HANDBOOK ON THE ECONOMICS OF CLIMATE CHANGE

Handbook on the Economics of Climate Change

Edited by

Graciela Chichilnisky

Professor of Economics and Mathematical Statistics, Columbia University, CEO and Co-Founder of Global Thermostat, New York, USA, and former US Lead Author of the UN IPCC

Armon Rezai

Professor, WU Vienna University of Economics and Business, and Senior Guest Research Scholar, International Institute for Applied Systems Analysis, Austria

Edward Elgar
PUBLISHING

Cheltenham, UK • Northampton, MA, USA

© Graciela Chichilnisky and Armon Rezai 2020

All rights reserved. No part of this publication may be reproduced, stored in a retrieval system or transmitted in any form or by any means, electronic, mechanical or photocopying, recording, or otherwise without the prior permission of the publisher.

Published by
Edward Elgar Publishing Limited
The Lypiatts
15 Lansdown Road
Cheltenham
Glos GL50 2JA
UK

Edward Elgar Publishing, Inc.
William Pratt House
9 Dewey Court
Northampton
Massachusetts 01060
USA

A catalogue record for this book
is available from the British Library

Library of Congress Control Number: 2019956769

This book is available electronically in the **Elgar**online
Economics subject collection
DOI 10.4337/9780857939067

Printed on elemental chlorine free (ECF)
recycled paper containing 30% Post-Consumer Waste

ISBN 978 0 85793 905 0 (cased)
ISBN 978 0 85793 906 7 (eBook)

Typeset by Servis Filmsetting Ltd, Stockport, Cheshire
Printed and bound in the USA

Contents

List of contributors	vii
Introduction to the *Handbook on the Economics of Climate Change* *Graciela Chichilnisky and Armon Rezai*	1

PART I THE POLITICAL ECONOMY OF CLIMATE CHANGE AND CLIMATE POLICY

1	Distributional issues in climate policy: air quality co-benefits and carbon rent *James K. Boyce*	12
2	Evaluating policies to implement the Paris Agreement: a toolkit with application to China *Ian Parry, Baoping Shang, Nate Vernon, Philippe Wingender and Tarun Narasimhan*	32
3	Bargaining to lose: a permeability approach to post-transition resource extraction *Natasha Chichilnisky-Heal*	68
4	Host–MNC relations in resource-rich countries *Natasha Chichilnisky-Heal and Geoffrey M. Heal*	83
5	Bargaining to lose the global commons *Natasha Chichilnisky-Heal and Graciela Chichilnisky*	106

PART II INTEGRATED ASSESSMENT MODELLING

6	Integrated Assessment Models of climate change *Chris Hope*	114
7	Climate change policy under spatial heat transport and polar amplification *William Brock and Anastasios Xepapadeas*	127
8	Progressive adaptation strategies in European coastal cities: a response to flood-risk under uncertainty *Luis M. Abadie, Elisa Sainz de Murieta, Ibon Galarraga and Anil Markandya*	167
9	Economic growth and the social cost of carbon: additive versus multiplicative damages *Armon Rezai, Frederick van der Ploeg and Cees Withagen*	199

10 Optimal global climate policy and regional carbon prices   224
    Mark Budolfson and Francis Dennig

11 Tipping and reference points in climate change games   239
    Alessandro Tavoni and Doruk İriş

PART III CLIMATE CHANGE AND SUSTAINABILITY

12 Climate change, Malthus and collapse   260
    Norman Schofield

13 Greenhouse gas and cyclical growth   281
    Lance Taylor and Duncan Foley

14 Growth and sustainability   296
    Robin Hahnel

15 Intergenerational altruism: a solution to the climate problem? 310
    Frikk Nesje and Geir B. Asheim

16 On intertemporal equity and efficiency in a model of global warming 326
    John M. Hartwick and Tapan Mitra

17 Transformational change: parallels for addressing climate and
    development goals   397
    Penny Mealy and Cameron Hepburn

18 Less precision, more truth: uncertainty in climate economics and
    macroprudential policy   420
    Cameron Hepburn and J. Doyne Farmer

Index   439

Contributors

Luis M. Abadie holds a PhD in economics from the University of the Basque Country and works in the fields of energy and climate economics. He is a member of the editorial board of Dyna.

Geir B. Asheim is Professor of Economics at the University of Oslo, Norway. His PhD is from the University of California, Santa Barbara. He has had research stays at several North-American universities, including Cornell, Harvard, Montréal, Northwestern and Stanford, and been resident at institutes for advanced study in Marseille and Paris. His research interests include game theory as well as the study of sustainable development in a world where environmental concerns and limited natural resources impose challenges.

James K. Boyce is a Senior Fellow at the Political Economy Research Institute and Professor Emeritus of Economics at the University of Massachusetts Amherst. His most recent books are *Economics for People and the Planet: Inequality in the Era of Climate Change* (Anthem Press, 2019) and *The Case for Carbon Dividends* (Polity Press, 2019).

William Brock is Vilas Research Professor of Economics Emeritus at the University of Wisconsin, Madison, Research Professor at the University of Missouri, Columbia, and Researcher at the RDCEP, University of Chicago. He is also a member of US National Academy of Sciences, distinguished Fellow of the American Economic Association, member of American Academy of Arts and Sciences and Fellow of the Econometric Society.

Mark Budolfson works on interdisciplinary issues in public policy, economics, and ethics, especially in connection with sustainable development goals and collective action problems. He is a Faculty Fellow at the Edmond J. Safra Center at Harvard, and Assistant Professor and a Fellow at the Gund Institute for Environment at the University of Vermont.

Graciela Chichilnisky is Professor of Economics and Mathematical Statistics and Director of the Columbia Consortium for Risk Management at Columbia University and the architect of the Kyoto Protocol Carbon Market. She has published 17 books and 319 scientific articles and taught at Harvard, Stanford, and Essex. She is also the CEO and Co-Founder of Global Thermostat, a company in which she co-invented a "Carbon Negative Technology"™ that captures CO_2 from the air and transforms it into profitable assets. In 2017, she was selected by the Carnegie Foundation for their prestigious Great Immigrant, Great American award.

Natasha Chichilnisky-Heal was a PhD candidate in political science at Yale University who has previously worked at the World Bank and the World Energy Forum. Her research focuses on natural resource wealth, global governance, and distributive justice and, in 2019, her work led to the creation of the Yale Program in Environmental Justice in Developing Countries in her honor. She was also a Senior Analyst at Developing World

Markets, a US-based fund manager specializing in environmentally responsible investments across the emerging markets, carried out policy work in Zambia and Mongolia, and created the concept of "permeable state".

Francis Dennig is an economist working on public policy questions relating to climate change, inequality, and their intersection. He is a NUS Fellow at the Center for Advanced Study in the Behavioral Sciences at Stanford University, and Assistant Professor at Yale-NUS College.

J. Doyne Farmer is Director of the Complexity Economics program at the Institute for New Economic Thinking at the Oxford Martin School, and is a Professor of Mathematics at the University of Oxford, as well as an External Professor at the Santa Fe Institute. His current research is in economics, including financial instability, sustainability, technological change and economic simulation in general.

Duncan Foley is Leo Model Professor of Economics at the New School for Social Research, and author of papers on political economy, mathematical economics, and climate change.

Ibon Galarraga is Research Professor at the Basque Centre for Climate Change (BC3) and was also Deputy Director for the period 2013–2015. He holds a PhD in economics from the University of Bath (UK) and was Deputy Minister for the Environment of the Basque Government, Executive Vice-president of the Environmental Public Society IHOBE, and Member of the Board of the Basque Energy Board (EVE).

Robin Hahnel is Professor Emeritus American University, Washington DC, and Teppola Distinguished Visiting Professor, Willamette University, Salem, Oregon. He is also Co-Director Economics for Equity and the Environment, the 3E Network.

John M. Hartwick is Professor of Economics at Queen's University, Kingston, Ontario. He has published a number of papers on "Solow Sustainability", before the chapter with Tapan Mitra in this volume on "Solow–Stollery Sustainability".

Geoffrey M. Heal is noted for contributions to economic theory and resource and environmental economics. He holds a doctoral degree from Cambridge University and an Honorary Doctorate from the Université de Paris Dauphine. He is Donald Waite III Professor of Social Enterprise at Columbia's Graduate School of Business and a Member of the National Academy of Sciences, a Fellow of the Econometric Society, Past President of the Association of Environmental and Resource Economists, recipient of its prize for publications of enduring quality and Life Fellow, and a Director of the Union of Concerned Scientists.

Cameron Hepburn is Director of the Smith School and an expert in environmental, resource and energy economics. He is a Professor of Environmental Economics at the Smith School and at the Institute for New Economic Thinking at the Oxford Martin School, and is also Professorial Research Fellow at the Grantham Research Institute at the London School of Economics and a Fellow at New College, Oxford.

Chris Hope is Emeritus Reader in Policy Modelling at Judge Business School, and Visiting Professor at University College, London. He was a Lead Author and Review Editor for the Third and Fourth Assessment Reports of the Intergovernmental Panel on Climate

Change, and an advisor on the PAGE model to the Stern review of the Economics of Climate Change. He has published extensively in books and peer-reviewed journals on the Integrated Assessment Modelling of climate change.

Doruk İriş is an Associate Professor in the School of Economics at the Sogang University (S. Korea) where he has been a faculty member since 2012. His main areas of research are applied microeconomic theory and behavioral economics, investigating both theoretically and experimentally how political and behavioral constraints affect the provision of public goods.

Anil Markandya is Professor Emeritus in the University of Bath (UK). He has been a Lead Author for the 2nd, 3rd, 4th and 5th IPCC Reports and recently contributed to the Special Report in 1.5°C. He has held researcher positions at Princeton, Berkeley and Harvard in the United States, and University College London in the United Kingdom. In 2008, he was acknowledged by Cambridge University as one of the 50 most influential global thinkers in the field of sustainability.

Penny Mealy is a Research Fellow at the Institute of New Economic Thinking at the Oxford Martin School and the Smith School of Environment and Enterprise. She is also a Research Associate at the Bennett Institute of Public Policy at the University of Cambridge. Her research focuses on economic development, technological evolution, transformational change, network science and agent-based modelling.

Tapan Mitra was Goldwin Smith Professor of Economics and a faculty member of the Center for Applied Mathematics at Cornell University. He was an Alfred P. Sloan Fellow, a Fellow of the Econometric Society, and a Fellow of the Society for the Advancement of Economic Theory. His research interests were economic dynamics, social choice and natural resource economics.

Tarun Narasimhan worked as a Research Analyst in the International Monetary Fund's Tax Policy Division from 2014 through 2016. His focus was on researching corporate income taxes and providing technical assistance. After receiving a Master's in statistics and data science from Stanford, he is working at a start up on improving US health care through machine learning.

Frikk Nesje is Research Fellow at the Department of Economics (the Alfred Weber Institute), Heidelberg University, Germany. His research interests include welfare economics, resource and environmental economics, and game theory. His current research is in intergenerational altruism, thresholds and natural disasters, and social discounting.

Ian Parry is Principal Environmental Fiscal Policy Expert in the IMF's Fiscal Affairs Department. He received a PhD in economics from the University of Chicago and prior to joining the IMF in 2010, he worked for 15 years at Resources for the Future. Parry has written numerous articles evaluating environmental, energy, and transportation policies in different countries, emphasizing the critical role of fiscal instruments for mitigating externalities.

Frederick van der Ploeg is Professor of Economics and Research Director of the Oxford Centre for the Analysis of Resource-Rich Economies (OxCarre). Also, affiliated with Vrije Universiteit Amsterdam, CEPR and CESifo. Formerly at Cambridge, LSE, EUI,

Tilburg and Amsterdam and also a Member of Parliament and Secretary of State in the Netherlands. Research interests are international finance, public economics and macroeconomics with special interests in the economics of climate and of natural resources.

Armon Rezai is Professor at WU Vienna University of Economics and Business, and Senior Guest Research Scholar at IIASA. His research interests are macroeconomics and political economy with special focus on employment and distribution, and the economics of natural resources and climate change.

Elisa Sainz de Murieta is a Post-Doctoral Researcher at the Basque Centre for Climate Change (BC3) and Visiting Fellow at the LSE's Grantham Research Institute for Climate Change and the Environment. Her current research focuses on climate change adaptation in cities, risk and decision-making under uncertainty.

Norman Schofield was Director of the Center in Political Economy, the William R. Taussig Professor of Political Economy, and Professor in the Departments of Economics and Political Science at Washington University, St Louis.

Baoping Shang is a senior economist at the Fiscal Affairs Department of the International Monetary Fund (IMF) and holds a PhD in policy analysis. At the IMF, his research has primarily focused on fiscal and expenditure policies. Prior to his current position, he worked at several leading research institutions in the United States, including RAND, the National Bureau of Economic Research (NBER), and the Urban Institute.

Alessandro Tavoni is an environmental economist based at the London School of Economics, where he leads the Changing Behaviour Group at the Grantham Research Institute. His research primarily relates to overcoming behavioral and political economy barriers to cooperation in the (climate) commons.

Lance Taylor is Arnold Professor Emeritus, New School for Social Research. His research work has spanned development economics, computable general equilibrium models, macroeconomics, and climate change.

Nate Vernon is a Senior Associate at IDinsight. He previously worked as a Research Analyst in the Fiscal Affairs Department of the International Monetary Fund (IMF). At the IMF, Nate focused on environmental and extractive industry fiscal policy.

Philippe Wingender is Economist in the Nordic Division of the European Department at the IMF having joined the Fund after obtaining his PhD in economics from the University of California, Berkeley. His research interests include inequality and policies to address it, the macroeconomic effects of fiscal policy and empirical methods for policy evaluation.

Cees Withagen is Emeritus Professor of Environmental Economics at Vrije Universiteit Amsterdam. He is Fellow of Tinbergen Institute and CentER (Tilburg University) and Research Professor at CESifo. His main research interest is in the economics of non-renewable resources and climate change, focusing on the various aspects of the Green Paradox.

Anastasios Xepapadeas is currently Professor of Economics at the Department of International and European Economic Studies of Athens University of Economics and Business, and the Department of Economics of the University of Bologna. He is a foreign associate at the US National Academy of Sciences, past President of the European Association of Environmental and Resource Economics and past Chair of the Board of Directors of the Beijer Institute of Ecological Economics of the Royal Swedish Academy of Sciences.

In loving and admiring memory of Natasha, Norman and Tapan.

Introduction to the *Handbook on the Economics of Climate Change*
Graciela Chichilnisky and Armon Rezai

This *Handbook* comes to light at a time when economic sciences start to recognize the inevitable emergence of climate change as the defining topic of our time. Economic thinking is evolving in front of our eyes, calling for reflection and reconsideration. The chapters contain ideas and policies to support and accelerate the change. We now know that climate change embodies and forecasts the future of human civilization and therefore its economic organization. It is the purpose of the *Handbook* to contribute to the transformation of economics in the midst of this momentous evolution.

The importance of climate change in economics should be no surprise. It is natural and to be expected because, as the traditional definition goes, economics is about the production, use, and distribution of resources, a definition that was famously proposed by T. Koopmans in the middle of the 20th century. Resources are at the core of economics, this much is clear. What is perhaps less clear is the transformation that has occurred in our perception of resources. Now, for the first time, we have 7.3 billion humans who have come to dominate the planet creating a new geological period that has replaced the Holocene and which geologists now call the Anthropocene. Only now that we dominate the geology of the planet have we come to recognize that the most important resources for human societies are the atmosphere of the planet, its bodies of waters and its biodiversity, namely the global environment. The definition of economics proposed by T. Koopmans has not changed: it is our understanding of resources that has fundamentally changed. To achieve its goal the book is divided into three sections that cover critical new areas and ideas about economics and climate change: The political economy of climate change and climate policy, integrated assessment modelling, and climate change and sustainability. For the convenience of the reader and using abstracts provided by the authors, the content is summarized in the following.

Part I examines issues of "The Political Economy of Climate Change and Climate Policy" and expands the conventional economic answer to climate change: "Make polluters pay." The externality associated with emitting harmful greenhouse gases needs to be internalized so that those reaping the benefits of emitting also bear its costs. This answer rests on the assumption that any distributional issue can be overcome by appropriate compensation of losers by winners of climate policy. That climate policy in fact poses net benefits and represents a "so-called" Pareto Improvement are deep insights from welfare economics. Decades of frail climate policy, however, reveal that there are powerful impediments in correcting price signals and unleashing market forces in a transition to carbon-free technologies. Increasingly, researchers and politicians realize that problems of political economy and distribution, of both carbon underground and in the atmosphere, are at the heart of the impediments to climate policy and their solutions are tantamount to successful climate policy. They are explored in the first section of this book.

In Chapter 1, entitled "Distributional issues in climate policy: air quality co-benefits and carbon rent", James K. Boyce tackles the implementation of a carbon tax that brings benefit to the current population through better air quality and the recycled tax revenue. He points out that the case for, and against, climate policy is typically made on grounds of inter-generational equity, assuming a tradeoff between future environmental well-being and present economic well-being. Boyce points out however that this framing of the problem is somewhat limited as it ignores the potential to design policies that mitigate climate change while yielding net benefits for most people who are alive today. This chapter considers two ways that climate policy can bring substantial benefits to the present generation: (i) air quality improvements from reduced burning of fossil fuels; and (ii) recycling of the rent created by carbon pricing. Both these considerations entail important issues of intra-generational equity that the chapter develops, which change our evaluation of climate policy.

In Chapter 2, entitled "Evaluating policies to implement the Paris Agreement: a toolkit with application to China", Ian Parry, Baoping Shang, Nate Vernon, Philippe Wingender, and Tarun Narasimhan discuss the recent policy developments and future policy options in China who, with its 1.3 billion human population, has become the world's largest annual emitter of carbon dioxide (CO_2). They propose a spreadsheet model for evaluating alternative fiscal and regulatory instruments that policy makers may consider for implementing the UN 2015 Paris Agreement's mitigation pledges, or national implementation targets. Various policies are evaluated against alternative metrics, including impacts on (CO_2) emissions, revenue, deaths from local air pollution, economic welfare, and economic incidence across households and industries. The model is applied to China but could be transferred to most other countries. For China, in the central case, they consider a carbon tax or coal tax that progressively rises to \$35 per ton of CO_2 cuts CO_2 emissions by about 20 per cent and raises well over 1 per cent of GDP in revenue by 2030 while, cumulated over the period 2017–30, saves approaching 2 million lives and generates discounted welfare gains equivalent to over 30 per cent of 2015 GDP. They show that an equivalently scaled emissions trading system applied to large emissions sources has roughly half the environmental and fiscal effectiveness, while other policies (e.g., incentives for energy efficiency and renewable sources of energy, and taxes on electricity and road fuels) are substantially less effective. The authors show that using around 5 per cent of the revenue from carbon/coal taxes can compensate low-income groups for increased energy prices, while 10 per cent of the revenues could compensate energy-intensive and trade-exposed firms.

Chapter 3, entitled "Bargaining to lose: a permeability approach to post-transition resource extraction", is authored by Natasha Chichilnisky-Heal who turns to the source of emissions, discussing resource extraction in resource-based developing economies and the rich political economy that comes with the multi-national extraction industry. The chapter is based on previous work entitled "Bargaining to lose: the permeability approach to post-transition resource extraction" where Natasha Chichilnisky-Heal introduced an original and fertile explanation for the resource curse. Her "permeability" approach questions the traditional treatment of the state as a decision maker having the public good as an objective, replacing it by the result of a bargaining game between the state and international organizations. Her new theory is illustrated with unique hands-on experience in the case of copper and gold mines in post-communist Mongolia and in post-socialist

Zambia, the largest in the world, and focuses on a bargaining game between the state and key financial organizations: the Bretton Woods Institutions (IMF, World Bank) and multinational corporations (MNCs) such as the resources conglomerate Rio Tinto. Permeability is the process by which external non-state actors such as the International Monetary Fund and Multinational corporations, by virtue of their relationships with cash strapped resource-rich governments, enter into crucial roles in the governance of these nations. This chapter goes beyond the traditional theory of the resource curse and proposes a relationship between permeability and a reduction in democratic accountability of these governments to their domestic constituencies, which Chichilnisky-Heal calls political underdevelopment. The argument runs as follows: external actors (multilateral and MNCs) bargain extensively with host governments over the regulation of extractive industries and offer development aid or loan packages over the regulation of extractive industries, and often tie development aid and loan packages to the satisfactory adjustment of regulations or conclusion of investment deals in the extractive sector. She argues, using data from the two cases mentioned, that this phenomenon skews the democratic process, providing the governments of these states with yet another constituency – the constituency of external actors. This perverts the democratic process not simply by making the government economically beholden to the external actors, as has been extensively argued, but by giving the external actors a permanent seat at the bargaining table of domestic politics. Permeability is not a binary process but rather an ordinal variable that measures the degree to which a democratic government and its processes have been "permeated" by actors other than its domestic constituent basis. The implications for the global environmental policies and the global environment itself are presented both theoretically and in practical terms, providing striking examples in the nations discussed.

Chapter 4, entitled "Host–MNC relations in resource-rich countries", Natasha Chichilnisky-Heal and Geoffrey M. Heal develop further the analysis of Chapter 3 within a theoretical context and from the perspective of the global commons and an individual firm's investment decision. The chapter discusses the relationship between a resource-rich developing country and a multinational corporation (MNC) that is developing the nations' resources for the international market. The authors model the connections between transparency, permeability (a term as well as a concept introduced by Natasha Chichilnisky-Heal and defined anew in this chapter as the amount of resource rent that leaves the country) and economic development, considering the polar cases of democracy and autocracy. It begins by considering the role of permeability in domestic politics, showing that a decrease in permeability will always benefit the incumbent, whether the country is a democracy or an autocracy. It then suggests that the relation between the host and the MNC has the features of a classical and quite intractable version of the hold-up problem, and that this may provide the MNC with incentives to influence political outcomes within the host country by whatever means are at its disposal. The hold-up problem can be overcome by the use of a Bilateral Investment Treaty that restricts the host country's ability to alter the terms of any agreement into which it has entered, and we investigate why a country might enter into a treaty that limits its freedom of action in this way. A possible answer is to be found in the capacity of a small number of poor countries to "tip" an equilibrium where none sign such treaties to one where all sign, in the process making all worse off. This chapter's analysis provides a micro foundation for the "obsolescing bargain model" of host–MNC relations.

In Chapter 5, entitled "Bargaining to lose the global commons", Natasha Chichilnisky-Heal and Graciela Chichilnisky apply the analysis of Chichilnisky-Heal in Chapter 3 to the global commons. Natasha Chichilnisky-Heal's "permeability" approach questions the treatment of the state as a decision maker having the public good as an objective and replaces it by the results of a bargaining game between the state and the International Organizations (IMF, World Bank). Her new theory is illustrated by her unique hands-on experience for the cases of copper and gold mines in Mongolia and in Zambia. Chapter 5 generalizes Chichilnisky-Heal's "bargaining to lose" approach to the resource curse providing economic models that validate the original conclusions and exploring its implications for the global commons: the atmosphere, the oceans and biodiversity. Chichilnisky-Heal "permeable state" is a transition to a new globalized society where the sovereign state – a relatively recent creation – is receding giving rise to a new set of global economic agents and institutions that better explain the dynamics of the destruction of the global commons in today's globalized world. We show that the permeable state theory is connected to explanations for the resource curse as a global market failure magnified by globalization and originating in the lack of property rights on natural resources during the pre-industrial period. We explore Chichilnisky-Heal's approach to the resource curse and its natural implications for the environmental crisis of the global commons. "Permeability" magnifies the losses from bargaining by the nation state. The results are ever-increasing overexploitation of resources and inefficient ever-increasing exports at increasingly lower prices. This complements the results arising from ill-defined property rights on resources in the private sector that drive Chichilnisky's (1994) explanation of the overexploitation of resources in developing nations. The latter are based on the inefficiency of market solutions that imply the (overexploitation of resources at market prices that are below replacement costs, due to the lack of well-defined property rights in developing nations prior to industrialization). Natural resources are inefficiently over-extracted, leading to the overexploitation of the global commons such as overuse of fossil fuels, minerals, and forests causing severe degradation of atmospheric water bodies and biodiversity degradation. Both state policies and private sector approaches lead to overexploitation of natural resources beyond what is efficient, with severe degradation of the global environment and the satisfaction of basic needs in the planet. The solutions that Chichilnisky-Heal proposes for permeability, for example, limiting the Bretton Woods' Institutions' "seat at the negotiation table" of resource extraction contracts, could help resolve the global environmental crisis, including climate change, which arise from the overextraction of global resources.

Part II of the *Handbook* covers "Integrated Assessment Modelling". To understand the connections between individual elements of the climate and the economy, climate scientists and economists have developed models for the integrated assessment of both. These help researchers and policy makers better understand the effects of climate change and how mitigation and adaptation policy can avert damages and make the economy more resilient. The chapters of this section introduce the most prominent models used in climate policy and show how they can be improved to include important complex geophysical phenomena, distributional aspects, and sectoral, institutional, and behavioral details or to provide simple policy rules for politicians.

In Chapter 6, entitled "Integrated Assessment Models of climate change", Chris Hope presents an overview of the development of the economic modelling of climate change by integrating impacts of climate on the economy and of the economy on the climate. The use of the models in policy-making has been rapid and influential, and the development and use of the models under the scrutiny of critics is strong and ongoing. As it becomes clearer that carbon pricing must be a large part of the solution to reducing greenhouse gas emissions, the importance and influence of Integrated Assessment Models (IAMs) may continue to increase. IAMs are tools that help translate the current knowledge about climate change, including its profound uncertainty, into policy advice. Chapter 6 describes three IAMs of climate change: The Dynamic Integrated Climate-Economy (DICE) model developed by Bill Nordhaus at Yale University, the Policy Analysis of the Greenhouse Effect (PAGE) model, developed by Chris Hope, at the University of Cambridge, UK, and the Climate Framework for Uncertainty, Negotiation and Distribution (FUND) model, developed by Richard Tol at Sussex University, UK. The vast majority of the independent impact and estimates of the social cost of carbon that appear in the peer-reviewed literature are derived from these three models. This chapter discusses their origins, influences, and shortcomings.

In Chapter 7, William Brock and Anastasios Xepapadeas introduce a spatial dimension to a standard model of integrated economic assessment by explicitly tracing the movement of heat and considering polar amplification. This chapter, entitled "Climate change policy under spatial heat transport and polar amplification", is the first instance in climate economics that considers a combination of spatial heat transport and polar amplification. It simplifies the problem by stratifying the Earth into latitude belts and assuming that the two hemispheres are symmetric. The results suggest that it is possible to build climate economic models that include the very real climatic phenomena of heat transport and polar amplification, and still maintain analytical tractability and show that the effect of heat transfer and polar amplification on climate policy depends upon the interaction of climate component dynamics with the distribution of welfare weights, population, and productive capacities across latitudes. The chapter discusses optimal fossil fuel taxes in a competitive environment with income effects and shows that optimal taxes have a spatial structure and are dependent on each latitude's output. In addition, it characterizes the interactions between spatial transport and the competitive equilibrium price path of tradable permits. Using general power utility functions, the authors show that an increase in the coefficient of relative risk aversion will reduce the social price of the climate externality.

In Chapter 8, the focus is shifted from global to regional modelling. Authors Luis M. Abadie, Elisa Sainz de Murieta, Ibon Galarraga, and Anil Markandya discuss in "Progressive adaptation strategies in European coastal cities: a response to flood-risk under uncertainty" strategies for European coastal cities of adaptation to climate change. The authors describe a novel stochastic model of sea-level rise and show how risk measures can be applied to such rising sea levels and the way in which sea-level rise and socio-economic development has been integrated. They present the estimated damage costs for different cities under different IPCC scenarios and for different time periods. The authors then turn to the concrete example of Glasgow and illustrate an application of real options analysis for this city. Using their approach a certain acceptable risk level can be determined above which no city is willing to go. This threshold allows estimating not only how much adaptation is needed but also when adaptation should start. This important piece of information complements other data used in decision making. Using the

Glasgow example, postponing a decision until more information on the impacts of sea-level rise or until other climate policy becomes available has to be weighed against other possibilities such as building a flexible defense that will allow a greater degree of protection in the future.

In Chapter 9, Armon Rezai and his co-authors Frederick van der Ploeg and Cees Withagen contribute in "Economic growth and the social cost of carbon: additive versus multiplicative damages" to recent debates on how fossil fuel emissions should be priced and whether there are simple rules which policy makers can follow. They therefore discuss optimal carbon pricing and derive simple policy rules for the cases of additive and multiplicative damage. In a calibrated integrated assessment model of Ramsey growth and climate change in the global economy, they investigate the differential impact of additive and multiplicative global warming damages for both a socially optimal and business-as-usual scenario. Fossil fuel is available at a cost that rises as reserves diminish and a carbon-free backstop is supplied at decreasing cost. If damages are not proportional to aggregate production and the economy is along a development path, the optimal carbon tax is smaller than with multiplicative damages. The economy switches later from fossil fuel to the carbon-free backstop and leaves less fossil fuel in situ. By adjusting climate policy in this way there is very little difference on the paths for global consumption, output and capital, and thus very little difference for social welfare despite the higher temperatures. They show that for all specifications the optimal carbon tax is not a fixed proportion of world GDP but rather follows a hump shape.

In Chapter 10, Mark Budolfson and Francis Dennig study how income inequalities across and within countries affect the optimal response to climate change. In "Optimal global climate policy and regional carbon prices" they find that, rather than impose one global price, heterogeneous prices are warranted with potentially large disparities if society's aversion to inequality is considerable. It is often stated that optimal global climate policy requires a single carbon price throughout the world. Chichilnisky and Heal (1994) have argued, however, that distributional issues or lump-sum transfers break this theorem and a policy in which different regions face different carbon prices can become superior to the uniform one. The chapter calculates utilitarian-optimal carbon prices under zero cross-regional lump-sum transfers in the multi-region IAM NICE, an adaptation of the DICE model introduced in Chapter 6. The resulting optimal global climate policy has differing regional carbon prices. Regions with the lowest levels of income start at lower prices, while richer regions face higher prices. This entails significant welfare gains over the standard single price optima commonly reported, which, as outlined briefly in the concluding remarks, can be improved upon still by allowing international trading in the corresponding emissions' allocations. Budolfson and Dennig show that the welfare gain from optimal differential prices is always positive for all their simulations and provides theoretical insights on monetary transfer necessary in concrete climate negotiations.

In Chapter 11, Alessandro Tavoni and Doruk İriş in "Tipping and reference points in climate change games" take a closer look at the dynamics of energy transition that carbon pricing can help usher. Once a tipping point for investment in low carbon technologies has been reached, and constituencies with stakes in the nascent markets have been formed, standard economic forces could sustain the transition to a carbon-neutral economy. The chapter reviews some of the recent literature that provides clues about when such reinforcing dynamics take place and the transition takes off. Given the wide scientific uncertainties surrounding the location of thresholds, the role of expectations and reference points are crucial for their formation and in supporting cooperation. The chapter reviews both theoretical and experimental literature featuring tipping points and reference dependence, with the aim of extending the understanding of potentially game-changing impacts on climate change cooperation. To this end, Tavoni and İriş examine the role of thresholds and reference levels in public goods and coalition formation games, which capture important features of dangerous climate change and its impacts on human behavior. The existence of ecological tipping points associated with abrupt and catastrophic, rather than gradual, climatic change has important behavioral repercussions in terms of the incentives to cooperate on mitigation efforts. Key insights emerging from the literature are that strong leadership in mitigation efforts induces cooperation by others; the reasons for why countries' high expectations about others' abatement efforts could have detrimental effects; and the reasons why developing countries have been relatively reluctant to exert even limited abatement efforts.

The last part of the *Handbook* combines "Climate Change and Sustainability" by embedding the climate crisis in the broader study of sustainability and contextualizing it either historically and in previous environmental disasters or within broader equity considerations, be they intertemporal across generations or using developmental approaches across regions. The final chapter argues for an incorporation of recent insights from finance on how not to model complex systems and offers potential ways forward for climate change economics.

In Chapter 12, entitled "Climate change, Malthus and collapse", Norman Schofield examines the impending deleterious and catastrophic effects of climate change in the context of sustainability and previous environmental disasters and presents a way forward by drawing on the logic of the Condorcet Jury Theorem. This chapter points out that the Agricultural Revolution of about 10 000 years ago triggered the rapid growth of world population, but also created Malthusian traps where population outgrew the availability of food resources. The Roman Empire was apparently constantly faced with such a trap. The British Empire, in contrast, was able to expand its resource base using the innovations of the Industrial Revolution. The author draws a parallel between Rome and our global economy to suggest that climate change could induce a Malthusian trap for us unless we pay heed to Pope Francis's call for us to "Care for Our Common Home". Since this presents us with a common goal, it is possible that the logic of the Condorcet Jury Theorem may give us hope of wiser choices over our future.

In Chapter 13, Lance Taylor and Duncan Foley also frame climate change in the context of the Malthusian theory. In "Greenhouse gas and cyclical growth", they develop a demand-driven integrated assessment model in which ever-increasing energy demand to power growing labor productivity, is the key to achieving sustained economic growth. To highlight the role of growing energy use in economic development, their growth model incorporates dynamics of capital per capita, atmospheric CO_2 concentration, and labor and energy productivity. Taylor and Foley show that in the "medium run" output and employment grow rapidly and are determined by effective demand in contrast to most models of climate change. In a "long run" of several centuries, the model converges to a stationary state with zero net emissions of CO_2. Properties of dismal and non-dismal stationary states are explored, with the latter requiring a relatively high level of investment in mitigation of emissions. Without such investment, under "business as usual", output dynamics are strongly cyclical in numerical simulations; there is strong output growth for about eight decades, followed by climate crisis, and output crash. Resources are the constraining factor in an economy generating endogenous growth as in the original essay of Malthus.

Chapter 14 by Robin Hahnel is entitled "Growth and sustainability" and discusses climate change, and more broadly, environmental sustainability in relation to economic production and labor productivity, arguing for a steady state economy and delineating what is necessary for achieving it. The relationship between economic growth and environmental sustainability has been the subject of much controversy. Hahnel argues that this has been in part due to lack of precision in defining terms and a failure to develop suitable models for studying the issue, both of which have been obstacles in developing understanding. The problem has been partly resolved in "What is sustainability?" by Graciela Chichilnisky (2011), which provided a rigorous definition and economic models to resolve this shortcoming. Following the earlier contributions, this chapter explains the difference between growth of production of goods and growth of environmental throughput, and it presents a model suitable for rigorously measuring changes in labor productivity and throughput efficiency. Hahnel formulates sufficient conditions for environmental sustainability, and helps to clarify how continued growth of production can be consistent with environmental sustainability. The chapter concludes with establishing linkages to the discussions on "steady-state" and "de-growth" economics, suggesting how the findings of this chapter can be extended to a world where nature is heterogeneous and parts of it are non-renewable, providing new observations about climate change policy and connecting it the broader questions of environmental sustainability.

In Chapter 15, Frikk Nesje and Geir B. Asheim discuss whether altruism is necessarily to the benefit of future generations, asking "Intergenerational altruism: a solution to the climate problem?" In an application of the theory of second-best, they show that in the presence of an uncorrected climate externality, greater utility weight on future generations can harm future generations by exacerbating the climate externality. Intergenerational altruism may induce more economic growth today to provide future generations with the means to weather the deleterious effects of climate change, thereby spurring climate change itself. Only when the climate externality is addressed by pricing carbon effectively, increased intergenerational altruism reduces the threat of climate change. Put differently, in a second-best setting with insufficient control of greenhouse gas emissions in the atmosphere, increased transfers to future generations through accumulation of capital might result in additional accumulation of greenhouse gases, and thereby aggravate the climate problem. In contrast, transfers to the future through control of greenhouse gas emissions will alleviate the climate problem. Whether increased intergenerational altruism is a means for achieving accumulation of consumption potential (through accumulation of capital) without increasing the climate threat depends on how it affects factors motivating the accumulation of capital and the control of emissions of greenhouse gases. Nesje and Asheim provide reasons for why increased intergenerational altruism aggravates the over-investment in brown capital and under-investment in green capital, that is, the atmosphere.

In Chapter 16, John M. Hartwick and Tapan Mitra also consider sustainability in the context of intergenerational justice in their essay "On intertemporal equity and efficiency in a model of global warming". They study equitable paths in a model where irreversible global warming is produced using an exhaustible resource. Global warming is assumed to affect both production and instantaneous welfare of society, both adversely. Global warming is generated by the use of an exhaustible resource, and they establish three equivalence results connecting the concepts of equity, Hartwick's Rule of investing the resource rents, and a suitably extended version of Hotelling's Rule, which takes into account the externalities caused by global warming and provides as an explicit solution an equitable path which satisfies Hartwick's Rule of investing resource rents. Consumption and global warming are bounded and the path is asymptotically similar to the maximin path obtained by Solow (1974) in a model without global warming. When the global warming function is strictly concave, they provide an explicit solution of an equitable path in which consumption and global warming exhibit unbounded quasi-arithmetic growth. This path follows an extended version of Hartwick's Rule of investment, and attain the maximum sustainable utility among all equitable paths that have a constant savings rate.

In Chapter 17, Penny Mealy and Cameron Hepburn approach sustainability in the context of the Sustainable Development Goals in "Transformational change: parallels for addressing climate and development goals", and focus on climate change and poverty alleviation which, as Stern (2016) has stated, are "the twin defining challenges of our century". Historically, efforts to address these two challenges have been conflicted. Adverse impacts of climate change are likely to hit the poorest of this world hardest, but traditional industrial routes out of poverty are dangerously emissions-intensive. At the same time, climate policy must not impose the same abatement burden on the poorest as argued by Budolfson and Dennig in Chapter 10. In fact, such tensions have been major impediments to emissions reduction in earlier climate negotiations and largely underpinned the failure of the 2009 Copenhagen COP to reach a global climate agreement. The authors argue that the UN Paris Agreement (UNFCCC, 2015), which has now been ratified by the large majority of countries, provides a promising new international platform to progress a unique collective framework for global climate cooperation. The confluence of these global agendas represents an historic opportunity to marry efforts on climate and development fronts and drive significant progress on sustainable development. Against this encouraging backdrop, Mealy and Hepburn draw attention to the twin climate and development challenges: both require societies to navigate and manage system-wide transformative change. Transformational change processes, particularly as they relate to climate and development contexts, need to be better understood and the authors attempt to draw these fields together, highlighting key commonalities and shared learning opportunities. There are clear advantages for both research and policy. In relation to research, climate and development economists have traditionally studied the process of transformative change in separate fields and with differing emphases. However, identifying key commonalities in respective change processes may not only improve shared learning outcomes, it could also illuminate a more generalized theory of transformational change. For policy, a lack of integration in climate and development initiatives can lead to outcomes that are myopic, at best, and detrimental to their intended objectives, at worst. In terms of methodological tools for analyzing and modeling transformational change, the chapter reviews four different approaches that have been used in both climate and development contexts. Network analysis provides a useful framework to investigate relationships across economic sectors and allows scholars and policy makers to better understand technological diffusion and industrial transition possibilities for a socially and environmentally more sustainable future.

In Chapter 18, Cameron Hepburn and J. Doyne Farmer in "Less precision, more truth: uncertainty in climate economics and macroprudential policy" draw on the existing knowledge of complex systems, particularly financial systems, to inform our understanding and modelling of climate change and climate policy. They argue that climate change economics is falling into the same traps that much of financial modelling did previously: parameters are expressed with too much confidence, important variables are omitted, and modelling of feedbacks, non-linearities, heterogeneity, and non-rational behavior are inadequately represented. Hepburn and Farmer, however, argue that while the climate and financial systems share various features – they are both "complex systems" – they also have important differences. The authors, therefore, carefully explore what lessons might be learned for climate system modelling from financial system modelling and macroprudential policy. Their key findings are: systematic data collection at a variety of scales is fundamental to properly understanding and modelling systemic risk; parameter values and the resulting model outputs need to treated with great skepticism due to the underlying uncertainty. In addition to parameter uncertainty, model uncertainty is likely to be large due to the fundamental difficulty of validation. Finally, since most economic models spring off the same set of narrow assumptions, significant variables, feedbacks, non-linearities, heterogeneity and non-rational behavior can, therefore, be easily overlooked. The authors argue for "conclusions that are less precise but more truthful" since they place emphasis on resilience as well as efficiency.

REFERENCE

Chichilnisky, G. (2011), 'What is sustainability?', *International Journal of Sustainable Economy*, **3**(2), 125–40.

PART I

THE POLITICAL ECONOMY OF CLIMATE CHANGE AND CLIMATE POLICY

1. Distributional issues in climate policy: air quality co-benefits and carbon rent*
James K. Boyce

1. INTRODUCTION

Climate change is often framed as posing a tradeoff between the welfare of the present and future generations. Policies that aim to mitigate climate change – most importantly, by reducing the use of fossil fuels – are assumed to require sacrifices on the part of those alive today for the sake of those who will follow. Invoking normative criteria of equity, efficiency or both, policy proponents maintain that the future gain from curtailing emissions will outweigh the present pain, while opponents make the opposite argument. Both sides agree, however, that the policies will require upfront costs. The public is left to weigh conflicting views on whether the benefits to future generations really justify the costs of taking action today.

The choice of an appropriate discount rate is a critical issue once this framework is accepted. For example, the UK government's Stern Review used a discount rate of about 1.4 per cent, whereas William Nordhaus of Yale University has used rates of 4 per cent or higher in his integrated assessment model.[1] The choice has major effects on what policies are deemed efficient: a lower discount rate justifies more aggressive policies to reduce emissions.

Inter-generational equity figures centrally in this debate. "Many would argue," the Stern Review noted, "that future generations have the right to enjoy a world whose climate has not been transformed in a way that makes human life much more difficult" (2007, p. 47). Citing a forecast that global per capita income will rise from $10000 today to $130000 (in today's dollars) in the next two centuries, Nordhaus (2008) countered: "While there are plausible reasons to act quickly on climate change, the need to redistribute income to a wealthy future does not seem to be one of them."[2]

Intra-generational equity has received less attention in climate policy debates. This reflects the prevalent assumption that climate policy necessarily will impose costs on the present generation. How these costs will be distributed has been seen as a secondary issue, overshadowed by the contentious tradeoff of present costs for future benefits.

---

* Research for this chapter was supported by Grant #INO15-00008 from the Institute for New Economic Thinking (INET). An earlier version was presented at the annual INET Conference held in Paris in April 2015. The author is grateful to conference participants and to the editors for thoughtful comments.

[1] Stern (2007); Nordhaus (2007). For a review of this debate, see Goulder and Williams (2012).

[2] This projected income may fail to account adequately for the potential adverse economic impacts of climate change itself (see Moore and Diaz, 2015; Stern, 2016).

This chapter challenges this conventional framing of the problem by examining the potential to design mitigation policies that yield substantial net benefits here and now.[3] These include benefits to the public at large, above and beyond the income and employment gains that would be generated by investments in energy efficiency and renewables.[4] They also are above and beyond the near-term benefits from mitigation itself, such as reduced risk from coastal flooding or extreme heat waves.

Two sorts of present-day benefits are considered here. The first are the improvements in air quality and public health that would come with reduced use of fossil fuels. The second is the net income gain that the majority of households would receive if the rent derived from a price on carbon emissions were to be recycled to the public as equal per capita dividends. Both sorts of benefits entail important issues of intra-generational equity. And both suggest that the assumed tradeoff between present and future generations can be overcome by climate policies that secure broad public support on the basis of present-day benefits.

2. AIR QUALITY CO-BENEFITS

In addition to reducing emissions of carbon dioxide, policies that curtail fossil fuel combustion reduce emissions of numerous air pollutants that damage human health, including particulate matter, sulfur dioxide, nitrogen oxides, and carbon monoxide. In fact, exercises that have assigned monetary values to the damages from carbon dioxide and these "co-pollutants" often put higher values on the latter. Co-pollutant intensity (the ratio of co-pollutant damages to carbon dioxide emissions) varies, however, across emission sources. For this reason, the spatial and sectoral composition of emissions reductions is important for efficiency and equity.

The World Health Organization (WHO) characterizes air pollution as "the world's largest single environmental health risk," calculating that outdoor and indoor air pollution together is responsible for one in eight premature deaths worldwide. Outdoor air pollution causes 3.7 million deaths annually. The WHO notes that more sustainable strategies in transport, energy and other sectors will be "more economical in the long term due to health-care cost savings as well as climate gains" (WHO, 2014).

Valuing the Health Costs of Air Pollution

Several studies have valued the costs of outdoor air pollution in monetary terms. A multi-country analysis by the Organisation for Economic Co-operation and Development (OECD) concluded that outdoor air pollution (specifically, particulate matter and ozone) was responsible for 2.45 million premature deaths annually in the OECD countries plus China and India in 2010 (see Table 1.1). China and India accounted for roughly 80 per cent of the total; among OECD countries the US ranked first with roughly 110 000 deaths.

---

[3] Here my focus is solely on mitigation. Climate change adaptation also poses deep distributional questions; for discussion, see Boyce (2014b).
[4] On the net employment gains of "green growth," see Pollin (2015).

Table 1.1 Costs of outdoor pollution in China, India and OECD countries, 2010

Country	Premature deaths (per year)	Value of a statistical life[1] (USD million)	Economic cost[2] (USD billion/year)
China	1 278 890	0.975	1371.4
India	692 425	0.602	458.4
US	110 292	4.498	545.8
Japan	65 776	3.068	222.0
Germany	42 578	3.480	163.0
Italy	34 143	2.995	112.5
Turkey	28 924	2.024	64.4
Poland	25 091	2.098	57.9
UK	24 064	3.554	94.1
Korea	23 161	3.027	77.1
Mexico	21 594	1.811	43.0
France	17 389	3.155	60.4
Other OECD	85 092	3.078	288.5
Total	2 449 419	1.321	3558.4

Notes:
[1] OECD calculation of the value of a statistical life (VSL) as a function of income per capita. For discussion, see text.
[2] Economic cost = Cost of mortality + morbidity.

Source: OECD, 2014, Tables 2.4, 2.7, 2.10, and 2.13–2.18.

To monetize these impacts, the OECD multiplied the number of deaths by the value of a statistical life (VSL, also sometimes termed the value of a prevented fatality), computed as a concave function of national income per capita, and then added 10 per cent for the costs of non-fatal illnesses. The total cost in these countries amounted to $3.5 trillion/year, with the OECD member countries accounting for about half of this and China and India for the rest (OECD, 2014).

The lower shares of China and India in monetary damages than total deaths are attributable to the OECD study's use of country-specific VSLs. Statistical lives in India and China were valued at $602 000 and $975 000, respectively, compared, for example, to $4.5 million in the US.[5] The OECD (2014, pp. 53–5) offers the following rationale for this procedure:

> A VSL value is meant to be an aggregation of individual valuations: an aggregation of individuals' WTP [willingness to pay], as communicated through WTP surveys, to secure a marginal reduction in the risk of premature death. In the world as we know it, individuals are differently endowed with the means with which to make such a trade-off; some work for their living for a dollar a day, some inherit a fortune yielding unearned income of a billion dollars a year. Human societies without exception have sought to socialise these risks to a greater or lesser extent in the

---

[5] Adopting a similar valuation procedure, a report of the World Bank and the Institute for Health Metrics and Evaluation (2016) used 2013 VSLs for India, China, and the US of $400 000, $978 000, and $5 million, respectively (purchasing power parity-adjusted 2011 dollars, calculated from data in Appendix B of the report).

form of public goods ... And it so happens that the level at which this socialisation of risks is executed today is the level of the nation-state. It is for this reason, and this reason alone, that it is appropriate to aggregate at the level of country-specific VSLs.

As discussed below, an alternative procedure would be to apply a uniform VSL to all countries, based on the ethical premise that all human lives are equally valuable regardless of individual wealth or per capita income in the country where the person happens to reside, or to use a poverty-weighted VSL that puts greater value on the lives of the most vulnerable. As Sunstein (2014, p. 89) remarks:

> If poor people are subject to a risk of 1/10 000, they do not have less of a claim to public attention than wealthy people who are subject to exactly the same risk. In fact they may have a greater claim, if only because they lack the resources to reduce that risk on their own.

Which of these valuation procedures is taken to be more appropriate depends on who foots the risk-reduction bill. When the poor must pay the cost of risk reduction themselves, a reasonable case can be made that they should not be compelled to spend as much as wealthier people would spend for protection against statistical risks. "Requiring poor people to buy Volvos," Sunstein (2014, p. 90) remarks, "is not the most sensible means of assisting them." On this basis, Sunstein contends that "for China or India, it would be disastrous to use a VSL equivalent to that of the United States or Canada." Whether the logic for individuals also applies to governments – with per capita income replacing individual income – is not obvious, however, since the distribution of risk-reduction costs and benefits is likely to vary across the national population. And as Sunstein himself notes, his argument "should not be taken to support the ludicrous proposition that donor institutions, both public and private, should value a risk reduction in a wealthy nation more than equivalent risk reduction in a poor nation."

Calculating the Co-Pollutant Cost of Carbon

The OECD's mortality data refer to outdoor air pollution from all sources, including not only fossil fuel use but also other sources such as wildfires, the burning of biomass, and construction dust. Reliable data on source-wise apportionment of air pollution are sparse, but for many pollutants in many countries fossil fuels are the most important source. Road transportation alone accounts for approximately half the outdoor air pollution in the EU24, according to the OECD (2014, p. 63), and for roughly one-third in the US where electric power generation (also from fossil fuel combustion) accounts for a higher share of the total than in Europe.

A study by MIT researchers estimates that 211 875 premature deaths (90 per cent confidence interval: 91 000–383 300) in the US in 2005 were attributable to particulate matter and ozone as a result of combustion emissions (Caiazzo et al., 2013).[6] Transportation (road, marine, rail and aviation) and electric power generation accounted for 60 per cent

---

[6] The difference between the total US deaths from outdoor air pollution as estimated by Caiazzo et al. (2013) and the OECD (2014) is likely attributable, in part, to emissions reductions between 2005 and 2010. See Fann et al. (2013).

Table 1.2 Premature deaths from outdoor air pollution in the US associated with combustion emissions from different sectors, 2005

Sector	Premature deaths	
	Number	%
Road transportation	58050	27.4
Electric power generation	53900	25.4
Industry	42550	20.1
Commercial/residential	42150	19.9
Marine transportation	8830	4.2
Rail transportation	5040	2.4
Aviation	1355	0.6
Total	211875	100

Source: Calculated from Caiazzo et al. (2013), Table 4.

of these, with the remainder due to other industrial, commercial and residential activities (Table 1.2).

An international analysis of premature mortality from outdoor air pollution reaches similar conclusions for the US, attributing 52 per cent to electric power generation and land traffic (Lelieveld et al., 2015). In other OECD countries included in this analysis, the authors estimate that the joint share of these two sectors ranges from 27.3 per cent (in Korea) to 35.5 per cent (in the UK). The joint shares in China and India are 20.8 per cent and 18.5 per cent, respectively, with residential and commercial energy use accounting for larger shares in those two countries. If we attribute all air pollution from transportation and the power sector to fossil fuel combustion, plus one-quarter of the air pollution from other sectors, these figures would imply that fossil fuel use accounts for roughly 65 per cent of premature mortality from outdoor air pollution in the US, 50 per cent in other OECD countries, and 40 per cent in China and India.

Applying these percentages to the OECD data in Table 1.1, we can calculate health impacts of co-pollutants per ton carbon dioxide emissions. I term this ratio the Co-Pollutant Cost of Carbon (CPCC). Three measures of the CPCC are reported in Table 1.3. The first, the number of premature deaths/ton CO_2, ranges from fewer than 13 in the US to more than 160 in India. The second, US dollars/ton using the OECD's valuation procedure (in which VSL varies with per capita income), ranges from $50/ton in Mexico to $134/ton in Italy. The final measure applies a uniform VSL to all countries, while holding unchanged the sum total of monetary damages according to the OECD study. By this measure, which is directly proportional to deaths/ton, India's CPCC exceeds $200/ton.

The CPCC for the US in 2010 based on the OECD valuation procedure was $64/ton. If, rather than the $4.5 million VSL used for the US in the OECD study, we were to apply the higher VSL used by the US Environmental Protection Agency (USEPA), the CPCC would increase correspondingly.[7]

---

[7] The official VSL used by the USEPA in 2013 was $9.7 million (USEPA, 2016, p. 2). For comparisons of the VSL used by USEPA and other US government agencies, see Robinson (2007).

Table 1.3   Co-pollutant cost of carbon, 2010

Country	Premature deaths from fossil fuel emissions	CO_2 emissions (million mt)	Co-pollutant cost of carbon (per mt CO_2)		
			US dollars		
			Deaths	OECD VSL	Equal VSL
China	511 556	7388.5	69.2	74.2	100.6
India	276 970	1714.9	161.5	106.9	234.6
US	71 690	5580.0	12.8	63.6	18.7
Japan	32 888	1177.3	27.9	94.3	40.6
Germany	21 289	797.0	26.7	102.3	38.8
Italy	17 072	419.8	40.7	134.0	59.1
Turkey	14 462	268.5	53.9	119.9	78.2
Poland	12 546	304.6	41.2	95.0	59.0
UK	12 032	529.5	22.7	88.8	33.0
Korea	11 580	584.0	19.8	66.0	28.4
Mexico	10 797	434.0	24.9	49.6	35.6
France	8 694	385.6	22.5	78.3	32.3
Other OECD	42 546	2588.3	16.4	55.7	23.6
Total	1 044 122	22 172.1	47.1	68.4	68.4

Sources: Premature deaths from fossil fuel emissions and co-pollutant cost of carbon: author's calculations (see text). CO_2 emissions (from consumption of coal, petroleum + natural gas): US Energy Information Agency, https://www.eia.gov/cfapps/ipdbproject/IEDIndex3.cfm?tid=90&pid=44&aid=8, accessed 11 February 2016.

It is instructive to compare the CPCC to the Social Cost of Carbon (SCC) that is used by the US government in regulatory analyses as a measure of climate damage. The SCC does not include the damages from co-pollutants that are released along with carbon dioxide. The average SCC in 2015 ranged from $11 to $56/ton CO_2 depending on the choice of the discount rate, with $105/ton used to test the sensitivity of cost–benefit analysis results to "the potential for higher-than-average damages" (USEPA, 2015).[8] The CPCC in the US thus is comparable to, and possibly even larger than, the government's average SCC.

Other studies have come to similar conclusions. An analysis of prospective air quality co-benefits from "deep decarbonization" policies in the US found that they would prevent approximately 36 000 premature deaths/year from 2016 to 2030 and concluded that the co-benefits would exceed the climate benefits valued on the basis of the official SCC (Shindell et al., 2016). For the European Union, a study by the Netherlands Environmental Assessment Agency concluded that the air quality co-benefits from a stringent climate policy would be large enough to offset the policy's costs "even when the long-term benefits of avoided climate impacts are not taken into account" (Berk et

---

[8]   As of 2014 the SCC had been used in more than 40 regulatory impact analyses by US government agencies (GAO, 2014). For details on how it was derived, see US Interagency Working Group on the Social Cost of Carbon (2013). For critiques, see Ackerman and Stanton (2012) and Foley et al. (2013).

al., 2006). Summarizing 37 studies from around the world, Nemet et al. (2010) found a mean value for air quality co-benefits of $49/ton of CO_2. An IMF study (Parry et al., 2014) concluded that in the top twenty CO_2-emitting countries the average nationally efficient carbon price based on domestic co-benefits alone would be $57.5/ton, without counting global climate benefits.

For climate policy the salience of air quality co-benefits may be even greater than these monetary valuations suggest. Air quality benefits are predominantly near-term and national, whereas climate benefits are predominantly long-term and global.[9] Greater emphasis on the magnitude of air quality co-benefits may therefore help to overcome the political impediments to climate policy that arise from myopia and concerns about international free riding.

Efficiency Implications

From an efficiency standpoint two conclusions follow:

First, inclusion of the air quality co-benefits justifies more stringent regulatory measures than if policy were based solely on damages from CO_2 emissions. Let us define the Full Social Cost of Carbon (FSCC) – total damages per ton of fossil CO_2 emissions – as the sum of the climate change cost of carbon (CCCC) and the co-pollutant cost of carbon (CPCC):

$$FSCC = CCCC + CPCC$$

Compared to the conventional SCC, which is based on the CCCC alone, the FSCC strengthens the efficiency case for curtailing use of fossil fuels, providing a yardstick for higher carbon prices and more ambitious emission reduction targets.[10]

Second, insofar as air quality co-benefits per ton of CO_2 vary across pollution sources and locations, efficiency can be enhanced by designing policies so as to achieve deeper emissions reductions where co-benefits are greater. The rationale for doing so can be illustrated by an example. Consider two facilities in California: a power plant located outside Bakersfield and a petroleum refinery located in metropolitan Los Angeles, each of which emits the same amount of CO_2, roughly 3 million tons per year (t/yr). The power plant also emits about 50 t/yr of particulate matter (PM) and has fewer than 600 residents living in a 6-mile radius, while the refinery emits about 350 t/yr of PM and has about 800 000 residents living within a 6-mile radius (Pastor et al., 2013). Clearly, the health co-benefits associated with a ton of carbon emission reductions will be greater at the refinery than at the power plant. Though this example is particularly dramatic, substantial variations in co-pollutant intensity are found across industrial sectors in the US (see, for example, Table 1.4).

---

[9] As Shindell (2015) observes, "near-term health impacts seem to typically be considered more important to citizens than longer-term impacts of any sort, consistent with the vastly greater sums spent on medical care and research than on long-term environmental protection."

[10] Shindell (2015) similarly proposes the term "Social Cost of Atmospheric Release" (SCAR) to refer to the combined climate and air quality damages from emissions of multiple pollutants. FSCC thus can be defined as SCAR per ton of CO_2 emissions from fossil fuel combustion.

Table 1.4 $PM_{2.5}$ intensity by industrial sector, United States

Industrial sector	Population-weighted $PM_{2.5}$ per ton CO_2	Minority share (%)
Primary metal manufacturers	19.7	47.5
Non-metallic mineral product manufacturers	8.6	39.8
Petroleum refineries	8.4	59.5
Chemical manufacturers	5.2	43.9
Power plants	3.0	38.8

Source: Boyce and Pastor (2013).

Equity Implications

From an equity standpoint as well, air quality co-benefits have significant implications for climate policy. In the US, for example, racial and ethnic minorities and low-income communities tend to bear disproportionate air pollution burdens (see, for example, Ringquist, 2005; Mohai, 2008; Morello-Frosch et al., 2011; Cushing et al., 2016).

Executive Order 12898, "Federal Actions to Address Environmental Justice in Minority Populations and Low-Income Populations," issued by President Bill Clinton in 1994, directs every US government agency to take steps to identify and rectify "disproportionately high and adverse human health or environmental effects of its programs, policies, and activities on minority populations and low-income populations." This directive made equity an explicit objective in federal environmental policy. Many US states now have similar environmental justice policies (Bonorris, 2010).

The extent of air pollution exposure disparities varies across facilities, industrial sectors and locations (Ash and Boyce, 2011; Ash et al., 2009; Zwickl et al., 2014). Nationwide, racial and ethnic minorities bear 59.5 per cent of the impact of particulate emissions from petroleum refineries, for example, compared to 38.8 per cent in the case of power plants, the latter figure being closer to their 34.2 per cent share in the national population in 2005–09 (see Table 1.4). An equity-oriented climate policy would aim to achieve greater emissions reductions not only from those sources where air quality co-benefits are greater, but also from sources where co-pollutant damages are more unequally distributed by race, ethnicity and income.

Incorporating Air Quality Co-Benefits into Climate Policy

Two broad categories of direct policy instruments can be used to reduce fossil fuel combustion: quantitative regulations, such as fuel economy standards for automobiles and renewable portfolio standards for power plants; and price-based policies, such as a carbon tax or marketed carbon permits. These are not mutually exclusive, since the public policy mix can and usually does include both. Anti-smoking policies, for example, combine restrictions on who can buy tobacco and where smoking is permitted with excise taxes to discourage smoking. Similarly, policies to cut sulfur dioxide emissions from US power plants have combined mandated technologies and emission standards with a cap-and-trade permit system.

One reason to include price-based tools in the policy mix is that they create incentives not only to adopt existing pollution control technologies but also to invest in research and development of new ones. Price-based policies often have encountered opposition, however, from environmental justice advocates on the grounds that they may allow co-pollutant "hot spots" to persist, and possibly even worsen, in overburdened communities. Environmental justice organizations in California, for example, filed a lawsuit in an attempt to block implementation of the state's CO_2 cap-and-trade program for this reason (Farber, 2012). Hot-spot concerns can be addressed both by quantitative regulations and by allowing prices to vary across sources, for example by a zone system with tighter caps (or, equivalently, higher prices) in priority locations (Boyce and Pastor, 2013).

Some economists have argued that co-pollutants should not be factored into climate policy design because they are best regulated separately (Schatzki and Stavins, 2009). Of course, pollution control technologies, such as scrubbers in smokestacks, can reduce co-pollutant emissions without reducing the use of fossil fuels (and hence without reducing CO_2 emissions). The adoption of such technologies by advanced industrialized countries is a major reason why their premature deaths from fossil fuel emissions per ton of carbon are much lower than in China and India, as seen in Table 1.3. But the fact that the CPCC remains high even in advanced industrialized countries with relatively stringent environmental regulations – ranging from \$64 to \$134 per ton CO_2 in Table 1.3 – means that it remains an important component of the full social cost of carbon, and hence relevant for assessing the full social benefits of carbon reduction.

The health costs of air pollution also remain large relative to the costs of pollution control, implying that actually existing environmental policies are far from what economists would characterize as efficient. A cost–benefit analysis for the European Union's Thematic Strategy on Air Pollution (TSAP) found current policies to be suboptimal in all EU member countries (Holland, 2012, 2014). The "extraordinary high net benefits and benefit-cost ratios" reported in the TSAP study, the OECD (2014, p. 76) remarks, "suggest that something has gone wrong with the decision-making process."

Unless and until one can reasonably assume that co-pollutant impacts are efficiently and equitably addressed by separate regulations, climate policy should take them into account. This approach to climate policy is consistent with the growing embrace of multi-pollutant strategies for air-quality management.[11] The authoritative US government document on regulatory impact analysis, Office of Management and Budget (OMB) Circular A-4, for example, explicitly directs federal agencies to consider co-benefits (also known as "ancillary benefits"):

> Your analysis should look beyond the direct benefits and direct costs of your rulemaking and consider any important ancillary benefits and countervailing risks. An ancillary benefit is a favorable impact of the rule that is typically unrelated or secondary to the statutory purpose of the rulemaking (e.g., reduced refinery emissions due to more stringent fuel economy standards for light trucks) . . . (OMB, 2003, p. 26)

---

[11] See National Academy of Sciences (2004) and McCarthy et al. (2010).

In a similar vein, a study by the European Environment Agency (2006) concluded that climate policies could significantly reduce both health damages from air pollution and the costs of controlling air pollutant emissions.

Administratively, it is not terribly difficult to incorporate air quality co-benefits into climate policy design, particularly in settings where co-pollutant damages are concentrated in a relatively small number of facilities, sectors or locations, as in the case of US industrial point source emissions (Boyce and Pastor, 2013). Policy options include the following:

1. *Monitor impacts on co-pollutants*: A minimalist option is simply to monitor co-pollutant emissions with a view to instituting remedial measures if the climate policy has unacceptable impacts, such as exacerbation of environmental disparities across racial, ethnic or income groups. This was the approach taken by the California Air Resources Board (2011) in its adaptive management plan for the state's cap-and-trade policy.
2. *Zonal tax or permit systems*: Carbon permit or tax systems can ensure emissions reductions in high-priority zones where potential public health benefits are greatest. Zone-specific caps were used, for example, in California's Regional Clean Air Incentives Market, which was initiated in 1994 to reduce emissions of NO_x and SO_2 in the Los Angeles basin (Gangadharan, 2004).
3. *Sectoral tax or permit systems*: Similarly, sector-specific permit or tax systems can be designed to ensure emissions reductions in those economic and industrial sectors with the highest co-pollutant intensities or the greatest disproportionate impacts on minority and low-income populations.
4. *Trading ratios*: In a tradable permit system where damages per unit of emissions vary across sources, the exchange rate at which permits are traded can serve as another policy instrument for achieving greater reductions from specific sources. For example, if the full social cost of carbon (CO_2 plus co-pollutant damages per ton CO_2) are twice as high in location A as in location B, the exchange rate ("trading ratio") would be 1:2 (Muller and Mendelsohn, 2009).
5. *Community benefit funds*: Finally, some fraction of the revenue obtained from carbon taxes or permit auctions (the "carbon rent" discussed below) can be channeled into community benefit funds to mitigate pollution impacts and protect public health in overburdened and vulnerable communities. This strategy has been enacted for revenues from permit auctions under California's climate policy.[12]

3. CARBON RENT ALLOCATION

Putting a price on carbon emissions by means of a cap or a tax is widely viewed as a central element of climate policy for good reason. Although "command-and-control" regulatory instruments often figure in the policy mix, too – as in the implementation of

---

[12] California Senate Bill 535, signed into law in 2012, mandates that 25% of the revenue from the state's carbon permit auctions is to be spent on projects that benefit disadvantaged communities. For discussion, see Truong (2014).

California's Global Warming Solutions Act – carbon pricing can be an effective, if not indispensable, instrument both to drive emissions reductions in the short run and to create incentives for technological innovation in the long run. In addition, carbon pricing offers an opportunity to build and sustain public support for climate policy if the revenue – here termed "carbon rent" – is allocated in a manner that is transparent and widely regarded as fair.

From an administrative standpoint, a carbon cap or tax is most easily and efficiently implemented "upstream" where fossil fuels enter the economy: at tanker terminals, pipeline hubs, coal mine heads, etc. For each ton of fossil carbon that a firm brings into the economy, it surrenders a permit or pays a tax. In the US, an upstream system would entail roughly 2 000 collection points nationwide, far fewer than the number of compliance entities that would need to be monitored in a downstream system (US Congressional Budget Office, 2001).

What is Carbon Rent?

The revenue generated by a carbon price is depicted in Figure 1.1. A cap reduces the quantity of fossil fuel from Q_0 to Q_1. A tax raises the price of fossil fuel from P_0 to P_1. A cap sets the quantity of carbon emissions and lets the price adjust, while a tax sets the price and lets the quantity adjust. Apart from this difference the two are equivalent. Carbon rent, represented by the shaded area in the diagram, is the product of the carbon price and the quantity of carbon in fossil fuel entering the economy.

The price elasticity of demand for fossil fuels is low, especially in the short run: the percentage change in price exceeds the percentage change in quantity. Hence the tighter the cap or higher the tax, the bigger the carbon rent.

Carbon rent is sometimes confused with the resource cost of reducing emissions, but the two are quite distinct. Investments in energy efficiency and alternative energy use real resources. A number of studies have concluded that the resource costs of emission reductions in response to the introduction of a carbon price will be fairly modest. In an analysis of the Waxman-Markey bill, an unsuccessful attempt to enact federal carbon pricing legislation in the US, the Congressional Budget Office (2009) estimated that the resource cost in the year 2020 would amount to only 18 cents per household per day.[13] Indeed, a study by McKinsey & Co. (2007) concluded that substantial emissions reductions can be achieved initially at negative marginal cost – that is, the investments would pay for themselves at market interest rates.

The difference between the resource cost of emission reductions and carbon rent is depicted in Figure 1.2. The horizontal axis is the quantity of emissions, starting from zero reduction (100 per cent of current emissions); the vertical axis is the price. The rising curve represents the marginal abatement cost, here shown beginning at zero (rather than in the negative range reported in the McKinsey study). The figure shows the effect of

---

[13] The 18 cents/day figure comes from dividing the CBO estimate of "net annual economywide cost" of $22 billion/yr by the US population (335 million). In addition to resource costs of energy efficiency and alternative fuels, the CBO's $22 billion estimate included costs for the purchase of international offsets and the production cost of domestic offsets (both of which would have been allowed under the bill) and overseas spending on adaptation and mitigation.

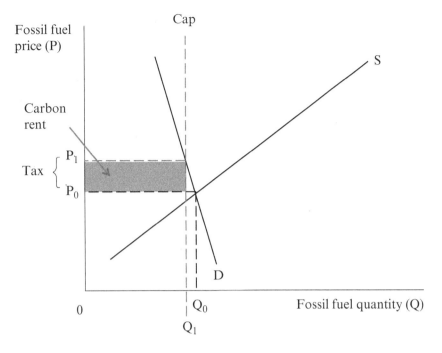

Figure 1.1   Carbon rent

capping emissions at 75 per cent of their current level, or equivalently, setting a carbon tax at the level needed to obtain this outcome. The resource cost triangle is the cost of preventing emissions. The carbon rent rectangle is the price paid by fossil fuel users for emissions that are *not* prevented. This is often termed "allowance value" in discussions of cap-and-permit systems, where an allowance is a synonym for a permit.

As Figure 1.2 shows, the carbon rent generated by pricing policies is likely to be substantially larger than the resource cost of reducing emissions. In economic terms, carbon rent is not a cost: it is a *transfer*. The carbon rent is not spent on retrofitting buildings or installing solar panels. It is a surcharge paid on fossil fuel resources that would have been produced even in the absence of the policy.

Who Pays Carbon Rent?

The tax or permit price is paid by the compliance entities – fossil fuel firms in an upstream system – and then passed through to final consumers, either directly in the market prices of gasoline, heating fuels and electricity, or indirectly in the market prices of food, manufactured goods, and everything else that is produced or distributed using fossil fuels.[14] The extra money paid by consumers is the main source of carbon rent (as

---

[14] Most economic analyses assume that 100% of the carbon price is passed through to consumers. In practice, it is possible that "pass-through" would be a little less than 100% (or even

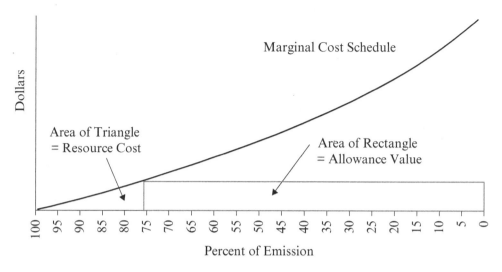

Source: Burtraw et al. (2009).

Figure 1.2 Resource cost versus allowance value (carbon rent)

noted below, some also comes from non-household final users of fossil fuels). Consumers pay in proportion to their direct and indirect consumption of fossil fuels, their "carbon footprints." Because upper-income households generally have bigger carbon footprints, they pay more in absolute terms than other households. As a percentage of their incomes, however, lower-income households may pay more. If so, carbon pricing is a regressive tax.

The incidence of carbon pricing can be analyzed by combining consumer expenditure survey data with input–output tables that provide information on the quantities of fossil carbon embodied in different goods and services. Figure 1.3 depicts the results of such calculations for US households. The relationship is concave: carbon footprints rise with total household expenditure, as expected, but decline as a percentage of expenditure. Similar patterns have been found in a number of other industrialized countries.[15]

There have been fewer studies of the distributional effects of carbon pricing in low and middle-income countries. In some of these countries, low-income households may have smaller carbon footprints than upper-income households not only absolutely but also in relative terms, as a percentage of total expenditure, by virtue of their very low consumption of fossil fuels. Figure 1.4 depicts the relationship between carbon emissions and household expenditure in China in the year 1995. The convex curve indicates that at that time, the incidence of carbon pricing in China would have been progressive.

---

a little more), if firms cut their profit margins in an effort to protect their market share (or use the policy as a pretext to increase profit margins). For discussion of the effects of the degree of pass-through on carbon rent, see Boyce and Riddle (2007).

[15] See, for example, Cramton and Kerr (1999); Symons et al. (2000); and Wier et al. (2005).

Distributional issues in climate policy 25

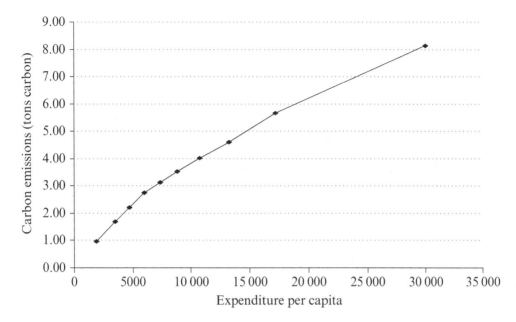

Source: Boyce and Riddle (2007).

Figure 1.3 Carbon emissions by expenditure class, United States

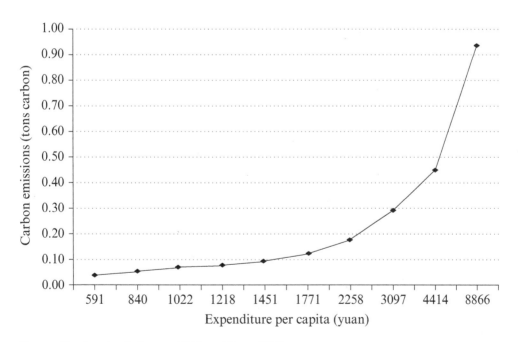

Source: Based on data in Brenner, Riddle and Boyce (2007).

Figure 1.4 Carbon emissions by expenditure class, China (1995)

Who Receives Carbon Rent?

The net impact of carbon pricing on income distribution, however, depends crucially on where the money goes: to whom the carbon rent is transferred. Broadly, there are three possibilities:

- *Cap-and-giveaway-and-trade* policies (usually called "cap-and-trade") distribute carbon permits to firms free of charge, allocating them on a formula based on historic emissions.[16] The European Union's Emissions Trading System for power plants and industrial point sources is an example. Much as OPEC increases profits for member countries by restricting oil supplies, the firms that receive free permits reap windfall profits by virtue of the cap: the supply of fossil fuels is reduced, prices go up, and suppliers keep the money. Under this policy the carbon rent ultimately flows to the shareholders and executives of firms that get free permits. Because these recipients generally are upper-income households, the effect of the rent distribution is regressive, compounding whatever regressivity exists in the incidence of the carbon price itself.
- *Cap-and-spend* policies auction carbon permits up to the limit set by the cap rather than giving them away. The auction revenue goes to the government. This equals the carbon rent, since what firms bid for permits is equal to what they will recoup in higher prices from buyers of fossil fuels. Government then can use the revenue for public expenditures, tax cuts (or "tax expenditures"), budget deficit reduction, or some combination of these. The Regional Greenhouse Gas Initiative for power plants in the northeastern US states is an example of such a policy. A carbon tax in which the revenues are deposited into the government treasury is another. Under this policy the net distributional impact of carbon pricing depends on how the government chooses to spend the carbon revenue.
- *Cap-and-dividend* policies auction the permits, too, but under this option the revenue is returned to the public as equal per capita dividends rather than being retained by the government. The underlying normative principle behind cap-and-dividend policy is that the scarce carbon absorptive capacity of the biosphere, or more precisely, a state's share of it, belongs in common and equal measure to all its people rather than to firms or the government.[17] Under this policy the net distributional impact is progressive, since high-income households pay more in absolute terms than low-income households (as shown in Figures 1.3 and 1.4), while all individuals receive the same dividends. A carbon tax in which the revenues are returned to the public as lump-sum payments (this is sometimes called "fee-and-dividend") has the same effect.

---

[16] Because the permits are given to them free of charge, rather than auctioned, some firms may find it profitable to sell permits to others who find it profitable to buy them. For this reason, permit trading is invariably allowed in the cap-and-giveaway policy.

[17] A similar principle can be applied to royalties from natural resource extraction. For example, oil revenues paid into the Alaska Permanent Fund provide annual dividends to all state residents (Barnes, 2014). For more on carbon dividends, see Boyce (2019).

Table 1.5 Net incidence of a carbon dividend policy in the United States

Expenditure decile	Net impact (USD/household/year)[1]	
	Scenario 1: 100% as dividends	Scenario 2: 75% as dividends[2]
1 (poorest)	289	190
2	253	154
3	225	126
4	201	102
5	175	76
6	148	49
7	117	18
8	77	−22
9	18	−81
10 (richest)	−109	−200

Notes:
[1] Net impact = dividend minus amount paid in higher prices for fossil fuels.
[2] Excludes impacts from the 25% of carbon rent allocated to other purposes.

Source: Author's calculations; for methods, see Boyce and Riddle (2011).

Table 1.5 shows the net impacts of carbon dividends in the United States under two scenarios. Both assume a modest carbon price of $25 per ton of CO_2. In the first scenario, 100 per cent of the carbon rent is recycled directly to the public as equal per capita dividends. This was proposed in a climate policy bill introduced by Congressman Chris Van Hollen in July 2014.[18] In the second scenario, 75 per cent of the carbon rent is recycled as dividends and 25 per cent is retained for public investment. This was proposed in the climate policy bill introduced by Senators Maria Cantwell and Susan Collins in December 2009.[19]

In either scenario, the majority of households would receive positive net benefits: their dividends would exceed what they pay as a result of higher fossil fuel prices.[20] There are two reasons for this result. First, because household income and expenditure are highly concentrated in the upper deciles, so are carbon footprints. The mean household carbon use is above the median, and dividends are based on the mean. Second, household consumption accounts for roughly two-thirds of fossil fuel use in the US. The remainder is consumed mainly by governments (federal, state, and local), and to a lesser extent by non-profit institutions and production of exports (Boyce and Riddle, 2008). If more than two-thirds of the total carbon rent is returned to households, they receive a transfer from these other sectors.

---

[18] See Boyce (2014a).
[19] For details, see Boyce and Riddle (2011).
[20] This result holds not only at the national level but also in each of the 50 states, although the percentage of households that would come out ahead varies depending, in particular, on the carbon-intensity of the state's electricity supply (Boyce and Riddle, 2009).

28 *Handbook on the economics of climate change*

A 2017 study by the US Treasury Department's Office of Tax Analysis reached similar conclusions. The study estimated that a carbon price of $49 per ton of CO_2 would generate annual dividends of $583 per person. Roughly 70 per cent of households would experience net income gains, and the distributional impact would be strongly progressive, ranging from an 8.7 per cent gain in net income for the poorest decile to a 1.0 per cent loss for the top decile. In contrast, using the carbon revenue to reduce corporate tax rates would have a regressive impact, resulting in net income losses for the lowest nine deciles and net gains only for the top decile (Horowitz et al., 2017, Table 6).

An Efficiency–Equity Tradeoff?

Some economists argue that the choice between carbon dividends and a revenue-neutral "green tax shift," in which carbon revenues are offset by tax cuts, poses a tradeoff between equity and efficiency (Burtraw and Sekar, 2014). Equal per capita dividends would be more progressive, as illustrated above, but cuts in income or sales taxes would boost output since these taxes reduce the supply of labor and capital.

The validity of this ostensible equity–efficiency tradeoff can be questioned on three grounds. First, in real-world contexts characterized by unemployed labor and underutilized capital, increases in the supply of labor and capital do not translate into increased output; instead they translate into more unemployment and excess capital. Second, from a social welfare standpoint the main problem may be that people work too much, as predicted by the relative income hypothesis, rather than too little (Wendner and Goulder, 2008). Third, higher output (as measured by GDP) is not synonymous with higher social welfare.

Apart from its appeal on equity grounds, an attractive feature of the carbon dividend option is that it could help to ensure durable public support for climate policy even in the face of rising fuel prices. Effective climate policy is not a one-shot game in which success is simply enacting legislation. Once in place, the policy must be able to continue over the decades needed to complete the clean energy transition. It must be popular enough to survive no matter what party controls the government. It is hard to imagine how robust public support can be secured in the face of rising fuel prices unless the carbon rent is returned to the population in a way that visibly and fairly protects the net incomes of most households. Additionally, it is hard to imagine any outcome that would be more inefficient than the failure to curtail the use of fossil fuels.

4. CONCLUSIONS

The distributional issues in climate policy are often posed in inter-generational terms, as a tradeoff between the welfare of present and future generations. This chapter has argued that climate policy has important distributional implications within the present generation, too, and that policies that take these into account can attenuate the myopia and free-rider problems that have impeded efforts to curtail the use of fossil fuels.

Two crucial intra-generational issues have been explored here. The first relates to the air quality co-benefits of reduced use of fossil fuels. By some calculations, these benefits are as large or larger than the climate benefits themselves. Because air quality co-benefits are local rather than global, and because co-benefits vary spatially and across pollution sources, the ways in which these benefits are (or are not) integrated into climate policy can have important distributional implications. Co-pollutant damages often are greatest in lower-income and politically disenfranchised communities. Therefore, designing climate policy to achieve greater emission reductions where they yield the greatest public health benefits can promote equity as well as efficiency.

The second distributional issue concerns the allocation of carbon rent. Climate change mitigation policy is a form of property creation, in that it converts the limited carbon-absorptive capacity of the biosphere from an open-access resource (where property rights are absent) into a resource governed by rights and responsibilities. When carbon pricing is in the policy mix – in the form of a carbon tax or cap-and-permit system – the new bundle of property rights includes the right to receive income from payments for use of the scarce resource. The allocation of this income, or carbon rent, again poses important distributional issues. These are illustrated by the choice among the cap-and-giveaway-and-trade, cap-and-spend, and cap-and-dividend policy options.

If climate policy addresses these distributional issues in an egalitarian fashion – based on the twin principles of equal rights to a clean and healthy environment and equal rights to carbon rent – the outcome can be positive net benefits for the majority of people in the present generation. These health and income benefits can attenuate or eliminate the ostensible tradeoff in climate policy between present and future welfare. In turn, this could help to overcome one of the greatest political obstacles to taking effective steps to safeguard the world's climate.

REFERENCES

Ackerman, Frank and Elizabeth A. Stanton (2012), 'Climate Risks and Carbon Prices: Revising the Social Cost of Carbon', *Economics: The Open-Access, Open-Assessment E-Journal*, **6**, 1–25.

Ash, Michael and James K. Boyce (2011), 'Measuring Corporate Environmental Justice Performance', *Corporate Social Responsibility and Environmental Management*, **18**, 61–79.

Ash, Michael, James K. Boyce, Grace Chang, Justin Scoggins and Manuel Pastor (2009), 'Justice in the Air: Tracking Toxic Pollution from America's Industries and Companies to our States, Cities, and Neighborhoods', Political Economy Research Institute, University of Massachusetts Amherst, MA and Program for Environmental and Regional Equity, University of Southern California, LA.

Barnes, Peter (2014), *With Liberty and Dividends for All*, San Francisco: Berrett-Koehler.

Berk, M., J. Bollen, H. Eerens, A. Manders and D.P. van Vuuren (2006), 'Sustainable Energy: Trade-offs and Synergies between Energy Security, Competitiveness, and Environment', Technical Report, Netherlands Environmental Assessment Agency (MNP), Bilthoven.

Bonorris, S. (ed.) (2010), *Environmental Justice for All: A Fifty State Survey of Legislation, Policies and Cases*, 4th edn, Berkeley, CA: American Bar Association and Hastings College of the Law, University of California.

Boyce, James K. (2014a), 'The Carbon Dividend', *New York Times*, 30 July.

Boyce, James K. (2014b), 'Amid Climate Change, What's More Important: Protecting Money or People?', *Los Angeles Times*, 22 December.

Boyce, James K. (2019), *The Case for Carbon Dividends*, London: Polity.

Boyce, James K. and Manuel Pastor (2013), 'Clearing the Air: Incorporating Air Quality and Environmental Justice into Climate Policy', *Climatic Change*, **120**(4), 801–14.

Boyce, James K. and Matthew Riddle (2007), 'Cap and Dividend: How to Curb Global Warming While Protecting the Incomes of American Families', Amherst, MA: Political Economy Research Institute, Working Paper No. 150.

Boyce, James K. and Matthew Riddle (2008), 'Keeping the Government Whole: The Impact of a Cap-and-Dividend Policy for Curbing Global Warming on Government Revenue and Expenditure', Amherst, MA: Political Economy Research Institute, Working Paper No. 188.

Boyce, James K. and Matthew Riddle (2009), 'Cap and Dividend: A State-by-State Analysis', Amherst, MA: Political Economy Research Institute; Portland, OR: Economics for Equity and Environment Network.

Boyce, James K. and Matthew Riddle (2011), 'CLEAR Economics: State-Level Impacts of the Carbon Limits and Energy for America's Renewal Act on Family Incomes and Jobs', Amherst, MA: Political Economy Research Institute.

Brenner, Mark, Matthew Riddle and James K. Boyce (2007), 'A Chinese Sky Trust? Distributional Impacts of Carbon Charges and Revenue Recycling', *Energy Policy*, **35**(3), 1771–84.

Burtraw, Dallas and Samantha Sekar (2014), 'Two World Views on Carbon Revenues', *Journal of Environmental Studies and Sciences*, **4**, 110–20.

Burtraw, Dallas, Richard Sweeney and Margaret Walls (2009), 'The Incidence of U.S. Climate Policy', Washington, DC, Resources for the Future, April.

Caiazzo, Fabio, Akshay Ashok, Ian A. Waitz, Steve H.L. Yim and Steven R.H. Barrett (2013), 'Air Pollution and Early Deaths in the United States. Part I: Quantifying the Impact of Major Sectors in 2005', *Atmospheric Environment*, **79**, 198–208.

California Air Resources Board (2011), 'Adaptive Management Plan for the Cap-and-Trade Regulation', 10 October, accessed 30 January 2017 at http://www.arb.ca.gov/cc/capandtrade/adaptive_management/plan.pdf.

Cramton, Peter and Suzi Kerr (1999), 'The Distributional Effects of Carbon Regulation: Why Auctioned Carbon Permits are Attractive and Feasible', in Thomas Sterner (ed.), *The Market and the Environment*, Cheltenham, UK and Northampton, MA: Edward Elgar, pp. 255–71.

Cushing, Lara J., Madeline Wander, Rachel Morello-Frosch, Manuel Pastor, Allen Zhu and James Sadd (2016), 'A Preliminary Environmental Equity Assessment of California's Cap-and-Trade Program', Los Angeles, University of Southern California, Program for Regional and Environmental Equity, September.

European Environment Agency (2006), 'Air Quality and Ancillary Benefits of Climate Change Policies', Luxembourg, Office for Official Publications of the European Communities.

Fann, Neal, Charles M. Fulcher and Kirk Baker (2013), 'The Recent and Future Health Burden of Air Pollution Apportioned Across U.S. Sectors', *Environmental Science & Technology*, **47**(8), 3580–89.

Farber, Daniel A. (2012), 'Pollution Markets and Social Equity: Analyzing the Fairness of Cap and Trade', *Ecology Law Quarterly*, **39**, 1–56.

Foley, Duncan K., Armon Rezai and Lance Taylor (2013), 'The Social Cost of Carbon Emissions: Seven Propositions', *Economics Letters*, **121**, 90–97.

Gangadharan, Lata (2004), 'Analysis of Prices in Tradable Emission Markets: An Empirical Study of the Regional Clean Air Incentives Market in Los Angeles', *Applied Economics*, **36**, 1569–82.

Goulder, Lawrence H. and Robert C. Williams III (2012), 'The Choice of Discount Rate for Climate Change Policy Evaluation', Washington, DC, Resources for the Future, Discussion Paper 12-43.

Holland, Mike (2012), 'Cost–Benefit Analysis of Scenarios for Cost-Effective Emission Controls after 2020', Version 1.02, Corresponding to IIASA Thematic Strategy on Air Pollution Report #7, November.

Holland, Mike (2014), 'Cost–Benefit Analysis of Final Policy Scenarios for the EU Clean Air Package', Version 2, Corresponding to IIASA Thematic Strategy on Air Pollution Report #11, Version 2a, October.

Horowitz, John, Julie-Anne Cronin, Hannah Hawkins, Laura Konda and Alex Yuskavage (2017), 'Methodology for Analyzing a Carbon Tax', Washington, DC, United States Department of the Treasury, Office of Tax Analysis, Working Paper 115, January.

Lelieveld, J., J.S. Evans, M. Fnais, D. Giannadaki and A. Pozzer (2015), 'The Contribution of Outdoor Air Pollution Sources to Premature Mortality on a Global Scale', *Nature*, **525**, 367–84.

McCarthy, James E., Larry Parker and Robert Meltz (2010), 'After the CAIR Decision: Multipollutant Approaches to Controlling Powerplant Emissions', Washington, DC, Congressional Research Service, 4 March.

McKinsey & Co. (2007), 'Reducing U.S. Greenhouse Gas Emissions: How Much at What Cost?', December.

Mohai, Paul (2008), 'Equity and the Environmental Justice Debate', in Robert C. Wilkinson and William R. Freudenberg (eds), *Equity and the Environment*, Research in Social Problems and Public Policy, vol. 15. Amsterdam, The Netherlands: Elsevier, pp. 21–50.

Moore, Frances C. and Delavane B. Diaz (2015), 'Temperature Impacts on Economic Growth Warrant Stringent Mitigation Policy', *Nature Climate Change*, **5**, 127–31.

Morello-Frosch, R., M. Zuk, M. Jerrett, B. Shamasunder and A.D. Kyle (2011), 'Understanding the Cumulative Impacts of Inequalities in Environmental Health: Implications for Policy', *Health Affairs*, **30**(5), 879–87.

Muller, Nicholas Z. and Robert Mendelsohn (2009), 'Efficient Pollution Regulation: Getting the Prices Right', *American Economic Review*, **99**(5), 1714–39.

National Academy of Sciences (2004), 'Air Quality Management in the United States', Washington, DC, National Academies Press.

Nemet, G.F., T. Holloway and P. Meier (2010), 'Implications of Incorporating Air-Quality Co-Benefits into Climate Change Policymaking', *Environmental Research Letters*, **5**, 1–9.

Nordhaus, William (2007), 'A Review of the *Stern Review on the Economics of Climate Change*', *Journal of Economic Literature*, **45**, 686–702.

Nordhaus, William (2008), 'The Question of Global Warming: An Exchange', *New York Review of Books*, 25 September.

OECD (2014), 'The Cost of Air Pollution: Health Impacts of Road Transport', OECD, Paris, France, http://dx.doi.org/10.1787/9789264210448-en.

Parry, I., C. Veung and D. Heine (2014), 'How Much Carbon Pricing is in Countries' Own Interests? The Critical Role of Co-Benefits', International Monetary Fund, Working Paper No. 174.

Pastor, M., R. Morello-Frosch, J. Sadd and J. Scoggins (2013), 'Risky Business: Cap-and-Trade, Public Health, and Environmental Justice', in C.G. Boone and M. Fragkias (eds), *Urbanization and Sustainability*, Dordrecht, Netherlands: Springer, pp. 75–94.

Pollin, Robert (2015), *Greening the Global Economy*, Boston, MA: Boston Review Books.

Ringquist, Evan J. (2005), 'Assessing Evidence of Environmental Inequities: A Meta-Analysis', *Journal of Policy Analysis and Management*, **24**(2), 223–47.

Robinson, Lisa A. (2007), 'How US Government Agencies Value Mortality Risk Reductions', *Review of Environmental Economics and Policy*, **1**(2), 283–99.

Schatzki, Todd and Robert N. Stavins (2009), 'Addressing Environmental Justice Concerns in the Design of California's Climate Policy', Comment submitted to the Economic and Allocation Advisory Committee, California Air Resources Board and California Environmental Protection Agency.

Shindell, Drew (2015), 'The Social Cost of Atmospheric Release', *Climatic Change*, **130**(2), 313–26.

Shindell, D.T., Y. Lee and G. Faluvegi (2016), 'Climate and Health Impacts of US Emissions Reductions Consistent with 2° C', *Nature Climate Change*, **6**(5), 503.

Stern, Nicholas (2007), *The Economics of Climate Change: The Stern Review*, Cambridge: Cambridge University Press.

Stern, Nicholas (2016), 'Current Climate Models are Grossly Misleading', *Nature*, **530**, 407–9.

Sunstein, C.R. (2014), *Why Nudge? The Politics of Libertarian Paternalism*, New Haven, CT: Yale University Press.

Symons, Elizabeth, Stefan Speck and John Proops (2000), 'The Effects of Pollution and Energy Taxes across the European Income Distribution', Keele University Economics Research Paper No. 2000/05.

Truong, Vien (2014), 'Addressing Poverty and Pollution: California's SB 535 Greenhouse Gas Reduction Fund', *Harvard Civil Rights–Civil Liberties Law Review*, **49**(2), 493–529.

US Congressional Budget Office (2001), 'An Evaluation of Cap-and-Trade Programs for Reducing U.S. Carbon Emissions', Washington, DC, June.

US Congressional Budget Office (2009), 'The Estimated Costs to Households from the Cap-and-Trade Provisions of H.R.', 2454, 19 June.

US Environmental Protection Agency (2015), 'EPA Fact Sheet: Social Cost of Carbon', December.

US Environmental Protection Agency (2016), 'Valuing Mortality Risk Reductions for Policy: A Meta-Analytic Approach', Prepared by the US Environmental Protection Agency's Office of Policy, National Center for Environmental Economics, for review by the EPA's Science Advisory Board, Environmental Economics Advisory Committee, February.

US Government Accountability Office (GAO) (2014), 'Regulatory Impact Analysis: The Development of Social Cost of Carbon Estimates', Washington, DC, July.

US Interagency Working Group on Social Cost of Carbon (2013), 'Technical Support Document: Technical Update of the SCC for Regulatory Impact Analyses Under Executive Order 12866', Washington, DC.

US Office of Management and Budget (OMB) (2003), Circular A-4, 17 September.

Wendner, R. and L. Goulder (2008), 'Status Effects, Public Goods Provision, and Excess Burden', *Journal of Public Economics*, **92**, 1968–85.

Wier, Mette, Katja Birr-Pedersen, Henrik Klinge Jacobsen and Jacob Klok (2005), 'Are CO_2 Taxes Regressive? Evidence from the Danish Experience', *Ecological Economics*, **52**, 239–51.

World Bank and Institute for Health Metrics Evaluation (2016), 'The Cost of Air Pollution: Strengthening the Economic Case for Action', Washington, DC, World Bank.

World Health Organization (WHO) (2014), '7 Million Premature Deaths Annually Linked to Air Pollution', accessed 6 January 2015 at http://www.who.int/mediacentre/news/releases/2014/air-pollution/en/.

Zwickl, Klara, Michael Ash and James K. Boyce (2014), 'Regional Variation in Environmental Inequality: Industrial Air Toxics Exposure in U.S. cities', *Ecological Economics*, **107**, 494–509.

2. Evaluating policies to implement the Paris Agreement: a toolkit with application to China*
Ian Parry, Baoping Shang, Nate Vernon, Philippe Wingender and Tarun Narasimhan

1. INTRODUCTION

One hundred and ninety parties submitted pledges to mitigate carbon dioxide (CO_2) and other greenhouse gases for the 2015 Paris Agreement on climate change. A typical pledge among advanced countries is reduce emissions by around 30 percent by 2030 relative to emissions in a baseline year[1] while China and India (for whom projected GDP growth is rapid but uncertain) pledged to lower the emissions intensity of GDP.[2] Under the Agreement, countries are required to report progress on meeting these pledges and to submit revised pledges every five years, which are expected to be progressively more stringent.[3] Current emissions pledges are in line with containing long-run, mean projected warming to 3.0–3.2°C above pre-industrial levels[4]—more aggressive targets would be needed to contain warming to the internationally agreed 1.5–2.0°C target. From a practical perspective, however, the immediate challenge is what specific policy actions are needed now in different countries and how obstacles to these actions might be overcome.

There is a wide array of alternative fiscal and regulatory CO_2 mitigation instruments—a non-exhaustive list includes carbon taxes; emissions trading systems (ETS); individual taxes on coal, road fuels, and electricity; renewables subsidies; measures to reduce the CO_2 intensity of power generation; and policies to increase energy efficiency in different sectors.

Although there is general agreement among economists, policymakers, and business leaders[5] that, ideally, comprehensive carbon pricing in line with environmental objectives would form the centerpiece of mitigation policy, there are practical reasons (e.g., opposition to large increases in energy prices) why policymakers may be under pressure to implement a more limited form of carbon pricing, perhaps in combination with other instruments. To make sound choices across instruments, to design the stringency

---

* We are grateful to Armon Rezai and a referee for helpful comments on an earlier draft.

[1] For some countries, the baseline is emissions in a previous year (e.g., 1990 or 2005) while for other countries it is projected emissions for 2030 in the absence of mitigation policy.

[2] Emissions pledges by country are listed at http://www4.unfccc.int/submissions/indc/Submission%20Pages/submissions.aspx.

[3] The Agreement came into force on November 4, 2016, following ratification by over 55 countries accounting for over 55 percent of emissions (see http://unfccc.int/paris_agreement/items/9444.php).

[4] UNEP (2016).

[5] See, for example, Krupnick et al. (2010) and www.carbonpricingleadership.org/carbon-pricing-panel.

of specific policies, and to communicate the case for policy actions to legislators and stakeholders, policymakers need an overarching quantitative framework for comparing options against a wide range of metrics including their effects on carbon dioxide (CO_2) emissions, revenue, premature deaths from local air pollution, incidence across household and industry groups, as well as their overall domestic economic benefits and costs.

While there has been plenty of valuable modelling of specific carbon mitigation policies in specific countries,[6] there has been little analysis comparing a broad range of policy options using a consistent modelling approach. Probably the most comprehensive study in this regard is Krupnick et al. (2010) who examined the effectiveness and cost-effectiveness of over 30 policies and policy combinations to reduce US CO_2 emissions and oil consumption using a variant of the National Energy Modelling System (NEMS), which contains considerable detail on energy sectors, regions, and adoption of specific technologies. Detailed computational models like NEMS provide considerable sophistication, but intuition underlying their results is not always transparent, the models are not easily replicated and transferred to other countries, they evaluate policies against a limited range of metrics (typically just impacts on CO_2 emissions), and the results may date quickly as energy systems evolve.

A far more streamlined model—one that is easily run for different countries by non-experts in a spreadsheet—would provide a useful, complementary toolkit, as it potentially facilitates transparent comparisons across instruments, across countries, and across a broad range of metrics, while enabling easy checks on the sensitivity of results to uncertain parameters (e.g., fuel price elasticities) and regular updating for the latest energy data.

This chapter discusses such a toolkit and applies it in the case of China, which accounted for 28 percent of global fossil fuel CO_2 emissions in 2014.[7] The model, which distinguishes power generation, road transport and other energy use in industry and homes, begins with energy flow data for China and projects this forward to 2030. The responses to different policies hinge on behavioral response assumptions for fuel use, energy efficiency, and so on in different sectors, which are based on empirical literature.[8]

To provide a flavor of the results for China, taxes on the carbon content of fossil fuels, or just coal (both of which are straightforward administratively), greatly outperform other policies across a broad range of metrics. In the central case, charges rising progressively to $35 per ton[9] of CO_2 by 2030 reduce CO_2 emissions by around 20 percent relative to baseline levels (while meeting China's main Paris pledge on CO_2 intensity) and raise

---

[6] See, for example, Aldy et al. (2016) and, for a discussion of modelling results for the United States, Fawcett et al. (2015).

[7] IEA (2016a), Table II.4.

[8] The discussion below is based on a larger working paper written for the 2016 Article IV consultation between the IMF and China. As regards prior analyses of carbon policies for China, Cao et al. (2013) analyze a carbon tax by integrating a detailed treatment of air pollution damages into an economic-energy model incorporating capital dynamics and disaggregating 33 different industries. With similar assumptions about the price responsiveness of coal use, our results on the carbon and local health benefits of carbon taxes are similar in a broad sense. The present analysis differs from Cao et al. (2013) by considering a wider range of carbon mitigation instruments, evaluating them against a broader range of metrics, and using a highly streamlined modelling approach. For a review of other recent carbon policy modelling exercises for China, see Karplus et al. (2016).

[9] Monetary figures are expressed in 2015 US$. To convert into Chinese Yen multiply by 6.5.

34 *Handbook on the economics of climate change*

revenues well over 1 percent of GDP in 2030. Cumulated over 2017–30, these policies also save close to 2 million air pollution deaths[10] while generating discounted domestic welfare gains (excluding climate benefits) equivalent to over 30 percent of 2015 GDP. These are extremely large benefits and imply substantial carbon pricing is in China's own interests due to the domestic environmental benefits alone. An equivalently scaled ETS applied to large industrial emissions sources has—roughly speaking—about half of the environmental and (with auctioned allowances) fiscal effectiveness of the carbon and coal tax policies, while other policies often have dramatically lower environmental, fiscal, and welfare benefits. The relative ranking of different mitigation instruments is generally robust to parameter uncertainty.

Incidence analysis can be conducted by linking the policy-induced impacts on energy prices from the spreadsheet tool to an input–output model to trace through price impacts on different industries and consumer products, and combining that with household survey data on spending for energy and other goods by different income groups. For China, carbon taxes are regressive but only moderately so—although the poor spend a greater share of their budget on electricity, natural gas, and heating than do wealthier households, the opposite applies for road fuels and other consumer products whose prices increase indirectly from higher energy costs. Recycling about 5 percent of the carbon tax revenues to lower social security contributions and stronger social safety nets—areas where China has been lagging[11]—could compensate the bottom two deciles. Carbon policies also complement broader efforts to re-balance the Chinese economy away from energy-intensive activities. Any transitory compensation for trade sensitive sectors should not involve large fiscal costs (around 10 percent of carbon pricing revenues) as these sectors are not disproportionately impacted by carbon mitigation policies.

The rest of the chapter is organized as follows. Section 2 describes the model and policy scenarios. Section 3 presents the main policy comparisons. Section 4 discusses policy incidence. Section 5 offers concluding remarks.

2. ANALYTICAL FRAMEWORK

A great deal about environmental, fiscal, and economic impacts of carbon mitigation policies can be learned from an aggregate-level (reduced form) model, parameterized so baseline energy projections are consistent with those from more disaggregated (structural) models and the responsiveness of fuels to policies is consistent with empirical evidence and other modelling results. So long as the model contains the key features for distinguishing among the baseline scenario and alternative mitigation policies, additional disaggregation

---

[10] There are more efficient policies for reducing local air pollution than carbon pricing, such as direct charges on smokestack emissions, or upfront charges on coal supply combined with credits for adoption of mitigation technologies (like sulfur dioxide scrubbers) at power plants. However, until these more efficient policies have comprehensively priced air pollution externalities (likely a long time), it is entirely appropriate to consider the unpriced air pollution co-benefits of (near-term) carbon pricing.

[11] Lam and Wingender (2015).

is not necessary for a first-pass policy assessment. These features include distinguishing the main energy sectors, fuel use within those sectors, and changes in energy efficiency from changes in the use of energy-consuming products. This section briefly outlines the model, with the details on equations and parameterization provided in Appendices 2A.1 and 2A.2, and then discusses the policies to be analyzed.

A. Energy Sectors

Five fossil fuels are distinguished, namely coal, natural gas, gasoline, road diesel, and an aggregate of other oil products (e.g., used in domestic aviation and petrochemicals). The model projects, out to 2030, annual fuel use in three sectors—power generation, road transport, and an 'other energy' sector, where the latter represents an aggregation of direct energy use by households, firms, and non-road transport.

Power sector

In the power sector, electricity demand in the 'business-as-usual' (BAU) scenario—that is, with no fiscal or regulatory policy changes to reduce fossil fuel use beyond those already implicit in recently observed fuel use and price data—increases over time with growth in GDP according to the income elasticity of demand for electricity. Higher electricity prices reduce electricity demand through improvements in energy efficiency and reductions in the use of electricity-consuming products. The efficiency of electricity-using products also improves over time with autonomous technological progress.

Power (in China) can be generated from coal, natural gas, oil, nuclear, hydro, and (non-hydro) renewables like solar and wind. Increases in the unit generation cost for one fuel lead to switching away from that fuel to other generation fuels. Unit generation costs also decline gradually over time with autonomous technological progress, where the rate of decline is assumed to be faster for renewables (a relatively immature technology). Changes in electricity demand result in proportional changes in generation from the different fuel types.

Road transport

The road transport sector distinguishes gasoline (i.e., light-duty) vehicles and diesel (primarily heavy-duty) vehicles. Again, future fuel use in the BAU varies positively with future GDP growth (through income elasticities for vehicle use), negatively with higher fuel prices (which promote use of more fuel-efficient vehicles and less driving) and autonomous improvements in vehicle fuel efficiency.

Other energy sector

The other energy sector reflects an aggregation of all other energy use and includes industry, non-road transport, and residences. In China, this sector uses coal, natural gas, (non-hydro) renewables, and (non-road) oil products.[12] Small-scale fuel users, namely households and small firms, are distinguished from large industrial users as this allows

---

[12] Given the focus on policies to reduce fossil fuels, the model does not capture (non-combustion) CO_2 emissions released, for example, during the cement-making process.

the modelling of downstream ETSs and regulations which can only cover the latter.[13] Again, baseline fuel use varies positively with GDP (through income elasticities) and negatively with fuel prices (through changes in energy efficiency and product usage) and autonomous improvements in energy efficiency.

Model simplifications and solution

One simplification in this model is that fuel use adjusts instantly and fully to changes in fuel prices within a period, whereas in reality the adjustment occurs progressively over time as capital turns over—in other words, the price responsiveness of fuel use is smaller in the shorter term than the longer term. However, given that policies are likely to be anticipated and phased in gradually, and the focus is on their longer-term impacts, there is less need to distinguish shorter-term responses (which would add considerable analytical complexity).

Full pass through of fuel prices into consumer prices for electricity and other energy products is assumed, though this may be a reasonable long-run approximation, even for China, given ongoing de-regulation of the energy sector.[14] Linkages with international trade are also ignored, given that fuel tax reforms are imposed on fuel consumption (from both domestic and imported sources) and the impacts of mitigation in other countries through changes in international fuel prices are beyond our scope.

The model is solved by first developing BAU fuel use by sector going forward to 2030, using equations of the model and projections of energy prices and GDP. The impacts of policy reform are then calculated by computing induced changes in fuel and electricity prices, and the resulting changes in energy efficiency, use of energy products, and hence fuel demand across the three sectors. The resulting change in air pollution deaths, carbon emissions, and revenue are calculated from the changes in fuel use and the deaths, CO_2 emissions, and prior taxes per unit of fuel use. Economic welfare costs and net benefits are calculated by applying standard formulas in the literature (Parry et al., 2016).

Data

The International Energy Agency's Extended World Energy Balances is used to aggregate fuel use by sector in China for 2013 (this data is available for around 160 countries). Fuel use is projected forward using GDP forecasts, income elasticities for energy products of between 0.25 and 1.25, and assumptions about autonomous technological change by fuel and sector taken from other studies.

Fuel prices and taxes are taken from IMF sources. Supply prices for fossil fuels are inferred from an international reference price (adjusted for transport and distribution costs), user prices are based on publicly available sources, and the difference between the two (after adjusting for general consumption taxes that should be applied to household fuels) is the specific fuel tax. Supply prices are projected forward using an average of (most recent) international projections from the US Energy Information Administration

---

[13] A threshold of 26,000 tons of CO_2 per year has been suggested for participation in the China ETS.

[14] In markets with regulated prices it can be very difficult to gauge what fraction of a new tax will be absorbed in losses for state owned enterprises. Alternatively, the new taxes might be interpreted as the nominal tax times the (generally large) fraction that is passed forward.

and the IMF (based on futures markets), while (real) pre-existing fuel taxes are taken as constant out to 2030.

Price elasticities (assumed constant) for electricity demand, road fuels, and fuels used in the other energy sector are taken to be –0.5, with half of the response coming from reductions in the use of energy-consuming products and half from improvements in energy efficiency.[15] Within the power sector, the coal price elasticity is –0.35.

B. Policies

This subsection discusses the different policy instruments under 'moderate' and 'aggressive' stringency scenarios.

Carbon tax: A comprehensive tax on fossil fuel CO_2 emissions promotes the full range of emissions mitigation opportunities (switching to cleaner fuels, improving energy efficiency, conserving on usage of energy-consuming products) across all sectors.

The best way to administer the tax would be to levy it upstream at the point of entry in the economy, for example, at the mine mouth for coal building off existing administrative structures for China's Resource tax[16] and for petroleum products at the refinery or gas processing plants, while imported fuel would be taxed at the border. There are currently in China about 11,000 coal mines (though restructuring will likely close around 4,000 over the next few years) and far fewer petroleum refineries and gas processing plants. This would contrast, by orders of magnitude, with the number of transactions the tax administration would have to monitor to collect a carbon tax downstream. Alternatively, the tax could be levied on large emitters though, besides missing a significant portion of emissions, measuring emissions is technically more challenging than measuring carbon content of fuel combustion, requiring a high level of technical expertise typically not found in tax administrations (Calder, 2015).

Two carbon tax scenarios are considered, including a moderate case with the tax rate increasing in equal yearly increments of $2.5 per ton from 2017 to reach $35 per ton by 2030 and an aggressive case with yearly tax rates twice as high.

Coal excise: This policy (and its stringency) is the same as for the carbon tax, with charges just applied to coal use.

ETS: This policy, building on regional pilot schemes, has been announced for China starting in 2020 (though the caps are yet to be determined) and is assumed here to apply to emissions from power generators, large industrial sources,[17] and domestic aviation, which

---

[15] Improvements in energy efficiency reduce unit operating costs for energy consuming products, hence increasing their demand, though the resulting extra energy use from this 'rebound effect' offsets only about 10 percent of the savings from higher efficiency.

[16] That is, adding a specific component to the ad valorem structure recently introduced. Alternatively, the tax could be set on coal processing plants which are far fewer in number than coal mines.

[17] Including petro and other chemicals, building materials, iron and steel, non-ferrous metals, and paper (see http://carbon-pulse.com/14353).

amounts to about 50 percent of current economy-wide CO_2 emissions. To facilitate policy comparisons, the ETS is modelled by its implicit tax, that is, the emissions price that would be established by the cap and (given equivalency between charges on carbon content and on emissions) with charges applying to fuels used in power generation and large industry for the same price trajectories as under the carbon tax.

Electricity excise: Excises on (mostly residential) electricity are applied in many countries (in part rationalized on environmental grounds), though their environmental effectiveness is limited as they do not promote switching to cleaner generation fuels or reductions beyond the power sector. Electricity taxes (applied to all uses) are considered, with the rates matched to the increase in electricity prices under the modest and aggressive carbon tax scenarios respectively.

Increased renewable generation subsidies: Here the focus is on renewables in power generation, given their greater potential for use in that sector (though intermittency and the geographic mismatch between sites and population centers limit scaling up). Renewable subsidies have limited effects on reducing CO_2 emissions as they do not promote some fuel switching possibilities (e.g., from coal to gas and from these fuels to nuclear), nor do they reduce electricity demand, or emissions beyond power. Subsidies for renewable power generation in China amounted to $0.03 per kWh in 2013 and moderate and aggressive scenarios are considered increasing the subsidy by 50 and 150 percent respectively above this level (higher subsidies than this start to imply negative generation costs).[18]

Reducing CO_2/kWh in power generation: Another policy to reduce power sector CO_2 is to impose a CO_2/kWh standard.[19] One fiscal analog of this policy is a tax/subsidy scheme involving taxes on relatively dirty generators—in proportion to the difference between the average CO_2 per kWh across their plants and a pivot point emission rate—and subsidies for relatively clean generators—in proportion to the difference between their CO_2 per kWh and the pivot point. A second fiscal analog is a carbon tax applied to the emissions content of power generation fuels with all the revenues recycled in a subsidy per unit of power generation (e.g., Bernard et al., 2007), which is how the policy is modelled here, with the carbon tax rates chosen to mimic those in the modest and aggressive carbon tax scenarios. All of these policies promote all opportunities for fuel switching to reduce power sector emissions, rather than just shifting to renewables (e.g., Krupnick et al., 2010), though they have a much weaker impact on electricity prices.

---

[18] In principle, there is an economic rationale for combining carbon pricing with renewables incentives if this addresses additional market failures (e.g., the inability of firms developing, or pioneering use of, technologies to capture spillover benefits to other firms from their own 'learning-by-doing' experiences) though whether they warrant substantial renewable deployment incentives is not entirely clear (e.g., Dechezleprêtre and Popp, 2016; Löschel and Schenker, 2016).

[19] Credit trading would be required for the policy to be cost effective in a model (unlike the present one) with differences in abatement cost schedules across power generators.

Increasing the efficiency of electricity-using capital: Regulations are commonly used to raise the efficiency of electricity-using capital.[20] The policy scenario considered here provides an upper bound on effectiveness and cost-effectiveness in the sense that it implicitly improves the efficiency of all electricity-using capital (appliances, lighting, buildings, heating and cooling equipment, etc.), and with equalized incremental costs per ton of CO_2 reduced across all products.[21] The policy is modelled by applying an implicit tax (with rates equal to those in the electricity tax scenarios) to reduce the electricity consumption rate, but not applying it to the demand for electricity-using capital, hence usage increases slightly from the rebound effect.

Higher road fuel taxes: Road fuel taxes in China were $0.16 and $0.13 per liter for gasoline and diesel respectively in 2013. These taxes are the most effective policies for reducing road fuel use as they promote both higher fuel economy and less driving. A modest scenario is considered where gasoline and diesel taxes are increased by the same amount as in the aggressive carbon tax and these tax increases are doubled for the aggressive scenario.

Fuel economy policies: Passenger vehicles have been regulated in China since 2005, the latest standards targeting new vehicle fuel consumption of 5 liters per 100 km (48 miles per gallon) by 2020 (UNEP, 2015). Heavy-duty vehicles (trucks and buses), which consume diesel, are not subject to regulation.[22] As the model does not distinguish vehicles of different vintages, an implicit policy raising average on road fuel economy progressively over time is considered. In the moderate scenario, the increase in fuel economy in each period matches that in the aggressive road fuel tax and in the aggressive fuel scenario, the increase in fuel economy in each period is twice that in the moderate fuel economy scenario.

Increasing efficiency in the other energy sector: The final policy increases the energy efficiency of fossil fuel-using capital for large users in the other energy sector (but not small users who are more difficult to regulate). As above, the policy is modelled by applying an implicit tax to reduce the consumption rate of coal, natural gas, and oil products but not applying it to the price in the demand for use of energy products. The implicit tax is chosen to mimic the increase in fuel price under the modest and aggressive carbon tax scenarios.

---

[20] Besides their environmental benefits, it is sometimes suggested that these policies address an additional market failure due to the private sector undervaluing the discounted energy savings from higher energy efficiency, though the evidence on this is mixed (e.g., Allcott and Wozny, 2013; Helfand and Wolverton, 2011). Allowing for this market failure could imply that, up to a point, policies to increase energy efficiency could have net economic benefits (before counting environmental benefits), though these benefits appear to be small relative to those from directly pricing emissions (e.g., Parry et al., 2014a).

[21] In reality, some capital may be difficult to regulate (e.g. smaller appliances, audio and entertainment equipment, industrial processes such as assembly lines) and without extensive credit trading incremental costs may differ substantially across different efficiency programs.

[22] Implementing regulations for heavy trucks, for example, is complicated given that fuel economy is very sensitive to the weight of freight (see Harrington, 2012).

3. RESULTS

This section discusses the baseline scenario, policy comparisons, sensitivity analyses, and a fully efficient pricing policy.

A. Baseline Projections

The baseline projections assume no new (or tightening of existing) policies beyond those implicit in observed data for 2013, aside from regulations resulting in a progressive reduction of local air pollution mortality rates and policies are then considered relative to this baseline. Inevitably, these projections are sensitive to different parameter assumptions, though this is less applicable to the relative policy comparisons which are the main focus of this chapter.

Figure 2.1 shows baseline and CO_2 emissions trends. GDP expands 131 percent between 2015 and 2030 (from IMF projections), while total energy consumption increases by 27 percent, implying a 45 percent decline in the energy to GDP ratio. CO_2 per unit of energy remains about constant however, implying a similar decline in the CO_2 to GDP ratio—although the productivity of zero-carbon fuels grows faster than for coal, this has little effect on reducing coal use per unit of energy, given lower future coal prices (see below) and limited substitution between fuels.[23] The CO_2 to GDP intensity in 2015 was 87 percent of the 2005 level[24] and declines to 45 percent of it by 2030, still substantially short of China's pledge for Paris to lower it by 60–65 percent below the 2005 level. Overall CO_2 emissions are 21 percent higher in 2030 compared with 2015.[25]

There is little change in the composition of primary energy out to 2030—the share of non-fossil fuel energy rises from 9 to 13 percent, while that for coal share falls from 66 to 63 percent. Given its high carbon intensity coal accounts for a disproportionately larger share, 82 percent in 2015, of CO_2 than for primary energy, while natural gas accounts for 3 percent and (road and non-road) oil products 14 percent. In terms of sectors, electricity accounts for 40 percent of CO_2 emissions in 2015, the transportation sector 7 percent, and the other energy sector 53 percent, and there is little change in these shares in the baseline.

---

[23] On the consumer demand side, there should be a high degree of substitution between electricity generated from fossil fuels and from renewables but evidence discussed in Appendix 2A.2 implies substitution possibilities on the supply side are more limited.

[24] Emissions intensity is 77 percent of the 2005 level in 2013 but then rises initially with the sharp decline in coal prices (and the immediate adjustment of the fuel mix in the model) to reach 87 percent by 2015 before declining.

[25] These energy demand and CO_2 projections are broadly consistent with those from the range of energy models for China summarized in Mischke and Karlsson (2014), Figures 2 and 3—GDP growth is moderately larger in the present model, though this is offset by a faster decline in the energy intensity of GDP. The projected decline in the energy intensity of GDP is about a third smaller than in Green and Stern (2016), with part of the difference due to the counteracting effect (in the current model) of lower future energy prices. In addition, the CO_2 intensity of energy is projected to fall by about 20 percent in Green and Stern (2016), based on extrapolating recent trends. Our projections for future coal use and CO_2 emissions are similar to those for China in IEA (2015), Current Policies Scenario (for the same future energy price projections).

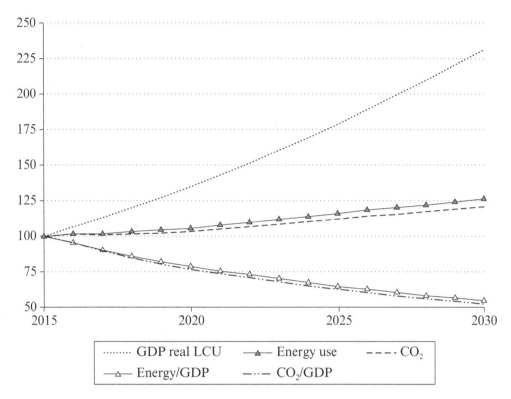

Note: The CO_2 intensity of energy changes very little in the baseline, implying almost identical trends in energy and CO_2, and in energy and CO_2 intensities, relative to GDP.

Source: Parry et al. (2016).

Figure 2.1 Energy use and CO_2: baseline scenario (2015 = 100)

Figure 2.2 indicates (real) energy price trends assumed in the baseline. All fossil fuel prices decline sharply between 2013 and 2016—by 62 percent for crude oil, about 31 percent for coal, 48 percent for natural gas, 45 percent for road fuels, and 12 percent for electricity—and thereafter rise slowly (or remain flat for electricity) but are still well below 2013 levels in 2030. Renewables prices, as proxied by power generation costs, fall by over 50 percent during the period (based on an assumed annual productivity growth rate of 4.5 percent).

Finally, estimated annual deaths from fossil fuel air pollution are 1.12 million in 2015[26] and rise to 1.3 million by 2030, with about half of the increase due to increased coal use and the other half rising population exposure due to urban migration (Parry et al., 2016). Initially, 53 percent of deaths are from coal combustion in the power sector and 42 percent

---

[26] This is significantly less than the estimate of outdoor air pollution deaths in the Global Burden of Disease project (Brauer et al., 2012), one potential explanation being that the latter also includes pollution from non-fossil sources (e.g., agriculture, plastics, refrigerants, landfills, mining).

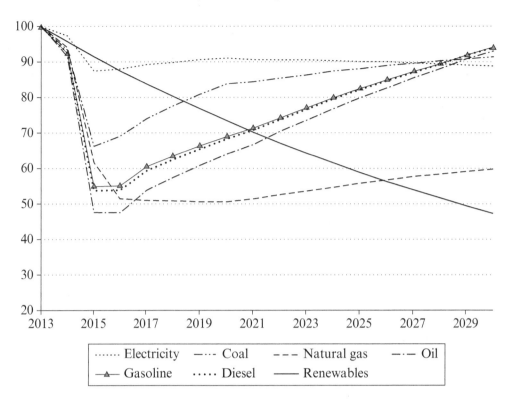

Notes: Gasoline and diesel prices follow identical trends. Renewables prices are renewable power generation costs. China recently introduced a domestic oil price floor of RMB 260 ($40) per barrel of oil but this is non-binding in our scenarios and is not reached in the baseline.

Source: IMF (2015) and authors' assumptions.

Figure 2.2 Energy prices: baseline scenario (2013 = 100)

from coal use in the other energy sector, however, the share of power sector coal in total deaths drops to 28 percent by 2030 due to greater deployment of control technologies at coal plants which roughly halves the industry average air pollution emission rate by 2030.

B. **Policy Comparison**

This subsection compares the impact of different policies on CO_2, revenue, air pollution deaths, and economic welfare, relative to the baseline outcome.

(i) CO_2 emissions: Figure 2.3 indicates the percent reduction (relative to the baseline level) in CO_2 emissions in 2020 and 2030 under each policy and stringency scenario. The carbon tax is the most effective policy, reducing CO_2 by 13 percent and 30 percent below baseline levels in 2020 and 2030 in the aggressive case, and by 7 percent and 19 percent in those years (both modest and aggressive taxes meet China's emissions intensity target) in

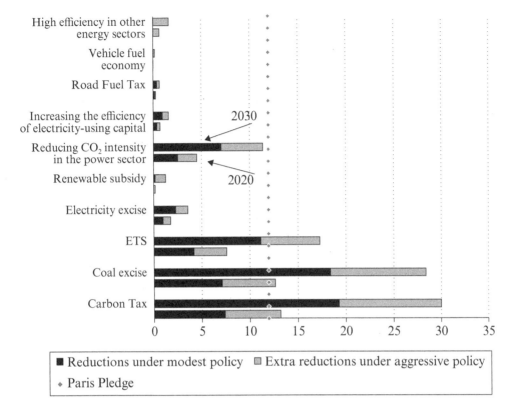

Note: Bars indicate percent reductions in fossil fuel CO_2 emissions relative to baseline emissions in that year.

Source: Parry et al. (2016).

Figure 2.3 Carbon emissions reductions in 2020 and 2030 (percent)

the moderate case. These results are driven by reductions in coal use, which account for 97 percent of the CO_2 reductions.[27] Doubling the carbon tax has a less than proportionate impact on CO_2 reductions, given the standard assumption that fuel demand curves are convex. The coal tax reduces CO_2 by 95–96 percent of the reductions under the carbon tax across years and stringencies, this small difference reflecting the relatively small emissions reductions forgone from failing to charge for CO_2 from natural gas and oil.

The ETS has intermediate effectiveness, reducing emissions by about 57 percent of the reductions under the carbon tax across years and stringency scenarios. The ETS produces the same CO_2 reductions from the power sector as under the carbon tax, but only a quarter of those from the other energy sector, as it does not cover small users, and none

---

[27] The moderate tax in 2020 raises coal prices by 23 percent and reduces coal use by 9 percent. In contrast, road fuel prices increase by only about 3 percent, and in any case these fuels only accounted for 7 percent of economy-wide emissions in the baseline for 2020.

from road transportation. The power sector CO_2/kWh intensity standard has about 30–40 percent of the effectiveness of the carbon tax across years and scenarios. The electricity excise has about 12 percent of the effectiveness of the carbon tax, that is, about 12 percent of the emissions reductions under the carbon tax come from reductions in electricity demand. This reduction is split about equally between improvements in energy efficiency and less usage of electricity-using capital—hence the policy to increase the efficiency of electricity-consuming products has about 5 percent of the effectiveness of the carbon tax. The road transportation policies have very limited effectiveness, and the same applies for the enhanced renewable generation subsidy (as this builds off a small base) and the efficiency policy for the other energy sector (which applies only to large firms).

Although combinations of policies are not explicitly modelled, in many cases they are essentially additive. For example, a regulatory combination to reduce the CO_2 intensity of power generation and improve energy efficiency across all three sectors has less than half of the effectiveness of the carbon tax.

(ii) Revenue: As indicated in Figure 2.4, the carbon tax also has the greatest fiscal benefit, raising revenues of 1.7 percent and 3.0 percent of GDP in 2030 in the modest and

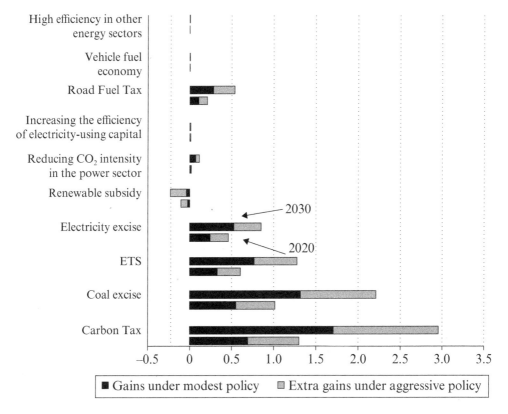

Source: Parry et al. (2016).

Figure 2.4 Fiscal gains in 2020 and 2030 (percent of GDP)

aggressive scenarios. Although carbon tax rates are 3.5 times as high in 2030 compared with 2020, revenues are only about a third higher relative to GDP because the baseline CO_2 to GDP ratio is 50 percent lower in 2030 and the higher carbon taxes have a bigger impact on eroding the tax base. Again, the coal tax is not far behind, raising revenues of 74–79 of those under the carbon tax across years and stringency scenarios. The ETS—if allowances are auctioned—and the electricity tax are intermediate cases, raising revenues of about 45 and 30 percent respectively as under the carbon tax. Road fuel taxes raise about 18 percent of the revenue from the carbon tax. Policies to reduce the CO_2 intensity of power generation and to improve energy efficiency in the power and other energy sectors have no revenue impacts. The renewable generation subsidy loses revenue, as does the vehicle fuel economy policy (which erodes the tax base of prior fuel taxes) but the losses are relatively small (less than 0.25 percent of GDP).

(iii) Local air pollution deaths: The percentage reduction in air pollution deaths in 2020 and 2030 for the major CO_2 mitigation policies are fairly similar to the percentage CO_2 reductions. For example, the modest carbon tax and coal tax both reduce deaths by about 9 percent in 2020 and 22 percent in 2030, while the aggressive versions of these taxes reduce deaths by 33 percent in 2030. The ETS reduces deaths between 5 and 20 percent across years and scenarios. More interesting perhaps is Figure 2.5 showing the

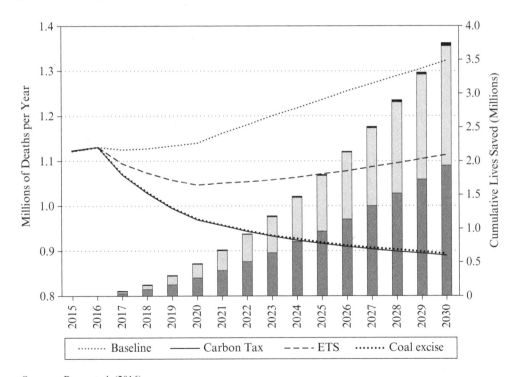

Source: Parry et al. (2016).

Figure 2.5 Pollution-related premature deaths: 2015–30 (millions, aggressive policy scenarios)

46   *Handbook on the economics of climate change*

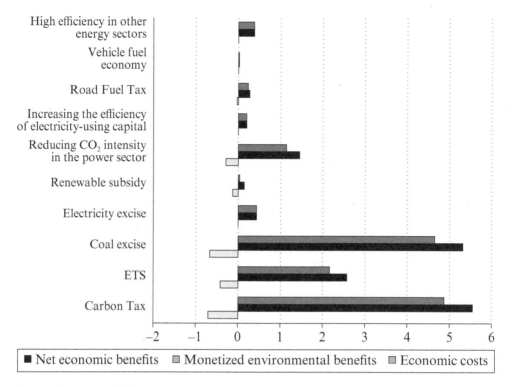

Source: Parry et al. (2016).

Figure 2.6 Domestic economic benefits and costs in 2030 (percent of GDP, aggressive policy scenarios)

time profile of air pollution deaths under selected policies in the aggressive scenarios. Lives saved (the difference between deaths in the baseline and under different policies) progressively increases over time as policies become more stringent. Cumulated over the 2017 to 2030 period, the carbon and coal taxes save about 3.7 million lives and the ETS about 1.9 million.

(iv) Domestic welfare benefits and costs: Figure 2.6 indicates the economic welfare costs, monetized domestic environmental benefits (essentially local air pollution benefits, excluding global climate benefits), and net welfare benefits (environmental benefits less economic costs) focusing on the aggressive policies for 2030. The carbon and coal tax perform far better than other policies, causing costs of about 0.7 percent of GDP, domestic environmental benefits approaching 6 percent of GDP, leaving a net welfare gain approaching 5 percent of GDP—a huge number.[28] Substantial carbon pricing

---

[28]   For comparison, Robert Lucas once estimated that replacing capital taxes with (less distortionary) labor taxes in the United States would generate an annual welfare gain approaching one

(beyond meeting its Paris pledge) is therefore in China's own domestic interest due to the local pollution benefits. Net welfare gains are smaller but still significant under the ETS at 2.2 percent of GDP, 1.2 percent under the policy to reduce the CO_2 intensity of power generation, and 0.5 percent of GDP or less under all other policies.

(v) Sensitivity analysis: Table 2.1 presents sensitivity analyses for the carbon and coal tax, ETS, and CO_2/kWh policy for the moderate scenarios (see Parry et al., 2016 for details).

The percent reduction in CO_2 emissions under different policies is obviously sensitive to fuel price elasticities—for example, if the magnitude of fuel price elasticities is 50 percent larger, the percent CO_2 reductions under different policies are increased by about 30–40 percent. Fossil fuel prices also matter—with (higher) International Energy Agency (IEA) (2015) price projections, the percent CO_2 reductions under different policies are about a third lower (in part because carbon charges have a smaller proportional impact on fuel prices). Changing income elasticities for energy products affects baseline CO_2 emissions but has essentially no effect on the policy-induced percent reductions in CO_2.

Revenue gains from fiscal policies as a percent of GDP are moderately sensitive to income and elasticities and productivity trends (which affect future tax bases relative to GDP).

Cumulative lives saved over the 2017–30 period are sensitive to parameter variations as they affect either baseline deaths and/or policy responsiveness. For example, with higher fuel price elasticities the carbon and coal taxes save 3.4 million lives, while lives saved from the ETS drops to just under 1 million under a lower scenario for baseline air pollution deaths. Welfare gains (discounted over the 2017–30 period and expressed as a percent of 2015 GDP) vary significantly in absolute terms but the relative welfare gains from policies are fairly robust—in all cases the ETS achieves about half of the welfare gains from the carbon and coal tax, and the CO_2/kWh policy achieves about 25–35 percent of these gains.

(vi) Comparison with a fully efficient policy: Parry et al. (2016) consider a fully efficient policy comprehensively charging fossil fuels for global and domestic environmental costs. This policy imposes much higher coal taxes (to reflect domestic air pollution costs) than the aggressive carbon tax. It reduces CO_2 emissions in 2030 by 58 percent, reduces air pollution deaths by 63 percent, raises revenue of 8.5 percent of GDP, and generates a net domestic welfare gain (again, excluding global climate benefits) of 6.2 percent of GDP.[29] Nonetheless, the aggressive carbon tax still performs reasonably well achieving 52 percent of the CO_2 reductions as under the fully efficient policy, 52 percent of the air pollution deaths, and 79 percent of the domestic welfare gains in 2030, though the shortfall in fiscal benefits is more pronounced at 35 percent of those under the fully efficient policy.

---

percent of GDP which he called 'the largest genuinely free lunch I have seen in 25 years' (Lucas, 1990, p. 314).

[29] A cautionary note here is that the uncertainties surrounding the effects of such dramatic policy changes are especially large.

Table 2.1 Sensitivity analysis: moderate policy scenarios in 2030

		CO$_2$ reduction (%)			Revenue gain (% of GDP)				Cumulative lives saved (millions)				PDV of welfare gain (% of 2015 GDP)			
		Carbon tax	Coal excise	ETS	Reducing CO$_2$/kWh for power	Carbon tax	Coal excise	ETS	Reducing CO$_2$/kWh for power	Carbon tax	Coal excise	ETS	Reducing CO$_2$/kWh for power	Carbon tax	Coal excise	ETS	Reducing CO$_2$/kWh for power
Central case		19.3	18.4	11.1	7.1	1.7	1.3	0.8	0.1	2.3	2.3	1.2	0.6	33.6	32.5	16.3	8.8
Income elasticities	Low	19.2	18.4	11.1	7.0	1.3	1.0	0.6	0.1	1.9	1.9	1.0	0.5	28.0	27.0	13.6	7.3
	High	19.3	18.5	11.2	7.1	2.2	1.7	1.0	0.1	2.8	2.7	1.4	0.8	40.4	39.1	19.6	10.5
Price elasticities	Low	10.0	9.7	6.2	4.2	1.8	1.4	0.8	0.1	1.1	1.1	0.6	0.4	15.7	15.3	8.2	4.9
	High	26.2	24.9	15.1	9.8	1.6	1.2	0.7	0.0	3.4	3.3	1.7	1.0	49.3	47.4	24.0	12.9
Productivity growth	Low	19.0	18.1	11.0	6.1	1.9	1.4	0.8	0.1	2.4	2.4	1.2	0.6	40.1	38.9	22.0	12.6
	High	19.6	18.9	11.4	7.3	1.5	1.2	0.7	0.1	2.2	2.2	1.1	0.6	32.0	30.9	15.5	8.4
Fossil fuel prices	Low	27.1	25.7	14.5	8.9	1.5	1.1	0.7	-0.1	3.4	3.4	1.6	0.8	49.5	47.5	21.4	10.4
	High	12.5	12.1	7.7	5.2	1.9	1.5	0.9	0.1	1.4	1.4	0.8	0.6	20.8	20.2	11.0	7.6
Mortality rates	Low	19.3	18.4	11.1	7.1	1.7	1.3	0.8	0.1	1.5	1.5	0.8	0.4	21.4	20.5	10.6	5.8
	High	19.3	18.4	11.1	7.1	1.7	1.3	0.8	0.1	2.7	2.7	1.6	1.0	40.0	38.9	22.7	13.8
Commodity prices	Low	24.7	23.5	13.7	9.7	1.9	1.5	0.9	0.2	3.5	3.5	1.7	1.1	51.7	49.9	24.1	15.3
	High	15.8	15.2	9.3	6.0	1.5	1.2	0.7	0.1	1.6	1.6	0.9	0.5	24.2	23.4	12.0	6.6

Source: Parry et al. (2016).

4. INCIDENCE ANALYSIS

Having a sense of the distributional incidence of carbon pricing across household and industry groups is critical to informing policy dialogue and aiding in the design of accompanying measures (e.g., compensation schemes). Policies perceived as broadly fair in this regard are not only desirable for their own sake but also may stand a better chance of being enacted and sustained. Household and industry incidence are discussed in turn below, again using China to illustrate the application of analytical tools.

A. Household Incidence

(i) Methodology: A first pass estimate of the incidence on household groups from higher prices of consumer products caused by carbon pricing can be obtained by calculating first-order consumer surplus losses using:

$$\sum_g \pi_t^{hg} \cdot \rho_t^{hg}$$

Here h denotes a household income group, $g = 1 \ldots G$ denotes major categories of consumer goods whose prices rise in response to carbon pricing, π_t^{hg} is the share of household h's budget spent on good g at time t and ρ_t^{hg} is the percent increase in the price of good g. According to this formula, if the budget share for a product is, say, 5 percent, a 10 percent increase in its price will decrease the household group's real income by the equivalent of 0.5 percent.

The budget shares for energy and non-energy products needed to implement this formula are available from household expenditure surveys, which are becoming increasingly common.[30] The direct price impacts on energy products are an output of the previously described spreadsheet tool or can be inferred from CO_2 emissions factors.[31] Indirect impacts on the prices of other consumer goods can be estimated, assuming full pass through, from input–output tables[32]—dividing fuel and electricity purchases by fuel prices, and applying CO_2 emissions factors, gives the embodied CO_2 per $ for each intermediate, and ultimately each consumer product, and multiplying by the CO_2 price gives the product price increase. Input–output tables are also becoming more widely available (e.g., Timmer et al., 2015).

There are a number of caveats to using the above formula as the basis for incidence analysis.

---

[30] These surveys are routinely conducted for many advanced countries and the World Bank's Living Standards Measurement Study compiles them for approximately 40 developing countries.

[31] Fuel price increases are the carbon price times the fuel's CO_2 emissions factor and the latter are well established and vary very little (per unit of energy for coal and natural gas and per liter for petroleum products) across countries. Approximate impacts on electricity prices can be inferred from the CO_2 emissions per unit of electricity, which are available by country from the International Energy Agency. Baseline fuel prices (needed to convert absolute increases into percent increases) are available from an IMF database (www.imf.org/external/np/fad/subsidies/data/subsidiestemplate.xlsx).

[32] See, for example, Coady and Newhouse (2006).

For one thing, the CO_2 emissions factor for power generation, and the embodied carbon in various energy products, will decline in response to carbon pricing, hence use of input–output tables leads to an overstatement of the consumer price increases from carbon pricing, though this overstatement is relatively modest for the level of carbon pricing envisioned for the near to medium term. Similarly, the formula overstates the loss of consumer surplus by ignoring reductions in demand for energy-intensive products in response to carbon pricing, though again the difference is relatively modest for non-dramatic carbon prices.[33]

Another caveat is that some (likely minor) fraction of the burden of carbon taxes may be passed backwards in lower producer prices, if fuel supply curves are upward sloping in the medium to longer term. To the extent this reduces the net of tax return to capital, some of the incidence of carbon pricing is borne by owners of capital, though if net of tax returns are largely determined in world markets the burden of lower producer prices is mainly borne by workers in the form of lower wages. The resulting incidence effects become difficult to estimate as they depend on whether energy-intensive firms disproportionately hire high- or low-wage workers, substitution elasticities between energy and other inputs, etc. (e.g., Fullerton and Heutel, 2011), though some studies suggest these incidence effects are not that large and may lower the regressivity of carbon pricing.[34]

Furthermore, household survey and input–output data is backward looking. Incidence effects could be projected forward based on assumptions about how household consumption patterns and the energy intensity of different industries might change over time, though these trends are likely gradual—below, we assume household spending patterns and industry structure in 2020 are the same as in 2012, the last year of available survey and input–output data.[35]

The appropriate definition of income against which carbon pricing burdens should be measured is also somewhat unsettled. Annual income is problematic given that many people with low annual income (e.g., students, people temporarily laid off or on maternity leave) are not poor in a life-cycle context, yet they contribute greatly to disparities in annual income across households.[36] This problem is partly (though, because of constraints on consumption smoothing across the life cycle, not fully) alleviated by measuring incidence against annual consumption expenditure rather than annual income.[37]

---

[33] For example, the first-order approximation (a rectangle) overstates the loss of consumer surplus (a trapezoid) by only about 5 percent when demand for a fuel product falls by 10 percent. Substantial carbon prices may be implemented over a longer time horizon, but incidence analyses can be periodically updated with new household survey data to account for the impact of previous carbon price increases on budget shares.

[34] For example, Rausch et al. (2011).

[35] As long as any trends reduce energy budget shares for all household groups in roughly the same proportion, the relative incidence of carbon pricing across households should be largely unaffected. One exception might be the prospects for rising budget shares for gasoline among middle and lower income households with potential for growth in vehicle ownership rates among these groups.

[36] Up to one-half of the inequalities in annual income across households might be attributed to variations in income over their life cycle rather than differences between life-cycle income (Lillard, 1977).

[37] See for example Poterba (1991), Hassett et al. (2009).

Yet another caveat is that here (as in other studies) the distributional incidence of the domestic environmental benefits of carbon pricing—principally the air pollution benefits—are not considered. Supposing (based on OECD, 2012) that the valuation of health risks is roughly proportional to income, then these benefits may be skewed to lower income households if these households are more likely to reside in severely polluted areas. Again, the effects become complex however, if they raise property values in areas with improving air quality with adverse effects for low-income renters.

Finally, the overall incidence effects of carbon pricing will critically hinge on how revenues are used. Regressive impacts from energy price increases might be offset through cuts in social security contributions or increases in welfare and social spending targeted at the poorest households[38] while the incidence implications of using revenues for cutting personal income, payroll, consumption, and corporate income taxes will depend on the specific parameters of a country's fiscal system and the degree of income inequality.[39]

(ii) Application to China: For China, household budget shares were obtained from the China Family Panel Studies (CFPS) data which provides household expenditures for 25 aggregated categories of goods and services. The latest year available for the survey is 2012 and includes information from a nationally representative sample of more than 13,000 households.[40] Households were first separated into income deciles using consumption as a proxy for permanent income and budget shares were calculated by dividing expenditure on individual goods and services by total household consumption.

As illustrated in Figure 2.7, the fraction of total expenditure spent on energy declines sharply with income from about 10 percent for the bottom income decile to about 4 percent for the top decile. The large reductions in the budget shares between the 1st and 2nd deciles, and between the 9th and 10th deciles, reflect the high degree of income inequality at the bottom and top end of the income distribution in China (the poor being especially poor and the rich being especially rich). Lower-income groups also allocate a relatively higher share of energy expenditure to coal, electricity, heating, and natural gas while higher-income groups spend disproportionately more on gasoline.[41]

We also used China's national input–output table for 2012, the latest version published by the National Bureau of Statistics, which disaggregates 139 sectors. We look at the incidence of a $10 per ton carbon tax for 2020 assuming the budget share and input–output data for 2010 is still applicable, and from the spreadsheet model this carbon price increases

---

[38] See Dinan (2015) for an extensive discussion of options in the United States.

[39] Another noteworthy point is that if allowances in an ETS are given away for free to polluting firms, as happened for several pilot programs in China, this transfers windfall gains to owners of capital in these firms (predominantly higher income households) which can greatly compound the regressive effects of carbon pricing (Parry, 2004).

[40] The CFPS is conducted by the Institute of Social Science Survey at Peking University and covers about 95 percent of the Chinese population in 25 provinces. Income distribution and poverty studies have found the CFPS to be consistent with other large-scale nationally representative household surveys in China, while Xi et al. (2014) find the sex–age structure of the 2010 CFPS survey closely tracks the 2010 Census.

[41] The declining budget share for electricity is often observed for advanced economies (e.g., Morris and Mathur, 2015), though not for low-income countries where the poor lack access to the power grid (e.g., Arze del Granado et al., 2012).

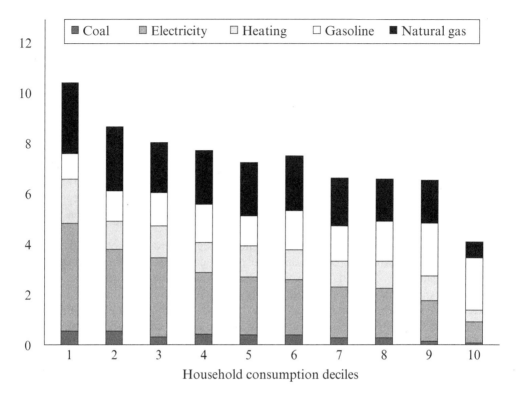

Source: IMF Staff estimates based on the CFPS 2012.

Figure 2.7 Composition of household energy expenditure by income group, 2012 (in percentage of total household expenditure)

the price of coal by 23 percent, natural gas by 6 percent, and gasoline by 3 percent. The products most impacted by these energy price increases include water, furniture, transport and communications, and cars, however the impacts on food, medical expenses, clothing, and cigarettes and alcohol turn out to be more significant given the larger budget shares for these products.

Overall, the carbon tax increases consumer prices on average by around 1.1 percent in 2020, with the total effect regressive, though only mildly so—the burden for the bottom income decile is 1.4 percent of income and that for the top income decile is 0.9 percent (Figure 2.8). This mainly reflects burdens from the indirect price effects which are substantial and increase moderately (as a share of income) for higher income households. The incidence analysis of the other policies discussed above (the results are not presented here) reveal broadly similar patterns of relative incidence across the income distribution, though the degree of regressivity is even less pronounced for the regulatory measures.

Figure 2.9 compares distributional incidence when (as previously, and in line with many economists' recommendations) households are grouped by consumption, with incidence when households are grouped by annual income. Previous studies (mainly for advanced

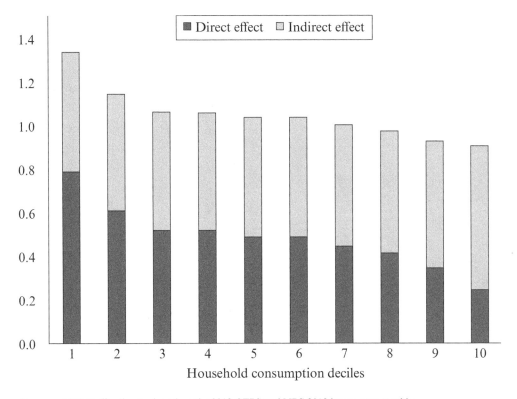

Source: IMF Staff estimates based on the 2012 CFPS and NBS 2012 input–output table.

Figure 2.8 Impact of a carbon tax in China, 2020 (in percent of total household consumption)

countries) find that carbon pricing is less regressive in the former case[42] but we find the opposite in the Chinese data—the burden of carbon pricing is approximately constant (1.1 to 1.3 percent of income) across different income groups. The difference in the degree of progressivity operates almost entirely through the direct impact of energy prices and is not due to any single source such as electricity or gasoline, but operates similarly for all energy sources.

A potentially important implication of this finding is that if the government wishes to use fiscal transfers to compensate the poorest—which would require using less than 5 percent of the carbon pricing revenues—care should be given in the identification of the neediest households. In particular using income as the basis for compensation might transfer more resources to income-poor households that in fact do not face such a large loss in purchasing power. On the other hand, identifying households based on aggregate consumption is quite data intensive and challenging. It might therefore be important

---

[42] See, for example, Morris and Mathur (2015).

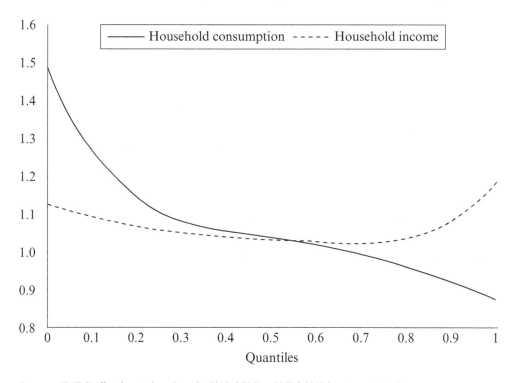

Source: IMF Staff estimates based on the 2012 CFPS and NBS 2012 input–output table.

Figure 2.9 Distributional impact of a carbon tax: consumption vs. income (in percent of total household consumption)

to complement eligibility requirements for targeted compensation programs with other indicators of financial deprivation.

B. Industry Incidence

Carbon pricing increases production costs across all industries, a particular concern being impacts on energy-intensive and trade-exposed sectors. Besides reducing competitiveness, these impacts can lead to emissions leakage, though estimated leakage rates are not that dramatic,[43] and some of the estimate reflects reductions in international fuel prices raising fuel demand in non-mitigating countries rather than migration of capital to these countries. Competitiveness and leakage are ameliorated to the extent there is global action on mitigation, which is the idea underpinning the 2015 Paris Agreement.

Measuring the incidence of carbon pricing on different sectors using input–output tables is less contentious than incidence analysis for households, the main issues being

---

[43] Typically, between about 5 and 20 percent, depending on the size and composition of the coalition of countries pricing carbon (e.g., Böhringer et al., 2012).

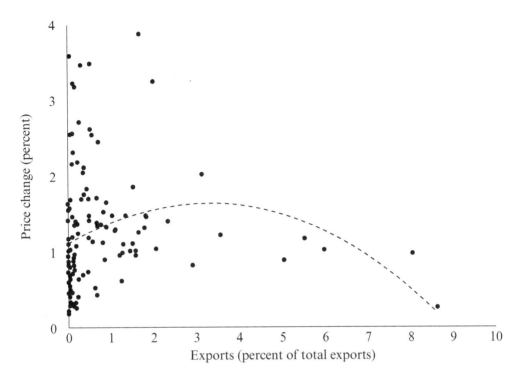

Source: IMF Staff estimates based on the NBS 2012 input–output table.

Figure 2.10   Cost increases from carbon tax and exports

whether there is full pass through of carbon pricing in energy prices, significant reductions in emissions or energy intensity in vulnerable sectors in response to carbon pricing, and possible changes in relative input prices from general equilibrium effects.

Figure 2.10 provides a preliminary sense of how the moderate carbon tax would increase costs across industries in China, by plotting the estimated increase in sectors' costs against their respective contribution to the country's total exports. While there is no clear pattern emerging, the sectors that contribute most to total exports are also among those that would face the smallest cost increase from higher energy prices—most of the hardest hit sectors are in fact those with small export shares.

Table 2.2 lists the 10 sectors most and least impacted by the introduction of a carbon tax based on energy intensity of production estimated from the input–output coefficient matrix. All the sectors in the top panel are heavy industries associated with China's 'old growth' model. In contrast, sectors that would experience the smallest cost increase are overwhelmingly in the service sector or in the consumer goods sectors such as tobacco and fishery. These results suggest that a carbon tax would also promote the rebalancing of the Chinese economy from heavy manufacturing, investment and real estate to services and consumption-led growth.

Similar results can also be easily estimated for sectors' share of value added and labor intensity. Panel a) in Figure 2.11 displays a strongly negative correlation between cost

56 *Handbook on the economics of climate change*

Table 2.2 Estimated cost increase from moderate carbon tax, 2020

	Cost Increase
Sectors Most Affected	
Basic Chemical Raw Materials	3.9
Cement, Lime and Gypsum	3.6
Brick, Stone and other Construction Materials	3.5
Fertilizer	3.5
Steel Flat-Rolled Products	3.3
Steel, Iron and Cast	3.3
Graphite and other Non-Metallic Mineral Products	3.2
Chemical Fiber Products	2.7
Composites	2.6
Ferroalloy Products	2.6
Sectors Least Affected	
Real Estate	0.19
Capital Market Services	0.21
Social Security	0.23
Wholesale and Retail	0.27
Insurance	0.27
Monetary and Financial and other Financial Services	0.29
Education	0.30
Tobacco Products	0.32
Entertainment	0.33
Fishery Products	0.34

Source: IMF Staff estimates based on the NBS 2012 input–output table.

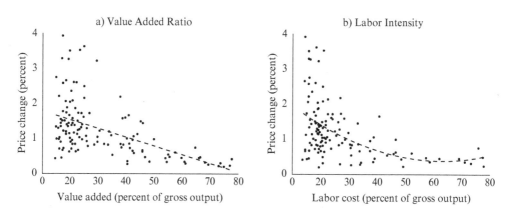

Source: IMF Staff estimates based on the NBS 2012 Input–output table.

Figure 2.11 Sectoral cost increase and rebalancing indicators

increases from a moderate carbon tax and the degree of value added by sector. This again suggests that a carbon tax would promote rebalancing by affecting relatively less the cost structure of sectors with high value added. We find a similar pattern for labor share of total output by sector in panel b). This suggests that labor intensive sectors would benefit more in relative terms, thereby promoting a rising share of labor and household income.

The bottom line is that competitiveness impacts of carbon pricing in China are not overly severe—in fact, even if China moves forward unilaterally with carbon pricing, trade-exposed firms could be fully compensated for higher energy costs using about 10 percent of the carbon pricing revenues.

5. CONCLUSION

This chapter describes a flexible model, implementable in a spreadsheet, for providing a first-pass comparison of a wide range of CO_2 mitigation policies across key metrics of concern to policymakers and applied it to China, though the tool can be readily applied to most other countries with publicly available data. While many of the qualitative insights from the above analysis might have been anticipated (e.g., applying the same charge to a wider range of emissions is obviously more effective at reducing emissions) it is still critical to understand the quantitative impacts of policies and the key factors determining them. This informs policymakers of the trade-offs in instrument choice, the specific actions needed to make headway on emissions commitments, and the broader (e.g., fiscal, public health, incidence) implications of these measures. More generally, future use of these sorts of tools for cross-country comparisons (based on consistent methodology and data) may be useful as countries revise their Paris mitigation commitments, taking into account the actions of others.

For China, the results underscore the drawbacks of a downstream ETS missing about half of coal use, and therefore forgoing roughly half of the emissions reductions, half of the lives saved, and half of the fiscal benefits of upstream charging systems covering all coal use. Upstream charges, moreover, could easily build off existing resource tax administration levied at the mine mouth. Although the government is committed to introducing an ETS in 2017, this should not preclude the simultaneous introduction of an upstream carbon or coal tax (perhaps with credits for firms required to obtain emissions allowances).

We would expect some of the results, such as the relatively limited environmental and fiscal effectiveness of road fuel and electricity taxes and incentives for renewable generation and energy efficiency, to carry over, in a broad sense, to other countries. Other results however, such as the strong environmental and fiscal advantages of carbon/coal taxes over ETS, will be less pronounced for other countries where coal is a minor share of CO_2 emissions or where (as in the United States) coal use is confined to the power and large industrial sectors.

There are multiple dimensions for refining the spreadsheet tool by incorporating additional policies (e.g., for biofuels) and country-specific factors (e.g., cross-border mobility of fuels), so long as transparency and accessibility to the non-specialist is preserved. One complementary direction for future research might be to develop a computable general equilibrium analogue with greater sophistication (e.g., capital dynamics, endogenous technological change) that can be collapsed to replicate the spreadsheet model, thereby informing on how additional modelling detail improves the accuracy of the policy results and providing guidance on possible 'offline' adjustments to the spreadsheet tool.

REFERENCES

Aldy, J., W. Pizer, M. Tavoni, L.A. Reis, K. Akimoto, G. Blanford, C. Carraro, L.E. Clarke, J. Edmonds, G.C. Iyer and H.C. McJeon, H.C. (2016), 'Economic Tools to Promote Transparency and Comparability in the Paris Agreement', *Nature Climate Change*, **6**(11), 1000.

Allcott, Hunt, and Nathan Wozny (2013), 'Gasoline Prices, Fuel Economy, and the Energy Paradox', NBER Working Paper No. 18583, National Bureau of Economic Research, Cambridge, MA.

Arze del Granado, J., D. Coady and R. Gillingham (2012), 'The Unequal Benefits of Fuel Subsidies: A Review of Evidence for Developing Countries', *World Development*, **40**(11), 2234–48.

Bernard, A.L., C. Fischer and A.K. Fox (2007), 'Is There a Rationale for Output-based Rebating of Environmental Levies?', *Resource and Energy Economics*, **29**, 83–101.

Böhringer, C., J.C. Carbone and T.F. Rutherford (2012), 'Unilateral Climate Policy Design: Efficiency and Equity Implications of Alternative Instruments to Reduce Carbon Leakage', *Energy Economics*, **34** (Supplement 2), S208–S217 (December).

Brauer, Michael, Markus Amann, Rick T. Burnett, Aaron Cohen, Frank Dentener, Majid Ezzati, Sarah B. Henderson, Michal Krzyzanowski, Randall V. Martin, Rita Van Dingenen, Aaron van Donkelaar and George D. Thurston (2012), 'Exposure Assessment for Estimation of the Global Burden of Disease Attributable to Outdoor Air Pollution', *Environmental Science and Technology*, **46**, 652–60.

Burnett, Richard T., C. Arden Pope, Majid Ezzati, Casey Olives, Stephen S. Lim, Sumi Mehta, Hwashin H. Shin et al. (2013), 'An Integrated Risk Function for Estimating the Global Burden of Disease Attributable to Ambient Fine Particulate Matter Exposure', *Environmental Health Perspectives*, **122**(4), 397–403.

Calder, Jack (2015), 'Administration of a U.S. Carbon Tax', in I. Parry, A. Morris, and R. Williams (eds), *Implementing a U.S. Carbon Tax: Challenges and Debates*, New York: Routledge, pp. 38–61.

Cao, Jing, Mun S. Ho and Dale W. Jorgenson (2013), 'The Economics of Environmental Policies in China', in C.P. Nielsen and M.S. Ho (eds), *Clearer Skies over China: Reconciling Air Quality, Climate, and Economic Goals*, Cambridge, MA: MIT Press, pp. 329–73.

Charap, Joshua, Arthur Ribeiro da Silva and Pedro Rodriguez (2013), 'Energy Subsidies and Energy Consumption—A Cross Country Analysis', Working Paper 13-112, International Monetary Fund, Washington, DC.

Coady, D. and D. Newhouse (2006), 'Evaluating the Distribution of the Real Income Effects of Increases in Petroleum Product Prices in Ghana', in A. Coudouel, A. Dani and S. Paternostro (eds), *Analyzing the Distributional Impacts of Reforms: Operational Experience in Implementing Poverty and Social Impact Analysis*, World Bank, Washington, DC.

Dahl, Carol, A. (2012), 'Measuring Global Gasoline and Diesel Price and Income Elasticities', *Energy Policy*, **41**, 2–13.

Dechezleprêtre, Antoine and David Popp (2016), 'Fiscal and Regulatory Instruments for Clean Technology Development in the European Union', Paper presented at a conference Energy Tax and Regulatory Policy in Europe: Reform Priorities and Research Needs, Ifo Institute, Munich, November 2014.

Dinan, Terry (2015), 'Offsetting a Carbon Tax's Burden on Low-Income Households', in I. Parry, A. Morris, and R. Williams (eds), *Implementing a US Carbon Tax: Challenges and Debates*, New York: Routledge, pp. 38–61.

Farid, Mai, Michael Keen, Michael Papaioannou, Ian Parry, Catherine Pattillo and Anna Ter-Martirosyan (2016), 'After Paris: Fiscal, Macroeconomic, and Financial Implications of Climate Change', Staff Discussion Note 16/01, International Monetary Fund, Washington, DC.

Fawcett, Allen A., Leon C. Clarke and John P. Weyant (2015), 'Carbon Taxes to Achieve Emissions Targets: Insights from EMF 24', in I. Parry, A. Morris, and R. Williams (eds), *Implementing a U.S. Carbon Tax: Challenges and Debates*, New York: Routledge, pp. 62–82.

Fullerton, Don and Garth Heutel (2011), 'Analytical General Equilibrium Effects of Energy Policy on Output and Factor Prices', *The B.E. Journal of Economic Analysis & Policy*, **10**, 1–26.

GCEC (2014), *China and the New Climate Economy*, Global Commission on the Economy and Climate, Tsinghua University, Beijing.

Green, Fergus and Nicholas Stern (2016), 'China's Changing Economy: Implications for its Carbon Dioxide Emissions', Centre for Climate Change Economics and Policy, Working Paper 258.

Grubb, M., F. Sha, T. Spencer, N. Hughes, Z. Zhang and P. Agnolucci (2015), 'A Review of Chinese CO_2 Emission Projections to 2030: the Role of Economic Structure and Policy', *Climate Policy*, **15**(S1), S7–S39.
Harberger, Arnold C. (1964), 'The Measurement of Waste', *American Economic Review*, **54**, 58–76.
Harrington, Winston (2012), 'Improving Fuel Economy in Heavy-Duty Vehicles', Discussion Paper 12-02, Resources for the Future, Washington, DC.
Hassett, K., A. Mathur and G. Metcalf (2009), 'The Incidence of a US Carbon Tax: A Lifetime and Regional Analysis', *Energy Journal*, **30**, 155–78.
Helfand, Gloria and Ann Wolverton (2011), 'Evaluating the Consumer Response to Fuel Economy: A Review of the Literature', *International Review of Environmental and Resource Economics*, **5**, 103–46.
IEA (2015), 'World Energy Outlook 2015', International Energy Agency, Paris, France.
IEA (2016a), 'CO_2 Emissions from Fuel Combustion', International Energy Agency, Paris, France.
IEA (2016b), 'World Energy Balances', International Energy Agency, Paris, France, accessed December 2016 at www.oecd-ilibrary.org/energy/data/iea-world-energy-statistics-and-balances/world-energy-balances_data-00512-en.
IMF (2015), 'People's Republic of China: Staff Report for the 2015 Article IV Consultation', International Monetary Fund, Washington, DC.
IMF (2016), 'World Economic Outlook', International Monetary Fund, Washington, DC.
Jamil, Faisal and Eatzaz Ahmad (2011), 'Income and Price Elasticities of Electricity Demand: Aggregate and Sector-Wise Analyses', *Energy Policy*, **39**(5), 519–27.
Karplus, Valerie J., Sebastian Rausch and Da Zhanga (2016), 'Energy Caps: Alternative Climate Policy Instruments for China?', *Energy Economics*, **56**, 422–31.
Krupnick, Alan J., Ian W.H. Parry, Margaret Walls, Tony Knowles and Kristin Hayes (2010), *Toward a New National Energy Policy: Assessing the Options*, Washington, DC, Resources for the Future and National Energy Policy Institute.
Lam, W. Raphael and Philippe Wingender (2015), 'China: How Can Revenue Reforms Contribute to Inclusive and Sustainable Growth?', IMF Working Paper 1566, Washington, DC.
Lillard, L.A. (1977), 'Inequality: Earnings versus Human Wealth', *American Economic Review*, **67**, 42–53.
Löschel, Andreas and Oliver Schenker (2016), 'On the Coherence of Economic Instruments: Climate, Renewables and Energy Efficiency Policies', Paper presented at a conference Energy Tax and Regulatory Policy in Europe: Reform Priorities and Research Needs, Ifo Institute, Munich, November 2014.
Lucas, Robert (1990), 'Supply-Side Economics: An Analytical Review', *Oxford Economic Papers*, **42**, 293–316.
Mischke, Peggy and Kenneth B. Karlsson (2014), 'Modelling Tools to Evaluate China's Future Energy System – A Review of the Chinese Perspective', *Energy*, **69**, 132–43.
Morris, Adele and Aparna Mathur (2015), 'The Distributional Burden of a Carbon Tax: Evidence and Implications for Policy', in I. Parry, A. Morris, and R. Williams (eds), *Implementing a US Carbon Tax: Challenges and Debates*, New York: Routledge, 38–61.
National Research Council (NRC) (2009), 'Hidden Costs of Energy: Unpriced Consequences of Energy Production and Use', Washington: National Research Council, National Academies.
Organisation for Economic Cooperation and Development (OECD) (2012), 'Mortality Risk Valuation in Environment, Health and Transport Policies', Paris: Organisation for Economic Cooperation and Development.
Parry, Ian (2004), 'Are Emissions Permits Regressive?', *Journal of Environmental Economics and Management*, **47**, 364–87.
Parry, Ian W.H. and Kenneth A. Small (2005), 'Does Britain or the United States Have the Right Gasoline Tax?', *American Economic Review*, **95**, 1276–89.
Parry, Ian W.H., David Evans and Wallace E. Oates (2014a), 'Are Energy Efficiency Standards Justified?', *Journal of Environmental Economics and Management*, **67**, 104–25.
Parry, Ian W.H., Dirk Heine, Shanjun Li and Eliza Lis (2014b), 'Getting Energy Prices Right: From Principle to Practice', International Monetary Fund, Washington, DC.
Parry, Ian W.H., Baoping Shang, Philippe Wingender, Nate Vernon and Tarun Narasimhan (2016), 'Climate Mitigation in China: Which Policies Are Most Effective?', Working paper 16-148, International Monetary Fund, Washington, DC.
Poterba, James M. (1991), 'Is the Gasoline Tax Regressive?', in David Bradford (ed.), *Tax Policy and the Economy*, Vol. 5, Cambridge, MA: National Bureau of Economic Research.
Rausch, S., G.E. Metcalf and J.M. Reilly (2011), 'Distributional Impacts of Carbon Pricing: A General Equilibrium Approach with Micro-Data for Households', *Energy Economics*, **33**(S1), S20–S33.
Sterner, Thomas (2007), 'Fuel Taxes: An Important Instrument for Climate Policy', *Energy Policy*, **35**(3), 194–202.
Timmer, M.P., E. Dietzenbacher, B. Los, R. Stehrer and G.J. de Vries (2015), 'An Illustrated User Guide to the World Input–Output Database: The Case of Global Automotive Production', *Review of International Economics*, **23**(3), 575–605.

UNEP (2015), 'The Chinese Automotive Fuel Economy Policy: February 2015 Update', United Nations Environment Program.
UNEP (2016), 'UN Environment Emissions Gap Report', UN Environment Programme, Nairobi, Kenya.
United States Inter-Agency Working Group (US IAWG) (2013), 'Technical Update of the Social Cost of Carbon for Regulatory Impact Analysis Under Executive Order 12866', Washington.
WB/SEPAC (2007), 'Cost of Pollution in China: Economic Estimates of Physical Damages', World Bank and State Environmental Protection Agency of China, Washington, DC, World Bank.

APPENDIX 2A.1 MODEL EQUATIONS

A discrete time period model is used where $t = 0 \ldots \bar{t}$ denotes a particular year. Fossil fuels are first discussed, followed by fuel use in the power, road transport, and 'other energy' sectors.[44]

(i) Fossil Fuels: Coal, natural gas, gasoline, road diesel, and the aggregate of non-road oil products are denoted by $i =$ COAL, NGAS, GAS, DIES, and OIL respectively. The consumer fuel price at time t, denoted p_t^i, is:

$$p_t^i = \tau_t^i + \hat{p}_t^i \tag{2A.1}$$

τ_t^i is the tax on fuel i including any excise or carbon charge. $\hat{p}_t^i$ is the pre-tax fuel price or supply cost which is exogenous, meaning fuel supply curves are perfectly elastic. For fuels used in multiple sectors, pre-tax prices and taxes are taken to be the same for all fuel users.

(ii) Power Sector – Electricity demand: Residential, commercial, and industrial electricity uses are aggregated into one economy-wide demand for electricity in year t, denoted Y_t^E, and determined by:[45]

$$Y_t^E = \left(\frac{U_t^E}{U_0^E} \cdot \frac{h_t^E}{h_0^E}\right) \cdot Y_0^E, \quad \frac{U_t^E}{U_0^E} = \left(\frac{GDP_t}{GDP_0}\right)^{\upsilon^E} \cdot \left(\frac{h_t^E p_t^E}{h_0^E p_0^E}\right)^{\eta^{UE}}, \quad \frac{h_t^E}{h_0^E} = (1+\alpha^E)^{-t} \cdot \left(\frac{p_t^E}{p_0^E}\right)^{\eta^{hE}} \tag{2A.2}$$

U_t^E is usage of electricity-consuming products or capital and h_t^E is the electricity consumption rate (e.g., kWh per unit of capital usage), or the inverse of energy efficiency. Product use increases with gross domestic product (GDP_t), where υ^E is the (constant) income elasticity of demand for electricity-using products. Product use declines with higher electricity costs (the user electricity price p_t^E times the electricity consumption rate), where $\eta^{UE} < 0$ is the (constant) elasticity of demand for use of electricity-consuming products. The electricity consumption rate declines (given other factors) at a fixed annual rate of $\alpha^E \geq 0$, reflecting autonomous energy efficiency improvements. Higher electricity prices increase energy efficiency, implicitly through adoption of more efficient technologies, where η^{hE} is the elasticity of the energy consumption rate with respect to energy prices.

Mix of power generation fuels: Power generation fuels used in China include coal, natural gas, oil, nuclear, hydro, and (non-hydro) renewables (wind, solar, biofuels), where the latter are denoted by $i =$ NUC, HYD, and REN. The share of fuel i in generation, denoted θ_t^{Ei}, is defined:

$$\theta_t^{Ei} = \theta_0^{Ei} \left\{ \left(\frac{g_t^i}{g_0^i}\right)^{\tilde{\varepsilon}^{Ei}} + \sum_{j \neq i} \theta_0^{Ej} \left[1 - \left(\frac{g_t^j}{g_0^j}\right)^{\tilde{\varepsilon}^{Ej}}\right] / \sum_{l \neq j} \theta_0^{El} \right\} \tag{2A.3}$$

---

[44] Cross-price effects across the three sectors are not modelled as they are likely small for the time horizon due to products being poor substitutes for one another (e.g., higher prices for transport vehicles will have a minimal effect on the demand for space heating fuels).

[45] All energy demand functions are taken to have constant income and price elasticities, which is a very common assumption in the literature.

where $i, j, l = COAL, NGAS, OIL, NUC, HYD, REN$. g_t^i is the cost of generating a unit of electricity using fuel i at time t and $\tilde{\varepsilon}^{Ei} < 0$ is the conditional own-price elasticity of generation from fuel i with respect to generation cost—these elasticities are chosen to be consistent with empirical evidence on fuel price responsiveness. From (2A.3) fuel i's generation share decreases in its own generation cost and increases in the generation cost of other fuels.

Use of fossil fuel i in power generation at time t, denoted F_t^{Ei}, is given by:

$$F_t^{Ei} = \frac{\theta_t^{Ei} Y_t^E}{\rho_t^{Ei}} \quad (2A.4)$$

Fuel use equals the generation share times total electricity output and divided by ρ_t^{Ei}, productivity of fuel use or electricity generated per unit of F_t^{Ei}. The total supply of power generation in each period is assumed equal to total electricity demand.

Prices and costs: Unit generation costs are given by

$$g_t^{Ei} = \frac{p_t^i + k_t^{Ei}}{\rho_t^{Ei}} - s_t^{Ei}, \quad p_t^i = 0 \quad \text{for } i = NUC, HYD, REN,$$

$$s_t^{Ei} = 0 \text{ for } i \neq REN, \quad \rho_t^{Ei} = (1 + \alpha^{\rho i})^t \rho_0^{Ei} \quad (2A.5)$$

k_t^{Ei} denotes non-fossil fuel costs (e.g., capital and labor) and s_t^{EREN} is a unit subsidy for renewables. Productivity of generation by fuel i increases at rate $\alpha^{\rho i} \geq 0$ per year implicitly from better production technologies and retirement of older, less efficient plants. Finally:

$$p_t^E = \sum_i q_t^{Ei} \theta_t^{Ei} + k_t^{ET} + \tau_t^E \quad (2A.6)$$

The consumer price of electricity is the product of the generation shares and unit generation costs summed over fuels, plus unit transmission costs, k_t^{ET}, and any excise tax on electricity consumption, τ_t^E (zero in the baseline). Upstream fuel taxes are therefore fully passed forward in higher electricity prices, implying an equivalency between taxing the carbon content of power generation fuels and taxing emissions at the point of combustion.

(iii) Road Transport Sector: In the road transport sector, gasoline and diesel vehicles are denoted $i = GAS, DIES$ respectively—the former (for China) represents cars and motorbikes while the latter trucks and buses. Analogous to (2A.1), gasoline and diesel fuel demand at time t, denoted F_t^{Ti} is:

$$F_t^{Ti} = \left(\frac{U_t^{Ti}}{U_0^{Ti}} \cdot \frac{h_t^{Ti}}{h_0^{Ti}}\right) F_0^{Ti}; \quad \frac{U_t^{Ti}}{U_0^{Ti}} = \left(\frac{GDP_t}{GDP_0}\right)^{v^{Ti}} \cdot \left(\frac{h_t^{Ti} p_t^i}{h_0^{Ti} p_0^i}\right)^{\eta^{UTi}}; \quad \frac{h_t^{Ti}}{h_0^{Ti}} = (1 + \alpha^{hTi})^{-t} \cdot \left(\frac{p_t^i}{p_0^i}\right)^{\eta^{hTi}} \quad (2A.7)$$

U_t^{Ti} is kilometers (km) driven by vehicles with fuel type i and h_t^{Ti} is fuel use per vehicle km (the inverse of fuel economy). Km driven in vehicle type i increases with real GDP, according to the income elasticity of demand v^{Ti} and inversely with proportionate changes in fuel costs per km $h_t^{Ti} p_t^i$, where $\eta^{UTi} < 0$ is the km driven elasticity with respect to per km fuel costs.[46] $\alpha^{Ti} \geq 0$ is an annual reduction in the fuel consumption rate due to autonomous fuel economy improvements. Higher fuel prices also reduce fuel consumption rates (e.g., through promoting engine efficiency increases or lighter weight materials) according to $\eta^{hTi} \leq 0$, the elasticity of the fuel consumption rate.

(iv) Other Energy Sector: Large and small energy users in the other energy sector are denoted by $q = LARGE, SMALL$, respectively. Use of fuel i in the other energy sector, by group q, at time t, denoted F_t^{Oqi}, is:

$$F_t^{Oqi} = \left(\frac{U_t^{Oqi}}{U_0^{Oqi}} \cdot \frac{h_t^{Oqi}}{h_0^{Oqi}}\right) F_0^{Oqi}; \quad \frac{U_t^{Oqi}}{U_0^{Oqi}} = \left(\frac{GDP_t}{GDP_0}\right)^{v^{Oi}} \cdot \left(\frac{h_t^{Oqi} p_t^i}{h_0^{Oqi} p_0^i}\right)^{\eta^{UOi}};$$

$$\frac{h_t^{Oqi}}{h_0^{Oqi}} = (1 + \alpha^{Oi})^{-t} \cdot \left(\frac{p_t^i}{p_0^i}\right)^{\eta^{hOi}} \qquad (2A.8)$$

where $i = COAL, NGAS, OIL$, and REN. The interpretation for (2A.8) is analogous to that for (2A.2) and (2A.7) with U_t^{Oqi} and h_t^{Oqi} denoting respectively, usage of products requiring fuel i at time t by group q and its fuel consumption rate. Parameters v^{Oi}, η^{UOi}, η^{hOi}, and α^{Oi} have analogous interpretations to previous notation and are taken to be the same across large and small users. Given the limited scope for substituting among different fuels used for very different products (compared with fuels producing a homogeneous product in the power sector), fuel switching possibilities are not modelled in the other energy sector.

(v) Initial metrics for comparing policies and model solution CO_2 emissions: CO_2 emissions from fossil fuel use at time t are:

$$\sum_{ji} F_t^{ji} \cdot \mu^{CO2i} \qquad (2A.9)$$

where $j = E, T, O$ denotes a sector and μ^{CO2i} is fuel i's CO_2 emissions factor (zero for non-hydro renewables, hydro, and nuclear). There is significant variation in CO_2 emissions factors among different coal types, but this is not really the case when (as here) emission rates are defined per unit of energy. The CO_2 emissions factors for fuels are fixed, that is, the model does not allow for the possibility of reducing them through carbon capture and storage technologies (given the high CO_2 prices needed for these technologies to be viable).

Revenue: Revenue from fuel and electricity taxes, less renewables subsidies, is:

$$\sum_{ji} F_t^{ji} \cdot \tau_t^i + Y_t^E \cdot \tau_t^E - s_t^{EREN} \cdot \theta_t^{EREN} \cdot Y_t^E \qquad (2A.10)$$

---

[46] The model abstracts from substitution between use of gasoline and diesel vehicles given the very different vehicle types and that the policy scenarios increase gasoline and diesel prices increase in roughly the same proportion as they emit similar amounts of carbon.

Air pollution mortality: Deaths from fossil fuel air pollution, at time t, is:

$$\sum_{ij} F_t^{ji} \cdot m_t^{ji} \qquad (2A.11)$$

m_t^{ji} is mortality per unit of (fossil) fuel i used in sector j, which may differ by sector due to differing use of control technologies and local population exposure to emissions.[47]

Economic welfare gains: Formulas for measuring the domestic welfare gains of policies are described in Parry et al. (2016), Appendix C. These are based on well-established second-order approximations (reflecting various triangles, rectangles, and trapezoids in fuel markets) from the literature (e.g., Harberger,1964) and capture domestic environmental benefits (principally reduced local air pollution and, far less importantly, reductions in congestion and other external costs of road vehicle use) less economic welfare costs (distortions in fuel markets created by, or exacerbated by, new policies). Climate benefits are excluded from the calculations given their global nature and dispute over their value.

APPENDIX 2A.2 PARAMETERIZATION

Here we discuss data to parameterize the model (mostly for 2013 to be projected forward).[48]

Fuel prices and taxes: Pre-tax prices for coal, natural gas, gasoline, diesel, and oil products for 2013 are from an IMF database on international prices. These prices are then projected forward to 2030 based on splitting the difference between IMF and IEA (2015) projections of international commodity price indices for coal, natural gas, and crude oil. Also from IMF sources, pre-tax excises for gasoline and diesel are $0.16 and $0.13 per liter and zero for other fossil fuels.

Power sector: From IEA (2016b), electricity consumption for China in 2013 was 386,971 kilotons of oil equivalent (ktoe).[49] Empirical studies for various countries suggest a range for the income elasticity of electricity-using products of around 0.5–1.5.[50] However, China is undergoing a structural rebalancing[51] that is lowering the energy intensity of GDP and a smaller value (0.5) is used so trends are in line with other projections (see text). Based on empirical evidence for a range of countries (see Parry et al., 2016), the price

---

[47] Local air pollution causes a range of other damages beyond mortality (morbidity, impaired visibility, building corrosion, crop damage, lake acidification, etc.) but previous studies suggest their combined damages are modest relative to mortality damages (e.g., NRC, 2009; WB/SEPAC, 2007).

[48] For a more in-depth discussion and ranges for sensitivity analyses, see Parry et al. (2016).

[49] Generation, rather than consumption, is what matters for fuel use and emissions, though the difference between them (reflecting electricity exports and imports) is less than 1 percent.

[50] For example, Jamil and Ahmad (2011), Table 1, report 26 estimates of long-run income elasticities for electricity from 17 studies, almost all of them lying within the above range.

[51] See, for example, GCEC (2014), Green and Stern (2016), Grubb et al. (2015), and IMF (2015).

elasticities for usage of electricity products and the electricity consumption rate are both taken to be –0.25 (implying an own-price elasticity for electricity of –0.5). The annual rate of efficiency improvement for electricity-using products (which is of modest importance for the results) is taken from Cao et al. (2013) to be 1 percent.

Generation shares are obtained from IEA (2016b) by electricity produced from a fuel divided by total electricity production. The price responsiveness of coal (in the power and other energy sector) is the most critical parameter determining the effectiveness of major CO_2 mitigation policies in China. A value of –0.35 is used based on empirical evidence for China and other counties (Parry et al., 2016), which—dividing by the share of fuel costs in coal generation costs—implies a generation cost elasticity of –0.6. Evidence to parameterize other generation cost elasticities is less solid and the same generation cost elasticity is used for other fuels, though the results (for China) are not especially sensitive to different values.

Fossil fuel inputs into the power sector for 2013 are from IEA (2016b) and fuel productivity is electricity produced by that fuel divided by fuel input. For coal plants annual average productivity growth during 2003 to 2013 was 2 percent, though IEA (2015) projects lower future growth of around 0.5 percent (as assumed here), not least because average coal plant efficiency in China has surpassed that in advanced countries. For natural gas and nuclear, there is likely more room for productivity improvements and a growth rate of 2 percent is assumed. For non-hydro renewables, annual productivity growth from 2003 to 2013 was the most striking at 6 percent, though this seems unlikely to be sustainable out to 2030—a productivity growth rate of 4.5 percent is used, and, lastly, for hydro and oil products, annual productivity growth of 0.005 percent is assumed.

For coal plants, non-fuel generation costs are taken to be two-thirds as large as 2013 fuel costs,[52] or $0.03 per kWh. For natural gas plants (which have low fixed and high variable costs), non-fuel generation costs are taken to be one quarter of those for coal plants. The power transmission cost is taken to be 60 percent of the electricity generation cost in 2013, or $0.05 per kWh.[53] The renewables subsidy is taken to be $0.03 per kWh (Parry et al., 2016).

Road transport: In the road transport sector, consumption of gasoline and diesel was 96,471 ktoe and 170,729 ktoe respectively in 2013 (IEA, 2016b). Estimates of the income elasticity for km driven are typically between about 0.35 and 0.8, although a few estimates exceed unity (Parry and Small, 2005) and a central value of 0.6 is used. Numerous studies have estimated motor fuel (especially gasoline) price elasticities for different countries and some studies decompose the contribution of reduced vehicle km from improvements in average fleet fuel efficiencies. Based on this literature, a value of –0.25 is used for each of these elasticities and for both gasoline and diesel—the total fuel price elasticities are therefore—0.5.[54] The annual rate of decline in vehicle fuel consumption rates (from technological improvements) are set at 1 percent (e.g., Cao et al., 2013).

---

[52] From Cao et al. (2013), pp. 341 (after accounting for differences in coal prices).
[53] This is approximately consistent with Cao et al. (2013), pp. 343.
[54] There is, however, significant variation among studies: for example, Sterner (2007) reports globally averaged (long-run) gasoline price elasticities of around –0.7 while individual country estimates in Dahl (2012) are closer to about –0.25 on average (see Charap et al., 2013 for further

Other Energy Sector: We assume 50 percent of fuel consumption (from IEA, 2016b) for mining and quarrying, iron and steel, chemical and petrochemical, non-ferrous metals, paper, pulp and print, and non-metallic minerals is by large firms and potentially covered by the ETS and regulation. Fuel consumption by small energy users is total fuel consumption less fuel use in power generation, road transport, and large other energy users. Evidence on income and price elasticities for fuels used in the industrial and residential sectors is more limited. Based on judgement, the same income elasticities for coal and oil products are used as for electricity, while a value of 1.0 is assumed for natural gas and renewables products. Values for the usage and efficiency price elasticities for all fuels are taken to be the same as those for road fuels and electricity. The annual rate of productivity improvement follows those for the same fuel as used in the power sector.

GDP growth: Projected GDP out to 2021 is from the IMF's World Energy Outlook. From 2022 onwards, real GDP growth is assumed to decrease linearly from 6 to 5 percent in 2030.

Mortality rates from fuel combustion: Coal accounts for the vast majority of air pollution deaths from fossil fuel combustion in China, primarily from fine particulates produced directly and formed indirectly (and in greater quantities) from sulfur dioxide (SO_2) and nitrogen oxide (NO_x). China requires new coal plants are fitted with flue-gas desulfurization (FGD) equipment, is closing small-scale (high polluting) plants, and is requiring other existing plants to retrofit with FGD. As of 2010, FGD equipment had been installed on around 80 percent of electric coal plants (Cao et al., 2013, pp. 343), though even with these technologies plants still emit some SO_2, in addition to NO_x and direct particulates.

Air pollution mortality is taken from Parry et al. (2014b), with some adjustments. Parry et al. (2014b) estimate that the average coal plant in China caused 10.4 air pollution deaths per petajoule (PJ), or 0.435 deaths per ktoe, and the average coal plant with control technologies caused 5.3 deaths per PJ. In the absence of other factors, we assume the mortality rate from coal combusted at power plants and large industrial sources would converge linearly from 10.4 to 5.3 deaths per PJ between 2010 and 2030. However, the share of the Chinese population residing in urban areas is projected to increase by about 25 percent between 2010 and 2030 (Cao et al., 2013) and it is the urban population that is mostly exposed to air pollution. We therefore make a linear upward adjustment in the mortality rate each year to account for this, where the upward adjustment reaches 25 percent by 2030. For small-scale coal emissions, we assume the mortality rate is 10.4 deaths per PJ in 2010, rising in proportion to the rising share of the urban population.

Also, based on Parry et al. (2014b), the 2010 mortality rates for natural gas, gasoline, diesel, and oil products are taken to be 1.1 per PJ, 36 per billion liters, 124 per billion liters, and 20 per million barrels of other oil products respectively, and again these are scaled

---

discussion). For a summary of evidence on the decomposition of the fuel price elasticities into the vehicle mileage and fuel efficiency responses see Parry and Small (2005). The responsiveness of fuel efficiency to taxation will be dampened in the presence of binding fuel economy regulations, though this issue is not so relevant for the present analysis which compares policies in isolation (rather than jointly).

up for the rising urban population (though these fuels contribute only a small share to total mortality).

One caveat is that some evidence suggests people's channels for absorbing air pollution become saturated at very high outdoor pollution concentrations implying, paradoxically, that the health benefits from incremental reductions in fuel combustion are smaller at high pollution concentrations than at more moderate concentrations.[55] In this regard, our analysis may overstate the domestic health benefits of carbon mitigation policies as it assumes incremental benefits are the same, regardless of pollution concentrations.

To value premature mortality, Parry et al. (2014b) use a value of $1.13 million per mortality for year 2010 based on extrapolating empirical evidence for advanced countries under an assumption that the income elasticity for this valuation is 0.8. This figure is first increased by 15 percent to update it to year 2015 based on the average increase in consumer prices between 2010 and 2015 (see IMF, 2016). And the 2010 figure is updated to future periods using future per capita income relative to 2010 raised to the power 0.8.

Other externalities: Parry et al. (2014b) estimate congestion, accident, and road damage externalities for gasoline and diesel vehicles at $0.86 and $0.56 per liter respectively for year 2010. This figure is updated to future periods in the same way as for the value of mortality. The additional parameter needed to compute fully efficient pricing policy is the social cost of carbon. For this we use the US IAWG (2013) value of $60 for emissions in 2030 (expressed in year $2015).

---

[55] That is, the relationship between mortality and pollution concentrations may be concave rather than linear (e.g., Burnett et al., 2013).

3. Bargaining to lose: a permeability approach to post-transition resource extraction
Natasha Chichilnisky-Heal

INTRODUCTION

This chapter will attempt to introduce a framework for considering the problem often identified as the resource curse. The framework will address the political economy problems that emerge as a result of the distinctive path experienced by many GDP-poor but resource-rich nations in the past 20 years, focusing specifically on the issue of quality of democracy.

The foundation of my argument rests on the development of the theory of permeability, which I will develop through the analysis of two cases: those of the copper mining sectors in post-socialist Zambia and post-Communist Mongolia. While the two cases may seem to bear little relation at first, I hope to show later in the chapter that the comparison is not only justifiable based on the similarities alone, but that by comparing these cases we can glean valuable insights on the changing nature of democracy in poverty-stricken resource-rich nations around the world. If a theory is developed by comparing a state in sub-Saharan Africa to one in the post-Soviet sphere, it should have some traction on other states in each region and possibly also in other regions as well, although discussion or proof of that point is beyond the scope of this chapter.

In defining permeability, I find it useful to describe the actors involved in the bargaining game I will describe. Firstly, there are the governments of the two countries in question. I refer not specifically to the various governments that have held power since each country transitioned to multi-party democracy, but rather to the institution of the state that has existed following transition. This institution, for the purposes of my argument, incorporates party leaders (Presidents and Prime Ministers as well as leaders of the opposition), major political parties, and various bureaucracies specifically related to resource management, extraction, and export. Secondly, we have the enfranchised portion of the population in each country. And finally, we have the multinational or multilateral external actors that I argue "permeate" the democratic government of these countries through their involvement in economic and resource-based activities.

Permeability is the process by which external non-state actors such as the International Monetary Fund and multinational corporations (the IMF, or other multilaterals, and MNCs), by virtue of their relationships with cash-strapped resource-rich governments, enter into crucial roles in the governance of these nations. The foundations of the theory of the resource curse, which has at this point seen many and varied proposed mechanisms, rest on explaining negative developmental outcomes in these nations. I will explore those in more detail below. This chapter is exclusively interested in proposing a relationship between permeability and a reduction in the democratic accountability of these governments to their domestic constituencies, or what I call political underdevelopment. The argument runs as follows. External actors (multilaterals and MNCs) bargain extensively with host governments over the regulation of extractive industries, and often tie development aid or loan packages to the satisfactory adjustment of regulations or conclusion of investment deals in the extractive sector. I argue, using the two cases I have mentioned, that this phenomenon skews the democratic process, providing the governments of these states with yet another constituency – the constituency of the external actors. This perverts the democratic process not simply by making the government economically beholden to the external actors, as has already been extensively argued, but by giving the external actors a permanent seat at the bargaining table of domestic politics. There are some exceptions in the cases I examine, which will be addressed during the case analysis.

The concept of permeability still holds value, however – it is not a binary but rather an ordinal variable, which measures the degree to which a democratic government and its processes have been "permeated" by actors other than its domestic constituent base. One example that I will cite below is the case of the Zambian constitutional redrafting, where despite the fact that Zambia, as a former British colony, does not operate under a Napoleonic legal system, the World Bank insisted that the government hire a Napoleonic legal system expert to lead the redrafting. Only the World Bank's permeation of the Zambian government, I claim, led to such an egregious error on the part of the Zambian government. Permeability functions as a concept opposed to that of sovereignty – while it is clear that in today's world few if any states exercise absolute sovereignty over their own territory and economy, states that exhibit permeation, or experience the phenomenon of permeability, necessarily exhibit lower levels of domestic sovereignty.

ROADMAP

In the rest of this chapter I will do the following. Firstly, I will provide the motivation for the chapter. Next, I will demonstrate the comparability of the cases of post-transition Zambia and Mongolia and the usefulness of the comparison to my theory. Thirdly, I will provide an overview of the bargaining game I (and others) claim takes place between external actors and developing resource-rich states, incorporating the domestic voting game into the picture. I will also describe my preliminary thinking on the incentive structures of different players in the permeability game I propose as a useful construct through which to view the resource curse problem. Finally, I will contextualize this bargaining game in the theoretical framework I have just proposed.

MOTIVATION FOR THE CHAPTER

The so-called resource curse is one of the greatest challenges facing developing nations today. Numerous cases of oil- and mineral-rich states stunted by civil war, large-scale corruption, hyperinflation, and limited state capacity, illustrate a tragic situation, where the wealth concentrated beneath the surface has not transformed itself into wealth in the hands of the citizenry. Political and economic actors the world over want to know: what is stopping the massive revenues flowing through the extractive industries from promoting economic development in some of the poorest countries in the world? This question is both critical and timely; the emergence of the Extractive Industries Transparency Initiative (which was written into Section 1504 of the Dodd-Frank Act) indicates the momentum gathering around this basic development issue. We must therefore ask: "What is preventing the development of this large set of resource-rich post-colonial states?" In this chapter, I develop a theoretical framework in which to study this question by conducting a comparative analysis of the post-transition development of the mining industries in Zambia and Mongolia – two developing states with significant mineral wealth (specifically copper) that experienced a massive exogenous economic (as well as political) shock in the early 1990s, and that subsequently have undergone similar processes impacting their permeability and thereby, I claim, their political and economic development. The purpose of this chapter is to use these two comparable but varying cases to deduce a basic theory of resource-driven underdevelopment.

Zambia and Mongolia are not the only countries to face the challenges of underdevelopment despite possession of significant reserves of natural resources. From Iraq to Bolivia, we find cases that seem to exemplify the concept of the resource curse: the idea that the existence of significant natural resources, particularly hydrocarbons or other minerals, within a nation's territorial boundaries is the primary cause of such diverse outcomes as civil war, brutal authoritarian regimes, ethnic conflict, corruption, and widespread poverty. However, what is often neglected in the literature and has only in the past several years been given a platform, is the fact that many of the wealthiest and most stable democracies around the world possess significant natural resources. The United States, Canada, Norway and Australia are but the most commonly cited counter examples to the resource curse theory. What makes these countries different? What has limited, in particular, the quality of democracy in resource-rich states in the third world? I propose that the negative outcomes clumped under the umbrella of the resource curse are in fact highly context-specific, and that there is a great need for each type of negative outcome (loss of democratic accountability, for example) to be studied individually. It is for this reason that I have chosen to compare the cases of Zambia and Mongolia, two countries that share sufficient points of comparability to demonstrate the workings of my theory as indicating a way in which resource wealth can translate into underdevelopment, and yet are different enough to indicate the generalizability of the permeability theory. These cases will be used to construct the theory presented here, and in future iterations of this project additional case studies will be developed for theory testing.

DEMONSTRATION OF COMPARABILITY OF CASES AND USEFULNESS OF COMPARISON

In order to show the usefulness of the comparison, I will first outline the nature of the "underdevelopment" that both Zambia and Mongolia have experienced. Underdevelopment can be broken down into economic, political, and social components – I will focus primarily on the political component. The per capita GDP ppp of Zambia in 2012 was $1700 (in 2011 US dollars), ranking it 202nd out of all the countries in the world according to the CIA World Factbook. Of the Zambian population, 64% lived below the poverty line in 2006, according to the same source. And yet, Zambia was once the world's second greatest producer of copper after Chile – copper, a highly traded and critical global commodity. Turning to Mongolia, the World Factbook states GDP per capita ppp in 2012 stood at $5400, placing it at rank number 151 in comparison to other nations. As of 2011, according to the same source, 29.8% of the population was below the poverty line. These facts of contemporary Mongolian life should be contextualized further, however, with the knowledge that this data from 2011 and 2012 comes eight and nine years respectively after discovery of the largest undeveloped copper-gold deposit in the world in the South Gobi desert. While Mongolia's GDP has been growing at a rate of between 6–7% per annum since 2010, this growth has not been reflected in the life of the average Mongolian.

The above statistics merely give a rough sketch of the economic situation of these two countries. What about the political situation? Political underdevelopment, as I operationalize it in this chapter, can be broken into corruption, unstable legal environment, ethnic fragmentation, and violent protest. These four components limit the capacity of the state to enact "positive sovereignty" – the formulation and implementation of policy over its population and territory, specifically of policy that benefits the domestic population rather than either the politicians themselves or the external actors referenced in the introduction. In Zambia, the Anti-Corruption Commission, a special agency put in place by the party elected to government in 2011 (the Patriotic Front), was recently accused of corrupt behavior in its ruling on several high-profile political cases.[1] The country has been in the process of drafting a new constitution since 2009, but had failed to make significant progress towards passing a draft even since the consolidation of government in 2011 up until late 2013 when a new draft was claimed to have been produced, although President Sata has refused to release it, leading to a dissolution of Parliament. The current President and head of the Patriotic Front, Michael Sata, ran his party on an anti-Chinese investment campaign in the mid-2000s, and has currently taken a note from his neighbor to the south in Zimbabwe in hinting at instituting land reforms stripping white Zambian nationals with British citizenship of ownership and labor rights.[2] While his campaign slogan leading up to the 2010 elections was highly palatable, "More money in your pocket," Sata has come under fire from opposing parties and civil society organizations for corruption and serious ineffectiveness in his government. Perhaps even more tellingly, during my preliminary field research in Zambia in January 2013 my interviews with citizens in the capital, Lusaka, the general attitude towards the government was succinctly summarized by one interviewee, who said, "Yes, President Sata made many promises, and yes, we have not yet seen any real progress on those promises. But it has only been two years, and here in Zambia we have learned to wait a long time for real change."[3]

What of Mongolia? The government, also a hybrid parliamentary-presidential system, has weathered a series of scandals in the past several years. One high-profile case was the sentencing to four years in jail of former President and former Prime Minister Nambar Enkhbayar in August 2012.[4] Accusations of Enkhbayar's Mongolian People's Revolutionary Party's involvement in vote-buying triggered riots in the capital,

---

[1] See http://www.lusakatimes.com/2013/02/20/anti-corruption-commission-clears-kabimba-and-gbm-of-any-corruption/.
[2] From personal communication with Hugh Carruthers.
[3] From personal communication with Sebastian Kopulande.
[4] See http://www.aljazeera.com/news/asia-pacific/2012/08/20128393741765865.html.

Ulaanbaatar, in July 2008 – the first time in Mongolia's history as a democracy where a transition of power was less than peaceful. More recently, Bayartsogt Sanjagav, former deputy speaker of Parliament, resigned his post in April 2013 after a report by the International Consortium of Investigative Journalists that he had secreted over $1 million in undeclared off-shore accounts.[5] Recent journalistic pieces on Mongolian politics have described both the government and the opposition as "billionaires' clubs." The generally dysfunctional nature of the Mongolian government's various coalitions between the three or four main parties, which has been ongoing since the early 2000s, is another indicator of the political challenges faced by the rapidly changing nation. Despite stable democratic practices, three of the four indicators of political underdevelopment are rampant in the country. Corruption is ever-present. The legal environment is not only destabilized by frequent changes to major legislation, but also by the ineffectiveness of regulations in the mining sector in particular. Violent protest, unheard of during the transition to multi-party democracy, is now a feature of everyday life in the capital Ulaanbaatar, and these protests take place over mining policy, either near corporations, or near the seat of government. Finally, while ethnic fragmentation is not a major concern for the primarily single-ethnicity state, Mongolia has experienced a resurgence of nationalist political parties since the floodgates opened to foreign mining and construction firms in the late 1990s and early 2000s, leading to the emergence of a neo-fascistic nationalist party known as the Blue Mongols, who support the ejection of all non-Mongolians from the country immediately – another point of commonality with recent trends in Zambian politics. One of this party's infamous slogans, which could easily be translated to Zambia, is, "Don't kill people – kill the Chinese." I saw it graffitied on lots of building walls during my many years in and out of Mongolia.

Zambia's troubles with corruption have been briefly described above. The legal environment in Zambia today is fundamentally unstable, as the country is in the process of revising its constitution but has made little to no progress on presenting, debating, or passing the new draft in the course of the past three years since the arrival of the Sata regime. Corruption and challenges to the rule of law and civil rights are united in a scandal that emerged in September 2013, wherein the president of one opposition party, Movement for Multi-party Democracy, Nevers Mumba, was to be arrested for illegitimate charges of defamation of President Michael Sata following a radio interview he gave in 2013 criticizing the government. The country has also interjected new mining investment laws (such as a windfall profits tax, which was exactly paralleled in Mongolia) into the regulatory environment and adjusted others frequently over the past decade. Violent protest has been relatively infrequent, although there was a rash of violent crimes against Chinese and Indian migrants following the death of several Zambian mineworkers during a protest at a Chinese-owned copper mine in 2005. In addition to anti-foreigner conflicts, there have been a number of high-profile protest movements taking place in the country over the past several years. In the spring of 2013, students at the national university consistently engaged in violent conflicts with the police. Both the ruling party and opposition factions have come into conflict with the police as well during political

---

[5] See http://www.business-mongolia.com/mongolia/2013/04/18/deputy-speaker-bayartsogt-s-explains-icij-offshore-releases-to-sgk/.

protests in the capital Lusaka. And, as mentioned above, while ethnic fragmentation among the main tribes in the country is very limited, it is growing, and the focus on removing foreigners and Sata's recent push to limit the rights of white Zambians represent strong sub-national movements.

I have yet to detail the conflict both the Mongolian and Zambian governments have generated internally (and indeed, across the global extractives sector) through their varied efforts to drastically change their mining laws since the commodities market upswing started in 2005. Both countries introduced windfall taxes (Mongolia in 2006 and Zambia in 2008). Both countries rewrote their mineral laws, and are today rewriting them again. Both countries have drawn the antagonism of mining firms and multilateral development agencies with vested interests in preventing resource nationalization from spreading like wildfire across the globe, as happened during the period from 2006 to 2009, from Mongolia to Latin America. This conflict with external non-state actors is at the heart of my theory of the underdevelopment of both countries. While some scholars of the resource curse argue that all the negative political phenomena can be attributed to what is known as the obsolescing bargain model, in which governments make bad deals when commodity prices are low and renege, with dramatic effects, when commodity prices are high, both Zambia and Mongolia have experienced such extreme interaction with external actors, particularly multilaterals, during these times that it is impossible to ignore the role of external actors in the negative political outcomes. When Mongolia goes through a downswing in commodity prices, it does not return to the MNCs on whose contracts it reneged – it goes to the World Bank, and the World Bank gives its advice and sits (literally) at the negotiating table.

Having now indicated that both Zambia and Mongolia suffer from low economic and political development despite significant mineral wealth (Mongolia possesses the largest undeveloped copper-gold deposit in the world, while Zambia remains Africa's largest copper producer and has 35 million tons of reserves left), I will now proceed to analyze the comparability of the two cases along other dimensions. After all, Zambia is a small landlocked country in south-east Africa with a population of roughly 13.5 million sharing territorial borders with Zimbabwe, Angola, Mozambique, the Democratic Republic of Congo, Namibia, Malawi, and Tanzania. Mongolia is the 19th largest country in the world and possibly the least densely populated (roughly three million inhabitants), with only two territorial neighbors: Russia and China. Zambia is a former British colony, while Mongolia was never technically a part of the USSR (despite often being called the sixteenth Soviet satellite). However, both countries have depended heavily on copper production and export for their economic survival for many decades. Zambia's northwestern region, known as the Copperbelt, has been producing raw copper since the 1930s. Mongolia's single largest revenue earner remains the massive Erdenet copper mine on the Russian border, which was developed by the Russians in the 1970s and remains jointly owned by the Russian and Mongolian governments. In this way, economic colonialism in the copper sector continues in both countries – one of the largest multinational mining companies operating in Zambia's copper sector today is Anglo American, which has been mining in the Copperbelt since 1928.

In addition to the political troubles that plague both nations today, they have a shared history of state socialism (communism in the Mongolian case) followed by a peaceful transition to democracy beginning in the late 1980s and early 1990s. Immediately following transition both countries were in dire economic distress. Zambia's sovereign debt topped $7 billion according to World Bank estimates at the time, while Mongolia's economy, which had mostly been propped up by Russian industry, essentially collapsed once the Russians left the country after the disintegration of the USSR. And while the copper sector in Zambia was significantly more developed in Zambia than in Mongolia, where large-scale production had only got underway in the 1970s, by the time of transition the Zambian copper sector was practically bankrupt, primarily due to lack of reinvestment during the socialist period from independence in 1964 to 1991.

Most critically for my theory, both countries were exposed to the same set of actors immediately post-transition with respect to their extractives industries. This set includes firstly multilateral development agencies, at that time ever ready with a Washington Consensus-approved set of structural adjustment programs that included rapid opening of the mining sector to international investors. The second group of actors in this set is the multinational mining companies that were invited into Zambia and Mongolia at the behest of, primarily, the World Bank and the IMF. This invitation took place through the restructuring of the countries' mineral laws, and in the Zambian case, the forced privatization and de-bundling of Zambian Consolidated Copper Mines, the state-owned enterprise that held all Zambia's mining assets and was responsible for the social welfare programs attached to them.

To summarize, Zambia and Mongolia are comparable along the key parameters of my argument. Table 3.1 lists the similarities and a brief discussion of the dissimilarities and their impact on the theory development.

The major dissimilarities come down to the following: ethnic fragmentation, regional environment, geopolitical significance, climate, population density, and nature of

Table 3.1 Relevant similarities and dissimilarities of state: external actor relations in Zambia and Mongolia

Zambia	Mongolia
Colonial occupation by the British	Pseudo-colonial occupation by Imperial Russia and then the USSR
State-run socialism from 1964 to 1991	State-run communism from 1921 to 1989
Dependence on copper mining revenues since 1920s	Dependence on copper mining revenues since 1970s
Collapse of copper mining sector by 1990s	Collapse of copper mining sector by 1990s
Small population in landlocked territory	Small population in landlocked territory
Extensive experience with World Bank, IMF, and multinational mining corporations	Extensive experience with World Bank, IMF, and multinational mining corporations
Survival of socialist party (UNIP) to present day	Survival of communist party (MPRP) to present day
Stable, hybrid parliamentary-presidential democracy since 1991	Stable, hybrid parliamentary-presidential democracy since 1991
Bargaining between government and external actors over mining investment	Bargaining between government and external actors over mining investment
Major cultural backlash against foreign investors (Chinese & Indians)	Major cultural backlash against foreign investors (Chinese & Canadians)

pre-democracy regime. Zambia is composed of several ethnic groups, one of which (Bemba) is predominant. Mongolia has a small Kazakh ethnic minority in the far west, but ethnic politics play no role in the national dialogue. However Zambian ethnic fragmentation is minimal – it mostly acts on a highly localized level of government that is not hugely relevant for the purposes of this chapter. The copper sector in Zambia is based in the Copperbelt, whose local politics are closely tied to national politics and which is predominantly Bemba. Regional environment and geopolitical significance are closely linked for these two countries; neither is significant on its own, so each state's geopolitical relevance is defined by its relations with its neighbors. Mongolia sits in the middle of the Sino-Russian region, has ties to Central Asia, and relies heavily on the US and EU to prop it up and keep it out of the clutches of the Russians and Chinese. Zambia is surrounded primarily by other small, relatively poor African nations, but it is a regional hub for trade and migration and is proximate to South Africa, the behemoth of the region. The main difference between Mongolia and Zambia along this dimension is that the West is more deeply invested in supporting Mongolia due to its location in a highly strategic region and its democratic regime in a world of authoritarianism. But when it comes to the mining sector, the impact of the West is felt equally in both countries. Climate is essentially irrelevant for copper mining, the issue under consideration. Variation in population density of course changes the composition of the economy and the particular developmental challenges, as well as the nature of domestic politics. But high urban concentration and almost identical population levels in Ulaanbaatar and Lusaka (1.5–2 million in each), and secondary population concentration around the major mining areas (the Copperbelt in Zambia, and Erdenet and the South Gobi in Mongolia) limit the relevance of differing population density.

The final obvious area of dissimilarity comes in the nature of the regimes prior to the establishment of multi-party democracy. This includes both the colonial or pseudo-colonial regimes and Zambia's first government post-independence. Zambia was a British colony from the late 19th century to 1964, after which Kenneth Kaunda's United National Independence Party took over in a single-party system and ruled till 1991. Mongolia was never officially colonized by Imperial Russia, but the White Russians fought against Mongolian revolutionaries from 1917–21, at which point Mongolia declared itself an independent communist nation under the Mongolian People's Revolutionary Party. That party ruled until 1989, and remains a factor in Mongolian politics to this day (like UNIP in Zambia). The key differences between the two countries are the nature of the colonizers, the timing of independence, and the nature of the first independent regime. For the purposes of this chapter, I am making the simplification that conditions prior to the 1970s will be excluded from the analysis. This decision is justified as follows. Firstly, copper mining in Mongolia did not become a significant revenue earner until the 1970s. Secondly, Zambia was a colony until 1964. Thirdly, the nature of the problem I am studying (bargaining between the state and external non-state actors over natural resource extraction) changed fundamentally in the 1970s with the advent of the New International Economic Order (NIEO), a set of proposals put forth by a group of developing states that highlighted developing nations' rights to:

- Regulate the activity of multinational corporations within their territory;
- Nationalize foreign property within their territory; and
- Obtain fair prices for exports of raw materials on the international market.

The NEIO, while relatively toothless in the end, changed the conversation about resource extraction in the developing world in ways that have lasted to this day. I argue that, for the above reasons, beginning the development of the comparative theoretical framework in the 1970s is justified.

In the rest of this chapter, I will outline the process by which these two sets of external non-state actors (multilaterals (IFIs) and MNCs), in their interactions with the post-transition Zambian and Mongolian governments, laid the groundwork for the underdevelopment we see today through their permeation of each government.

OVERVIEW OF BARGAINING THEORY

The permeability theory I propose is best understood through the form of a bargaining game. Governments bargain with multilateral development agencies over the terms of economic assistance, and those terms include concessions regarding the legislative framework for extractive industries. Governments also bargain with multinational mining corporations over the terms of individual mining agreements. There is, of course, another bargaining game that occurs, between the government and the electorate, but a key point is that in both Zambia and Mongolia, despite the electorate's efforts, through national elections, to improve each country's relative bargaining position with respect to the external non-state actors, initial conditions in the post-transition period and then subsequent conditions following each iteration of the bargaining games have made it so that no matter what party or coalition came to power, the host country remained at a severe disadvantage and was ultimately forced to choose, at various points from 1991 to today, between acceding to the terms of the external non-state actors or stalling investment and ensuring further economic stagnation. This is the essence of the "permeation" I have identified, and that I claim is a key concern in the operation of democratic politics. Each case shows this mechanism at work, where increased permeability initiated during transition to multi-party democracy and economic crisis leads to a "permeability trap" wherein regardless of which party comes to power, politicians find themselves unable to effect true change in the structure of the bargaining game as well as reliant upon external actors, and in many cases turn to graft as an alternative way of satisfying their desires while in power.

I propose a new model, building on a bargaining framework, to study the relationship between extractive industries and development outcomes: the permeability approach. While in this chapter I will not formalize this model, for the purposes of discussion I define permeability as a continuous variable that represents the degree to which positive and negative sovereignty have been compromised, and argue that it enters the process under study through bargains that take place between the host country (HC) and the various types of external non-state actors. Permeability is normalized to one, where a value of zero represents complete positive and negative sovereignty and a value of one represents complete failed statehood. Obviously, there are no states that exist on the ends of this open interval, but my purpose in developing the permeability concept comes from the recognition that most post-colonial states have higher levels of permeability than developed, colonizing nations, and those forced to open their political and economic borders to external non-state actors approach a value of one. One of the next steps in this chapter project will be to select key quantifiable indicators that can be used to measure permeability.

I have identified four primary mechanisms through which permeability, as it is increased, can lead to political underdevelopment. After listing each mechanism, I will provide a brief explanation of the process, and some examples of the process from the Zambian and Mongolian cases.

Mechanism 1: Higher levels of permeability cause lower levels of political (and economic) development by creating opportunities for and incentivizing corruption.

The argument here runs as follows. When permeability increases (e.g., when Mongolia's economic situation in the early 1990s forced it to approach the IFIs and other sources of funding for economic assistance), as a result of the legislative changes that were demanded by the IFIs, Mongolia was forced to accept a number of suboptimal bargains for mining contracts. Over time, when the Mongolian government wished to renege on or oppose such contracts, politicians were able to seek high levels of graft from the MNCs as well as other interested players, such as Russia. In addition, as I have described briefly above, in both the Zambian and Mongolian cases politicians who enter office with the desire to effect change in the mining sector find themselves unable to do so, and are therefore more likely to accept bribes. As a secondary result of this effect, politicians who are corruptly motivated rather than ideologically motivated are more likely to seek office in future.

Mechanism 2: Higher levels of permeability cause lower levels of political and economic growth through regulatory entrenchment of bad bargains.

This is a fairly straightforward claim. Mongolia has been forced into signing a number of "stability agreements" with mining companies, which guarantee protection from any tax reform or legal reform of other kinds during the life of the agreement. Their tenure usually runs from 10–30 years, which is unheard of outside of the developing world. Peru is another example of a nation plagued by stability agreements, which leave essentially only the option of nationalization as the government's last resort, destroying international credibility and causing capital flight. The governments in both Zambia and Mongolia instituted windfall profit taxes on copper and gold in the mid-2000s, another attempt to redress the suboptimal terms of initial bargains. These led to reduction in production levels and increased conflict with the MNCs, requiring the assistance of the World Bank in both cases as mediator, further solidifying the Bank's role IN government in both countries. Therefore, permeability close to 1 in the mid-1990s led to bad bargains which led both to increased entrenchment of external constituencies in each country's government and to suboptimal economic growth rates.

Mechanism 3: Higher levels of permeability cause lower levels of political and economic growth by *causing the policy space in the country's political sphere to collapse and become uni-dimensional.*

This is an argument that requires further exploration, and is as yet primarily a theoretical proposal. I have witnessed during my fieldwork in both countries, as well as other developing resource-rich states, the degree to which national political campaigns focus on the extractive sector and public discussion of politics focuses on how to use mineral wealth to reduce poverty more effectively. I hypothesize that this reduction of political debate to one issue – whether or not to nationalize natural resources, most often – reduces the opportunity to hold politicians accountable for their actions in other, perhaps more important, spheres such as education, healthcare, infrastructure, and other sources of economic growth. This explanation has not been explored at all in the literature, and I have another working paper in which I am currently expanding the concept and identifying appropriate theory-testing mechanisms.

Mechanism 4: Higher levels of permeability cause lower levels of political and economic development through *misdirection of bureaucratic growth*.

There are two primary explanations for this mechanism, and they are intertwined. The first is a secondary effect of Mechanism 1: corruption. This is very much evident in Nigeria, where the Minister of Oil is one of the wealthiest and most powerful women in the country. MNCs chalk bribes to the appropriate bureaucratic agency up to the cost of doing business in the process of competing for a new contract or extending an existing one. Opportunities for rent-seeking may draw more and more high-level bureaucrats or rent-seekers to the agencies responsible for dealing with the extractive sector, such as the Mineral Resource and Petroleum Authority of Mongolia (MRPAM) or the Ministry of Mining in Zambia. There are other agencies that may experience this effect as well, such as the Revenue Authority in Zambia, which deals with collecting customs from mining companies exporting copper along Zambia's many borders. Bureaucrats are likely to expand their own opportunities for rent-seeking by expanding the authority or reach of their own agencies wherever possible.

The second explanation for this mechanism, while it can coexist with the first, is slightly different. It deals with the idea that there is an ideal division of the government's budget between all bureaucratic agencies, and that permeability of the country's government, combined with Mechanism 3, create an inequitable division of resources. The Ministries of Education, Health, and Transportation are likely to suffer significantly as time, energy, political clout and resources are distributed away from them and toward the agencies that deal with what is perceived to be the most critical issue in the country: the extractive sector. Permeability can also cause this effect by forcing the government to develop its own "expert task force" on resource issues to stand on the other side of the table from the external advisors with whom the government negotiates over critical policy. Underinvestment in health, education, and infrastructure is a clear path to underdevelopment.

The value of the permeability approach is that it is synthetic. It allows us to examine the mechanisms and symptoms identified by other approaches, but grounds the analysis in a study of the evolution of the political and economic institutions of developing nation-states through a bargaining framework. It emphasizes the importance of understanding where there is autonomy or control in the process of development, and where it can be taken away. It also points us to impacts we might not otherwise have considered, such as the phenomenon, which I have observed, of a government increasing the size of its Ministry of Mining simply because it feels outnumbered by external technical experts. The permeability approach explicitly links development outcomes with the degree of sovereign statehood and thereby the degree of a government's democratic accountability downward, towards its population, versus upwards, towards external actors.

In order to develop the permeability approach using the cases of Zambia and Mongolia, I have outlined a basic timing framework for the bargaining model under discussion:

- Time t = 1: State undergoes exogenous economic shock (e.g. decolonization, decline in commodities prices).
- Time t = 2: State requires financial assistance and seeks it from external actors (IFIs, other governments, MNCs).
- Time t = 3: In the case of resource-rich states, a bargain (Bargain 1) is struck regarding investment in extractive industries. Although this process can often take years, I choose to represent it as a single point on the timeline. This bargain comprises: a) an extractive industries investment protection regime (regulatory framework); and b) an initial investment contract for at least one project.
- Time t = 4: Investment occurs. This is also a drawn-out point on the timeline.
- Time t = 5: Election occurs.
- Time t = 6: Bargain 2 is negotiated. This may or may not include renegotiation of Bargain 1.

ANALYSIS OF INCENTIVE STRUCTURES OF THE ACTORS IN THE BARGAINING GAME

Host Country Government (HC)

Although the incentive structure of the HC might seem to be the most straightforward in the bargaining game, in fact the HC is strictly confined in its set of strategy profiles by exogenous circumstances and by the strategy profiles of the other players in the game. The HC is also not a unitary actor – it can be broken down into parties, into bureaucracies versus executives, into individual legislators, into distinct ministries, and so forth. The HC is not only a non-unitary actor at any given time t, but whatever aspect of it that is under study changes as we shift from one game to the next. Elections in time t = 5 are a stand-in for the possible replacement of the HC actor, whatever actor or actors we are examining. Even bureaucrats can be replaced when elections come around.

So, what are the incentives of the different types of HC actors? The executive, for example, obtains utility over several terms: his level of power/effectiveness during his time in office, his likelihood of re-election, his payoffs, his ability to redistribute payoffs, his enactment of his programmatic policy preferences, and his historical legacy/reputation. There are additional factors at play, but from my field research in Zambia and Mongolia these seem to be the most critical elements. Many of these terms are interdependent and probabilistic in nature.

A legislator has much the same utility profile as an executive, except perhaps that he will place higher weight on re-election than an executive. A legislator has made a career out of politics that he hopes will last him through each iteration of the game, whereas the executive, while he may have been a legislator in a past iteration of the game, he expects that having reached the position of executive he is unlikely to be able to continue in politics much beyond one additional iteration of the game (i.e., election). Therefore, the executive may, based on political possibilities at the time and his own preferences, emphasize either implementing his preferred programmatic policies or obtaining payoffs.

Bureaucrats represent a slightly different type of player again. Career civil servants look more like legislators except that they are likely to have less control over policy formation and to place more weight on small, consistent payoffs as large payoffs can be relatively easily detected when made to relatively low-level civil servants. Ministers, on the other hand, are political appointees and are therefore interested in their own careers as functions of the careers of their party, the executive who appointed them, and are heavily invested in obtaining payoffs – in both Zambia and Mongolia this seemed to be considered as one of the main reasons to be a Minister, as well as to enable one to develop a patronage network of one's own. However, in Zambia, more weight may be placed on remaining in office by pleasing the executive than in Mongolia – during my fieldwork in Zambia in January 2013, I heard many stories of ministers appointed by the Patriotic Front government being replaced multiple times just since the PF's election to government in 2011.

Multilateral Development Agencies (IFIs)

IFIs can be viewed as unitary actors, principal–agent actors, network actors, or all of these options. When viewed as unitary actors, their incentives are relatively simple; they have both a policy agenda that they wish to promote and a desire to be granted access to the decision-makers in the HC in the future. These incentives create something like a voting game, in which the IFIs present an initial policy, the HC government responds, and the IFIs adapt their policies until a compromise is reached. But viewing IFIs as unitary actors is overly simplistic. Both principal–agent problems and social network issues come into play. The principal–agent concerns have to do with distance, misinformation, and possibly conflicting incentives. For example, the World Bank in Washington, DC may have certain preferences and mandates that have to be translated by the representatives on the ground in Mongolia. And, of course, after a certain number of years away from the home office, many World Bank specialists I have known have exhibited a tendency to become mandates unto themselves, especially in the more remote corners of the world.

The social network issues can be even more pressing in some cases. Normally, the association of an agent on the ground with an IFI is a strong link to the international reputation of the IFI, and that link restrains principal–agent problems of mismatched incentives, at least. But as Keck and Sikkink's work on transnational advocacy networks and Stiglitz's work on the social networks within the World Bank and IMF demonstrate, network connections can easily pervert intentions, and the safety of acting with others allows transgressions against the principal and its stated aims that solitary agents might not be engaged in otherwise. In one piece of anecdotal evidence from Mongolia, I observed the in-country World Bank Country Director and member of the WB's Oil, Gas & Mining Group spend five years pressuring the Mongolian government to accede to a particular bargain over the largest undeveloped copper-gold deposit in the world, located in the South Gobi. Once the government, its bargaining power dramatically reduced by the global financial crisis and commodity price downswing, finally agreed to the bargain, the Country Director quit his job at the WB and promptly accepted a position on the executive board of Rio Tinto, the Australian mining major with the controlling stake in the copper-gold project. While conducting field research into the social networks present in the Mongolian mining sphere in 2009, I came across a number of strong personal, informal ties between the WB Country Director and several representatives of Rio Tinto and its local partner. In addition, on many occasions during the year I lived in Mongolia, I saw this group seated around a table in the local Irish bar and restaurant, forming the bonds that would influence the fate of an entire nation.

Overlapping Incentives

A key element of future analysis will be fleshing out the overlapping incentives between different sets of actors. For example, the executive in Mongolia in 1997 may want very badly to please the World Bank in order to keep grant and loan funding flowing. The World Bank wants to maintain its credibility as an impartial advisor to governments, and yet it faces the challenge that the majority shareholders are Western nations who may choose to use the Bank to promote their own agenda, often in relation to creating a favorable business environment for Western MNCs. And MNCs may be (and frequently are) at the local level made up of both host country nationals and expats, leading to potential conflicts of interest in preference formation.

THE BARGAINING GAME IN THE PERMEABILITY CONTEXT

I am only one of many scholars to view the resource curse as essentially a bargaining problem. I hope to contribute to the field of scholarship on this issue by utilizing the framework of permeability theory, and proposing the four mechanisms outlined as ways in which permeability interferes in the bargaining process and leads to suboptimal outcomes that get iteratively worse for the host country and reduce the democratic accountability of the government. Let us examine the bargaining game through the lens of permeability. Permeability enters into the equation at time $T = 2$, where the government negotiates with IFIs and then MNCs. By $T = 3$, the government has been "permeated." Many of the four mechanisms identified above are self-reinforcing – that is, the negative development outcomes that emerge from permeability reinforce permeability. That is why this is written as an iterative bargaining game, in which, even if commodity prices skyrocket, the government almost always loses. One solution to this problem that I have seen enacted in several countries occurs when the government hires competent experts of its own, who have knowledge of how the IFIs and MNCs work and are capable of holding their ground and helping the government to do the same. Of course, the ideal solution would be for citizens of these countries to start being trained to fulfill this capacity, but in the meantime, we have few options. One would be to enhance the role of the EITI in these nations, reducing the risk of corruption or at the very least bringing it into the light. Another would be to encourage, as far as possible, domestically owned refinement of raw materials being extracted and exported in their unrefined form at the present moment. This is not always economically feasible, however, and often the technical capacity to build and run such an operation is not present in the country. A final (and obvious) option would be to dramatically restrict the role of entities like the World Bank and the IMF, and to do so for unilateral development agencies and MNCs as well. If host country governments were suddenly told that the WB experts could not sit in on governmental meetings, not because the government had denied them but because the WB charter had been amended, this could dramatically change the permeability factor. It would, of course, change the nature of the WB's operations fundamentally, but perhaps it would be for the better.

CONCLUSION

This chapter has presented a new model of the resource curse. It proposes that a lack of scholarship on the relationship between governments and external actors has led to under-theorization of the role of permeability in national politics, and has led to theorists overlooking the impact on democratic accountability (and therefore on many other development outcomes) of "permeation."

4. Host–MNC relations in resource-rich countries*
Natasha Chichilnisky-Heal and Geoffrey M. Heal

1. INTRODUCTION

Our aim is to explore issues relating to the political economy of many income-poor but resource-rich countries. This is hardly a new topic: there is an extensive literature on both the political and economic aspects of this, focusing on the resource curse, and noting that in many resource-rich countries, the population are poor in spite of the country's wealth: wealth beneath the surface has not been transformed into purchasing power in the hands of the citizenry. Additionally, and perhaps related to this, resource-riches are often associated with bad governance. Nigeria is a wealthy country whose population is mostly poor, and which is politically unstable and badly governed: Saudi Arabia and other Gulf countries are more prosperous but autocratically ruled. There are, however, resource-rich countries that are prosperous, stable and democratic. Botswana is both prosperous and democratic (as are a number of developed resource-rich countries such as the US, Canada and Australia). Such dramatic differences in outcomes have naturally attracted the attention of researchers (see Ross (2018) for a review). Sachs and Warner (1999) were amongst the first to point out that resource-rich countries appeared to grow less, rather than more, than similar resource-poor countries, and Karl (1997) noted that the massive resource transfer in favor of oil-producers associated with the oil shocks of the 1970s seemed to lead rather counter-intuitively to economic deterioration and political decay. A series of papers on the resource curse can be found in Humphries et al. (2007), and Dunning (2008) contains an interesting analysis of the political aspects of being resource rich, suggesting how oil dependence can lead either to democracy or to autocracy, depending on the details of the domestic political configuration. Ross (2012) makes the case that oil resources predispose to corruption.

The particular dimension of concern here is rather narrower and is the nature of the deal struck between a resource-rich country and the multinational corporation (MNC) which develops its resources, and the way in which this deal interacts with domestic politics. We focus on what we call the "permeability" of this deal, which here we quantify as the fraction of the total resource rent that leaves the country; on the transparency of resource-related transactions; and on the level of economic development. Chichilnisky-Heal (2014)[1] presents a broader interpretation of the idea of permeability, interpreting it as a measure of the degree to which the democratic government and its processes have been "permeated" by actors other than its domestic constituent base. As we shall see in

---

* The first author left a draft of this paper at her tragic death in November 2014. The second author then completed it. We are grateful to Susan Hyde, Jisung Park, John Roemer and Susan Rose-Ackerman for valuable comments.

[1] http://politicalscience.yale.edu/sites/default/files/ncheal_-_003.pdf.

84 *Handbook on the economics of climate change*

section 3, there are compelling reasons why a resource-producing corporation might wish to intervene in the domestic politics of the host country, making the host "permeable" in this broader sense. We are interested in how these factors interact with other issues and in particular with domestic political processes. International financial institutions such as the IMF and the World Bank are often parties to these interactions, suggesting or even emphasizing certain preferred approaches and exerting influence over the choices made by the host country government. We simplify here by focusing only on the interactions between the host country and the MNC, but studies of the interactions between host governments and international financial institutions in the cases of Mongolia and Zambia are reported in Chichilnisky-Heal (2014).

We explore these issues in the context of a simplified version of the model used by Dunning (2005) and Dunning (2008), which is in turn an adaptation of the Acemoglu-Robinson model of political transitions to resource-dependent countries (Acemoglu and Robinson (2001)). (We call this the ARD model.) The main simplification is that we work with a static version of the model, which can be thought of as representing the stationary equilibria of the system. There are two classes, the elite and the masses, who are vying for control of the society. Whichever is in control is threatened by the possibility that the other will mount a successful challenge and take over the government. There is an exogenous source of non-resource income, plus income from the exploitation of natural resources. Non-resource income accrues to both groups, whereas only the elite benefit from resource rent. Income of both types accruing to the elite is taxed to provide public goods, which benefit only the masses.

Whichever group is not in control can mount a challenge to the ruling group, with a certain probability of success, and with known payoffs to successful and unsuccessful outcomes. The ruling group can ward off this threat by ensuring that the out-group reach a welfare level at least equal to the expected outcome of an attempted overthrow – we call this a "consent constraint." This captures one of the main ideas of the ARD model.

Resource rent is divided two ways, between the elite and the MNC developing the resource. The fraction P accruing to the MNC, and so leaving the country, is called permeability. Not all of the resource rent is visible to the masses, only a fraction t where t stands for transparency. The fraction invisible to the masses cannot be taxed to fund the public good. So both groups have an interest in reducing permeability, though in this they are opposed to the MNC, but the two domestic groups have opposed interests with respect to transparency.

The host country is dependent on the MNC for the development of the resource, but this dependence depends on and falls with the level of the hosts' development. The MNC's incentive to develop the resource, and so generate rent for use in the host country, depends on the permeability: the more money it can extract, the greater its incentive to develop the resource base and the greater the total rent – but also the greater the fraction flowing out of the country.

Within this context we study the interactions between transparency, permeability and development, and how these affect the domestic political equilibrium. We show that whichever party is in control has a strong incentive to reduce permeability, as the host country's transparent (visible to the ruling group) benefits from doing so accrue exclusively and directly to the ruling group, and only the ruling group benefits from a reduction in permeability. This could go some way toward explaining the focus of domestic politics on resource-related issues in resource-dependent developing countries. If the masses are in control they gain from an increase in transparency, and vice versa, so that we expect democracies to be more transparent than autocracies.

We then investigate the nature of the deal struck between the host country and the MNC. The relationship here is complex, as it contains both cooperative and non-cooperative dimensions. Up to a certain level of permeability, both host country and MNC stand to gain from an increase in permeability, as it gives the MNC greater incentives to invest in developing the resource base, benefiting both parties, and the gain from a greater resource base more than outweighs the loss to the host country from greater permeability. However beyond this critical point interests diverge: the MNC gains from more permeability while the host country loses. The critical level of permeability decreases as the level of economic and political development increases, as more developed host countries are less in need of the skills and resources offered by MNCs. So the potential for conflict between the host and the MNC increases with development, and it is reasonable to assume that both parties anticipate this when negotiating contractual relationships. We show that the relationship between the host and the MNC has the classic features of a hold-up problem (Tirole (1986)). Many features of the deals seen in practice, such as stability agreements or some of the features of bilateral investment treaties (BITs), can be explained by this anticipation of increasing conflicts, as can some features of the roles played by international financial institutions. We show that although widely adopted by developing countries, BITs may make them worse off, in an argument which extends those of Guzman (1997) and Bubb and Rose-Ackerman (2007).

In section 2 we set out the basic model developing the relationship between transparency, permeability and domestic political equilibrium. In particular we look at the comparative statics of the equilibrium with respect to changes in permeability, showing that the benefits of a reduction in permeability, or the costs of an increase, accrue exclusively or largely to the ruling group. A decrease in permeability is a very attractive strategy for a ruling group seeking to increase its income. When the masses are in control, the elite can share in the gains from a reduction in permeability if transparency is less than complete, because a lack of transparency shields some of the gains that accrue to them from taxation by the ruling masses.

In section 3 we investigate the nature of the bargaining process between the host country and the MNC: it is this process which actually determines the permeability. We investigate the conflict between the host and the MNC. An agreement that the host reaches with the MNC today will generally seem to the host to involve an unacceptably high level of permeability in the future. This built-in and growing conflict leads to a hold-up problem and explains many of the features found in host–MNC agreements, so-called "stability agreements," which make it extremely difficult to revise the agreement. Section 3 offers some conclusions, and two appendices present mathematical details of some of the arguments.

2. A FORMAL MODEL

The total population is taken to be distributed uniformly on the unit interval $[0,1]$. A fraction r of the population is rich and constitutes the political and economic elite. The remainder are the masses. There are two sources of income in this society: income from producing goods and services, which totals I, and rent from the production of extractive resources, totaling R. Resource rent is defined as the difference between market price p and marginal extraction cost, so that assuming a given market price the total of resource rent is

$$R = \int_0^Q (p - mec[q])\,dq$$

where Q is the output of the resource and $mec[q]$ is the marginal extraction cost of the $q-th$ unit extracted.

A fraction $f > r$ of the income from goods and services accrues to the elite. Some of the resource rent accrues to foreign entities, namely the MNC. This fraction is denoted P for "permeability." The remainder of the resource rent goes entirely to the elite, whose total income per capita before taxes is therefore

$$I_e = I\frac{f}{r} + R\frac{(1-P)}{r} \tag{4.1}$$

Only a fraction t (for transparency) of this resource rent is visible to the masses: the remainder is hidden. All of the income from production is visible, and visible income is taxed at a rate τ, with the proceeds of the tax used to finance the provision of public goods, the benefits of which accrue entirely to the masses. The after-tax income of the elite I_e^a is

$$I_e^a = (1-\tau)I\frac{f}{r} + (1-\tau)t\frac{R}{r}(1-P) + (1-t)\frac{R}{r}(1-P) \tag{4.2}$$

The income per capita of the masses is correspondingly

$$I_m = I\frac{(1-f)}{1-r} + g \tag{4.3}$$

where

$$g = \tau I f + \tau t R(1-P) \tag{4.4}$$

The public good is not divided by population as being a public good it is fully available to all. Each group's preferences are represented by a smooth strictly concave utility function of per capita income, denoted U_e and U_m respectively.

2.1 Democracy

To understand the role of changes in permeability on the equilibrium we consider two governance cases. In the first we have a democracy, the median voter is a member of the masses, and the masses control the country to maximize their utilities. However, in doing this they face a constraint: the elite may be able to overthrow the democracy and take command, and will attempt to do so if their wellbeing falls below a certain critical level. So the masses must maximize their wellbeing subject to the constraint that the welfare of the elite is high enough to prevent them from trying to overthrow the democracy. We denote the welfare level needed to prevent the elite from attempting a coup by U_e^c.

We can think of U_e^c as being calculated as follows. If the elite carries out a coup successfully then their wellbeing is U_e^*. The probability of a coup being successful is s_e. The welfare level after an unsuccessful coup is U_e^l. So if the elite attempts a coup, then their expected welfare level is

$$U_e^c = s_e U_e^* + (1 - s_e) U_e^l \tag{4.5}$$

and this determines the welfare level that the masses have to allow the elite to attain to prevent a coup.

We assume that the constraint $U_e \geq U_e^c$ will bind so that

$$U_e = U_e^c \tag{4.6}$$

This is equivalent to a participation constraint in a principal–agent problem: the elite must be given a sufficiently attractive outcome that they willingly participate in the economy rather than trying to overthrow the system. We call this the "consent constraint" – hence the c in U_e^c. The objective of the government in a democracy is to max U_m, $U_e = U_e^c$.

2.2 Autocracy

Next we assume that the elite have successfully executed a coup and are in control of the country. They manage economic affairs so as to maximize their own welfare, subject to ensuring that the masses are sufficiently well off that they do not carry out a revolution. This is again a participation or consent constraint, this time to ensure the participation of the masses. The participation constraint for the masses is set by a similar equation to that used for the elite:

$$U_m^c = s_m U_m^* + (1 - s_m) U_m^l \tag{4.7}$$

where s_m is the success probability for a revolution initiated by the masses and U_m^*, U_m^l are respectively the welfare levels attained by the masses after successful and unsuccessful revolutions. The economic parameter controlled by the elite is again the tax rate τ, chosen so that the masses achieve welfare level U_m^c. (In principle, we have an inequality constraint here, $U_m \geq U_m^c$, but we assume it to be binding.) The tax rate affects the welfare of the masses via government spending, given as before by: $g = \tau If + \tau tR(1 - P)$. Again the objective is to max U_e, $U_m = U_m^c$.

2.3 Equilibrium Analysis

In this section we show that the concepts of the previous sections can all fit together consistently. Recall that U_m^c and U_e^c represent the welfare levels associated with the consent constraints of the masses and the elite respectively: these are the welfare levels they have to attain if they are not to try and overthrow the system. These were defined in equations (4.5) and (4.7) above as $U_e^c = s_e U_e^* + (1 - s_e) U_e^l$ and $U_m^c = s_m U_m^* + (1 - s_m) U_m^l$. Here s_i is the exogenously given probability of group i, $i = e, m$ overthrowing the system, and U_i^* and U_i^l are respectively the maximum welfare that party i can attain given the consent

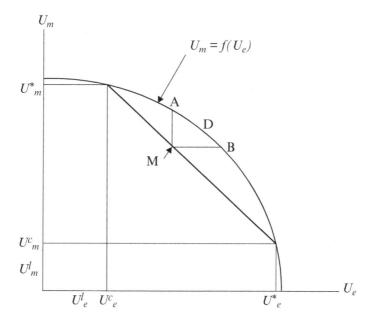

Figure 4.1 The variables, U_m^c, U_m^*, U_e^c, U_e^* and their relationships. $U_m^* = F(U_e^c)$ and $U^* = F^{-1}(U_m^c)$

constraint on the other, and the welfare level resulting from a failed attempt to overthrow the system.

We let $U_m = F(U_e)$ be the formula for the utility possibility frontier relating the efficient welfare levels of typical members of the masses and the elite (Figure 4.1). Clearly $U_m^* = F(U_e^c)$ and $U_e^* = F^{-1}(U_m^c)$. We also assume that the welfare level resulting from a failed attempt to overthrow the system is a known fraction $\lambda \in [0,1]$ of the welfare level from a successful coup, so that $U_m^l = \lambda U_m^*$ and $U_e^l = \lambda U_e^*$. Pulling all these observations together we have the following four equations

$$s_e U_e^* + (1 - s_e) U_e^l = F^{-1}(U_m^*) \tag{4.8}$$

$$s_m U_m^* + (1 - s_m) U_m^l = F(U_e^*) \tag{4.9}$$

$$U_m^l = \lambda U_m^* \tag{4.10}$$

$$U_e^l = \lambda U_e^* \tag{4.11}$$

which we can simplify to

$$s_e U_e^* + (1 - s_e) \lambda U_e^* = F^{-1}(U_m^*) \tag{4.12}$$

$$s_m U_m^* + (1 - s_m) \lambda U_m^* = F(U_e^*) \tag{4.13}$$

These are two equations in two unknowns, U_e^*, U_m^*. A solution will of course be a function of the parameters of the system, $s_e, s_m, \lambda, R, P, I, \tau, r, f, t$. Figure 4.1 shows these concepts and their relationships. While we cannot assert that this system of equations always has a positive solution, simulations for the cases of linear and circular utility frontiers indicate that it does for open sets of parameter values.

2.4 Comparative Statics

2.4.1 With respect to P

We are now in a position to do some comparative statics analysis, looking at how the welfare levels of the two groups change with changes in permeability P, the fraction of resource rent accruing to the MNC. First, consider the case of **democracy**, with the masses in control and facing the consent constraint (2.5). Recall that $g = \tau I f + \tau t R(1 - P)$. Suppose that P changes and that the welfare of the elite is kept constant at the initial constraint level U_e^c: this of course means holding the income of the elite constant. There is a choice to be made here: we can hold constant either total income or visible income. Holding total income constant is of course what the masses really want to do, but as not all income is visible they cannot be sure of doing this: they can however be sure of holding visible income constant. We investigate both cases, total income first. The total derivative of the (total) income of the masses I_m with respect to P is

$$\frac{dI_m}{dP} = \frac{\partial I_m}{\partial P} + \frac{\partial I_m}{\partial \tau} \frac{\partial \tau}{\partial P} \tag{4.14}$$

In this expression

$$\frac{\partial I_m}{\partial P} = -\tau t R, \quad \frac{\partial I_m}{\partial \tau} = If + tR(1 - P) \tag{4.15}$$

and the derivative of the tax rate with respect to P holding elite welfare constant is

$$\left.\frac{\partial \tau}{\partial P}\right|_{U_e} = -\left\{\frac{-(1-\tau)tR/r - (1-t)R/r}{-\frac{tR}{r}(1-P) - I_r^f}\right\} \tag{4.16}$$

so that

$$\frac{dI_m}{dP} = -R \tag{4.17}$$

Hence with elite welfare constant at its initial level the rate of change in government spending with permeability P equals the resource rent. We return to the implications of this below. If instead of keeping total income constant the rulers keep visible income constant, then we have

$$\frac{dI_m}{dP} = -tR \tag{4.18}$$

Next, we investigate the other governance case, with the **elite in control** and facing a consent constraint from the masses (4.7). As before we can investigate the effect of a change in the fraction of resource rent going to the MNC. In response to such a change, the elite change the tax rate so as to keep the welfare level of the masses constant, so that $\tau\{If + (1 - P)tR\}$ is kept constant. The total derivative of elite after-tax income with respect to P is

$$\frac{dI_e^a}{dP} = \frac{\partial I_e^a}{\partial P} + \frac{\partial I_e^a}{\partial \tau}\frac{\partial \tau}{\partial P} \tag{4.19}$$

and

$$\frac{\partial I_e^a}{\partial P} = -(1-\tau)\frac{R}{r} \tag{4.20}$$

$$\frac{\partial \tau}{\partial P}\bigg|_{U_m} = \frac{\tau t R}{r} \tag{4.21}$$

From this it follows that, when g is held constant,

$$\frac{\partial I_e^a}{\partial P} = -\frac{tR}{r} \tag{4.22}$$

There is an r in the denominator here because income released from taxation when net rent rises, given that the welfare of the masses is held constant, has to be expressed in per capita terms: this is not true in the democracy case as the public good, which is funded by taxation of the rent (amongst other sources) is available equally to all masses.

2.4.2 With respect to $(1-P)R$

An alternative approach that gives slightly more intuitive and cleaner results is to conduct comparative statics with respect to the amount of resource rent that remains in the host country, $(1 - P)R$. To do this we follow exactly the steps of the previous subsection, but instead of differentiating with respect to P we do so with respect to $(1 - P)R = \bar{R}$. We can skip the technical details: the results are

$$\frac{dI_m}{d\bar{R}} = -1 \text{ or } \frac{dI_m}{d\bar{R}} = -t \tag{4.23}$$

$$\frac{dI_e^a}{d\bar{R}} = -1 \tag{4.24}$$

In the first expression, giving the derivative of income of the masses, we have either -1 or $-t$ depending on whether actual income or visible income is held constant for the elite. Obviously if we have total transparency about income then $t = 1$ and these are the same. So these equations are saying that with full transparency every extra dollar that does not leave the country goes to the ruling group, whether we have democracy or autocracy. Otherwise, it is every extra visible dollar that goes to them.

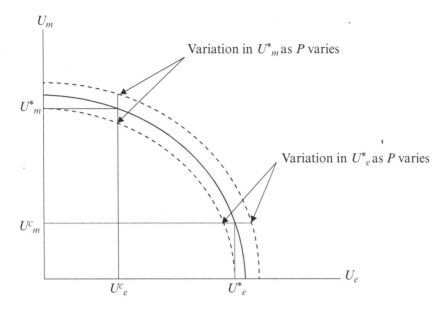

Figure 4.2   The utility possibility frontier moves in response to changes in permeability

2.5   Intuition

The rate at which the per capita income of the ruling group changes as permeability changes is essentially the same whichever group is in power: it is $-tR$, the total (visible) resource rent, in one case expressed in per capita terms. The rate at which the rent available to the home country changes with P is of course $-R$. In the autocracy case the elite are able to appropriate all of this extra rent for themselves. In the case of democracy, the masses also extract all the extra rent available to the home country themselves via an increase in government spending.

Why is the ruling group able to appropriate the entire extra rent available to the home country? Because of the participation constraints (4.5) and (4.7). These effectively specify income levels for the out-groups. These incomes do not change as permeability P changes, leaving all the extra resources available to the controlling group. Figure 4.2 illustrates this: the solid line shows a utility possibility frontier for representative members of the elite and the masses. In the case of democracy, the elite (whose welfare is on the horizontal axis) have to be given a welfare level of at least U_e^c, and clearly the best that the masses can do for themselves subject to this constraint is to attain their welfare level at the point on the frontier corresponding to U_e^c on the horizontal axis, shown as U_m^*. As permeability P changes, the utility possibility frontier moves in or out, as shown by the dotted lines, and the welfare attainable by the masses changes in response but the constraint on elite welfare is taken to be constant so that the masses absorb all the gains or losses from a change in permeability.

In the autocratic case the constraint is on the welfare of the masses, at the level shown as U_m^c, and the best that the elite can do is to pick the horizontal coordinate at the corresponding point on the frontier. Again, changes in the frontier clearly affect the welfare of the ruling group and not that of the ruled, that of the elite and not that of the masses.

This gives us insights into "how resource rents may shape economic conflict between elites and masses and thereby the emergence and persistence of democratic and authoritarian regimes." (Dunning (2008) p. 63 of the Kindle version.) Changes in the amount of retained resource rent – permeability – feed directly into the income of the ruling class, whichever class this may be, giving them a very strong interest in this variable, and explaining why the political debate in resource-rich developing countries frequently revolves mainly around the negotiation and renegotiation of the terms on which MNCs exploit the country's resources, whoever is in power. By way of illustration, in Mongolia from 2007 to 2013 the top item of news in the English-language newspaper each week has been either the redrafting of the minerals law or the renegotiation of major mining agreements. (For more details and an illustration of this in the case of Zambia see Chichilnisky-Heal 2014.)

An obvious extension of the analysis would be to allow the participation constraints to alter with changes in permeability: we could imagine that the out-group's expectations from switching to a controlling situation rise as the available resource rent rises. In terms of Figure 4.2, U_e^c and U_m^c would rise as the frontier moves outwards. Mathematically we would no longer differentiate with the out-group's welfare constant but with it in some functional relationship to resource rents available to the home country. In equations (4.5) and (4.7) the terms U_m^c, U_e^c would vary with the permeability.

We have taken transparency t to be exogenous, but it might be interesting and realistic to endogenize it. A first step could be to set $t = 1$ in the case of rule by the masses, reflecting the fact that once they are in charge they are well-placed to discover the full extent of resource-related income, but to allow $t > 1$ in the case of rule by the elite.

2.6 Pluralistic Governance

We have focused on two extreme cases, rule by the elite or the masses. There is another possibility, and in this case the effect of a change in permeability is quite different: incremental funds no longer accrue to the ruling group and there is no longer an incumbent advantage to a drop in permeability or a rise in the resource price. To understand this other possibility, consider Figure 4.1 again. Suppose that on average the society is ruled by the elite a fraction τ of the time and by the masses for the remaining fraction $1 - \tau$. Then for τ of the time it is at the point (U_e^*, U_m^C) and the rest of the time it is at (U_e^C, U_m^*), and let $M = \tau(U_e^*, U_m^C) + (1 - \tau)(U_e^C, U_m^*)$. The point M represents the average welfare levels of the two groups over time as control switches from one to the other. M is clearly Pareto inefficient:[2] both could be made better off at any point on the utility possibility frontier between A and B. The two groups should be able to bargain to reach a point such as D in this interval, as both gain by such a move. In this case there is no presumption that as the frontier moves in or out because of changes in permeability or prices or other external parameters, all benefits will accrue to one group: the distribution of gains will depend on the details of the arrangement used to select and support the point D.

---

[2] The inefficiency arises from risk aversion and not knowing what the welfare level will actually be at any point in time.

3. THE HOST AND THE MNC

To understand the relationship between the host country and the MNC and to generate any interesting predictions about the value of P we need a framework with more structure. To provide this we model the relationship between them as the equilibrium of a non-cooperative game (in the next subsection we investigate the implications of a bargaining approach).

We assume that the total rent available is a function of two variables, the permeability P, which is chosen by the host country, and the investment I in developing the resource deposits, which is chosen by the MNC. The outcome also depends on the stage of development of the host country, which is exogenous and is represented by D. The investment in resource development is chosen by the MNC in response to the financial terms offered to it, represented by permeability P, so $I = I(P)$. And the rent available is a function of the investment and the stage of development, $R = R(I(P), D)$. For a given level of development D, we assume the total rent available R is an increasing concave function of the permeability, reflecting the fact that greater payments to the MNC give it more incentive to prospect for resources and to invest in their discovery and extraction, but that there are diminishing returns to this process. So the curve $R(I(P), D)$ shows the rent resulting from the MNC's best response to the level of permeability P offered by the host. And for a given level of permeability, i.e. a given incentive structure, the total rent available increases with the level of development. More developed countries have better infrastructure, a better qualified labor force, and can manage the resource extraction process more effectively.

Figure 4.3 shows this set of relationships: permeability P is plotted horizontally and total rent R vertically, and the three labelled brown lines show how total rent available $R(I(P), D)$ increases with P for a given level of D, and also increase with D for a given P. The rent accruing to the host country is $(1 - P) R(I(P), D)$. We simplify the expression for rent to $R(P, D)$.

It seems natural to think of the host as initiating discussions with an MNC and trying to anticipate how this will react to terms offered by the host, so we use a Stackelberg leader-follower framework, with the host as the leader.[3] The host proposes a value for permeability P, and the MNC responds with a value for investment I which maximizes the MNC's net revenues and which for a given stage of development D implies a rent $R(I(P), D)$: this function is the MNC's reaction function to the host's choice of strategy P. The host is assumed to know this reaction function and so will choose P so as to maximize its receipts from the resource given the MNC's reaction function. That is, it chooses P to maximize $(1 - P) R(I(P), D)$. Maximizing this with respect to P leads to

$$-R(P,D) + (1 - P)\frac{\partial R}{\partial P} = 0 \; or \; \frac{R}{1 - P} = \frac{\partial R}{\partial P} \qquad (4.25)$$

as a first order condition. Here $\partial R/\partial P$ is the slope of the curve $R(P, D)$ and $R/(1 - P)$ is the slope of the rectangular hyperbola on which host revenues are constant $(1 - P) R$, which has its horizontal axis increasing to the left. These are the convex curves in

---

[3] An analysis based on the more symmetric Nash equilibrium concept does not lead to interesting insights into this problem.

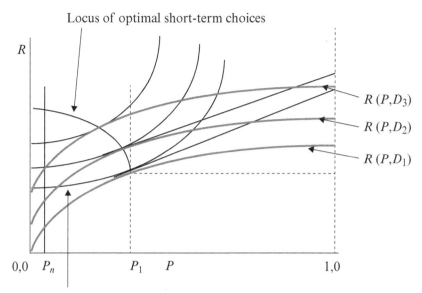

Figure 4.3 The host's optimal bargain changes with the level of development and follows the locus of points of tangency

Figure 4.3. The tangencies shown meet this condition, and the locus of such points of tangency will move upwards and to the left as D increases, as shown. These are optima for the host and represent a tradeoff between the loss of rent from greater permeability and the increase in rent because greater permeability provides stronger incentives to the MNC. Each point of tangency between a contour of the host country's payoff and a rent-permeability curve for a given level of development is a Stackelberg equilibrium in the game between the host and the MNC for a given level of development D.

If the host country's development level were D_1 then the best level of permeability for the host country would be P_1, but if the level of development were to increase, as it presumably will over time, the best choice of P will fall, as shown by the locus of optimal short-term choices. This poses a dilemma for the host country: at development level D_1 should it choose permeability P_1, which is currently optimal, or should it choose a lower level which will be optimal at some future level of development? If so, optimal at which future stage of development? The core of the problem the host country faces is that it knows that any choice which seems optimal under current circumstances will seem suboptimal at future dates with higher development levels. A sequence of choices of P, each of which is the best choice at the time it is made, is not time consistent: all these choices will seem wrong at later dates and the host country will wish to revise them. If the host country chooses P_1 and then stays with this, it will move over time up the vertical line through P_1, and at future dates the chosen level of permeability will seem too high.

A way of resolving this conundrum is to think of the host country's problem as a dynamic optimization problem, in which it seeks to choose permeability at each point in time to maximize the present value of revenues:

$$Max_{P(t)} \int_0^\infty (1 - P(t)) \, R(P(t) \, D(t)) \, e^{-\delta t} dt \qquad (4.26)$$

where $\delta \geq 0$ is a discount rate, the level of development $D(t)$ is an exogenous function of time, and $P(t)$ is the variable whose time path is to be chosen. The solution to this problem simply requires that P be chosen at each date to satisfy the first order condition (4.25), which means following the locus of points of tangency upwards and to the left in Figure 4.3. If the level of development $D(t)$ is made endogenous, so that D increases with the rent received, then the solution is qualitatively the same, except that the rate of movement may be different. (For a full mathematical treatment see Appendix 4A.1.)

Suppose then that the host country seeks to do what is optimal for it and revise the level of permeability at each stage of development, following the locus of short-term optima; what is the outcome? The problem here is that there are two parties to any bargain, the host country and the MNC. The MNC will wish to hold the host to the initial high level of permeability, and will not agree to revisions that worsen the terms it faces. So there is a fundamental conflict between what the host and the MNC want, which in fact becomes sharper over time as the host moves to higher levels of development and seeks to move to the left in Figure 4.3 and reduce the payments to the MNC.

Many agreements between hosts and MNCs anticipate precisely this conflict. For example, in 1997 Mongolia introduced the Minerals Law, described by the United States Geological Survey as the most liberal mineral law in the world at the time. This law made provision for "stability agreements," which guarantee that the terms negotiated between the host and the MNC will not be changed for some specified period – 20 to 60 years are typical. According to an authority on these agreements,

> Mining and petroleum agreements governing the exploration and development of natural resources frequently include contractual assurances of stability. These stability clauses are intended as legally binding commitments by the host country's government. The commitment may be for an initial period of years or for the length of the agreement. They may cover a broad-range of host country laws or be limited to fiscal laws or even certain provisions in the fiscal laws, such as tax and royalty rates. . . . Fiscal stability clauses are generally justified by: (1) the large size and the sunken nature of the initial investment, and (2) often a long period required to recover investment and earn a reasonable return, taken together with (3) a lack of credibility on behalf of the host country to abstain from changing the fiscal rules – possibly singling out high rent petroleum or mining operations – once the investment is sunk (the time inconsistency problem).[4]

In fact, it is not unusual for mining contracts to be negotiated in an ad hoc fashion outside of the country's mineral laws, with dispute settlement clauses that disempower the host country's domestic courts, and with the MNC being explicitly exempted from any subsequent changes in the host country laws. Bilateral investment treaties institutionalize this idea as does the Energy Charter Treaty in the case of petroleum-exporting countries (van Benthem and Stroebel (2013)). A particular source of conflict has been the introduction of environmental laws in countries where there were initially none, with mining MNCs –

---

[4] See https://www.international-arbitration-attorney.com/wp-content/uploads/arbitrationlaw1394930.pdf#page=422, Daniel and Sunley (2010).

often the main polluters – being exempt from these. The issue here is that the host country cannot credibly commit to an agreement that it will clearly wish to change, and has to resort to invoking authorities outside the country.[5]

This conflict of interests describes precisely the situation reflected in Figure 4.3: it is clear that the host country will wish to revise any constant P agreement that it signs early in its economic development, to the disadvantage of the MNC, because a choice that looks optimal today will not seem so in a decade. Although Mongolia introduced a pro-MNC legal regime in 1997, it has since drastically changed its mining laws, introducing a windfall profits tax in 2006. (The agreement that Mongolia initially signed with Centerra Gold provided for no income tax liability for ten years.) Zambia, another resource-rich poor country, followed suit in 2008: both were attempting to capture some of the extra rents generated by the resource price boom of the early 2000s. It has often been the case that international financial institutions, when negotiating with the host country about economic assistance programs, seek concessions concerning the legislative framework for extractive activities, promoting stability agreements amongst others and, in accord with the Washington Consensus, promoting the interests of the MNCs.

The quote above also hints at another aspect of the bargaining situation, namely the fact of large sunk costs in mining operations. The MNC has to make substantial irreversible investments before any revenues flow from the project, and so is vulnerable to exploitation by the host via a strategy that offers attractive terms before the investment is made and revises these once it is in place and cannot be undone.

These observations are all consistent with the "obsolescing bargain model,"[6] which suggests that host countries naturally strike bargains that become sub-optimal for them, although the mechanism through which sub-optimality emerges is rather different here: the obsolescing bargaining literature focuses on changes in rents due to changes in prices as a driver of dissatisfaction with existing deals (as in the cases of Mongolia and Zambia with windfall profits taxes cited above). The model presented here can be thought of as providing an analytical underpinning for and an extension of the obsolescing bargain model. The discussion of bargaining and the hold-up problem in the next subsection provides further insight into the idea of an obsolescing bargain between host and MNC.

4. THE HOLD-UP PROBLEM IN HOST–MNC RELATIONS

The central problem in reaching a stable and efficient relationship between the host and the MNC is an illustration of the hold-up problem (see Tirole (1986) or Hart (1995)). The host country is selling to the MNC the right to produce its resource and sell it on world markets. To take advantage of this offer the MNC must make a substantial investment that is specific to this trading relationship, and that has no value in any other context. The two parties agree terms before the investment is made, and on the basis of this

---

[5] See Radon (2007), who refers to "contractual colonialism."
[6] See for example Moran (1978). Thomas and Worral (1994) develop the idea of obsolescing bargains within a game-theory framework.

agreement the MNC executes the investment, generally in the multi-billion dollar range. Once the investment is made, the host may reopen the bargaining process and the MNC's bargaining power is then greatly reduced, as it already has a financial commitment to the relationship, something that both parties can anticipate in advance. A simple model illustrates this point.

Assume that the MNC can invest an amount I in the host country to generate profits of $\pi(I)$ and that the investment costs $kI, k > 0$. Obviously the optimal investment level I^* satisfies the first order condition $k = \frac{\partial \pi}{\partial I}$. Before any investment is made, the two parties bargain about how to share the profits generated, and agree that the MNC will receive a share P, with $(1 - P)$ going to the host country. Maximizing $P[\pi(I) - kI]$ is the same as maximizing $\pi(I) - kI$ so the MNC will still choose the socially optimal investment level. Assume the bargaining process that leads to this agreement on how to divide the profits is a Nash bargaining process, so that P is the sharing rule that maximizes the product of the differences between the host's and MNC's welfare levels and their fall-back points, the welfare levels they attain in the event of no agreement. In the event of no agreement, neither makes any profit or incurs any costs, so the fall-back point is at the origin. If the utility functions are identical then the outcome will be symmetric and we will have $P = 0.5$.

Now suppose that the parties reach an ex ante agreement on profit-sharing, a value of P, and on the basis of this the MNC makes the investment I^*. Once the investment is made and the associated costs are sunk, the host reopens the bargaining process and there is a further round of Nash bargaining. In this case in the event of no agreement the MNC loses its investment I^*, and the host country gains this. So the fall-back positions are no longer equal. A new round of Nash bargaining will lead to a smaller share of profits going to the MNC: this is the essence of the hold-up problem (see Appendix 4A.2 for detailed arguments). The stability agreements discussed above are a response to this, but they are generally inadequate. Bilateral investment treaties are a more sophisticated attempt to grapple with the same issues and are more successful, but have not prevented expropriations (Stroebel and van Bentham (2013)).

A response to hold-up problems widely considered in the industrial organization literature is to integrate the two parties to the problem. It is clearly not possible to integrate the host country and a multinational corporation, and the nearest one could come to this in the present case would be to set up a new company whose sole purpose is to produce and sell the resource from the host country. This company would be jointly owned by the host and the MNC, and profits would be shared between them according to their shareholdings (for more discussion on integration see Che and Sákovics (2004) and Bolton and Whinston (1993)). However, in the present context this would not overcome the possibility of hold-up. The nature of such a deal would be that the host would give the jointly owned company mineral development rights in exchange for shares, and the MNC would provide capital and technology in exchange for its shares. Once such a deal was in place and the investment had been carried out by the MNC, the host would as before be in a position to reopen negotiations and threaten the MNC with the loss of its right to operate and so of its investment if it failed to agree a larger shareholding for the host.

It seems that the potential for hold-up is very deeply embedded in the structure of host–MNC relations, leaving the MNC vulnerable to a loss of anticipated revenues from the venture. One reason for this is that the relationship is between a sovereign state and a foreign corporation. They can choose which legal system will govern their relationship – New York, London, etc. – but none of these jurisdictions have power to inflict substantial penalties on the host in the event that it violates the agreement. So even elaborate agreements with complex stability clauses, as discussed above, cannot ensure that the host will not at some stage reopen negotiations. A possible way to make the relationship more symmetrical is noted by Janeba (2000), who observes that the MNC's bargaining power is increased if it has the possibility of entering into a deal with a second host country that can also supply the resource at issue. The MNC can now construct capacity in both countries and threaten to pull its activities away from whichever offers least favorable terms. Of course, this solution comes at the cost of excess capacity. This point is developed further by Kessing and Hefeker (2014).

There are some events which tend to trigger a reopening of negotiations, and one way of avoiding such reopening is to specify in advance how terms of the contract may be modified if these triggers occur. For example, the price of the commodity being mined may rise, making the initial contract far more profitable for the MNC than the host had anticipated. Several authors studying the hold-up problem have shown that hold-ups triggered by unanticipated events can be avoided by specifying how the contract can be modified if such events occur (Bolton and Whinston (1993), van Benthem and Stroebel (2013)).

One option that must appeal to the MNC is to try to ensure that the host government is friendly and not disposed to reopen negotiations. There are clearly several routes to this. One is intervention in the domestic affairs of the host to influence political outcomes. This might involve supporting a party or politician seen as well-disposed, or assembling a group of supporters by what in game theory are described as side-payments, and in this context would surely be seen as bribes. This could be a factor contributing to the observation that political processes in many resource-rich countries are corrupt (Ross (2018)). Another route to a favorable political orientation in the host country is to persuade third parties to use influence over the host to ensure that it does not exploit its bargaining power. Natural allies would be international financial institutions, particularly those on which the host country might have some dependence such as the World Bank or a regional development bank, or the MNC's home government, which could be lobbied to make aid disbursements conditional on a suitable financial settlement between the host and the MNC. There is anecdotal evidence at least that all these options are used by resource-extraction MNCs: Chichilnisky-Heal (2014) refers to the use of World Bank influence in Mongolia and Zambia to attempt to prevent the government from reopening contractual terms. The second author has personal experience of a similar situation in Papua New Guinea. This issue continues to be topical: a headline in the *Financial Times* recently stated "Cash-strapped Nigeria to renegotiate contracts with oil majors," showing that the hold-up problem does not diminish over time.[7]

---

[7] *Financial Times*, posted online at 7.14pm, 15 September 2015. According to the *FT*, "The NNPC said it would overhaul its contracts with companies such as Shell, Chevron, Eni and ExxonMobil 'in the weeks and months ahead . . . to extract as much benefit as possible for Nigeria.'"

4.1 Adoption of Bilateral Investment Treaties

As we have mentioned, an increasingly common response to the hold-up problem implicit in host–MNC relations is for the host country to sign a Bilateral Investment Treaty. BITs are a way for a host country to commit itself to following certain procedures and staying within the scope of agreements. BITs are typically between a developing country that is potentially the host of investment and the home government of an MNC that might invest in the host, or the government of a country in which the MNC does business. These treaties involve an intergovernmental agreement that covers all aspects of investment, and typically includes provisions for international arbitration in the event of a dispute. Such provisions mean that when a host country that has signed a BIT with the home country of an MNC also signs an agreement with the MNC, this represents a commitment that cannot readily be altered or disregarded. BITs have become widespread in the last few decades, and a natural question is why a developing country would sign a treaty that limits its ability to exploit its bargaining power in the future. A part of the answer is clearly that this is seen as a prerequisite to obtaining more foreign investment, but this possible increase comes at a cost.

One of the most interesting observations on this question was suggested by Guzman (1997), who makes the argument that poor countries sign bilateral investment treaties (BITs) because of competition with other poor countries. One country signs initially and this diverts investment from others to it, giving the others an incentive to sign too, and this competition then leads to them all signing and makes them all worse off in the end. Dixit (2003) refers to a process like this as "entrapment." Guzman sees the structure of the game between them as something like a prisoners' dilemma game, with the best outcome for all poor countries being that none signs a BIT, but each having an incentive to defect from this equilibrium if they assume all others are going to continue in it.

Bubb and Rose-Ackerman (2007) model this idea slightly differently: they assume that the gains to a poor country from signing and not signing are both decreasing functions of the number of poor countries that have already signed, and then look for Nash equilibrium patterns of signing/not signing. They assume that all countries signing is a Nash equilibrium, that all not signing is not, and analyze when there may be intermediate equilibria.

Here we work with a rather different framework. We assume that there are at least two possible Nash equilibria to the game between poor countries when they are choosing whether or not to sign. One is that *all sign*, as with Bubb and Rose-Ackerman: the other is that *none sign*. So we are making an assumption *that all not signing is an equilibrium*, an assumption which neither Guzman nor Bubb and Rose-Ackerman make: in this sense we are making not signing more stable, making it more difficult for some countries signing to force others to sign. The point is not that this is a good description of reality, but rather to emphasize that even if not signing were to be a stable situation, it would still be the case that some countries deciding to sign can force all others to sign and that all will be worse off as a result of this. Countries that sign exercise a lot of leverage over those that don't.

Assume there are N developing countries indexed by $i = 1, \ldots, N$ trading with a set of rich countries and considering signing bilateral investment treaties with them. Each developing country has two strategies – to sign a BIT with the rich countries $S_i = 1$, or to decide not to sign, $S_i = 0$. Payoffs are a function of all agents' strategies and we write them $U_i(S_i, S_{-i})$ where S_{-i} is the vector of strategies of all agents other than agent i. We order strategy vectors $S = (S_i, S_{-i})$ by the usual vector ordering, so that $S' > S$ means that $S'_i \geq S_i \forall i \: \& \: \exists j : S'_j > S_j$.

We assume the payoffs show what Heal and Kunreuther call uniform strict increasing differences, which means that $\exists \varepsilon > 0$ such that if $S'_{-i} > S_{-i}$ then

$$U_i(1_i, S'_{-i}) - U_i(0_i, S'_{-i}) \geq \varepsilon + U_i(1_i, S_{-i}) - U_i(0_i, S_{-i}) \quad (4.27)$$

where $0_i, 1_i$ represent a zero or a one as the $i - th$ component of the vector. In intuitive terms this just says that decisions to sign BITs are strategic complements. The payoff to changing from not signing 0_i to signing 1_i increases as the number who have already signed gets larger. This is consistent with the case argued by Guzman and Bubb and Rose-Ackerman: every time another country signs a BIT it takes investment away from those who have not signed and increases their incentive to sign. This could be because the payoff to signing increases, or that to not signing decreases, or some combination.

In order to explain our result, we need some additional definitions. A tipping set is a subset of the developing countries, which we will denote $T \subset S$, with the property that if these countries choose to sign a BIT, then all others find this to be their best move also. Formally, if $S_i = 1 \forall i \in T$, then $\forall i \notin T$, $U_i(1_i, S_{-i}) \geq U_i(0_i, S_{-i})$. So if all poor countries are initially not signing BITs and then those in T change and decide to do so, the best move for all others is now to sign too. In this sense, the members of T can "tip" the system from no countries signing BITs to all doing so. The result we are able to prove is that provided there are enough developing countries (N is large enough), there is always a tipping set with fewer than N countries in it.

Proposition 1 *Under the assumption (4.27) of increasing differences, and with a large enough number N of developing countries, there is a tipping set T with fewer than N–1 members that can tip the equilibrium where no country signs a BIT to one where all sign.*

This is an immediate application of Heal and Kunreuther (2010).

Intuitively this result is driven by the increasing differences condition, which ensures that choices to sign BITs are strategic complements. A numerical example will perhaps be helpful in conveying how this works. Suppose that there are ten players ($N = 10$), each of whom may sign or not ($S_i = 1 \: or \: 0$). The payoffs to these choices are $U_i = 9 \: if \: S_i = 1$, $U_i = \#(0) \: if \: S_i = 0$. Here $\#(0)$ is the number of countries choosing strategy 0, i.e. the number choosing not to sign. So if all countries do not sign, they all get a payoff of 10. And if one were to change to signing and the others remained as before, the payoff to the signer would be 9, so there is no incentive for anyone to change from 0 to 1 when all are at 1: all choosing 1 is a Nash equilibrium. And if all choose to sign, then they all get a payoff of 9: any country that moves from 1 to 0 when the rest are at 1 will get a payoff of 1, so all signing is also a Nash equilibrium. Now suppose that all countries are not signing, $S_i = 0 \forall i$, and then two – any two – choose to change to signing. Then the payoff to any other country that joins them is 9 and the payoff to remaining a non-signer is 8. Hence, all countries will now join the initial two and sign. The first two to sign have in effect forced the remainder to sign, not forced in a legal or physical sense, but forced in the sense of changing the incentives facing them so that their best move is now to sign.

When all sign they will all get a payoff of 9: when none signed they all got a payoff of 10. So the first movers have forced the rest into a move which results in all being worse off.

In the context of the game, there is no reason why any country should move from the equilibrium where all are not signing BITs, which is an equilibrium – this is a difference between our model and that of Bubb and Rose-Ackerman. But if we place the game in a broader geopolitical context there are many explanations. For example, the leaders of the first countries to sign may gain personally from the existence of a BIT, or some of their supporters may. Alternatively, the rich countries, seeking to break an implicit cartel of non-signing countries, may offer some incentives outside the game, such as aid, military assistance or access to advanced military hardware, etc.

4.2 Investment in Resource-Rich LDCs

The work of Guzman and Bubb and Rose-Ackerman gives us some fundamental insights into the driving forces behind the expansion of BITs. Central to these models is the idea of competition between LDCs for investment from rich countries. Such competition is a reality in many cases: if all an LDC has to offer to potential investors is low wages, then the terms on which FDI is managed will be an important factor in locational choices of investors. There are many poor countries offering low wages to labor-intensive industries and none will have particular market power. But the same is not true of resource-rich countries: there are few resource-rich countries and many MNCs that would be happy to develop their resources. If a country really has rich reserves of a valuable mineral, MNCs are not so sensitive to the terms on which they can invest. As an example, consider the willingness of Western oil companies to invest in Russia, a country where the risks of expropriation are above the average. In the case of investment in resource-rich countries we are closer to a bilateral monopoly when it comes to thinking about the relations between host and MNC. Indeed the situation may be closer to a monopoly supplier of the resource facing competing buyers of the right to develop it. Most oil-rich countries choose to develop their reserves through national oil companies, sometimes working in collaboration with international oil majors, and obtain very favorable terms for their resources (van Benthem and Stroebel (2013)). The literature on BITs probably needs to differentiate between resource-rich and other developing countries.

5. CONCLUSIONS

We have investigated various aspects of the relationships between resource-rich developing countries and multi-national corporations that develop their resources. We began by considering the role of permeability, in the sense of the fraction of resource rent leaving the country, in domestic politics, showing that a decrease in permeability will always benefit the ruling group, whether the country is a democracy or an autocracy.

We then modeled the determination of permeability through games between the host and the MNC, and noted that there is a fundamental problem in reaching a stable agreement that both parties are happy with. In a non-cooperative framework the host will always view an agreement reached at an earlier date as sub-optimal and want to revise it and reduce permeability – as indicated by the obsolescing bargain literature, though for rather different and more fundamental and structural reasons. The obsolescing bargain literature sees the need for revision as coming from changes in external parameters such as prices: we indicate that this need is built into the structure of the problem. The discussion of bargaining and the hold-up problem in host–MNC relations emphasizes the point that the obsolescence of an initial bargain is guaranteed even if there are no changes in external parameters. The fundamental role that the possibility of hold-up plays in the relationship between the host and the MNC suggests strong incentives for the MNC to influence domestic politics in the host, possibly making the host's political system permeable in a broader sense.

A way of resolving the hold-up problem is for the host and the home country of the MNC to sign a BIT that places restrictions on the host's ability to alter the terms of an investment deal once it has been agreed. The signing of a BIT means that the host is committing itself when it makes a deal with an MNC: it is throwing away some of its bargaining power. Permeability becomes unchangeable. One might expect that this would happen only when there is a clear net gain from doing this, perhaps in terms of more investment being available because of the security implied by the BIT. However, Guzman has suggested that countries can be "entrapped" (using Dixit's words) into making agreements that in the end make them worse off, and we have developed that point as an application of the theory of tipping sets in games of strategic complementarity. If a subset of poor countries sign BITs with investor countries, they may in effect force their peers to follow suit, and in the end produce an equilibrium where all poor countries are worse off, with higher permeability, than if none signed BITs. The relationship between a resource-rich country and an MNC is a locus of conflicting pressures, with the domestic ruling group seeking lower permeability and external pressures from the MNC and the host country's peers acting in the opposite direction.

REFERENCES

Acemoglu, D. and Robinson, J. A. (2001). A theory of political transitions. *American Economic Review*, **91**(4): 938–963.
Bolton, P. and Whinston, M. (1993). Incomplete contracts, vertical integration and supply assurance. *Review of Economic Studies*, **60**: 121–148.
Bubb, R. J. and Rose-Ackerman, S. (2007). Bits and bargains: Strategic aspects of bilateral and multilateral regulation of foreign investment. *International Review of Law and Economics*, **27**(3): 291–311.
Che, Y.-K. and Sákovics, J. (2004). A dynamic theory of holdup. *Econometrica*, **72**: 1063–1103.
Chichilnisky-Heal, N. (2014). Bargaining to lose: Post-transition resource extraction. Unpublished manuscript, Department of Political Science, Yale.
Daniel, P. and Sunley, E. M. (2010). *The Taxation of Petroleum and Minerals*, chapter 14. Contractual assurances of fiscal stability. Abingdon, UK: Routledge.
Dixit, A. K. (2003). Clubs with entrapment. *American Economic Review*, **93**(5): 1824–1836.
Dunning, T. (2005). Resource dependence, economic performance and political stability. *Journal of Conflict Resolution*, **49**(4): 451–482.
Dunning, T. (2008). *Crude Democracy*. Cambridge: Cambridge University Press.
Guzman, A. T. (1997). Why LDCS sign treaties that hurt them: Explaining the popularity of bilateral investment treaties. *Virginia Journal of International Law*, **38**: 639–688.
Hart, O. (1995). *Firms, Contracts, and Financial Structure*. Oxford: Clarendon Press.
Heal, G. and Kunreuther, H. (2010). Social reinforcement: Cascades, entrapment and tipping. *American Economic Journals: Microeconomics*, **2**(1): 86–99.
Humphries, M., Sachs, J., and Stiglitz, J. (2007). *Escaping the Resource Curse*. New York: Columbia University Press.

Janeba, E. (2000). Tax competition when governments lack commitment: Excess capacity as a countervailing threat. *American Economic Review*, **90**(5): 1508–1519.

Karl, T. L. (1997). *The Paradox of Plenty: Oil Booms and Petro-States*. Berkeley, CA: University of California Press.

Kessing, S. and Hefeker, C. (2014). Competition for natural resources and the hold-up problem. Beitrage zur Jahresrtagung des Vereins fur Socialpolitik 2014, http://hdl.handle.net/10419/100361.

Moran, T. H. (1978). Multinational corporations and dependence: A dialogue for dependistas and non-dependistas. *International Organization*, **32**(1): 79–100.

Radon, J. (2007). *Escaping the Resource Curse*, chapter 4. How to negotiate an oil agreement. New York: Columbia University Press.

Ross, M. L. (2012). *The Oil Curse: How Petroleum Wealth Shapes the Development of Nations*. Oxford: Princeton University Press.

Ross, M. L. (2018). *Handbook of the Politics of Development*, chapter 12. The politics of the resource curse: a review. Oxford: Oxford University Press.

Sachs, J. and Warner, A. (1999). The big rush, natural resource booms and growth. *Journal of Development Economics*, **59**(1): 43–76.

Stroebel, J. and van Benthem, A. (2013). *Critical Issues in Taxation and Development*, chapter 5. Investment treaties and hydrocarbon taxation in developing countries. Cambridge, MA: MIT Press.

Thomas, J. and Worral, T. (1994). Foreign direct investment and the risk of expropriation. *Review of Economic Studies*, **61**: 81–108.

Tirole, J. (1986). Procurement and renegotiation. *Journal of Political Economy*, **94**: 235–259.

van Benthem, A. and Stroebel, J. (2013). Resource extraction contracts under threat of expropriation: Theory and evidence. *Review of Economics and Statistics*, **95**(5): 1622–1639.

APPENDICES

Appendix 4A.1

Above, we considered the problem

$$Max_{P(t)} \int_0^\infty (1 - P(t)) R(P(t) D(t)) e^{-\delta t} dt$$

which involves choosing permeability at each date so as to maximize the present value of revenues accruing to the host country. Here $\delta \geq 0$ is a discount rate and $D(t)$ is the time path of development, which we take as exogenous for the present. To solve this problem we choose $P(t)$ so as to maximize the Hamiltonian

$$H = [1 - P(t)] R[P(t) D(t)] e^{-\delta t}$$

at each time t, which gives as a first order condition

$$R(t) = (1 - P(t)) \frac{\partial R}{\partial P}$$

which is the same as equation (4.25) above. A more complex version of this approach is to endogenize the level of development as a function of cumulative revenues generated from the resource. This would give a state variable

$$D(t) = \int_0^t [1 - P(\tau)] R(\tau) d\tau \qquad (4A.1)$$

and its rate of change

$$\frac{dD}{dt} = [1 - P(t)] R(t) \qquad (4A.2)$$

The Hamiltonian is now

$$H = = [1 - P(t)] R[P(t) D(t)] e^{-\delta t} + \lambda e^{-\delta t} [1 - P(t)] R[P(t) D(t)] \qquad (4A.3)$$

The first order condition is again

$$R(t) = (1 - P(t)) \frac{\partial R}{\partial P} \qquad (4A.4)$$

so that the analysis of Figure 4.3 again applies and an optimal path involves following the locus of points of tangency.

Appendix 4A.2

Let $U_h((1 - P) R)$ and $U_m(PR)$ be the utilities of the host country and the MNC respectively, with $P \in [0,1]$ the fraction of total revenue accruing to the MNC. The two parties engage in a Nash bargaining process to choose P. Ex ante, before any investment is made, their fall-back positions, their payoffs in the event of no agreement, are both zero: neither party makes or loses money in the event of no agreement. Hence, the Nash bargaining solution satisfies

$$\underset{P}{Max}\{U_h((1-P)R) - 0\}\{U_m(PR) - 0\} \qquad (4A.5)$$

the solution to which requires that

$$\frac{U'_h}{U_h} = \frac{U'_m}{U_m} \qquad (4A.6)$$

If both payoff functions are identical, an assumption we make to allow us to focus on the factors essential to our argument, then this implies that $P = 0.5$.

Now assume that parties are bargaining again after the MNC has made an investment of I, perhaps because the host has forced a reopening of the terms of their agreement. In this situation, the MNC stands to lose its investment in the event of no agreement. The Nash solution therefore now satisfies

$$\underset{P}{Max}\{U_h((1-P)R) - 0\}\{U_m(PR) - I\} \qquad (4A.7)$$

and the first order conditions are now

$$\frac{U'_h}{U_h}\left[1 - \frac{I}{U_m}\right] = \frac{U'_m}{U_m} \qquad (4A.8)$$

We know that I represents an amount that is lost, so $I < 0$ and

$$\frac{U'_h}{U_h} < \frac{U'_m}{U_m} \qquad (4A.9)$$

We can also show that

$$\frac{d}{dx}\left[\frac{U(x)'}{U(x)}\right] < 0$$

under standard concavity conditions on the payoff functions, so it follows that $P < 0.5$.

5. Bargaining to lose the global commons
Natasha Chichilnisky-Heal and Graciela Chichilnisky

1. INTRODUCTION

In "Bargaining to Lose: The Permeability Approach to Post Transition Resource Extraction" Natasha Chichilnisky-Heal (2013) tackles the *resource curse* in an original and fertile way. Her "permeability" approach questions the standard treatment of the state as a single decision maker having the public good as an objective. She offers instead a new analytical and policy-oriented methodology, with hands-on experience in Mongolia and on Zambia, putting on the table the state's bargaining game with key global organizations. The state bargains with the Bretton Woods Institutions (IMF, World Bank) and with MNCs in the creation of contracts for exploiting the nations' resource extraction; these are the institutions that led the globalization of the world economy since the post-WWII period when colonialism receded and was replaced by market oriented international financial institutions. Chichilnisky-Heal argues that the developing nation loses from the bargaining – and the loss is the resource curse.

This article extends Chichilnisky-Heal's work in two directions: one, providing a mathematical model that formalizes her arguments about the impact of the state's permeability on the exploitation of extractive resources and showing that the model validates her results; and two, showing that the resource course takes a particular form: accelerating and increasing the quantities of resources extracted and lowering their prices in international markets. Through Hotelling's formalization of extractive resources as financial assets, it leads to higher discount rates on the future, which in turn affects negatively savings and investment within resource-rich GDP poor nations. This result links the resource curse analysis of Chichilnisky-Heal with other work (Chichilnisky, 1994) that views the resource curse in developing nations as originating from the lack of private property rights on resources in developing nations, and from the Bretton Woods' institutions role in magnifying the corresponding tragedy of the commons into the global environmental crisis of our time.

The resource curse explains the failure of resource-rich developing nations to use their resources as needed to achieve economic growth. It is highly applicable today to poor nations. About 70% of the exports of Latin America today are raw materials and that number is 90% in Africa. Ultimately, the resource curse impacts the entire global economy and underlies the global environmental crisis. This crisis is based on the overextraction of resources such as coal and petroleum, causing dangerous CO_2 emission levels, the biodiversity loss due to overextraction of forest resources and overextraction of biomass from the oceans. Most of the natural resources consumers in the world today are extracted in poor nations (LA, Africa, post-transition economies) and they are extracted for exports, leading to today's overconsumption in the OECD world and to worldwide environmental losses.

This piece explores the assumptions in "Bargaining to lose" comparing it with the literature, and generalizes its results to explore its implications for what is usually known as the *global commons*: the atmosphere, the oceans and the planet's biodiversity. The results address the following issues: (1) whether Chichilnisky-Heal's "permeable state" represents a transition to a new globalized society where the sovereign state – a relatively recent creation – is receding giving rise to a new set of global economic agents and institutions that better explains the dynamics of the global commons, (2) how the permeability approach substitutes and complements other explanations for the resource curse (Chichilnisky, 1994) as a global market failure greatly magnified by globalization and based on the lack of well defined property rights on natural resources in poor nations during the pre-industrial period, (3) provides an extension and generalization of Chichilnisky-Heal's "bargaining to lose" approach to enhance the understanding of the environmental crisis on the global commons, and the solutions that she proposes (2003), e.g. limiting Bretton Woods Institutions' "seat at the negotiation table" of resource extraction contracts, thus helping resolve the environmental crisis based on overextraction of global resources.

2. THE RESULTS

A simple mathematical model extends the original piece (Chichilnisky-Heal, 2013) under stereotyped conditions that capture its key aspects, and establish the validity of its conclusions. Both models interpret in mathematical terms the excellent policy examples and recommendations provided by Natasha Chichilnisky-Heal's original article (2013), as well as her game theoretical approach to a bargaining game between the state and IFIs (International Financial Institutions). We establish how the permeability of the state causes inferior outcomes in terms of undermining the nation's economic growth and inducing poverty. Within this simplified framework the model expands on Natasha's piece by showing how the resource curse causes environmental degradation by generally leading to overextraction of extractive resources in developing nations. The oversupply of resources leads to prices that are lower than would be optimal without the permeable state, ultimately inducing overconsumption in the industrial nations as well as global loss of biodiversity, overexploitation of the oceans, and overemission of CO_2 into the atmosphere.

3. THE "PERMEABLE" STATE: A FORMAL ECONOMIC MODEL

The model and their results are presented in a summarized form as appropriate for the Yale conference "Extractive Resources and Global Governance: Distributive Justice and Institutions" held on 23 October 2015. It formalizes the "permeability" approach by defining a continous parameter λ with values between 0 and 1 that measures the extent of "weakened sovereignity" by the state – this is the methodology proposed in Chichilnisky-Heal (2013). The model studies the optimal intertemporal allocation problem of a developing nation that extracts natural resource, such as for example, the optimal exploitation of the largest copper-gold mine in the world that is located in the Gobi Desert, Mongolia. The results show that there is overextraction in the short run and a lowering of the prices, due to the permeability of the state. In a permeable state, the intertemporal allocation in the developing nation is compromised and becomes suboptimal, undermining economic growth and producing poverty while the nation increases its extraction and exports of resources. This is the *resource curse* that Chichilnisky-Heal focused on in her original piece. The parameter λ that measures permeability of the state can be interpreted as the degreee of control by the state of its resources, including the state's ability to exercise property rights on the nation's territory and on its extractive resources, where $\lambda = 1$ implies full control and $\lambda = 1$ implies complete loss of control or lack of property rights – such as, for example, during the period of colonization. This formulation follows the approach suggested to a continuum of degrees of permeability as indicated in Chichilnisky-Heal's original piece (2013). Within this formulation we compare the intertemporal allocation of resources that occurs when the parameter $\lambda = 1$ and when $\lambda < 1$. The state negotiates a contract with a MNC for extraction of resources as explained in the original piece (Chichilnisky-Heal, 2013) – and the model shows how a low amount of control, namely a lower value of the parameter λ, leads to inefficient intertemporal allocation of resources. In particular, it leads to the overextraction of resources that are offered to the export markets at prices that are lower than would be optimal when $\lambda = 1$ and there is no permeability. The result is that the developing nation overextracts its resources, and that extractive resources are excessively exported, traded in international markets, and thus undervalued. This, in turn, produces a negative effect on the nation's economy, leading to poverty and to suboptimal economic growth. It also leads to negative effects on the world as a whole, a massive overconsumption of extractive resources. The former is the "resource curse" and the latter is the origin of the environmental crisis in the world economy. Therefore, we thus find an explanation based on the "permeability approach," for the resource curse and the global environmental crisis. The environmental crisis includes, for example, the overextraction of trees from developing nations' forests leading to biodiversity destruction and destruction of carbon sinks that recycle the world's carbon, as well as the overextraction of petroleum and coal leading to increased CO_2 emissions and therefore to climate change.

The model we propose for the bargaining between the state and the ICIs is based on contract theory and emerges from the original Chichilnisky-Heal piece (2013). It explains the process by which a developing nation negotiates a contract with a MNC – such as Rio Tinto and the Mongolian government in the example of the copper and gold mine in Mongolia – with the participation of the World Bank as a third party that participates in the process. In the example of Chichilnisky-Heal (2013), after five years of World Bank recommendations in favor of an (unfavorable) contract with Rio Tinto, the person who led the World Bank team joined Rio Tinto in a senior role once the contract was signed by the Mongolian government, see Chichilnisky-Heal (2013). We show that "hidden contracts" – which are typical in contract theory – can emerge in this context, and use the policy example in Chichilnisky-Heal to illustrate with a real world case study. Hidden contracts are known to cause inferior outcomes in terms of the allocation of resources in the developing nation, once again explaining the resource curse as originating from the permeability of the state, as Natasha set out to do in her original piece. This contract model is not presented formally here and will be developed mathematically elsewhere.

4. PERMEABILITY, THE RESOURCE CURSE AND THE GLOBAL ENVIRONMENT

This section examines the intertemporal allocation of resources in a nation with a permeable state. A sovereign nation N has a total amount M of an extractive resource such as gold, copper, coal, a forest, or a fishery. Developing nations typically lack private property rights for natural resources, which can be either treated as common or a free access asset or may be owned by the government, for a discussion see Chichilnisky (1994). This is different for developed nations such as USA and Australia, where resources are typically privately owned. As examples of developingn nations, consider Zambia and its copper mines, Mongolia and its Gobi Desert's gold-copper mines that are the largest in the world, Brazil and its rainforests, and Nigeria and Mexico and their petroleum. In these cases, the resource is either owned by the government, or in the case of Brazil the resource (the Amazon forest) is often used as a common or free access property. In all cases, there are no well defined private property rights. This dichotomy between industrial and developing nations along the lines of property rights on resources was observed in Chichilnisky (1994). In industrial nations where natural resources are often privately owned such as the US and its EXXON-Mobile petroleum, Australia and Rio Tinto, Canada and its tar sands.

Focus, therefore, on a developing nation N. Lacking *private* property rights, the role of N's sovereign state is to optimize over time the extraction and sale of M, and to do so optimally for the welfare of its citizens. In general terms the value of the resource to the nation N can be represented as

$$V(M)$$

where V represents welfare that the nation N can obtain from exploiting – extracting, producing exporting and selling – the resource stock M.

This section will offer a very simple economic model that validates, in standard economic terms, the results in Chichilnisky-Heal on how the *permeability* of the state in N diminishes the optimal value the nation can achieve from its extractive resource. It shows how the nation ends up overextracting the resource in the short run with respect to what would be an optimal pattern of resource extraction over time. This in turn causes lower resource prices when the nation has an impact on global markets – for example, in the case on Mongolia, who has the largest copper and gold mines in the world in its Gobi Desert. In the context of nation N, the loss of welfare and the economic losses produced by the misallocation of its extractive resources undermines economic performance and can be called the *resource curse*. Obviously there may be other contributing factors, but this model suffices to indicate the extent to which the *permeability of the state* inevitably leads to the misallocation of economic resources.

The following is a simple intertemporal allocation model for nation N. From the above, the optimal intertemporal allocation problem of N can be simply summarized as the following optimal resource extraction problem over time:

$$V(M) = Max_x(x + \beta V(M - x))$$

where M is the total amount of the resource in N, x is the amount extracted in the first period, β is a time discount factor indicating the value to the nation of increasing value today rather than tomorrow, and $V(M - x)$ represents the value from extracting the remaining stock after is extracted today. As famously indicated by Hotelling, an extractive resource can be considered a financial asset, so that its optimal allocation has to take into consideration the financial gains of extracting today and investing the money in financial markets, obtaining a financial rate of return r. The extraction plan must reflect N's indifference between a dollar from extraction today and a dollar times $(1 + r)$ tomorrow. Therefore β satisfies the equation

$$\beta = \frac{1}{1 + r}$$

where r is the financial return that can be obtained from extracting the resource and investing the money in a financial institution. The optimality condition is therefore

$$1 = \beta V'(M - x)$$

where V' is the derivative of V with respect to x. So far, this represents in a highly simplified form the optimal extraction plan for nation N over time.

Now introduce the "permeability factor" λ that is defined in Chichilnisky-Heal (2013), as discussed above. In that case the optimization problem changes to

$$V_\lambda(M) = Max_x(x + \lambda \beta V(M - x))$$

where the parameter λ represents the degree of permeability as defined in Chichilnisky-Heal (2013) and described above. The degreee of permeability is between 0 and 1, *i.e.* $\lambda \in [0,1]$. When the state is permeable, for any firm that exploits the resource, the value of the second period of the contract is more uncertain, since the state cannot fully ensure a future contract for extracting the resource. The extent of this uncertainty varies with the permeability parameter, between 0 and 1, with 0 representing no property rights by the state and 1 representing full control. This decreases the firm's economic value of a contract for extraction in period 2 with respect to the value of extraction in period 1. The loss of certainty is represented here by multiplying the second term by the parameter λ. We therefore have the new modified optimality condition:

$$V_\lambda(M) = Max_x(x + \lambda \beta V_\lambda(M - x))$$

where $V_\lambda(M)$ represents the optimal welfare that can be obtained under the permeability approach with permeability factor λ.

Proposition 1 *When the state is permeable, any firm that exploits the resource faces an uncertain future about its contract. The extent of this uncertainty varies with the perme-*

ability parameter and more permeability decreases further the firm's economic value of a contract for extraction in period 2 with respect to the value of extraction in period 1. The result is acceleration and magnification of extraction today, and a corresponding lowering of resource 6 prices in international markets, leading to the resource curse. The larger is the permeability, the larger is the inefficiency.

Proof: The loss of certainty is represented here by multiplying the second term by the parameter λ. We therefore have the new modified optimality condition:

$$V_\lambda(M) = Max_x(x + \lambda\beta V_\lambda(M - x))$$

where $V_\lambda(M)$ represents the optimal welfare that can be obtained under the permeability approach with permeability factor λ. Since $\lambda \in [0,1]$ we have the following inequality:

$$V_\lambda(M) = Max_{x_\lambda}(x_\lambda + \lambda\beta V_\lambda(M - x))$$

$$< V(M) = Max_x(x + \lambda\beta V(M - x))$$

so when the state is permeable, namely $\lambda<1$, the welfare of N decreases with the permeability of the state and the extraction in the first period X_λ increases, in particular with a permeable state the extraction today increases over what is optimal leading to lower international prices for the resource. Recall the example of Mongolia, whose copper and gold mines in the Gobi Desert are the largest in the world:

$$x_\lambda > x$$

This completes the proof.

Proposition 2 *Permeability leads to higher discount factors in financial markets and to lower rates of savings and investment in a developing nation that would be optimal.*

Proof: This follows directly from Hotelling's observation that extractive resources are equivalent to financial assets, and from the fact that more permeability is equivalent to a higher discount factor for the future.

Proposition 3 *Permeability leads to global overextraction of resources and to lower resource prices. The overextraction in developing nations and overconsumption in OECD nations causes the global environmental crisis in the Global Commons: Overuse of the atmosphere (overconsumption of fossil fuels as extractive resources), the Oceans (overextraction of biomass from the oceans) and Biodiversity extinction (overextraction of forests, destruction of ecosystem and landscapes).*

Proof: This follows from Propositions 1 and 2, see also Chichilnisky (1994).

REFERENCES

Chichilnisky, Graciela (1994). "North-South Trade and the Global Environment," *American Economic Review*, **84**(4), 851–74.

Chichilnisky-Heal, Natasha (2013). "Bargaining to Lose: A Permeability Approach to Post-Transition Resource Extraction," Department of Political Sciences, Yale University, Invited presentation at ISA Annual Meetings, March 27 2014, Washington, DC.

PART II

INTEGRATED ASSESSMENT MODELLING

6. Integrated Assessment Models of climate change
Chris Hope

ORIGINS

Integrated Assessment Models (IAMs) of climate change are so called because they integrate two things. Scientific information about the effects of greenhouse gases on climate. Economic information about the impacts of climate on welfare, including the costs of reducing the emissions of greenhouse gases. So IAMs all share the defining trait that they incorporate knowledge from more than one field of study, although different models vary greatly with regard to their scope (Weyant et al., 1996).

There are several other types of model that are superficially similar to IAMs, but which actually do something slightly different, such as measure the impacts of climate change in physical terms (Stehfest et al., 2014), or model the introduction of technologies that could be used to tackle climate change (Manne et al., 1995).

I shall not discuss these other types of model here, as they are not designed to do the main thing that has marked out IAMs in the policy process, which is to estimate the marginal impact of one more tonne of greenhouse gas emissions, specifically carbon dioxide. This is given the name of the social cost of CO_2 ($SCCO_2$), and is, in certain circumstances, the correct tax rate to be applied to all emissions of CO_2 (Hepburn, 2006).

There are three IAMs of climate change. The Dynamic Integrated Climate-Economy (DICE) model developed by Bill Nordhaus at Yale University, USA. The Policy Analysis of the Greenhouse Effect (PAGE) model, developed by the author, Chris Hope, at the University of Cambridge, UK. The Climate Framework for Uncertainty, Negotiation and Distribution (FUND) model, developed by Richard Tol, now of Sussex University, UK. The vast majority of the independent impact and $SCCO_2$ estimates that appear in the peer-reviewed literature are derived from these three models (Tol, 2008).

The first version of DICE was published in 1993 in Resource and Energy Economics (Nordhaus, 1993). The first version of PAGE was published in the same year, 1993, in Energy Policy (Hope et al., 1993). The first version of FUND appeared slightly later, in 1997, in Environmental Modelling and Assessment (Tol, 1997).

Another early model, the Carbon Emissions Trajectory Assessment (CETA), was created by Stephen Peck and Thomas Teisberg and was used in the early 1990s to explore optimal emissions and carbon taxes, but appears to have been subsequently abandoned (Peck and Teisberg, 1992).

All three models have been through several versions. At least six for DICE (Nordhaus, 1993; Nordhaus, 1994; Nordhaus and Boyer, 2000; Nordhaus, 2007; Nordhaus, 2010; Nordhaus, 2013), four for PAGE (Hope et al., 1993; Plambeck and Hope, 1996; Hope, 2006; Hope, 2013), and three major versions for FUND (Tol, 1997; Tol, 2002; Anthoff and Tol, 2009).

All three models have had the same lead developers throughout. David Popp, Zili Yang and Joseph Boyer have collaborated on DICE with Bill Nordhaus. Erica Plambeck and

Stephane Alberth have helped the author to develop versions of the PAGE model. David Anthoff has been a co-developer of FUND with Richard Tol since 2006.

Initial versions of the three models had different strengths. Optimizing models endogenously compute the optimal level of control of greenhouse gases, within their restricted framework. Simulation models evaluate the effect on the economy and the environment of a particular policy to control climate change. Some early versions of the models limited their treatment of uncertainty to a simple sensitivity analysis of an essentially deterministic model, others incorporated uncertainty at their heart, and carried uncertainties through the calculation in a systematic way (Hope, 2005).

DICE integrated in an end-to-end fashion, the economics, carbon cycle, climate science, and impacts in a highly aggregated model that allowed a weighing of the costs and benefits of taking steps to slow greenhouse warming. It was mainly used to calculate an optimal emissions pathway and the associated price on CO_2 emissions (Nordhaus, 1993).

PAGE included regional detail and expressed all the main inputs as probability distributions so that the risks of climate change could more easily be appreciated (Hope et al., 1993).

FUND was originally set up to study the role of international capital transfers in climate policy, but it soon evolved into a test-bed for studying impacts of climate change in a dynamic context. It had more regional and sectoral detail than the other two models (Tol, 1997).

Over time, developments of the three models have added capabilities they originally lacked. Regional details and some ability to run under uncertainty, were added to DICE in the RICE and PRICE models (Nordhaus and Yang, 1996; Nordhaus and Popp, 1997). An optimizing version of the PAGE model was produced (Hope, 2008). Probability distributions for some inputs have been added to FUND (Tol, 2003).

The three models were created independently and have continued to be developed separately. The lead developers have not collaborated on joint model-development projects. If they were to do so, there would be a danger that what appears to be three independent models would actually simply be variants of a single model. This reduction in diversity would be dangerously misleading to policymakers. All three models are still in active use today.

INFLUENCE

The PAGE model was the main input to the impact calculations in the Stern Review on the Economics of Climate Change, commissioned by the UK Government in 2005 and published in 2007 (Stern, 2007). Although not the first economic report on climate change, the Stern Review is the most widely known and most influential report of its kind (Sterner and Persson, 2008). The authors of the Stern Review chose the PAGE model because of its comprehensive treatment of risk and uncertainty, and the ease of replacing default input assumptions with customized values (Dietz et al., 2007). The Stern review estimated a mean $SCCO_2$ of $85 per tonne of CO_2, in the year 2001 (year 2000 US dollars), a higher value than nearly all previous estimates.

The fourth assessment report of the Intergovernmental Panel on Climate Change in 2007 highlighted the impact and $SCCO_2$ estimates from the three models in Chapter 20

116 *Handbook on the economics of climate change*

of working group two, including the recently published estimates from the Stern review (Parry et al., 2007).

In the United States, an interagency working group was convened by the Council of Economic Advisers and the Office of Management and Budget in 2009–10 to design an $SCCO_2$ modelling exercise and to develop estimates for use in rulemakings. The interagency group was composed of scientific and economic experts from the White House and federal agencies. The interagency group identified a variety of assumptions, which the Environmental Protection Agency (EPA) then used to estimate the $SCCO_2$ using the three integrated assessment models, DICE, PAGE, and FUND.

The interagency group recommended a set of four $SCCO_2$ estimates for use in regulatory analyses. The first three values are based on the average $SCCO_2$ from the EPA's runs with the DICE, PAGE and FUND models, at discount rates of 5, 3, and 2.5 per cent per year. The fourth value is the 95th percentile of the $SCCO_2$ from the three models at a 3 per cent discount rate, and is intended to represent the potential for higher-than-expected damages (Interagency Working Group, 2010).

The interagency group updated these estimates, using new versions of each integrated assessment model and published the updated estimates in May 2013. Minor technical corrections to the estimates were published in July 2015. The four $SCCO_2$ estimates are $16, $51, $76, and $150 per tonne of CO_2 emissions in the year 2025 (year 2014 US dollars) (Interagency Working Group, 2013). The EPA has used the $SCCO_2$ to analyse the carbon dioxide impacts of a large number of rulemakings since the interagency group first published estimates in 2010 (EPA, 2015).

The interagency working group calculations have been picked up and used by several other leading organizations around the world.

The International Monetary Fund (IMF) has developed a practical methodology, and associated tools, to show how the major environmental damages from energy can be quantified for different countries and how these estimates can be used to design an efficient set of energy taxes. It uses the interagency working group calculations to estimate the climate change damages (IMF, 2013; IMF, 2015). Its managing director, Christine Lagarde, says that carbon taxes should be as routine a part of finance ministers' toolkits as VAT or income tax (Lagarde, 2014).

The World Bank Group has also called on governments and corporations around the world to support carbon pricing to bring down emissions and drive investment into cleaner options (World Bank, 2015). The carbon price needs to be high enough to send a meaningful signal to investors and consumers (PWC, 2014), and the mean $SCCO_2$ estimates from IAMs would achieve this.

SCRUTINY

In the wake of their impact in the policy process, there have been several attempts to scrutinize and critique the three IAMs.

Following the Stern review, there were claims that the PAGE model had estimated a high mean $SCCO_2$ because it had consistently picked the most pessimistic option for every choice that was made (Tol, quoted in Cox and Vadon, 2007). In response, the Stern team showed that the high mean $SCCO_2$ was actually caused by the ethical choice of a low pure time preference rate and by the serious attention paid to the full range of risks from climate change (Dietz et al., 2007).

Other research reached the opposite conclusion that the model runs used in the Stern Review may well have underestimated US and global damages (Ackerman et al., 2009).

Taking a different tack, Weitzman argued that concern about climate change should be driven mainly by low probability, high consequence events with the potential to essentially wipe out human civilization, which he felt were not adequately represented in the Stern review's modelling (Weitzman, 2007).

He went on to argue that such fattened tails have strong implications for situations, like climate change, where a catastrophe is theoretically possible because prior knowledge cannot place sufficiently narrow bounds on overall damages, and that the economic consequences of fat-tailed structural uncertainty (along with unsureness about high-temperature damages) can outweigh the effects of discounting in climate-change policy analysis (Weitzman, 2009). He suggested adopting a statistical value of civilization as a way of bounding the influence of these catastrophes in models. This suggestion was adopted in the next version of the PAGE model, PAGE09 (Hope, 2013), while Nordhaus observed that societies do not appear to behave as if catastrophic outcomes have infinite disutility, and that climate change, by contrast, is a situation where we can learn as we go along, so the relevance of Weitzman's critique to climate change policy is unclear (Nordhaus, 2011).

The scrutiny of the models redoubled after their use by the EPA in the polarized political atmosphere of the United States (Interagency Working Group, 2015).

Stern observed that the scientific models, because they omit key factors that are hard to capture precisely, appear to substantially underestimate the risks of climate change. Many economic models add further gross underassessment of risk because the assumptions built into the economic modeling on growth, damages and risks, come close to assuming directly that the impacts and costs will be modest and close to excluding the possibility of catastrophic outcomes (Stern, 2013). He acknowledged that the PAGE model was partly exempt from this latter criticism. His recommendations included focusing more strongly on climate change as a risk-management problem and investigating the ways in which the long-lasting effects of damage might be incorporated in formal modelling.

The most trenchant criticism claimed that IAMs have crucial flaws that make them close to useless as tools for policy analysis: some inputs, such as the discount rate, are arbitrary, but have huge effects on the $SCCO_2$ estimates the models produce; the models' descriptions of the impact of climate change are completely ad hoc, with no theoretical or empirical foundation; and the models can tell us nothing about the most important driver of the $SCCO_2$, the possibility of a catastrophic climate outcome. Perhaps the best we can do is come up with rough, subjective estimates of the probability of a climate change sufficiently large to have a catastrophic impact, and then some distribution for the size of that impact (in terms, say, of a reduction in GDP or the effective capital stock) (Pindyck, 2013).

Pindyck's criticism seems too extreme. For instance, he failed to acknowledge that PAGE includes a representation of catastrophic outcomes almost exactly in the form he recommends (Hope, 2006). Weyant (2014) states that Pindyck's conclusion that IAMs are useless fundamentally misconceives the enterprise. IAMs and the $SCCO_2$ are conceptual frameworks for dealing with highly complex, non-linear, dynamic, and uncertain systems.

118 *Handbook on the economics of climate change*

The human mind is incapable of solving all the equations simultaneously, particularly under the conditions of profound uncertainty that actually prevail. The climate sensitivity is uncertain, as are the impacts for a given temperature rise (IPCC, 2007; Stern, 2007). Finding a way to take this uncertainty into account is one of the main tasks facing the designer of an IAM.

Uncertainty is a common problem for policymaking, particularly in the environmental area. But it is almost the main defining characteristic here. Parson and Fisher-Vanden (1997, p. 609) state that 'Uncertainty is central to climate change' (Hope, 2005).

The models have a simple enough form to confront this uncertainty head-on and have provided important insights into many aspects of climate change policy.

REPRESENTATIVE RESULTS UNDER UNCERTAINTY

We can see this by looking in a bit more detail at one result that is of most interest for setting prices on CO_2 emissions, the amount by which the Net Present Value (NPV) of impacts increases if one more tonne of CO_2 is emitted, or decreases if one less tonne is emitted – the social cost of CO_2 ($SCCO_2$). In the PAGE09 model, this is calculated by changing the emissions of CO_2 in a particular year (2009 in the default model) by 100 Gt, and dividing the difference in the NPV of impacts by 100 billion. This may seem like a non-marginal change, but tests with changes in emissions of 1 Gt, 10 Gt and 100 Gt give results within 1 per cent of each other for mean values of the inputs, so 100 Gt can actually be considered a marginal change in emissions of the stock pollutant CO_2.

Figure 6.1 shows the full probability distribution of the $SCCO_2$ in 2009 from the current default PAGE09 model, version 1.7. The mean value is $103 per tonne of CO_2, with a 5 per cent to 95 per cent range of about $11 to $283, all in $US (2005). This estimate assumes

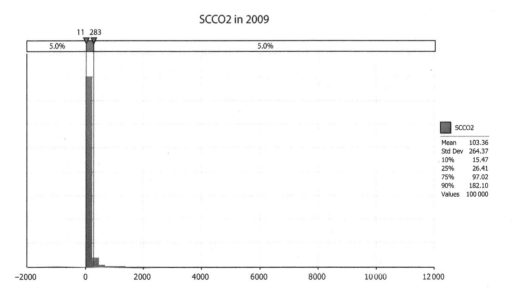

Figure 6.1 The $SCCO_2$ from the PAGE09 default model in 2009

GDP, population and emissions follow the IPCC's A1B scenario (Nakicenovic and Swart, 2000). The standard deviation of the result is larger than the mean, about $265, implying that with 100 000 runs the standard error of the mean NPV is about $0.75; 90 per cent of the time another 100 000 runs would give a mean NPV within about $1.5 of the $103 per tonne of CO_2 shown here.

The standard deviation is so large because the PAGE09 model keeps track of whether a discontinuity, such as the melting of the Greenland or West Antarctic ice sheets, has actually been triggered in each run by the temperature exceeding the tolerable level by enough to increase the probability of a discontinuity beyond the value of a uniform [0,1] random number in that run. The very high values for the $SCCO_2$ result when the small increase in emissions brings forward the date at which a discontinuity has been triggered.

The $SCCO_2$ is so hard to pin down accurately because of the possibility that even a small amount of extra emissions, such as 100 Gt of CO_2 might lead to an earlier discontinuity, in, say, 2075 rather than 2100. On average this happens in about 3000 of the 100 000 runs, and this is what produces the very long right tail in the distribution, giving a few $SCCO_2$ values of up to $10 000 or so. With mean values for all the inputs to the model, the $SCCO_2$ comes to only about $50, showing how important the proper treatment of risk is to understanding the $SCCO_2$.

The skewness of the result is so extreme that it is difficult to see the shape of the distribution when all the values are included. So, Figure 6.2 shows the same result but with the top 1 per cent of values omitted.

Comparing the mean value of $82 per tonne with the mean value of $103 when all runs are included shows that the top 1 per cent of runs contribute $21, or about 20 per cent, to the mean $SCCO_2$ value in the default PAGE09 model. This confirms the importance of properly representing risk and uncertainty when making estimates of the $SCCO_2$.

Another useful aid to understanding the variation in the $SCCO_2$ is shown in Figure 6.3.

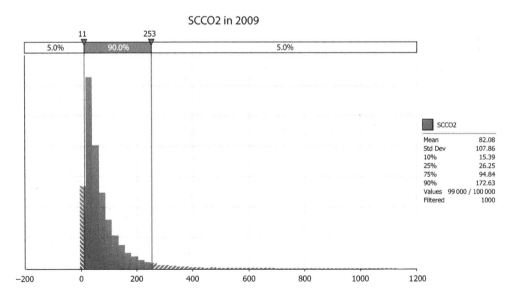

Figure 6.2 The $SCCO_2$ from the PAGE09 default model, top 1% of values omitted

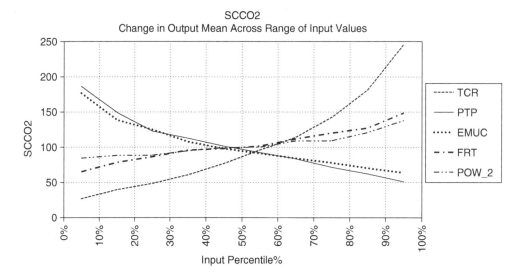

Figure 6.3 Major influences on the $SCCO_2$ from the default PAGE09 model in 2009

This shows the amount by which the mean $SCCO_2$ varies if the five most important influences on the $SCCO_2$ are varied across their input ranges, with all other inputs taking their full default ranges.

The most important influence is one of the components of the climate sensitivity, the transient climate response (TCR), defined as the temperature rise after 70 years, corresponding to the doubling-time of CO_2 concentration, with CO_2 concentration rising at 1 per cent per year. In the default PAGE09 model, the TCR has a triangular distribution with minimum value 1, mode 1.3 and maximum value 2.8°C (Andrews and Allen, 2008). A decrease in the TCR to the bottom 10 per cent of its range decreases the mean $SCCO_2$ to about $30 per tonne of CO_2. An increase in TCR to the top 10 per cent of its range increases the mean $SCCO_2$ to about $250 per tonne of CO_2.

Next are the pure time preference (PTP) rate and the equity weights (EMUC), and this time the slope of the influences are negative, with an increase in PTP or EMUC reducing the mean $SCCO_2$. A higher PTP rate means that impacts that occur in the future have a lower NPV, and the range for this input used in the default PAGE09 model is a triangular distribution with minimum value of 0.1, mode 1 and maximum value 2 per cent per year. The implication of these results is that a PTP rate of 0.1 per cent per year, as used in Stern (2007), would increase the mean $SCCO_2$ to about $180 per tonne of CO_2. Using a PTP rate of 2 per cent per year, as in Nordhaus (2007), would decrease the mean $SCCO_2$ to about $50 per tonne of CO_2. A higher EMUC means that impacts that occur in the future, when consumption per capita is on average higher than today's consumption per capita in the EU, are weighted less.

The feedback response time (FRT) is a measure of how long the Earth takes on average to respond to higher radiative forcing. The range for this input used in the default PAGE09 model is a triangular distribution with minimum value of 10, mode 30 and maximum value 65 years (Andrews and Allen, 2008). An increase in the FRT from its lowest to its highest decile increases the mean $SCCO_2$ from about $70 per tonne of CO_2 to about $150 per tonne of CO_2. It might be thought that the sign of this influence should be negative, as a higher FRT means the Earth takes longer on average to respond to higher radiative forcing, but in fact, if the TCR is fixed, a higher value for FRT means a higher value for the climate sensitivity, and so a larger response to higher concentrations of CO_2 overall.

The shape of the non-economic impact function is represented by the exponent linking non-economic impacts to temperature (POW_2), which in the default PAGE09 model, has a triangular distribution with minimum value 1.5, mode 2 and maximum value 3. Increasing POW_2 from its lowest to its highest decile increases the mean $SCCO_2$ from about $80 per tonne of CO_2 to about $140 per tonne of CO_2. Further definitions and details of all the inputs and $SCCO_2$ results are in Hope (2013).

REGIONAL RESPONSIBILITIES AND IMPACTS

If a tonne of CO_2 emitted in 2009 causes mean extra impacts of $103, it is also possible to use PAGE09 to answer the question: What is the regional split of the mean extra impacts caused by the marginal tonne of CO_2?

This is an important issue, as scientific assessments show that poor countries are likely to bear the brunt of changing temperatures (Harrington et al., 2016). Integrated Assessment Models allow us to see whether this is also true of the full range of climate change impacts.

Combining this regional output with the regional split of emissions in 2009 allows a matrix to be drawn up showing the mean extra impacts in region 'i' that are caused by emissions from region 'j', and vice versa. This helps to inform debates around whether the developed world must now take more responsibility for the consequences of climate change (WWF, 2012).

Table 6.1 shows the results for the regional distribution of the mean $SCCO_2$ from the default PAGE09 model in the A1B scenario. The first row shows that 1.0 per cent of the mean $SCCO_2$ is contributed by extra impacts in the EU from emissions in the EU, 0.8 per cent of the mean $SCCO_2$ is contributed by extra impacts in the US from emissions in the EU, and so on. Globally, 11.3 per cent of the mean $SCCO_2$ is contributed by emissions from the EU.

The first column shows that 1.0 per cent of the mean $SCCO_2$ is contributed by extra impacts in the EU from emissions in the EU, 1.3 per cent of the mean $SCCO_2$ is contributed by extra impacts in the EU from emissions in the US, and so on. Globally, 8.5 per cent of the mean $SCCO_2$ is contributed by extra impacts in the EU.

Table 6.2 summarizes the results by combining the first four regions which make up the annex 1 regions, and the last four regions which make up the rest of the world (RoW), largely developing countries.

Less than 10 per cent of the mean $SCCO_2$ comes from extra impacts in annex 1 from annex 1 emissions, while over 45 per cent comes from extra impacts in RoW from RoW emissions. About one third of the mean $SCCO_2$ comes from extra impacts in the RoW caused by emissions in Annex 1, while just over 10 per cent comes from extra impacts in annex 1 caused by emissions in the RoW. In total, annex 1 country emissions are on

Table 6.1 *Regional distributions of the mean SCCO₂, A1B scenario*

% of SCCO₂	Mean extra impacts in								
	EU	US	OT	EE	CA	IA	AF	LA	Global
From emissions in									
EU	1.0	0.8	0.5	0.3	1.2	4.0	2.9	0.8	**11.3**
US	1.3	1.1	0.7	0.4	1.7	5.6	4.0	1.1	**15.9**
OT	0.5	0.4	0.3	0.1	0.7	2.2	1.6	0.4	**6.3**
EE	0.7	0.6	0.4	0.2	0.9	3.0	2.1	0.6	**8.4**
CA	1.1	0.9	0.6	0.3	1.4	4.7	3.4	0.9	**13.4**
IA	1.9	1.5	1.0	0.5	2.3	7.7	5.6	1.5	**22.0**
AF	1.0	0.8	0.6	0.3	1.3	4.3	3.1	0.9	**12.3**
LA	0.9	0.7	0.5	0.2	1.1	3.7	2.7	0.7	**10.5**
Global	**8.5**	**6.8**	**4.5**	**2.2**	**10.5**	**35.1**	**25.5**	**7.0**	**100.0**

Note: EU: European Union; US: USA; OT: Other OECD; EE: Former Soviet Union and rest of Europe; CA: China and centrally planned Asia; IA: India and SE Asia; AF: Africa and Middle East, LA: Latin America.

Table 6.2 *Summary regional distributions of the mean SCCO₂, A1B scenario*

% of SCCO₂	Mean extra impacts in		
	Annex 1	RoW	Global
From emissions in			
Annex 1	9.2	32.6	**41.8**
RoW	12.8	45.4	**58.2**
Global	**22.0**	**78.0**	**100.0**

average responsible for about 40 per cent of the mean SCCO₂, while suffering about 20 per cent of the extra impacts.

FLOURISHING RESEARCH

There is a substantial research effort in several locations around the world to adapt and use the three IAMs. A few recent examples follow of this flourishing research.

The DICE model has been modified by researchers at Stanford University to incorporate recent empirical findings suggesting that climate change could slow economic growth rates, particularly in poor countries (Moore and Diaz, 2015).

Researchers at the London School of Economics have modified DICE to allow a wider range of values to be considered for climate sensitivity, and potential climate impacts (Dietz and Stern, 2015). These new estimates use Nordhaus's own discount rate assumptions, which are much higher than those Stern has previously used. Higher discount rates make impacts that appear in the future seem less worrying. The discount rate used in Dietz and Stern (2015) is 4.5 per cent per year. And yet the $SCCO_2$ from the model in 2015 goes up to between \$32 and \$103 per tonne of CO_2.

The PAGE model has been adapted by researchers at the Institute for Prospective Technological Studies, Seville to allow for the possibility of a thin, intermediate or fat tail for both the climate sensitivity parameter and the damage function exponent (Pycroft et al., 2011), and to include explicit consideration of extreme sea-level rise (Pycroft et al., 2014).

The FUND model has been used at the Pacific Northwest National Laboratory's Joint Global Change Research Institute in College Park, MD to calculate the social costs of four greenhouse gases – carbon dioxide, methane, nitrous oxide, and sulphur hexafluoride (Waldhoff et al., 2014).

There is also an on-going systematic investigation by a committee led by the US National Academy of Sciences (NAS) to assess the technical merits and challenges of a narrowly focused update to the $SCCO_2$ estimates, and to examine potential approaches, along with their relative merits and challenges, for a more comprehensive update to the $SCCO_2$ estimates to ensure the estimates reflect the best available science. The study was requested of the NAS by the Inter-agency Working Group that constructed the $SCCO_2$ and by the White House. Inter alia the committee is assessing the available science and how it would impact the choice of integrated assessment models and damage functions; climate science modelling assumptions; socio-economic and emissions scenarios; presentation of uncertainty; and discounting (National Academy of Sciences, 2015).

As well as their main use to date of calculating the mean and distribution of the $SCCO_2$, the three IAMs can be, and are being, used for a variety of other tasks: To find the mean and distribution of the appropriate price on non-CO_2 greenhouse gas emissions (Waldhoff et al., 2014); to find the mean and distribution of the extra impacts from arctic feedbacks (Hope and Schaefer, 2015); to find the mean and distribution of the net benefit of a low emissions scenario, and the influence of each input on the net benefit; to find the optimal path of emissions over time with moderate adaptation, and the appropriate price on CO_2; to find how the optimal path varies if we get better information about the Transient Climate Response (TCR); to find the value of better information about TCR; to find how the value of better information about TCR declines the later we get it (Hope, 2015). Each of these uses is a flourishing research area.

The three models were developed independently, have all withstood the test of time and have all been influential on policy. The DICE model has been written up in book form (Nordhaus and Boyer, 2000). The PAGE model has had most influence on policy through its use in the Stern review and by the US interagency working group. The FUND model has acted as a counterweight to the other two models, consistently producing lower estimates for the $SCCO_2$, partly because of its inclusion of possible benefits for the first degree or so of climate change (Tol, 2002).

The field is vibrant, the use of the models in policymaking has been rapid and influential, and the development and use of the models under the scrutiny of critics is strong and on-going. As it becomes clearer that carbon pricing must be a large part of the solution to reducing greenhouse gas emissions, the importance and influence of IAMs can only continue to increase (Lund et al., 2015). IAMs are the best tools we have for honestly translating the current knowledge about climate change, including its profound uncertainty, into policy advice.

REFERENCES

Ackerman, F., E.A. Stanton, C. Hope and S. Alberth (2009), 'Did the Stern Review Underestimate US and Global Climate Damages?', *Energy Policy*, **37**(7), 2717–21.
Andrews, D.G. and M.R. Allen (2008), 'Diagnosis of Climate Models in Terms of Transient Climate Response and Feedback Response Time', *Atmospheric Science Letters*, **9**(1), 7–12.
Anthoff, D. and R.S.J Tol (2009), 'The Impact of Climate Change on the Balanced Growth Equivalent: An Application of FUND', *Environmental and Resource Economics*, **43**(3), 351–67.
Cox, S. and R. Vadon (2007), 'Running the Rule over Stern's Numbers', *BBC News*, 26 January 2007.
Dietz, S. and N. Stern (2015), 'Endogenous Growth, Convexity of Damage and Climate Risk: How Nordhaus' Framework Supports Deep Cuts in Carbon Emissions', *Economic Journal*, **125**, 574–620. doi:10.1111/ecoj.12188.
Dietz, S., C. Hope and N. Patmore (2007), 'Some Economics of "Dangerous" Climate Change: Reflections on the Stern Review', *Global Environmental Change*, **17**(3–4), 311–25.
EPA (2015), 'EPA Fact Sheet Social Cost of Carbon', accessed March 2016 at http://www3.epa.gov/climatechange/Downloads/EPAactivities/social-cost-carbon.pdf.
Harrington, L.J., D.J. Frame, E.M. Fischer, E. Hawkins, M. Joshi and C.D. Jones (2016), 'Poorest Countries Experience Earlier Anthropogenic Emergence of Daily Temperature Extremes', *Environmental Research Letters*, **11**(5), 055007.
Hepburn, C. (2006), 'Regulation by Prices, Quantities or Both: An Update and an Overview', *Oxford Review of Economic Policy*, **22** (2), 226–47.
Hope, C. (2005), 'Integrated Assessment Models', chapter 4 in D. Helm (ed.), *Climate-Change Policy*, Oxford: Oxford University Press, p. 81.
Hope, C. (2006), 'The Marginal Impact of CO_2 from PAGE2002: An Integrated Assessment Model Incorporating the IPCC's Five Reasons for Concern', *Integrated Assessment*, **6**(1), 19–56.
Hope, C. (2008), 'Optimal Carbon Emissions and the Social Cost of Carbon over Time under Uncertainty', *Integrated Assessment*, **8**(1), 107–22.
Hope, C. (2013), 'Critical Issues for the Calculation of the Social Cost of CO_2: Why the Estimates from PAGE09 are Higher than those from PAGE2002', *Climatic Change*, **117**(3), 531–43.
Hope, C. (2015), 'The $10 Trillion Value of Better Information about the Transient Climate Response', *Philosophical Transactions of the Royal Society A*, **373**(2054), 20140429.
Hope, C. and K. Schaefer (2015), 'Economic Impacts of Carbon Dioxide and Methane Released from Thawing Permafrost', *Nature Climate Change*, **6**(1), 56.
Hope, C., P. Wenman and J. Anderson (1993), 'Policy Analysis of the Greenhouse Effect', *Energy Policy*, **21**(3), 327–38.
Interagency Working Group on Social Cost of Carbon, United States Government (2010), 'Technical Support Document: Social Cost of Carbon for Regulatory Impact Analysis Under Executive Order 12866', accessed March 2016 at http://www3.epa.gov/otaq/climate/regulations/scc-tsd.pdf.
Interagency Working Group on Social Cost of Carbon, United States Government (2013), 'Technical Support Document: Technical Update of the Social Cost of Carbon for Regulatory Impact Analysis Under Executive Order 12866', accessed March 2016 at https://www.whitehouse.gov/sites/default/files/omb/inforeg/social_cost_of_carbon_for_ria_2013_update.pdf.
Interagency Working Group on Social Cost of Carbon, United States Government (2015), 'Response to Comments: Social Cost of Carbon for Regulatory Impact Analysis Under Executive Order 12866', accessed March 2016 at https://www.whitehouse.gov/sites/default/files/omb/inforeg/scc-response-to-comments-final-july-2015.pdf.
International Monetary Fund (2013), 'Energy Subsidy Reform: Lessons and Implications', accessed March 2016 at http://www.imf.org/external/np/pp/eng/2013/012813.pdf.
International Monetary Fund (2015), 'Factsheet: Climate, Environment, and the IMF', accessed March 2016 at http://www.imf.org/external/np/exr/facts/enviro.htm.
IPCC (2007), 'Climate Change 2007. The Physical Science Basis. Summary for Policymakers', Contribution of Working Group I to the Fourth Assessment Report of the Intergovernmental Panel on Climate Change, IPCC Secretariat Switzerland.
Lagarde, C. (2014), 'Moving Ahead with Carbon Pricing: A Call to Finance Ministers', accessed March 2016 at http://www.imf.org/external/np/fad/environ/pdf/121214.pdf.
Lund, H., B. Dudley, C. Descalzi, B. van Beurden, E. Saetre and P. Pouyanné (2015), 'Widespread Carbon Pricing is Vital to Tackling Climate Change', Letter to the *Financial Times*, 1 June 2015, accessed March 2016 at http://www.ft.com/cms/s/0/682898fe-07e4-11e5-9579-00144feabdc0.html#axzz3ı ıBuXg9H.
Manne, A., R. Mendelsohn and R. Richels (1995), 'MERGE: A Model for Evaluating Regional and Global Effects of GHG Reduction Policies', *Energy Policy*, **23**(1), 17–34.

Moore, F.C. and D.B. Diaz (2015), 'Temperature Impacts on Economic Growth Warrant Stringent Mitigation Policy', *Nature Climate Change*, **5**(2), 127–31.

Nakicenovic, N. and R. Swart (2000), 'Special Report on Emissions Scenarios (SRES): A Special Report of Working Group III of the Intergovernmental Panel on Climate Change', Cambridge: Cambridge University Press, accessed March 2016 at http://www.grida.no/climate/ipcc/emission/index.htm and http://sres.ciesin.org/final _ data.html.

National Academy of Sciences (2015), 'Assessing Approaches to Updating the Social Cost of Carbon', accessed March 2016 at http://sites.nationalacademies.org/DBASSE/BECS/CurrentProjects/DBASSE_167526.

Nordhaus, W.D. (1993), 'Rolling the "DICE": An Optimal Transition Path for Controlling Greenhouse Gases', *Resource and Energy Economics*, **15**(1), 27–50.

Nordhaus, W.D. (1994), *Managing the Global Commons: The Economics of Climate Change* (Volume 31), Cambridge, MA: MIT Press.

Nordhaus, W.D. (2007), 'A Review of the *Stern Review on the Economics of Climate Change*', *Journal of Economic Literature*, **45**(3), 686–702.

Nordhaus, W.D. (2010), 'Economic Aspects of Global Warming in a Post-Copenhagen Environment', *Proceedings of the National Academy of Sciences*, **107**(26), 11721–6.

Nordhaus, W.D. (2011), 'The Economics of Tail Events with an Application to Climate Change', *Review of Environmental Economics and Policy*, **5**(2), 240–57.

Nordhaus, W.D. (2013), *The Climate Casino: Risk, Uncertainty, and Economics for a Warming World*, New Haven, CT: Yale University Press.

Nordhaus, W.D. and J. Boyer (2000), *Warming the World: Economic Models of Global Warming*, Cambridge, MA: MIT Press.

Nordhaus, W.D. and D. Popp (1997), 'What is the Value of Scientific Knowledge? An Application to Global Warming using the PRICE Model', *The Energy Journal*, **18**(1), 1–45.

Nordhaus, W.D. and Z. Yang (1996), 'A Regional Dynamic General-equilibrium Model of Alternative Climate-change Strategies', *The American Economic Review*, **86**(4), 741–65.

Parry, M.L., O.F. Canziani, J.P. Palutikof, P.J. van der Linden and C.E. Hanson (eds) (2007), *Contribution of Working Group II to the Fourth Assessment Report of the Intergovernmental Panel on Climate Change*, Cambridge and New York: Cambridge University Press.

Parson, E.A. and K. Fisher-Vanden (1997), 'Integrated Assessment Models of Global Climate Change', *Annual Review of Energy and the Environment*, **22**(1), 589–628.

Peck, S.C. and T.J. Teisberg (1992), 'CETA: A Model for Carbon Emissions Trajectory Assessment', *Energy Journal*, **13**(1), 55.

Pindyck, R.S. (2013), 'Climate Change Policy: What Do the Models Tell Us?', *Journal of Economic Literature*, **51**(3), 860–72.

Plambeck, E. and C. Hope (1996), 'PAGE95: An Updated Valuation of the Impacts of Global Warming', *Energy Policy*, **24**(9), 783–94.

PWC (2014), 'Why Putting a Price on Carbon is Becoming a Business and Economic Reality', PriceWaterhouseCoopers, accessed March 2016 at http://pwc.blogs.com/sustainability/2014/09/why-putting-a-price-on-carbon-is-becoming-a-business-and-economic-reality.html.

Pycroft, J., L. Vergano and C. Hope (2014), 'The Economic Impact of Extreme Sea-level Rise: Ice Sheet Vulnerability and the Social Cost of Carbon Dioxide', *Global Environmental Change*, **24**, 99–107.

Pycroft, J., L. Vergano, C. Hope, D. Paci and J.C. Ciscar (2011), 'A Tale of Tails: Uncertainty and the Social Cost of Carbon Dioxide', *Economics: The Open-Access, Open-Assessment E-Journal*, **5**(22), 1–29.

Stehfest, E., D. van Vuuren, T. Kram, L. Bouwman, R. Alkemade, M. Bakkenes, H. Biemans, A. Bouwman, M. den Elzen, J. Janse, P. Lucas, J. van Minnen, M. Muller and A. Gerdien Prins (2014), 'Integrated Assessment of Global Environmental Change with IMAGE 3.0: Model Description and Policy Applications', Report no. 735, PBL Netherlands Environmental Assessment Agency.

Stern, N. (2007), *The Economics of Climate Change: The Stern Review*, Cambridge and New York: Cambridge University Press.

Stern, N. (2013), 'The Structure of Economic Modelling of the Potential Impacts of Climate Change: Grafting Gross Underestimation of Risk onto Already Narrow Science Models', *Journal of Economic Literature*, **51**(3), 838–59.

Sterner, T. and U.M. Persson (2008), 'An Even Sterner Review: Introducing Relative Prices into the Discounting Debate', *Review of Environmental Economics Policy*, **2**(1), 61–76.

Tol, R.S.J. (1997), 'On the Optimal Control of Carbon Dioxide Emissions: An Application of FUND', *Environmental Modelling and Assessment*, **2**(3), 151–63.

Tol, R.S.J. (2002), 'New Estimates of the Damage Costs of Climate Change, Part I: Benchmark Estimates', *Environmental and Resource Economics*, **21**(1), 47–73.

Tol, R.S.J. (2003), 'Is the Uncertainty about Climate Change Too Large for Expected Cost–Benefit Analysis?', *Climatic Change*, **56**(3), 265–89.

Tol, R.S.J. (2008), 'The Social Cost of Carbon: Trends, Outliers and Catastrophes', *Economics: The Open-Access, Open-Assessment E-Journal*, **2**(25).
Waldhoff, S., D. Anthoff, S. Rose and R.S.J. Tol (2014), 'The Marginal Damage Costs of Different Greenhouse Gases: An Application of FUND', *Economics: The Open-Access, Open-Assessment E-Journal*, **8**(31), 1–33.
Weitzman, M.L. (2007), 'A Review of the *Stern Review on the Economics of Climate Change*', *Journal of Economic Literature*, **45**(3), 703–24.
Weitzman, M.L. (2009), 'On Modelling and Interpreting the Economics of Catastrophic Climate Change', *Review of Economics and Statistics*, **91**(1), 1–19.
Weyant, J. (2014), 'Integrated Assessment of Climate Change: State of the Literature', *Journal of Benefit Cost Analysis*, **5**(3), 377–409.
Weyant, J., O. Davidson, H. Dowlabathi, J. Edmonds, M. Grubb, E.A. Parson, R. Richels, J. Rotmans, P.R. Shukla, R.S.J. Tol and W. Cline (1996), 'Integrated Assessment of Climate Change: An Overview and Comparison of Approaches and Results', Chapter 10 of IPCC Working Group III report, Economic and Social Dimensions of Climate Change, Cambridge and New York: Cambridge University Press.
World Bank (2015), 'Pricing Carbon', accessed March 2016 at http://www.worldbank.org/en/programs/pricing-carbon.
WWF (2012), 'Tackling the Limits to Adaptation', accessed March 2016 at http://wwf.panda.org/wwf_news/?206889/Governments-must-get-serious-about-Loss-and-Damage-caused-by-climate-change-inaction#.

7. Climate change policy under spatial heat transport and polar amplification*
William Brock and Anastasios Xepapadeas

1. INTRODUCTION

While spatial heat transport and polar amplification are well-established phenomena in the science of climate change, they have been largely ignored in the economic modeling of climate change. This chapter introduces spatial heat transport and polar amplification in a simple spatial climate economics model, in which the climate model is based upon the work of Alexeev et al. (2005) and Langen and Alexeev (2007).

In this work the strength of spatial poleward heat transport from the lower latitudes to the higher latitudes depends upon the level of global mean average temperature. The spatial transport effect causes polar amplification due to increased meridional latent heat transport, as discussed by Alexeev et al. (2005) and further developed by Alexeev and Langen (2007) and Alexeev and Jackson (2012).

In order to exhibit the economic and climatic effects of spatial heat transport in the clearest and simplest possible way, we stratify the Earth into latitude belts and model the change in damages to each latitude belt from increased CO_2 into the atmosphere resulting from fossil fuel use in economic production activities located at each latitude. In order to focus completely on spatial climatic heat transport, we assume that total production at latitude belt x is given by $y(x, t)E(x, t)^\alpha$, $0 < \alpha < 1$, where $y(x, t)$ grows exogenously and $E(x, t)$ denotes emissions from fossil fuel inputs, or fossil fuel use by an appropriate choice of units in total production. This simplification and abstraction away from the allocative effects of other inputs to production on the economic side of the model enables us to keep a tight focus on climatic heat transport effects. In this context our approach and contributions can be explained in the following way.

First, consider the usual welfare optimization problem in which a social planner chooses the latitude emissions to maximize the integral over latitudes and time of discounted weighted utilities of consumption per capita where the climate dynamics are modeled by an energy balance model with spatial heat transport. Progress in this kind of modeling of more realistic climate representations in Integrated Assessment Models (IAMs) has been hindered by the analytical difficulties in dealing with more realistic climatic heat and moisture transport dynamics across a continuum of locations, and the modeling of the carbon cycle under anthropogenic forcing.

Seeking more realistic climate representations, we follow recent research suggesting that global temperature change can be described by a cumulative carbon emission budget. In this context, global mean warming is linearly proportional to cumulative

---

* The authors thank Joan Stefan for technical editing.

carbon[1] with the slope of the linear relationship referred to as the transient climate response to cumulative CO_2 emissions (TCRCE).[2] We show that by using the cumulative carbon emission budget approach, and by expanding the climate dynamics of the latitude temperature field $T(x, t)$ into an infinite series of even numbered Legendre polynomials, the optimization model can be solved to any desired degree of accuracy for usual specifications of utility functions and latitude climatic damages. This analytical contribution enables economists to introduce climate effects of spatial transport and still retain some useful analytical tractability in climate economic models at this level of aggregation. We believe that this theoretical contribution is important for advancing analytically tractable IAM modeling because it introduces more realistic climate dynamics than, for example, simple three box carbon cycle models and two box temperature dynamics models, for the climate component of IAMs. Analytic tractability enables us to understand how the climate and economic components of an IAM interact to produce outcomes.

As an example, consider the case of zero income effects when it is optimal for the tax function to be uniform across latitudes. In this case, we show that an increase in the strength of poleward amplification of transport of heat energy $\hat{r}$ from $\hat{r} = 0$ to a small positive number causes the optimal tax function to shift upward or downward, depending on the interaction of the distribution of welfare weights, population, and damages per capita across latitudes with the distribution of the temperature anomaly $T(x, t)$ across latitudes at each point in time. We illustrate this marginal distribution effect of increased polar amplification with data and plots of population distribution data across latitudes for potentially plausible per capita damage distributions across latitudes.[3]

It is worth noting that our analysis indicates that ignoring heat transport and polar amplification in the standard economic models of climate change implies a potential bias in the calculation of emission taxes and emission path for policy purposes. The present work on the distributional impacts of climate change suggests that it is worthwhile to generalize IAMs to include marginal distributional impacts of spatial heat and moisture transport across latitudes and longitudes. We believe that isolating this marginal impact of polar amplification of heat transport on the optimal tax function is new in our chapter.

Desmet and Rossi-Hansberg (2015) have studied spatial effects in IAM models at the

---

[1] As stated by Pierrehumbert (2014, p. 346),

CO_2 radiative forcing is concave downward as a function of concentration. However, the air fraction nonlinearity makes CO_2 concentration concave upward as a function of cumulative emissions, and it is a somewhat fortuitous consequence of the nature of carbonate chemistry nonlinearities that these two nonlinearities very nearly cancel for cumulative emissions up to several thousand gigatonnes of carbon. As a result, the value of a change in radiative forcing ΔF at the end of a given time period is linearly proportional to the cumulative emissions during that interval.

[2] See Matthews at al. (2009), Pierrehumbert (2012–13; 2014), Matthews et al. (2012), MacDougall and Friedlingstein (2015), Leduc et al. (2016).

[3] See e.g., we might expect that poorer latitudes will experience larger per capita damages, all other things equal (Burgess et al. (2014), Dell et al. (2012)). For example, Dell et al. (2012) have stressed the damaging effects of climatic changes in temperature and precipitation upon not only output levels but also growth rates of poorer countries. Burgess et al. (2014) document increased death rates due to high temperature extremes among the poor who do not have access to adaptation strategies such as air conditioning.

level of disaggregation into latitude belts as we do in this chapter. They do not include heat transport effects across latitudes and do not include polar amplification effects. On the other hand, they include the important adaptation response of migration to negative climate change while we do not include this response. Their paper shows the importance of removing, or at least reducing, restrictions on the adaptive response of migration.

Second, in a world where compensatory transfers are not possible, the usual result – that emission taxes should be uniform – fails because of income effects. We show that "poorer" latitudes should be taxed less than "richer" latitudes due to income effects. Furthermore we conduct comparative dynamics of optimal emissions taxes w.r.t. parameters, e.g., the strength of heat transfer, the strength of polar amplification $\hat{r}$, due to increased poleward latent heat transport, and more. Our comparative dynamics indicate that the optimal tax function depends not only on socioeconomic factors, but also on the interactions of these factors with climate dynamics as they are reflected in the heat transport process.

Third, our decomposition of the temperature field $T(x, t)$ into modes enables us to rank the modes by response times with the higher numbered modes responding faster than the lower numbered modes. This decomposition allows us to show that optimal paths may induce polar amplification.

The remainder of this chapter is organized as follows. Section 2 develops the basic analytical framework used in the chapter. Section 3 conducts welfare analysis and derives optimality conditions for the unified spatial climate and economic model. Section 4 studies the impact and exhibits the importance of heat transfer and polar amplification in the welfare analysis of climate change, and in particular on the social price of the climate change externality. This section shows how the comparative dynamics of heat transfer strength and polar amplification strength depend upon the interaction of climate component dynamics with the distribution of welfare weights, population, and productive capacities across latitudes. Section 5 discusses optimal fossil fuel taxes in a competitive environment with income effects; we show that optimal taxes have a spatial structure and are dependent on each latitude's output. We view our contribution here to be a type of "second best" analysis that moves the discussion of optimal taxation in climate economics towards adding more realistic institutional constraints on compensatory transfers across different sovereigns, as well as moving the discussion of optimal taxation towards more realistic climate models, but still preserving analytical tractability. Chichilnisky and Sheeran (2009) and Chichilnisky (2015) argue that efficiency and equity in a world of independent sovereigns could be tackled in climate economics by first bargaining over the allocation of emission permits, and then opening a trading market in these permits. We show how spatial transport phenomena interacting with the permit allocation impacts the competitive equilibrium price path of tradable permits. We conduct an analysis of permit prices as well as permit wealth ratios across latitudes and show how spatial transport impacts these quantities of economic interest. Section 6 conducts the same type of analysis as was done for logarithmic utility in earlier sections, but for general power utility functions. We show that an increase in the coefficient of relative risk aversion will reduce the social price of the climate externality. Section 7 includes a short summary, conclusions, and suggestions for future research. The Appendix contains the proofs of the propositions.

2. TEMPERATURE DYNAMICS AND HEAT TRANSPORT

To study the evolution of local temperature and its impact on climate policy when heat transport across the globe is taken into account, we build and extend the standard one-dimensional energy balance model (EBM) developed by North (1975a;b), North et al. (1981), and Wu and North (2007). We also substantially extend the work of Brock et al. (2013; 2014) which, for the first time to our knowledge, introduced into an one-dimensional EBM with spatial heat transport, the anthropogenic influence on local temperature resulting from the accumulation of carbon in the atmosphere and conducted economic optimization analysis in this type of model.

Let x denote the sine of the latitude. For simplicity we will just refer to x as "latitude", and let $T_{total}(x, t)$ denote surface (sea level) temperature measured in °C at latitude x and time t. We assume constant albedo across latitudes. The simplifying assumption of constant albedo allows us to cancel out the solar input and the constant in the outgoing radiation term of North et al. (1981) and decompose $T_{total}(x, t)$ into two parts: a baseline part and the temperature anomaly which is associated with human actions. Thus we define surface temperature as:

$$T_{total}(x, t) \equiv T_b(x, t) + T(x, t), \tag{7.1}$$

where the baseline temperature $T_b(x, t)$ is what the temperature at (x, t) would have been if humans were not increasing the carbon content of the atmosphere beyond the pre-industrial levels, and $T(x, t)$ is the temperature anomaly, which is the temperature increase attributed to the anthropogenic emissions of greenhouse gasses (GHGs).[4]

The basic energy balance equation for the surface temperature with human input added can be written as:

$$C\frac{\partial T_{total}(x, t)}{\partial t} = Q(x, t) - A - BT_{total}(x, t) + D\mathcal{L}T_{total}(x, t) + \Delta F \tag{7.2}$$

$$T_{total}(x, 0) = T_b(x, 0), \text{ given}$$

$$\mathcal{L}T_{total}(x, t) \equiv \frac{\partial}{\partial x}\left[\frac{(1 - x^2)\partial T_{total}(x, t)}{\partial x}\right], \tag{7.3}$$

where $x = 0$ denotes the Equator, $x = 1$ denotes the North Pole and $x = -1$ denotes the South Pole, and the heat capacity parameter "C" of North et al. (1981) is absorbed into the other parameters of (7.2). That is, we put $C = 1$ by absorbing it into the other parameters in (7.2). In (7.2), $Q(x, t)$ is the solar forcing, and D is a heat transport coefficient which is an adjustable parameter measured in W/(m²)(°C) which has been calibrated to match observed temperatures across latitudes. North et al. (1981, equation 21) state that D may

---

[4] At this stage, we consider the most general case where the baseline temperature depends on time in order to allow for cases in which t is defined at time scales where there are quasi-periodic regularities like El Nino and La Nina. In such cases, the baseline temperatures would not be constant in t. When we move to the economic analysis and since dealing with quasi periodicity of $T_b(x, t)$ is beyond the scope of this chapter we will assume that T_b is independent of time.

be a function of x and must be thought of as a free parameter to be adjusted empirically, but in their expansions they treat D as constant independent of x. We follow Alexeev et al. (2005) and specify the transport coefficient as proportional to the temperature anomaly. Finally, ΔF denotes radiative forcing associated with anthropogenic emissions $E(t)$ of GHGs. The diffusion operator $\mathcal{L}$ is a way of modeling the mostly uni-directional transport of heat and moisture from the Equator towards the Poles. Without this transport temperate zones would be colder.

Alexeev et al. (2005) specify the heat transport coefficient as a function of the temperature anomaly as

$$D(T_M(t)) = D_{ref}(T_M(t) - T_{ref}), \tag{7.4}$$

where $T_{ref} = 15°C$ and T_M is global mean temperature. Setting $T_M(t) = T_b + \overline{T}(t)$ and $T_b = T_{ref}$ we obtain[5]

$$D(T_M(t)) = D_{ref}[1 + \hat{r}(T_b(t) + \overline{T}(t) - T_{ref}(t))] = D_{ref}[1 + \hat{r}\overline{T}(t)] \tag{7.5}$$

$$D_{ref} = 0.445, \hat{r} = 0.03/K. \tag{7.6}$$

Alexeev et al.'s (2005) results suggest that there is some evidence for the diffusion coefficient to be increasing in global mean yearly average temperature and this effect is captured in equations (7.4)–(7.6).

The operator $\mathcal{L}$ is a linear operator on the space of functions of x with the property that the nth Legendre polynomial $P_n(x)$ is an eigenfunction of $\mathcal{L}$, i.e. $\mathcal{L}P_n(x) = -\lambda_n P_n(x)$, $\lambda_n = n(n+1)$.[6] We use this property in the solution of the model. The term $D(T(t))\mathcal{L}T(x, t)$ therefore models the heat flux associated with the temperature anomaly.

To enhance the tractability of the optimized model, since in (7.2) dynamics are described by partial differential equations (PDEs) with a nonlinear diffusion term, we introduce two approximations, one from North et al. (1981) and the other from Matthews et al. (2009).

North et al. (1981) note that $T(x, t)$ can be written in a series expansion in terms of Legendre polynomials, or

$$T(x, t) = \sum_{n=0, \text{ even}} T_n(t) P_n(x) \tag{7.7}$$

where $P_n(x)$ is the nth Legendre polynomial. They approximate $T(x, t)$ by truncating the expansion at some finite N.[7] This implies that the average global temperature anomaly (7.7) can be defined as:

---

[5] Alexeev et al. (2005) specify $\hat{r}$ to be 3% per degree Kelvin.

[6] $P_n(x) = 2^n \sum_{k=0}^{n} x^k \binom{n}{k}\binom{(n+k-1)/2}{n}$, $P_0(x) = 1, P_2(x) = \frac{1}{2}(3x^2 - 1)$.

[7] Note that $\sum_{\infty}^{n=0} T_n(0) P_n(0)$ does not imply that all $T_n(0)$'s are zero. Indeed, if all $T_n(0)$'s are zero, then the solution of (7.2) would be independent of x and all spatial effects would vanish for the anomaly. As one might expect, if one is dealing with differential equations in an infinite dimensional space, an infinite number of initial conditions must be specified.

$$\overline{T}(t) \equiv \frac{1}{2}\int_{-1}^{1} \sum_{n=0}^{\infty} T_n(t)P_n(x)dx = T_0(t). \tag{7.8}$$

Then, following Alexeev et al. (2005) the heat transport coefficient can be defined as:

$$D(T_M(t)) = D[1 + \hat{r}T_0(t)]. \tag{7.9}$$

Following Matthews et al. (2009) and MacDougall and Friedlingstein (2015), Leduc et al. (2016),[8] the full response of the Earth System to anthropogenic emissions is approximately linearly proportional to cumulative emissions at date t. Brock and Xepapadeas (2017) provide a simulation that adapts MacDougall and Friedlingstein's work (2015) which shows that the linear approximation can be quite good. Therefore, in modeling temperature dynamics we use the approximation

$$\Delta F \approx \lambda E(t), \; E(t) = \int_{x=-1}^{x=1} E(x,t)dx \tag{7.10}$$

Using North's approximation for the baseline local temperature and the local temperature anomaly which is

$$T_b(x,t) = \sum_{n,\,\text{even}} T_{bn}(t)P_n(x), \; T(x,t) = \sum_{n,\,\text{even}} T_n(t)P_n(x), \tag{7.11}$$

respectively, we can write the total temperature dynamics (baseline plus anomaly) as:

$$\frac{\partial (T_b(x,t) + T(x,t))}{\partial t} = \sum_{n=0}^{\infty} (\dot{T}_n(t) + \dot{T}_{bn}(t))P_n(t) = \tag{7.12}$$

$$Q(x,t) - \left(A + B\sum_{n=0}^{\infty}(T_n(t) + T_{bn}(t))P_n(x)\right) +$$

$$[D(1 + \hat{r}(\overline{T}_b(t) + \overline{T}(t) - 15))]\mathcal{L}(T + T_b)(x,t) + \Delta F$$

The baseline temperature dynamics are given by

$$\frac{\partial T_b(x,t)}{\partial t} = \sum_{n=0}^{\infty} \dot{T}_b(t)P_n(t) = \tag{7.13}$$

$$Q(x,t) - \left(A + B\sum_{n=0}^{\infty} T_b(t)P_n(x)\right) + [D(1 + \hat{r}(\overline{T}_b(t) - 15))]\mathcal{L}(T_b)(x,t)$$

$$= Q(x,t) - \left(A + B\sum_{n=0}^{\infty} T_{bn}(t)P_n(x)\right) + D\mathcal{L}(T_b)(x,t),$$

using $\overline{T}_b(t) = T_{ref} = 15$. Taking the difference between total and baseline temperature dynamics, we obtain the temperature anomaly dynamics as:

---

[8] Leduc et al. (2016) show that regional temperatures also respond approximately linearly to cumulative CO_2 emissions. We thank Victor Zhorin of RDCEP, University of Chicago, for bringing this paper to our attention.

$$\frac{\partial T(x,t)}{\partial t} = \sum_{n=0}^{\infty} \dot{T}_n(t) P_n(t) = -B \sum_{n=0}^{\infty} T_n(t) P_n(x) \quad (7.14)$$

$$+ D(1 + \hat{r}\overline{T}(t))\mathcal{L}(T + T_b)(x,t) - D\mathcal{L}(T_b)(x,t) + \Delta F$$

$$= -B \sum_{n=0}^{\infty} T_n(t) P_n(x) + D(1 + \hat{r}T_0(t))\mathcal{L}T(x,t) + D(1 + \hat{r}T_0(t))\mathcal{L}T_b(x,t)$$

$$- D\mathcal{L}T_b(x,t) + \Delta F.$$

Using $\mathcal{L}P_n(x) = -\lambda_n P_n(x)$, $\lambda_n = n(n+1)$ for the eigenvalues of $\mathcal{L}(\cdot)$, we obtain:

$$\frac{\partial T(x,t)}{\partial t} = \sum_{n=0}^{\infty} \dot{T}_n(t) P_n(x) = -B \sum_{n=0}^{\infty} T_n(t) P_n(x) \quad (7.15)$$

$$- D\left[\sum_{n=0}^{\infty} \lambda_n T_n(t) P_n(x)\right] + \Delta F - \hat{r} D T_0(t) \left(\sum_{m=0}^{\infty} \lambda_m [T_{bm}(t) + T_m(t)] P_m(x)\right)$$

$$T_0(x,0) = \sum_{n=0}^{\infty} T_n(0) P_n(x) = 0, \text{ given.}$$

We can simplify (7.15) by using the property that the Legendre polynomials are orthogonal with respect to the inner L^2 product on the interval $x \in [-1,1]$, which implies that

$$\langle P_n, P_m \rangle = \int_{x=-1}^{x=1} P_n(x) P_m(x) dx = \frac{2}{2n+1} \delta_{nm} \quad (7.16)$$

$$\delta_{nm} = \begin{cases} 1 & \text{if } n = m \\ 0 & \text{if } n \neq m \end{cases}. \quad (7.17)$$

Therefore, multiplying both sides of (7.15) by $P_n(x)$, integrating over $x \in [-1,1]$ (7.16–7.17), using the inner product notation $\int_{x=-1}^{x=1} F(x) G(x) dx = \langle F, G \rangle$, noting that $P_0(x) = 1$, $\int_{-1}^{1} P_0(x) dx = 2$, $\int_{-1}^{1} P_n(x) dx = 0$, $n = 2, 4, 6, \ldots$, and dropping the higher order term

$$\hat{r} D T_0(t) \sum_{m=0}^{\infty} \lambda_m T_m(t) P_m(x) = o(\|T(\cdot)\|) \quad (7.18)$$

in the norm of the function $T(\cdot)$, we obtain:[9]

$$\dot{T}_n(t) = [-B - D\lambda_n] T_n(t) + \Delta F \frac{\langle 1, P_n \rangle}{\langle P_n, P_n \rangle} - \hat{r} D T_0(t) \lambda_n T_{bn}(t) \quad (7.19)$$

$$n = 0, 2, 4, \ldots.$$

---

[9] Since the anomalies in the modes, $T_0(t)$, $T_2(t)$, ... are small, we expect the products to be small enough that the optimal paths ignoring the second order term (7.18) are workably close to the optimal paths when (7.18) is not dropped. However, numerical work is needed to actually verify how large the error in the optimal path is when (7.18) is dropped.

A two-mode approximation of (7.19), for example, results in the following system of ordinary differential equations:

$$\dot{T}_0(t) = -BT_0(t) + \lambda E(t), \; P_0(x) = 1, \langle P_0, P_0 \rangle = 2 \tag{7.20}$$

$$\dot{T}_2(t) = (-B - 6D)T_2(t) - 6\hat{r}DT_0(t)T_{b2}(t) \tag{7.21}$$

$$\langle 1, P_2 \rangle = 0, \; P_2(x) = \frac{1}{2}(3x^2 - 1). \tag{7.22}$$

We will use temperature dynamics (7.19) to derive the optimal emission paths and the corresponding optimal spatial taxes, under the additional assumption that the baseline temperature T_b does not depend on time.[10]

3. WELFARE MAXIMIZATION UNDER HEAT TRANSFER

To study optimal emissions paths in the context of the one-dimensional climate model described above, we consider a simple welfare maximization problem with logarithmic utility, where world welfare is given by:[11]

$$\int_{t=0}^{\infty} e^{-\rho t} \left[\int_{x=0}^{x=1} v(x) L(x) \ln[y(x,t) E(x,t)^\alpha (1 - A(x,t)) e^{-\phi_T(x)(T_{total} - bA)}] \right] dx dt, \tag{7.23}$$

where $y(x,t)E(x,t)^\alpha$, $0 < \alpha < 1$, $E(x,t)$, $T(x,t)$, $L(x)$ are output per capita, fossil fuel input, temperature anomaly, and fully employed population at location (or latitude) x at date t, respectively. The term $yE^\alpha(1 - A(x,t))$ stands for the fraction of output available for consumption after adaptation. The term $e^{-\phi_T(x)[T_{total}(x,t) - bA(x,t)]}$ reflects damages to output per capita in location x from an increase in the temperature anomaly at this location, which is the term $e^{-\phi_T(x)T_{total}(x,t)}$ net of reduction in damages due to adaptation, which is the term $e^{\phi_T(x)bA(x,t)}$. The damage coefficient ϕ_T may depend upon time, i.e., $\phi_T = \phi_T(x,t)$. Note that damages depend upon total temperature at (x,t) which is defined by the sum of baseline temperature which would have occurred if there were no human emissions into the system, $T_b(x)$, and the temperature anomaly, $T(x,t)$, caused by human emissions into the atmosphere. We assume that $y(x,t), L(x)$ are exogenously

---

[10] It should be noted that working at time scales where there are quasi-periodic regularities like El Nino and La Nina that baseline temperatures would not be constant in t. Dealing with quasi periodicity of $T_b(x,t)$ is, however, beyond the scope of this chapter. The assumption of constant $T_b(x,t)$ in equation (7.1) and in the work that follows, provides expositional simplicity. The constancy of $T_b(x,t)$ follows if we assume that in the temperature dynamics $T_b(x,t)$ is "differenced out" along with the solar forcing, so in the dynamics of the anomalies $T(x,t)$ the temperature dynamics are stationary if D is constant. In the Appendix we show that our results can be extended to time dependent $T_b(x,t)$ at the cost of expositional simplicity.

[11] For the rest of the chapter we follow North (1975a, b) and North et al. (1981) and consider the northern hemisphere only, i.e. $x \in [0,1]$.

given and fixed. That is, we are abstracting away from the problem of optimally accumulating capital inputs and other inputs in order to focus sharply on optimal fossil fuel taxes under transport effects. Finally, $v(x)$ represents welfare weights associated with location x

Formulation (7.23) allows the incorporation of another very important aspect of spatially distributed damages from climate change, namely, damages from precipitation. Defining total precipitation as the sum of baseline precipitation and the precipitation anomaly or $P_{total}(x, t) = P_b(x) + P(x, t)$, Castruccio et al. (2014) suggest the following approximation for the precipitation anomaly:[12]

$$P(x, t) = \psi(x)T(x, t). \tag{7.24}$$

Assuming exponential precipitation damages of the form $\exp(-\varphi(x, t)(P_b(x) + P(x, t)))$ and using Castruccio et al.'s (2014) approximation, we can write a welfare function, dropping adaptation to simplify, that contains both temperature impacts and precipitation damages as:

$$\int_0^\infty e^{-\rho t} \int_0^1 [v(x)L(x)\ln(y(x, t)E(x, t)^\alpha) \times \tag{7.25}$$

$$(e^{-\phi_T(x, t)[T_b(x) + T(x, t)]} e^{-\varphi(x, t)[P_b(x) + \psi(x)T(x, t)]})] dxdt.$$

In the case where we are assuming logarithmic utility and exponential damages to output both from temperature and precipitation, we can add a baseline temperature $T_b(x)$ and a baseline precipitation $P_b(x)$ to the corresponding anomalies and still be able to assert that (7.25) with adaptation included can be replaced for optimization purposes by the equivalent problem,

$$\max_{E(x, t), A(x, t)} \int_0^\infty e^{-\rho t} \int_0^1 v(x)L(x)[\alpha \ln E(x, t) + \ln(1 - A) \tag{7.26}$$

$$- \phi(x)T(x, t) - bA]dxdt,$$

where $\phi(x, t) = \phi_T(x, t) + \varphi(x, t)\psi(x)$. In the definition of $\phi(x, t)$, the term $\phi_T(x, t)$ accounts for temperature damages, while the term $\varphi(x, t)\psi(x)$ allows for precipitation damages. It should be noted that to the best of our knowledge, this is the first time that climate economics in terms of a spatial one-dimensional EBM that incorporates the important climate science phenomenon of heat transfer is combined with the spatial characteristics of damages from temperature and precipitation.[13] This combination results in a model of climate economics capable of determining the impact on the social

---

[12] We ignore the conditional variance since we are working with a deterministic model.
[13] It should be noted that there could be important interactions between temperature and precipitation. This is because precipitation interacts with cloud cover and this interaction could have an important effect on energy balance and temperature dynamics. Our equation (7.2), does not capture these potential important interactions, which could be an important area for further research.

136 *Handbook on the economics of climate change*

cost of climate externality of including spatial heat transport and, hence, the impact of spatial heat transport on optimal fossil fuel taxes.

The problem of a social planner would be to choose fossil fuel paths $E(x, t)$ or equivalently, by an appropriate change in units, emissions paths $E(x, t)$ to maximize (7.26) subject to climate dynamics given by (7.19), and an additional constraint reflecting the potential exhaustibility of global fossil fuel reserves.

$$\int_{t=0}^{\infty} E(t)dt \leq R_0, \ E(t) = \int_{x=0}^{x=1} E(x,t)dx, \ \int_{x=0}^{x=1} R_0(x)dx \qquad (7.27)$$

where R_0 denotes global fossil fuel reserves, and $R_0(x)$ fossil fuel reserves in location x.

Constraint (7.27) implies that the social planner is altruistic and treats fossil fuels reserves as a common property which can be transferred across locations. The alternative polar case is to assume that no transfers are possible and that each location is constrained by local fossil fuel reserves, or

$$\int_{t=0}^{\infty} E(t,x)dt \leq R_0(x) \text{ for all } x \in [0,1], \ \int_{x=0}^{x=1} E(x,t)dx = E(t). \qquad (7.28)$$

We start with the welfare maximization problem of the altruistic planner, making the simplifying assumption that the damage parameter $\phi(x, t)$ is independent of $s(x)$. The current value Hamiltonian for this problem is:[14]

$$H =$$

$$\int_{x=0}^{x=1} \left\{ v(x) L(x) \left[\ln[E^\alpha(1-A)] - \varphi(x) \left[\sum_{n=0,2,\ldots} T_n(t) P_n(x) - bA \right] \right] \right. \qquad (7.29)$$

$$\left. - \lambda_R(t) E(t,x) dx \right\} +$$

$$\sum_{n=0,2,\ldots} \lambda_{T_n} \left[[-B - D\lambda_n] T_n(t) + \lambda \left(\int_0^1 E(x,t) dx \right) \delta_{n0} - \hat{r} D T_0(t) \lambda_n T_{bn}(t) \right],$$

where $\delta_{n0} \equiv \frac{\langle 1, P_n \rangle}{\langle P_n, P_n \rangle}$, Note that $\delta_{n0} = 0, n \neq 0$, $\delta_{n0} = 1, n = 0$. The two-mode approximation, for example, would result in the following current value Hamiltonian:

$$H = \qquad (7.30)$$

$$\int_{x=0}^{x=1} \{v(x)L(x)[\ln[E^\alpha(1-A)] - \phi(x)[T_0(t) + T_2(t)P_2(x) - bA]]$$

---

[14] Note that with the logarithmic utility and exponential damage formulation used in this chapter, the baseline quantities for temperature and precipitation do not affect the optimization.

$$-\lambda_R(t)E(t,x)\}dx + \lambda_{T_0}(t)[-BT_0(t) + \lambda E(t)] +$$

$$\lambda_{T_2}(t)[(-B-6D)T_2(t) - 6\hat{r}DT_0(t)T_{b2}].$$

The first order necessary conditions (FONC) resulting from the maximum principle, after suppressing the (x, t) arguments to ease notation when necessary, can be obtained as follows. The optimal emission (or fossil fuel) $E^*(x, t)$ path and optimal adaptation $A^*(x)$ satisfy:

$$\frac{\alpha v(x)L(x)}{E^*(x,t)} = \lambda_R(t) - \lambda\lambda_{T_0}(t) \Rightarrow \quad (7.31)$$

$$E^*(x,t) = \frac{\alpha v(x)L(x)}{\lambda_R(t) - \lambda\lambda_{T_0}(t)}, \quad (7.32)$$

$$\frac{1}{1 - A^*(x)} = b\phi(x) \Rightarrow A^*(x) = \frac{b\phi(x) - 1}{b\phi(x)}. \quad (7.33)$$

In (7.31), $\xi_C(t) = -\lambda\lambda_{T_0}(t)$ is the social price of the climate externality, λ_R is the scarcity rent of the (essential) fossil fuel input, and $\xi_F(t) = \lambda_R(t) - \lambda\lambda_{T_0}(t)$ is the social price of fossil fuels. Here, we define social price of the climate externality to allow it to be negative, which it usually will be since it is typically a "bad". Furthermore (7.33) implies that for all latitudes and all dates where $b\phi(x) - 1 > 0$, there will be adaptation expenditures. The condition for positive adaptation stems from the fact that the marginal benefit of adaptation has to be bigger than its marginal cost for its use to be positive.

Setting $d(x) = v(x)L(x)\phi(x)$ to simplify the exposition, the costate variables evolve according to

$$\dot{\lambda}_{T_0} = (\rho + B)\lambda_{T_0} + \sum_{n=0,2,\ldots} \lambda_{T_n}n(n+1)\hat{r}DT_{bn}(t) + \langle 1, d(x)\rangle \quad (7.34)$$

$$\dot{\lambda}_{T_n} = (\rho + B + Dn(n+1))\lambda_{T_n} + \langle P_n(x), d(x)\rangle, n = 2, 4, \ldots \quad (7.35)$$

$$\dot{\lambda}_R(t) = \rho\lambda_R(t), \quad (7.36)$$

while temperature dynamics are given by

$$\dot{T}_0 = -BT_0(t) + \lambda E^*(t) \quad (7.37)$$

$$\dot{T}_n = -[B + D\lambda_n]T_n(t) - \hat{r}DT_0(t)\lambda_n T_{bn} \quad n = 2, 4, \ldots \quad (7.38)$$

and the fossil fuel constraint satisfies

$$E^*(t) = \int_{x=0}^{x=1} E^*(x,t)dx, \int_{t=0}^{\infty} \langle 1, E^*(x,t)\rangle dt = R_0. \quad (7.39)$$

If we assume that each location is constrained by local fossil fuel reserves $R_0(x)$ and that no transfers are possible, condition (7.36) should be replaced by

$$\dot{\lambda}_R(x, t) = \rho \lambda_R(x, t). \tag{7.40}$$

Then

$$E^*(x, t) = \frac{\alpha v(x) L(x)}{\lambda_R(x, t) - \lambda \lambda_{T_0}(t)}, \tag{7.41}$$

while the fossil fuel constraint becomes

$$\int_{t=0}^{\infty} E^*(x, t) \, dt = R_0(x). \tag{7.42}$$

3.1 The Two-Mode Solution

For the two-mode approximation the costate variables evolve according to

$$\dot{\lambda}_{T_0} = (\rho + B)\lambda_{T_0} + 6\hat{r} D T_{b2} \lambda_{T_2} + \langle 1, d \rangle \tag{7.43}$$

$$\dot{\lambda}_{T_2} = (\rho + B + 6D)\lambda_{T_2} + \langle P_2, d \rangle. \tag{7.44}$$

Since T_{b2} is fixed at all dates and independent of t, then (7.43), (7.44) imply that in the optimal control problem the values for $\lambda_{T_0}, \lambda_{T_2}$ should be fixed at their steady-state values $(\dot{\lambda}_{T_0}, \dot{\lambda}_{T_2}) = (0, 0)$, by setting the constants of the ODE solution equal to zero, otherwise the costates will explode to infinity.

Therefore,

$$\lambda^*_{T_2} = \frac{-\langle P_2, d \rangle}{(\rho + B + 6D)}, \quad \lambda^*_{T_0} = \frac{-6\hat{r} D T_{b2} \langle P_2, d \rangle}{(\rho + B)(\rho + B + 6D)} - \frac{\langle 1, d \rangle}{(\rho + B)}, \tag{7.45}$$

which implies that the optimal emissions will be constant

$$E^*(x, t) = \frac{\alpha v(x) L(x)}{\lambda_R(x, 0) e^{\rho t} - \lambda \lambda^*_{T_0}}. \tag{7.46}$$

With infinite reserves $\lambda_R(x, t) = 0$ and

$$E^*(x, t) = \frac{\alpha v(x) L(x)}{-\lambda \lambda^*_{T_0}} \tag{7.47}$$

independent of t. Temperature dynamics in the two-mode approach will be linear, or

$$\dot{T}_0 = -BT_0 + \lambda \left[\frac{\alpha v(1) L(1)}{-\lambda \lambda^*_{T_0}} + \frac{\alpha v(2) L(2)}{-\lambda \lambda^*_{T_0}} \right]$$

$$\dot{T}_2 = (-B - 6D) T_2 - 6\hat{r} D T_{b2} T_0$$

This is a linear system with negative eigenvalues $(-B, -B - 6D)$, with corresponding eigenvectors $(-1/\hat{r}T_{b2}, 1)'$ and $(0, 1)'$, so the solution paths for the anomalies are:

$$T_0(t) = \frac{-C_1}{\hat{r}T_{b2}} e^{-Bt} + T_0^*, \quad T_0(0) = 0 \tag{7.48}$$

$$T_2(t) = C_1 e^{-Bt} + C_2 e^{-(B+6D)t} + T^*_2, \quad T_2(0) = 0 \tag{7.49}$$

$$T_0^* = \lambda \left[\frac{\alpha v(1) L(1)}{-\lambda \lambda^*_{T_0}} + \frac{\alpha v(2) L(2)}{-\lambda \lambda^*_{T_0}} \right] / B \tag{7.50}$$

$$T_2^* = \frac{6\hat{r}DT_{b2}T_0^*}{(-B - 6D)} \tag{7.51}$$

Equations (7.45) and (7.48)–(7.51) completely specify the solution of the welfare maximization problem in the two-mode approximation.

3.2 Welfare Maximization when Heat Transfer is Ignored

To understand the impact of heat transfer on optimal fossil fuel paths (or emission paths) and optimal climate policy, it is helpful to consider at the beginning welfare optimization where heat transfer is ignored, or $D = 0$. The optimality conditions (7:34)–(7.36) with $D = 0$ become:[15]

$$\dot{\lambda}_{T_0} = (\rho + B)\lambda_{T_0} + \langle 1, \mathrm{d} \rangle \tag{7.52}$$

$$\dot{\lambda}_{T_n} = (\rho + B)\lambda_{T_n} + \langle P_n, \mathrm{d} \rangle, \; n = 2, 4, \ldots \tag{7.53}$$

while temperature dynamics in (7.37)–(7.38) are independent of D. Taking the forward solutions for the costate variables we obtain:

$$\lambda_{T_0} = -\int_{s=t}^{\infty} e^{-(\rho+B)(s-t)} \langle 1, \mathrm{d} \rangle ds \tag{7.54}$$

$$\lambda_{T_n} = -\int_{s=t}^{\infty} e^{-(\rho+B)(s-t)} \langle P_n, \mathrm{d} \rangle ds. \tag{7.55}$$

Then the optimal fossil fuel path for the log utility case is given by

$$E^*(x, t) = \frac{\alpha v(x) L(x)}{\lambda_R(0) e^{\rho t} - \lambda \lambda_{T_0}(t)}. \tag{7.56}$$

---

[15] Note that $D > 0$ implies that spatial heat transport occurs. On the other hand, we need $\hat{r} > 0$ so that spatial heat transport generates impacts in the context of our model, that is, to create asymmetric effects such as polar amplification. Note that D appears in mode 2's costate, but $\hat{r} = 0$ removes the effect of mode 2's costate on the dynamics of mode zero's costate, as can be seen from (7.43)–(7.44).

Using the assumption that population and the damage parameter do not change with time, the steady-state values for the costate variables implied from (7.52)–(7.53) are:

$$\lambda^*_{T_0} = -\frac{\langle 1, d \rangle}{(\rho + B)}, \lambda^*_{T_n} = -\frac{\langle P_n, d \rangle}{(\rho + B)} \quad (7.57)$$

$$\langle 1, d \rangle = \int_{x=0}^{x=1} v(x) L(x) \phi(x) dx \quad (7.58)$$

$$\langle P_n, d \rangle = \int_{x=0}^{x=1} P_n(x) v(x) L(x) \phi(x) dx. \quad (7.59)$$

This means that the steady-state costate variables are independent of location x. The resource constraint implies

$$R_0 \geq \int_{t=0}^{\infty} \int_{x=0}^{x=1} E(x,t) dx dt = \int_{t=0}^{\infty} \int_{x=0}^{x=1} \left(\frac{\alpha v(x) L(x)}{\lambda_R(0) e^{\rho t} - \lambda \lambda_{T_0}} \right) dx dt. \quad (7.60)$$

The initial value $\lambda_R(0)$ can be obtained by solving (7.60) for this initial value for any given value of total reserves R_0. Conditions (7.57)–(7.59) and (7.60) completely determine the optimal emission path for each location with the population kept constant at each location. Since the steady-state costate variables are independent of location x, the social price of the climate externality and the social price of fossil fuels are independent of location x. If we consider the case in which each location is constrained by local fossil fuel reserves $R_0(x)$, and assume that no transfers are possible, then the local resource constraint implies

$$\int_{t}^{\infty} E^*(x,t) dt = \int_{t=0}^{\infty} \left(\frac{\alpha v(x) L(x)}{\lambda_R(x,0) e^{\rho t} - \lambda \lambda_{T_0}} \right) dt = R_0(x). \quad (7.61)$$

This constraint can be used to determine the initial value $\lambda_R(x, 0)$ for any given value of total local reserves $R_0(x)$. In this case, although the social price of the climate externality does not depend on the location, the social price of fossil fuels depends on location through local reserves. This result is similar to an analogous result in Brock et al. (2014).

4. HEAT TRANSPORT AND CLIMATE CHANGE POLICY

We move now to one of the main objectives of this chapter, which is the characterization of the impact of heat transport towards the Poles on the social price of climate externality $\xi(t) = -\lambda \lambda_{T_0}$ and consequently on optimal fossil fuel paths and fossil fuel taxes. Since λ is a fixed parameter, the impact of heat transport should be realized through the costate variable λ_{T_0}. Under the assumption of fixed baseline temperature this costate is fixed at its optimal steady state and determines the optimal tax for the correction of the climate externality. We determine the impact of spatial heat transport by the ratio

$$\psi \equiv \frac{\tau(\hat{r} = 0)}{\tau(\hat{r} > 0)} = \frac{\lambda_{T_0}(\hat{r} = 0)}{\lambda_{T_0}(\hat{r} > 0)}, \quad (7.62)$$

which is computed at the steady state of λ_{T_0}. Using $d(x) = v(x) L(x) \phi(x)$, we obtain from (7.52)

$$\lambda_{T_0}(\hat{r} = 0) = -\frac{\langle 1, d(x) \rangle}{(\rho + B)}, \quad (7.63)$$

while from (7.34) we obtain

$$\lambda_{T_0}(\hat{r} > 0) = -\frac{\langle 1, d(x) \rangle + \sum_{n=2,\ldots} \lambda_{T_n} n(n+1)\hat{r} DT_{bn}(t)}{(\rho + B)} \quad (7.64)$$

$$\lambda_{T_n} = -\frac{\langle P_n(x), d(x) \rangle}{(\rho + B + Dn(n+1))}, \quad n = 2, 4, \ldots$$

If we take the two-mode approximation $T_b(x, t) \cong T_{b0} + T_{b2} P_2(x)$ with fixed baseline temperatures (North et al., 1981, equation (7.31)) with $T_{b0} = 14.97°C$ and $T_{b2} = -28.0°C$, then using the steady-state solutions for (7.43) and (7.44),

$$\lambda^*_{T_0}(\hat{r} > 0) = -\frac{\langle 1, d(x) \rangle + 6\lambda_{T_2} \hat{r} DT_{b2}}{(\rho + B)}, \quad (7.65)$$

and the ratio (7.62) becomes

$$\frac{\lambda_{T_0}(\hat{r} = 0)}{\lambda_{T_0}(\hat{r} > 0)} = 1 \Big/ \left(1 + \frac{6\lambda^*_{T_2} \hat{r} DT_{b2}}{\langle 1, d(x) \rangle}\right) \quad (7.66)$$

$$\lambda^*_{T_2} = -\frac{\langle P_2(x), d(x) \rangle}{(\rho + B + 6D)} = -\frac{\int_0^1 P_2(x) v(x) L(x) \phi(x) dx}{(\rho + B + 6D)}. \quad (7.67)$$

The above analysis suggests the following proposition.

Proposition 1 *Ignoring spatial heat transport, i.e. setting $\hat{r} = 0$ when $\hat{r} > 0$, will lead to underestimation of the optimal climate externality tax if $J > 0$, $J \equiv \frac{6\lambda^*_{T_2} \hat{r} DT_{b2}}{\langle 1, d(x) \rangle}$. The optimal climate externality tax will be overestimated if $-1 < J < 0$.*

The proof follows from the calculations above.

The value of the quantity of interest J depends on the distributions across latitudes of the welfare weights $v(x)$, the population $L(x)$, the damages $\phi(x)$ from an increase in global temperature, and $P_2(x)$ that reflects the dynamics of Nature on the spatial distribution of temperature. J can be written as

$$J \equiv \frac{6\hat{r} D(-T_{b2}) \int_0^1 P_2(x) v(x) L(x) \phi(x) dx}{(\rho + B + 6D) \int_0^1 v(x) L(x) \phi(x) dx} \quad (7.68)$$

$$\phi(x) = \phi_T(x) + \psi(x) \varphi(x), \quad (7.69)$$

or

$$J = \frac{6\hat{r}D(-T_{b2})}{(\rho + B + 6D)}\int_0^1 P_2(x)s(x)dx \qquad (7.70)$$

$$P_2(x) = \frac{1}{2}(3x^2 - 1), \ s(z) = \frac{v(z)L(z)\phi(z)}{\int_0^1 v(x)L(x)\phi(x)dx}, \qquad (7.71)$$

where $s(x)$ can be interpreted as the share of weighted (by welfare weights $v(x)$ and population) damages at location x. Since $T_{b2}<0$, the sign of J is the sign of $\psi = \int_0^1 P_2(x)s(x)dx$. In the case where $s(x)$ is a constant independent of x, then $J = 0$ since $\int_0^1 P_2(x)dx = 0$ and $\psi = 1$. Let $\hat{r} = 0.03$, $D = 0.0445$, $T_{b2} = -28$, $\rho = 0.02$, $B = 2$ so that $\frac{6\hat{r}D(-T_{b2})}{(\rho + B + 6D)} = 0.478$. If $s(x) = x$, $x \in [0,1]$, then $\psi = 0.05978$, and $J = 0.94359$. Thus when the share of damages is higher in higher latitudes, ignoring heat transport underestimates the optimal climate externality tax. On the other hand, if $s(x) = 1 - x$, $x \in [0,1]$, then $\psi = -0.05977$ and $J = 1.06358$. Thus, in this example, when the share of damages is higher in lower latitudes, ignoring heat transport overestimates the optimal climate externality tax.

To obtain more insights regarding the potential values of J, we consider the general function

$$s(x) = \frac{\gamma(1-x)^{\alpha_0}(1+x)^{\beta_0}(\gamma_0 + \delta_0 x^2)}{\int_{x=0}^{x=1}(1-x)^{\alpha_0}(1+x)^{\beta_0}(\gamma_0 + \delta_0 x^2)dx}, \qquad (7.72)$$

as an approximate distribution of $s(x)$ in $x \in [0, 1]$. From (7.71) the share $s(x)$ depends on the distribution of welfare weights $v(x)$, population $L(x)$, and marginal damages $\phi(x)$. Work by Mendelsohn et al. (2006) or Burgess et al. (2014), for example, suggest that climate change is expected to be most severe in poor countries surrounding the equator, with a skew towards southern latitudes. If we follow the usual approach of setting welfare weights equal to Negishi-type weights, then these weights can be set according to GDP per capita across latitudes. See Kummu and Varis (2011) for data by latitudes. However equal weights across locations, or weights where the most importance is given to latitudes around the equator, are also possibilities. Finally, for the population, evidence suggests that roughly 88% of the world's population lives in the northern hemisphere, and about half the world's population lives north of 27°N.[16] The actual distribution of $s(x)$ is an empirical issue that requires further research. In order, however, to focus on the possible over- or under-estimation of the externality tax, we consider the alternative distributions shown in Figure 7.1.

In Figure 7.1, distributions 1, 2 and 5 assign more damages to locations in the North while distributions 3, 4 and 6 assign more damages to locations around the equator.

The results of Table 7.1 show that ignoring heat transport will cause an underestimation of the optimal externality tax when the distribution of the weighted share of climate change damages is skewed towards the southern latitudes. The bias and its direction depends on natural parameters reflected in $P_2(x)$, but also on socioeconomic parameters reflected in

---

[16] See http://visual.ly/worlds-population-2000-latitude-and-longitude.

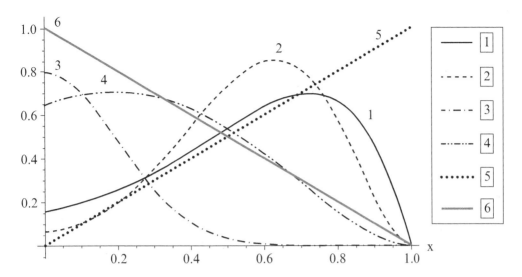

Figure 7.1 Possible s(x) distributions

Table 7.1 Externality tax comparison

Distribution s(x)	J	$\psi = \dfrac{\lambda_{T_0}(\hat{r} = 0)}{\lambda_{T_0}(\hat{r} > 0)}$
1	0.01567	0.98456
2	0.00636	0.99368
3	−0.04208	1.04393
4	−0.052657	1.05558
5	0.05978	0.94359
6	−0.05978	1.06358

the distribution of population, climate change damages and welfare weights. The important message, however, is that taking into account the spatial dynamics of nature emerging because of heat transport across latitudes, which is a well-documented natural process, changes the social price of the climate externality and the corresponding optimal tax relative to the case where heat transport is ignored. Since in general $\lambda_{T_0}(\hat{r} = 0) \neq \lambda_{T_0}(\hat{r} > 0)$, optimality conditions (7.32) or (7.41) suggest that optimal paths for fossil fuel use when heat transport across latitudes is ignored will in general either overestimate or underestimate the true optimal fossil fuel paths. That is, $E^*(x, t; \hat{r} > 0) \neq E^*(x, t; \hat{r} = 0)$.

4.1 Spatial Heat Transport and Cross Latitude Effects

Having established that taking into account that heat transport across latitudes affects the social price of the climate externality and consequently optimal emission paths and taxes through socioeconomic and natural factors, our next step is to examine in more detail the impacts of heat transport across locations on socially optimal fossil fuel use, the social price of fossil fuels, the socially optimal temperature paths and the discount rate for future costs and benefits. Since heat transport is captured by the parameter $\hat{r}$, the impacts are determined by taking the appropriate derivatives with respect to $\hat{r}$.

4.1.1 Heat transport and fossil fuel use

We examine first the impacts on fossil fuel use by calculation the derivatives of $E^*(x, t)$ and $\lambda_{T_0}^*$ with respect to $\hat{r}$. From (7.46) we obtain:[17]

$$\frac{\partial E^*(x, t)}{\partial \hat{r}} = \frac{-\alpha v(x)L(x)(\lambda_R'(x, 0)e^{\rho t} + \lambda(\lambda_{T_0}^*)')}{(\lambda_R(x, 0)e^{\rho t} - \lambda\lambda_{T_0}^*)^2}, \qquad (7.73)$$

where from (7.45)

$$\frac{\partial \lambda_{T_0}^*}{\partial \hat{r}} = \frac{6\hat{r}DT_{b2}\langle P_2, d\rangle}{(\rho + B)(\rho + B + 6D)}, \qquad (7.74)$$

Assuming infinite reserves this expression can be further simplified to

$$\frac{\partial E^*(x, t)}{\partial \hat{r}} = \frac{-\alpha v(x)L(x)}{(-\lambda\lambda_{T_0}^*)^2} \frac{6\hat{r}DT_{b2}\langle P_2, d\rangle}{(\rho + B)(\rho + B + 6D)}. \qquad (7.75)$$

The following proposition can then be stated.

Proposition 2 *Under the two-mode approximation of temperature dynamics, the impact on the shadow value of the mean global temperature from an increase in the heat transport across locations is given by (7.74) and the impact on fossil fuel use by (7.73), or (7.75).*

Since, following North et al. (1981) $T_{b2} < 0$, the sign of $\frac{\partial \lambda_{T_0}^*}{\partial \hat{r}}$ and consequently the sign of $\frac{\partial E^*(x, t)}{\partial \hat{r}}$, depend on the sign of $\langle P_2, d\rangle = \int_0^1 P_2(x)v(x)L(x)\phi(x)dx$, which reflects the interaction of Nature dynamics with the socioeconomic factors. An approximation of this derivative with numerical estimations was provided in the previous section.

4.1.2 Heat transport and the social price of fossil fuels

We turn now to the derivative $\frac{\partial \lambda_R(x, 0)}{\partial \hat{r}}$ which reflects the impact of heat transfer on the social price of finite fuel reserves. This is characterized in the following proposition.

Proposition 3 *Let $\xi = -\lambda\lambda_{T_0}^*$ be the social price of the climate externality which is independent of heat transfer when $\hat{r} = 0$. Then the sign of $\frac{\partial \lambda_R(x, 0)}{\partial \hat{r}}$ is opposite to the sign of $\xi' = \lambda\frac{\partial \lambda_{T_0}^*}{\partial \hat{r}}$.*

For the proof see Appendix.
When the social price of the climate externality changes because we account for heat

---

[17] To denote derivatives with respect to $\hat{r}$, we use either the straightfoward notation $\partial y/\partial \hat{r}$, or y_z', where y denotes a function like E and z is an index like R. This is to avoid complex notations.

transfer, the social price of the finite fuel reserves will move towards the opposite direction. Thus, if ξ goes down, the social price of a finite fossil fuel reserve should go up, because there is a tendency to extract more and vice versa.

4.1.3 Temperature paths and polar amplification

The identification of such a potential impact is important since our spatial model allows us to determine the characteristics of the temperature anomaly at the Poles, i.e., at $x = \pm 1$. An increase in the temperature anomaly at the Poles is related to the phenomenon of polar amplification (PA), which increases the loss of Arctic Sea ice relative to the case where PA is not present. This in turn has consequences for melting land ice and other effects. There is growing evidence suggesting a link between more rapid Arctic warming relative to the warming of the Northern hemisphere mid-latitudes when Global Mean Yearly Temperature (GMT) increases. This phenomenon has been called Arctic amplification and is expected to increase the frequency of extreme weather events (Francis and Vavrus, 2014). Melting land ice associated with a potential meltdown of Greenland and West Antarctica ice sheets due to polar amplifications might cause serious global sea level rise. It is estimated that the Greenland ice sheet holds an equivalent of 7 metres of global sea level rise, while the West Antarctica ice sheet holds the potential for up to 3.5 metres of global sea level rise (see Lenton et al., 2008).[18] On the other hand, the loss of Arctic Sea ice due to Arctic amplification may generate economic benefits by making possible the exploitation of natural resources and fossil fuel reserves which are not accessible now because of the sea ice. Thus, any polar or Arctic amplification implied by welfare maximization in the context of the spatial climate model should be taken into account.

Proposition 4 *Assuming infinite fossil fuel reserves, an increase in $\hat{r}$ in the neighborhood of $\hat{r} = 0$ is associated with an increase in PA at the North Pole for the socially optimal temperature path if the increase in $\hat{r}$ reduces the social price of the climate externality. If the increase in $\hat{r}$ increases the social price of the climate externality, then there is no association with PA. The impact from an increase in $\hat{r}$ on the Equator's temperature ($x = 0$) is ambiguous.*

For proof see Appendix.

The impact of heat transport on the social price of the climate externality depends on socioeconomic as well as natural factors. Therefore, PA may emerge from an optimization model as a result of specific choices like welfare weights or existing conditions, such as the distribution of population or production damages from climate change across latitudes. It should be noted that if we assume symmetry between the two hemispheres, the result of this proposition can be extended to the South Pole.

The potential generation of extra costs and benefits to mid-latitudes due to PA resulting from the optimizing model should be taken into account by fine tuning the spatial damage function. A damage function which includes damages to latitude x caused by spillovers

---

[18] In the discussion about tipping points, it has been stressed that the time scale of melting of the Greenland ice sheet is much longer than Arctic Sea ice melting. However, the Antarctic ice sheet could melt very fast once it gets started, but it will take an increase of 5°C of surface temperature for a serious destabilization.

146 *Handbook on the economics of climate change*

from temperature increases at other latitudes z, e.g., melting of land ice and potential indirect effects caused by melting of sea ice, can be written as:

$$\phi(x; \{T_0(t) + T_2(t)P_2(z)\}_{z=0}^{z=1})[T_0(t) + T_2(t)P_2(x)]. \tag{7.76}$$

Damages from increased melting of land ice is a flow variable rather than a stock variable, so the flow of damages should depend upon the flow of melted land ice which depends, in turn, on the volume of available ice to melt. Consider the following high-latitude belt temperature index:

$$I(T_0(t), T_2(t); z_c) \equiv \int_{z > z_c} (T_0(t) + T_2(t)P_2(z))dz = \tag{7.77}$$

$$\int_{z \in [z_c, 1]} (T_0(t) + T_2(t)P_2(z))dz = (1 - z_c)\left[T_0(t) + \frac{T_2(t)}{2}\right] z_c(1 + z_c).$$

Then, the damage function (7.76) where the high-latitude temperature anomaly affects mid-latitude damages can be specified as:

$$\phi(I(T_0(t), T_2(t); z_c); x)[T_0(t) + T_2(t)P_2(x)]. \tag{7.78}$$

It is plausible to assume that $\phi(\cdot)$ is positive and increasing in the index $I(T_0(t), T_2(t); z_c)$ for latitudes in the set $\{z: z \leq z_c\}$. Note that $\phi(\cdot)$ might even be negative for some high latitudes because of the potential opening of new shipping lanes and the potential opening of access to previously inaccessible natural resources and fossil fuel reserves. PA effects could become substantial if warming continues, i.e. $T_0(t)$ continues to increase.

Using (7.78), the current value Hamiltonian (7.30) for the two-mode approach becomes

$$H = \int_{x=0}^{x=1} v(x)L(x)\{\ln[E^a(1 - A)]$$

$$- \phi(I(T_0(t), T_2(t); z_c); x)[T_0(t) + T_2(t)P_2(x) - bA]] - \lambda_R(t)E(t, x)\}dx +$$

$$\lambda_{T_0}(t)[-BT_0(t) + \lambda E(t)] +$$

$$\lambda_{T_2}(t)[(-B - 6D)T_2(t) - \hat{r}D\lambda_2T_0(t)T_{b2}].$$

PA affects the costate variables for the two temperature modes T_0, T_2 which are now modified, relative to (7.34)–(7.35) and evolve according to

$$\dot{\lambda}_{T_0} = (\rho + B)\lambda_{T_0} + 6\hat{r}DT_{b2}\lambda_{T_2} + \left\langle vL, -\frac{\partial I}{\partial T_0}\right\rangle \tag{7.79}$$

$$\frac{\partial I}{\partial T_0} = \phi(I) + \frac{\partial \phi}{\partial I}z_c(1 - z_c)(1 + z_c)T_0$$

$$\dot\lambda_{T_2} = (\rho + B + 6D)\lambda_{T_2} + \left\langle vL, -\frac{\partial I}{\partial T_2}\right\rangle \qquad (7.80)$$

$$\frac{\partial I}{\partial T_2} = \phi(I)P_2 + \frac{\partial \phi}{\partial I}z_c(1-z_c)(1+z_c)P_2T_2.$$

The impact of PA is captured by the terms

$$\left\langle vL, -\frac{\partial I}{\partial T_0}\right\rangle, \left\langle vL, -\frac{\partial I}{\partial T_2}\right\rangle. \qquad (7.81)$$

Although it is difficult to provide analytical results at this stage, it is clear that the PA will affect the shadow values of the two temperature modes and, through them, the social price of the climate externality and the optimal temperature path. It is worth noting that PA effects are determined by socioeconomic factors and nature dynamics. Calibration might provide a quantification of all these effects but the insight obtained is clear.

4.1.4 Growth effects

Recent work by Moyer et al. (2014), Dietz and Stern (2015), Moore and Diaz (2015), and Hof (2015) has stressed the potentially large impacts of climate change on the growth of economic output as well as on the level of economic output especially in poorer economies (Moore and Diaz (2015), Hof (2015)). We take this effect into account as follows. In (7.23) we assume the $y(x, t)$ component of output is given by

$$y(x,t;T(x,t)) = y_0(x,t)\exp[(g_0(x) - g_1(x)T(x,t))t]\exp[-\phi(x)T(x,t)]. \qquad (7.82)$$

Thus, the temperature anomaly reduces local growth rate $g_0(x)$ by $g_1(x)T(x,t)$. The relevant Hamiltonian for optimization using the two-mode approximation is:

$$H = \qquad (7.83)$$

$$\int_{x=0}^{x=1} \{v(x)L(x)[\alpha\ln E(x,t) + \ln(1-A) - g_1(x)[T_0(t) + T_2(t)P_2(x)]t]$$

$$-\phi(x)[T_0(t) + T_2(t)P_2(x) - bA] - \lambda_R(t)E(t,x)\}dx +$$

$$\lambda_{T_0}(t)[-BT_0(t) + \lambda E(t)] +$$

$$\lambda_{T_2}(t)[(-B-6D)T_2(t) - 6\hat{r}DT_0(t)T_{b2}(t)].$$

The costate equations for the temperature modes are now given by

$$\dot\lambda_{T_0} = (\rho + B)\lambda_{T_0} + 6\hat{r}DT_{b2}\lambda_{T_2} + \langle 1, d_g\rangle \qquad (7.84)$$

$$\dot{\lambda}_{T_2} = (\rho + B + 6D)\lambda_{T_2} + \langle P_2, d_g \rangle \tag{7.85}$$

$$d_g = \phi(x)v(x)L(x) + v(x)L(x)g_1(x)t, \tag{7.86}$$

$$\langle 1, d_g \rangle = \int_{x=0}^{x=1} [\phi(x)v(x)L(x) + v(x)L(x)g_1(x)t]dx \tag{7.87}$$

with forward solutions

$$\lambda_{T_0}(t) = -\int_{s=t}^{\infty} e^{-(\rho+B)(s-t)}[6\hat{r}DT_{b2}\lambda_{T_2}(s) + \langle 1, d_g \rangle]ds \tag{7.88}$$

$$\lambda_{T_2}(t) = -\int_{s=t}^{\infty} e^{-(\rho+B+6D)(s-t)}[\langle P_2, d_g \rangle]ds. \tag{7.89}$$

From (7.88) and (7.89), it can be seen that the impact of heat/moisture transport on the costate $\lambda_{T_0}(t)$ for T_0, which determines the social cost of climate change externality, is "magnified" through the channel $6\hat{r}DT_{b2}\lambda_{T_2}(s)$. The evolution of the costate $\lambda_{T_2}(s)$, which determines this impact along with the climate parameters $\hat{r}DT_{b2}$, is determined by the socioeconomic factors $\phi(x)v(x)L(x)$ and the growth effect $v(x)L(x)g_1(x)t$. The growth effects of climate could therefore be important in characterizing optimal paths for fossil fuel emissions and policy instruments. Their quantitative impact is undoubtedly an interesting area for further research.

5. OPTIMAL CLIMATE CHANGE POLICIES

5.1 Fossil Fuel Taxes

The solution of the welfare maximization problem allows us to obtain some insight into the structure of optimal fuel taxes, or equivalently, optimal carbon emission taxes. A representative firm produces output using emissions or, equivalently, fossil fuels according to the production function $y(x,t)E(x,t)^\alpha$, and faces a fossil fuel tax (or carbon tax) $\tau(x,t)$.[19] Then the profit maximizing path of fossil fuel use $E(x, t)$ is determined by

$$E^0(x, t) = \underset{E(x,t)}{\operatorname{argmax}} \{y(x, t)E(x, t)^\alpha - \tau(x, t)E(x, t)\}, \tag{7.90}$$

with

$$E^0(x, t) = \left(\frac{\tau(x, t)}{\alpha y(t, x)}\right)^{\frac{1}{\alpha-1}} \text{ and} \tag{7.91}$$

---

[19] To simplify things, we assume that competitive markets exist so that output is sold at a competitive world price normalized to one, while fossil fuels are bought at a competitive world price p_F that satisfies the arbitrage condition $(\dot{p}_F(t)/p_F(t)) = r(t)$, where $r(t)$ denotes the world interest rate. Thus τ should be interpreted as including the exogenously determined fossil fuel price.

Consider now the problem of the social planner whose objective is to maximize

$$\int_{x=0}^{x=1} v(x)L(x)[\ln C(x,t) - \phi(I(T_0(t), T_2(t); z_c); x)[T_0(t) + T_2(t)P_2(x)]]dx, \quad (7.92)$$

subject to climate and resource availability constraints, where $C(t, x)$ is per capita consumption at latitude x and time t. The planner chooses an emission tax $\tau(x, t)$ for each latitude and then the representative firm in each latitude takes this tax as parametric and determines fossil fuel use to maximize latitude payoff according to (7.91). Taxes collected are given by $\tau(x, t)E^0(x, t)$. In a competitive equilibrium, the lump sum transfers from the social planner back to the consumers at latitude x at date t are equal to the taxes collected at this latitude and are given by

$$T_{tran}(x, t) = \tau(x, t)E^0(x, t) = \tau(x, t)\left(\frac{\tau(x,t)}{ay(t,x)}\right)^{\frac{1}{a-1}}. \quad (7.93)$$

Hence in equilibrium, consumption at latitude x is

$$C(t, x) = [y(x, t)E^0(x, t)^a - \tau(x, t)E^0(x, t)] + T_{tran}(x, t) \quad (7.94)$$

$$C(t, x) = y(t, x)\left(\frac{\tau(x,t)}{ay(t,x)}\right)^{\frac{1}{a-1}} = y(t, x)^{\frac{1}{1-a}}\tau(x, t)^{\frac{a}{1-a}}a^{\frac{a}{1-a}}. \quad (7.95)$$

With consumption determined in terms of the fossil fuel tax by (7.95), the social planner acting as a Stackelberg leader chooses the spatiotemporal path for the fossil fuel tax $\tau(x, t)$ to maximize the integral of discounted values of optimized objectives (7.92), subject to climate and resource availability constraints.[20] The current value Hamiltonian function for this problem is defined as:

$$H = \int_{x=0}^{x=1} v(x)L(x)\left\{\ln\left[y(t, x)^{\frac{1}{1-a}}\tau(x, t)^{\frac{a}{1-a}}a^{\frac{a}{1-a}}\right] - \right. \quad (7.96)$$

$$-\phi(I(T_0(t), T_2(t); z_c); x)[T_0(t) + T_2(t)P_2(x)] - \lambda_R(t)\left(\frac{\tau(x,t)}{ay(t,x)}\right)^{\frac{1}{a-1}}\right\}dx +$$

$$\lambda_{T_0}(t)\left[-BT_0(t) + \lambda\int_{x=0}^{x=1}\left[\left(\frac{\tau(x,t)}{ay(t,x)}\right)^{\frac{1}{a-1}}\right]dx\right] +$$

$$\lambda_{T_2}(t)[(-B - 6D)T_2(t) - \hat{r}D\lambda_2 T_0(t)T_{b2}].$$

To provide a first insight into the optimal tax, we consider the simplest possible case where there are infinite reserves and damages are independent of the high-latitude index $I(T_0(t), T_2(t); z_c)$. In this case the optimal tax is determined as

$$\tau^*(x, t) = \arg\max_{\tau}\left\{v(x)L(x)\ln\left[y(t, x)^{\frac{1}{1-a}}\tau(x, t)^{\frac{a}{1-a}}a^{\frac{a}{1-a}}\right]\right\} \quad (7.97)$$

---

[20] To simplify the exposition, we do not consider adaptation expenses.

$$+ \lambda_{T_0}(t)\lambda\left(\frac{\tau(x,t)}{ay(t,x)}\right)^{\frac{1}{\alpha-1}}\Bigg\}, \qquad (7.98)$$

which results in

$$\tau^*(x,t) = \alpha^\alpha (v(x)L(x))^{\alpha-1} y(x,t) (-\lambda\lambda_{T_0}(t))^{1-\alpha}. \qquad (7.99)$$

For the simplest case where $\hat{r} = 0$, the optimality conditions from (7.96) imply that at a steady state, $\lambda_{T_0} = -\frac{\langle 1,d\rangle}{(\rho+B)}$ and therefore

$$\tau^*(x,t) = \Lambda(x) y(x,t) \qquad (7.100)$$

$$\Lambda(x) = \left[\alpha^\alpha (v(x)L(x))^{\alpha-1}\left(\frac{\lambda\langle 1,d\rangle}{(\rho+B)}\right)^{1-\alpha}\right]. \qquad (7.101)$$

Hence although the steady-state social price of the climate externality, i.e., $-\frac{\langle 1,d\rangle}{(\rho+B)}$, is independent of location, the optimal steady-state fossil fuel tax is linear in $y(x,t)$ which can be interpreted as the output-productivity component of location x. Thus, there are two sources of spatial dependence for the optimal fossil fuel tax. The first is through the proportionality factor $\Lambda(x)$ of $y(t,x)$, which depends on different welfare weights and population across latitudes. The second is the output-productivity component $y(x,t)$. Note that even if welfare weights and population differences across latitudes are ignored, e.g. $v(x)L(x) = 1$, the spatial differentiation of the fossil fuel tax is introduced by spatial differences in the output-productivity component.

In the more general case in which spatial heat transport is taken into account and damages depend on the high-latitude index, i.e., we have the case $\phi(x,I)$, then, using (7.79) and (7.80), the steady-state values for λ_{T_0} and λ_{T_2} are

$$\bar{\lambda}_{T_0} = -\frac{6\hat{r}DT_{b2}\bar{\lambda}_{T_2} + \left\langle vL, -\frac{\partial I}{\partial T_0}\right\rangle}{(\rho+B)}, \bar{\lambda}_{T_2} = -\frac{\left\langle vL, -\frac{\partial I}{\partial T_2}\right\rangle}{(\rho+B+6D)}. \qquad (7.102)$$

When (7.102) is used to determine the optimal fossil fuel tax given by (7.99), it is clear that the fuel tax will be adjusted both for spatial heat transport, by the term $6\hat{r}DT_{b2}\bar{\lambda}_{T_2}$, and optimal PA effects, by the terms $\frac{\partial I}{\partial T_0}, \frac{\partial I}{\partial T_2}$.

It should be noted that for any given distribution of welfare weights $v(x)$ and population $L(x)$, poorer latitudes, i.e., latitudes with a relatively lower output-productivity component $y(t,x)$, are taxed less per unit emissions than richer latitudes. This result should be contrasted with the result derived under the standard assumption of compensatory transfers which indicates that a unit of emissions is taxed the same no matter which latitude belt emitted it.

In the finite reserve case it can easily be seen that the optimal tax will be

$$\tau^*(x,t) = \alpha^\alpha (v(x)L(x))^{\alpha-1} y(x,t) (\lambda_R(x,t) - \lambda\bar{\lambda}_{T_0}\lambda)^{1-\alpha}. \qquad (7.103)$$

Furthermore the exact impact of spatial heat transport on the the optimal fossil fuel tax is given by

$$\frac{\partial \tau^*(x, t)}{\partial \hat{r}} = \qquad (7.104)$$

$$(1 - \alpha)\alpha^\alpha (v(x)L(x))^{\alpha-1} y(x, t) (\lambda_R(x, t) - \lambda\bar{\lambda}_{T_0})^{-\alpha} (\lambda'_R(x, t) - \lambda\bar{\lambda}'_{T_0}).$$

5.2 Equilibrium Price of Permits

Chichilnisky and Sheeran (2009) have written an important book on carbon markets which stresses the "two sided coin" feature of carbon markets: (i) Efficiency objectives can be achieved by competitive equilibrium pricing on a world market of emissions permits, e.g., a uniform world market price on such markets helps prevent "carbon leakage" and other problems caused by different carbon prices/taxes in different locations; and (ii) Equity can be achieved by allocating more permits to more deserving countries.[21]

We explore their type of carbon market in our model where permits are allocated to latitudes and latitudes are treated as sovereigns. While latitudes are not countries, data on income distribution by latitude can be used to illustrate effects of allocation of permits to latitudes by income of latitudes and also to illustrate the effects of heat/moisture transport on the optimal number of emissions permits. Since uncertainty is absent, we can't address most issues raised in the debate between carbon taxes and carbon permit markets, e.g., Weitzman (2014), or say anything about how well markets will perform in the real world (Schmalensee and Stavins, 2015). Here, we just take a look at the issue of implementing Pareto Optima that are desired by a welfare optimizing planner who, perhaps, assigns higher welfare weights to more "deserving" latitudes, e.g., rapidly industrializing poorer latitudes that have not emitted much in the past relative to industrialized latitudes.

Let $P(x, t)$ denote the number of emission permits allocated to latitude x at date t. We choose units so that one permit is equivalent to emissions into the atmosphere by one unit of input of fossil fuels, E.

Assume latitude x at date $t = 0$ chooses emissions that maximize the Lagrangian,

$$\int_{t=0}^{\infty} e^{-\rho t}\alpha \ln E(x, t)\, dt + \mu_x \left(\int_{t=0}^{\infty} p(t)P(x, t)\, dt - \int_{t=0}^{\infty} p(t)E(x, t)\, dt \right), \qquad (7.105)$$

where $p(t)$ is the world market price of an emissions permit at date t. The FONC of optimization imply that

$$E(x, t) = \frac{\alpha e^{-\rho t}}{\mu_x p(t)}. \qquad (7.106)$$

We assume emissions markets are working well enough and that there are no impediments or obstructions that get in the way of permits trading at a uniform price at all latitudes. The total supply of permits at each date t is given by

---

[21] See also Chichilnisky's (2015) discussion of carbon markets.

152 *Handbook on the economics of climate change*

$$\int_{x=0}^{x=1} L(x)P(x)dx, \qquad (7.107)$$

where $L(x)$ is the population of latitude x which is assumed to be constant for simplicity, and to avoid notation clutter. It is easy to generalize the treatment here to growing populations and changing populations. Total demand by latitude x at date t is given by

$$L(x)E(x,t) = L(x)\left[\frac{\alpha e^{-\rho t}}{\mu_x p(t)}\right], \qquad (7.108)$$

while global demand at date t is given by

$$\int_{x=0}^{x=1} L(x)E(x,t)dx = \int_{x=0}^{x=1} L(x)\left[\frac{\alpha e^{-\rho t}}{\mu_x p(t)}\right]dx. \qquad (7.109)$$

Suppose $\{E^*(x,t), x \in [0,1], t \in [0,\infty]\}$ is a desired solution by a planner, e.g. a solution to a welfare optimization problem where poor latitudes are weighted more heavily than rich latitudes, or latitudes that have emitted more relative to others in the past during industrialization are weighted less heavily by appropriately defined weights $v(x)$.

We investigate here whether the desired solution can be implemented by choosing an allocation $\{P(x,t), x \in [0,1], t \in [0,\infty)\}$ of permits and opening a world market for permits as in Chichilnisky and Sheeran (2009). We try the allocation $P(x,t) = E^*(x,t)$ for all (x,t). Equating demand and supply at date t gives us

$$\int_{x=0}^{x=1} L(x)\left[\frac{\alpha e^{-\rho t}}{\mu_x p(t)}\right]dx = \int_{x=0}^{x=1} P(x,t)dx = \int_{x=0}^{x=1} E^*(x,t)dx, \qquad (7.110)$$

recalling that the planner's optimal solution for emissions by each latitude is given by

$$E^*(x,t) = \frac{v(x)L(x)}{\lambda_{R_0}^* e^{\rho t} - \lambda \lambda_{T_0}^*}. \qquad (7.111)$$

Hence,

$$p(t) = \frac{\alpha e^{-\rho t}\int_{x=0}^{x=1}(L(x)/\mu_x)dx}{\int_{x=0}^{x=1} E^*(x,t)dx} = \qquad (7.112)$$

$$\frac{\alpha[\lambda_{R_0}^* - \lambda \lambda_{T_0}^* e^{-\rho t}]\int_{x=0}^{x=1}(L(x)/\mu_x)dx}{\int_{x=0}^{x=1} v(x)L(x)dx}. \qquad (7.113)$$

Since, $\lambda_{T_0}^* < 0$ for $\hat{r} = 0$, we see that in this case $p(t)$ decreases to the asymptotic value, $\lambda_{R_0}^*$ as $t \to \infty$. When "space matters", i.e. $\hat{r} > 0$, we see from (7.43), (7.44) that the forward solutions for the costate variables are

$$\lambda_{T_0}^*(t) = -\int_{s=t}^{\infty} e^{-(\rho+B)t}[6\hat{r}DT_{b2}\lambda_{T_2}^*(s) + \langle 1, d(x)\rangle]ds \qquad (7.114)$$

$$\lambda_{T_2}^*(t) = -\int_{s=t}^{\infty} e^{-(\rho+B+6D)t}[\langle P_2, d(x)\rangle]ds \qquad (7.115)$$

$$d(x) = \phi(x)v(x)L(x)\phi(x). \qquad (7.116)$$

Hence, in this model, spatial transport impacts the equilibrium trading price of emissions permits in the Chichilnisky/Sheeran market in a way that can be computed in closed form, once the marginal damage function $d(x)$ is known.

It is of interest to calculate the equilibrium level of wealth, W_x, of each latitude under this allocation scheme. The wealth of latitude x is given by

$$W_x \equiv \int_{t=0}^{\infty} p(t)P(x,t)dt = \qquad (7.117)$$

$$\int_{t=0}^{\infty} \left\{ \frac{a[\lambda_{R_0}^* - \lambda\lambda_{T_0}^*(t)e^{-\rho t}]\int_{x=0}^{x=1}(L(x)/\mu_x)dx}{\int_{x=0}^{x=1} v(x)L(x)dx} \right\} P(x,t)dt =$$

$$\int_{t=0}^{\infty} \left\{ \frac{a[\lambda_{R_0}^* - \lambda\lambda_{T_0}^*(t)e^{-\rho t}]\int_{x=0}^{x=1}(L(x)/\mu_x)dx}{\int_{x=0}^{x=1} v(x)L(x)dx} \right\} \left\{ \frac{v(x)L(x)}{\lambda_{R_0}^* e^{\rho t} - \lambda\lambda_{T_0}^*(t)} \right\} dt =$$

$$\int_{t=0}^{\infty} e^{-\rho t} \left\{ \frac{a\int_{x=0}^{x=1}(L(x)/\mu_x)dx}{\int_{x=0}^{x=1} v(x)L(x)dx} \right\} \{v(x)L(x)\} dt =$$

$$\left(\frac{a}{\rho}\right) \left\{ \left[\int_{x=0}^{x=1}(L(x)/\mu_x)dx\right] \frac{v(x)L(x)}{\int_{x=0}^{x=1} v(x)L(x)dx} \right\}.$$

Thus, we see that this particular allocation scheme results in a total wealth ratio, between locations x and x'

$$\frac{W_x}{W_{x'}} = \frac{v(x)L(x)}{v(x')L(x')} \qquad (7.118)$$

and a per capita wealth ratio,

$$\frac{w_x}{w_{x'}} = \frac{v(x)}{v(x')}. \qquad (7.119)$$

6. CLIMATE EXTERNALITY PRICE AND "SAFETY FIRST" UTILITY

The results obtained above were based on the tractability advantages of the logarithmic utility function. In this section we seek to identify the impact on the social price of climate externality and the socially optimal use of fossil fuel under a more general utility function. In particular, we investigate the class of utilities where marginal disutility increases very fast relative to the logarithmic utility as consumption goes towards zero.

A more general utility function results in the following welfare function:

$$\int_{t=0}^{\infty} e^{-\rho t} \left[\int_{x=0}^{x=1} v(x) L(x) U\left[\frac{yE^{\alpha}e^{-\phi(x)T_{total}(x,t)}}{L(x)} \right] dx \right] dt = \quad (7.120)$$

$$\int_{0}^{\infty} e^{-\rho t} \left[\int_{-1}^{1} v(x) L(x) U\left[\frac{y(x,t)E(x,t)^{\alpha}e^{-\phi(x,t)[T_b(x)+T(x,t)]}}{L(x)} \right] dx \right] dt,$$

which is maximized by choosing the optimal path $E(x, t)$, subject to the constraints imposed by Nature dynamics and fossil fuel exhaustibility. Using the two-mode approximations and the approximations of the radiating forcing term employed above, the current value Hamiltonian for the problem is:

$$H = \int_{x=0}^{x=1} \left\{ v(x) L(x) U\left[\frac{y(x,t)E(x,t)^{\alpha}e^{-\phi(x)[T_0(t)+T_2(t)P_2(x)]}}{L(x)} \right] \right.$$

$$\left. - \lambda_R(t) E(t,x) \right\} dx +$$

$$\lambda_{T_0}(t) [-BT_0(t) + \lambda E(t)] +$$

$$\lambda_{T_2}(t) [(-B - 6D) T_2(t) - \hat{r} D \lambda_2 T_0(t) T_{b2}].$$

The FONC resulting from the maximum principle, after suppressing the (x, t) arguments to ease notation when necessary, are presented below. The optimal emission path $E^*(x, t)$ satisfies

$$\frac{\alpha v(x) L(x) [U'(\widetilde{C}(x,t)) \widetilde{C}(x,t)]}{E^*(x,t)} = \lambda_R(t) - \lambda \lambda_{T_0}(t) \quad (7.121)$$

$$\widetilde{C}(x,t) = y(x) E^*(x,t)^{\alpha} e^{-\phi(x)[T_0(t)+T_{b0}+(T_2(t)+T_{b2})P_2(x)]}. \quad (7.122)$$

We use $\widetilde{C}(x, t)$ to denote the output of the economy. We assume that this output is consumed, but the consumption value has been damaged by climate damages reflected in the exponential term. The costate variables evolve according to:

$$\dot\lambda_{T_0} = (\rho + B)\lambda_{T_0} + \langle vL, \phi U'(\tilde C)\tilde C\rangle + \lambda_{T_2} 6\hat r DT_{b2} \tag{7.123}$$

$$\dot\lambda_{T_2} = (\rho + B + 6D)\lambda_{T_2} + \langle vL, \phi P_2 U'(\tilde C)\tilde C\rangle \tag{7.124}$$

$$\dot\lambda_R(t) = \rho\lambda_R(t). \tag{7.125}$$

The optimality conditions for temperature dynamics, externality dynamics and the fossil fuel constraints are the same as (7.37)–(7.40).

If the heat transport is ignored, i.e. $D = 0$, the costate variables evolve according to:

$$\dot\lambda_{T_0} = (\rho + B)\lambda_{T_0} + \langle vL, \phi U'(\tilde C)\tilde C\rangle \tag{7.126}$$

$$\dot\lambda_{T_2} = (\rho + B)\lambda_{T_2} + \langle vL, \phi P_2 U'(\tilde C)\tilde C\rangle \tag{7.127}$$

$$\dot\lambda_R(t) = \rho\lambda_R(t),$$

with forward solutions

$$\lambda_{T_0} = -\int_{s=0}^{\infty} e^{-(\rho+B)(s-t)}\langle v(x)L(x), \phi(x)U'(\tilde C)\tilde C\rangle ds \tag{7.128}$$

$$\lambda_{T_2} = -\int_{s=0}^{\infty} e^{-(\rho+B+6D)(s-t)}\langle v(x)L(x), \phi(x)P_2(x)U'(\tilde C)\tilde C\rangle ds.$$

Conditions (7.123)–(7.124) indicate that the neat property of the log utility function obtained above is lost because of the term $U'(\tilde C)\tilde C$ which emerges when general utility functions are used. In order to obtain some analytical results, we consider the class of utility functions

$$U(C) = \frac{C^{1-\gamma}}{1-\gamma}, \tag{7.129}$$

where γ is both the coefficient of relative risk aversion and (minus) the elasticity of marginal utility with respect to consumption, while the log utility function is the special case $\gamma = 1$. For $\gamma > 1$, we call the class of utilities "safety first" because in this case, when the consumption value is damaged due to climate change, the disutility increases faster than the logarithmic utility for which $\gamma = 1$. In the same context, an increase of γ from the value of one implies an increase in the relative risk aversion. For this class of utility functions, we have $U'(\tilde C)\tilde C = \tilde C^{1-\gamma}$. The main question is whether an increase in the coefficient of relative risk aversion from the value of one will have an impact on the social price of the climate externality and the socially optimal fossil fuel path.

Using (7.129), the optimality condition for the optimal choice of fossil fuel use becomes

$$\frac{av(x)L(x)\tilde C(x,t;\hat r,\gamma)^{1-\gamma}}{E(x,t;\hat r,\gamma)} = \lambda_R(t) - \lambda\lambda_{T_0}. \tag{7.130}$$

Differentiating (7.130) with respect to γ, evaluating the derivatives at $(\hat{r}, \gamma) = (0, 1)$ and using $\frac{\partial \bar{C}^{1-\gamma}}{\partial \gamma} = -\ln \bar{C}$, and suppressing $(\hat{r}, \gamma)$ to ease notation, we obtain

$$\alpha v(x) L(x) \left[-\frac{1}{E(x, t)^2} \frac{\partial E(x, t)}{\partial \gamma} - \frac{\ln \bar{C}}{E(x, t)} \right] = \frac{\partial \lambda_R(t)}{\partial \gamma} - \frac{\lambda \lambda_{T_0}}{\partial \gamma}. \quad (7.131)$$

To identify the impact of increasing γ from the value $\gamma = 1$ on the social price of the climate externality, we consider expansions of any endogenous variable $\zeta(t; \hat{r}, \gamma)$ of our model with respect to $(\hat{r}, \gamma)$ around the point $(\hat{r}, \gamma) = (0, 1)$, or

$$\zeta(t; \hat{r}, \gamma) = \zeta(t; 0, 1) + \frac{\partial \zeta(t; 0, 1)}{\partial \hat{r}} \hat{r} + \frac{\partial \zeta(t; 0, 1)}{\partial \gamma} (\gamma - 1) + o(\hat{r}, |\gamma - 1|). \quad (7.132)$$

Since we are interested in the social price of the climate externality and the use of fossil fuels, we consider the following expansions

$$\lambda_{T_0}(t; \hat{r}, \gamma) = \lambda_{T_0}(t; 0, 1) + \frac{\partial \lambda_{T_0}(t; 0, 1)}{\partial \hat{r}} \hat{r} + \frac{\partial \lambda_{T_0}(t; 0, 1)}{\partial \gamma} (\gamma - 1) + \quad (7.133)$$

$$o(\hat{r}, |\gamma - 1|)$$

$$E(t; \hat{r}, \gamma) = E(t; 0, 1) + \frac{\partial E(t; 0, 1)}{\partial \hat{r}} \hat{r} + \frac{\partial E(t; 0, 1)}{\partial \gamma} (\gamma - 1) + \quad (7.134)$$

$$o(\hat{r}, |\gamma - 1|),$$

which approximate the climate externality price and the fossil fuel use. Using these expansions, we can state the following result.

Proposition 5 *Assuming no serious poverty at any location at any time, so that $\ln \bar{C}(x, t) > 0$ for all (x, t) and $\lambda_R(t) = 0$ for all t, then a small increase in the coefficient of relative risk aversion γ from $\gamma = 1$ will reduce the social price of the climate externality $-\frac{\partial \lambda_{T_0}(v Y; 0, 1)}{\partial \gamma} = \int_{s=v}^{\infty} v e^{-\rho(s-v)} \langle v L, \phi \ln \bar{C} \rangle(s) ds < 0.$*

For proof see Appendix.

Since in the safety-first class of utilities, climate damages in the utility function are realized through damages in the value of consumption, and recalling that $C = y E^u e^{-\phi T}$ and $U(C) = C^{1-\gamma}/(1-\gamma)$, it is reasonable to expect that an increase in γ from $\gamma = 1$ will reduce the price of the climate externality and the corresponding fuel tax when the stock of fossil fuels is assumed to be infinite. It should also be noted that the impact of the safety-first utility, as quantified by the derivative $\frac{\partial \lambda_{T_0}(v; 0, 1)}{\partial \gamma}$, depends on the socioeconomic factors $v(x), L(x), \phi(x)$ and the value of consumption $\bar{C}$ adjusted for climate change damages.

6.1 Consumption Discount Rates under Spatial Heat Transfer

The previous discussion made clear that allowing for spatial heat transfer has an impact on the social cost of the externality and the associated policy instruments. A question that emerges in this context is what the impact of heat transfer is on the discount rate used for discounting future flows of consumption costs and benefits. As is well known (e.g. Arrow et al. (2014) or Gollier (2007)) the consumption rate of interest, r_t, is defined in the context of the Ramsey rule as:

$$r_t = \rho - \frac{d}{dt} \ln \frac{\partial U(C(t))}{\partial C(t)}. \tag{7.135}$$

Consider the case where each location x is regarded as a "closed economy" in which case the consumption rate of interest can be a local equilibrium rate, that is,

$$r_t(x) = \rho - \frac{d}{dt} \ln \frac{\partial U(C(x,t))}{\partial C(x,t)}. \tag{7.136}$$

Define consumption after climate change damages have been accounted for by

$$\tilde{C}(x,t) = C(x,t) e^{-\phi(x)\hat{D}(x,t)}, \tag{7.137}$$

$$\hat{D}(x,t) = T_{b0}(t) + T_{b2}(t) P_2(x) + T_0(t) + T_2(t) P_2(x), \tag{7.138}$$

and consider the utility function

$$U(\tilde{C}(x,t)) = \frac{1}{1-\gamma}(C(x,t) e^{-\phi(x)\hat{D}(x,t)})^{1-\gamma}. \tag{7.139}$$

Using (7.136) we obtain:

$$r_t(x) = \rho + \gamma g(x,t) + (1-\gamma)\phi(x) \frac{d\hat{D}(x,t)}{dt} \tag{7.140}$$

$$g(x,t) = \frac{\dot{C}(x,t)}{C(x,t)}. \tag{7.141}$$

From (7.138),

$$\frac{d\hat{D}(x,t)}{dt} = \dot{T}_0(t) + \dot{T}_2(t) P_2(x), \tag{7.142}$$

since it is reasonable to assume that baseline temperature modes remain constant, i.e. $\dot{T}_{b0}(t) = \dot{T}_{b2}(t) = 0$. From the optimality conditions (7.37), (7.38)

$$\dot{T}_0 = -B T_0(t) + \lambda E^*(t) \tag{7.143}$$

$$\dot{T}_2 = -[B + 6D] T_2(t) - 6\hat{r} D T_0(t) T_{b2} \tag{7.144}$$

where $E^*(t)$ is the optimal aggregate path for fossil fuel use determined as

$$E^*(x, t) = \frac{\alpha v(x) L(x)}{\lambda_R(t) - \lambda \lambda_{T_0}^*}, \quad E^*(t) = \int_0^1 E^*(x, t) dx. \tag{7.145}$$

Conditions (7.140)–(7.145) indicate that the climate change adjustment to the discount rate reduces to zero for a logarithmic utility function. But it is not zero for the most often considered values of γ between 1.5 and 3 (Dasgupta, 2008). For $\gamma > 1$, the adjustment depends on the paths of the anomaly which are determined by socioeconomic conditions, i.e., $v(x), L(x), \phi(x)$ and Nature's spatial dynamics reflected in $P_2(t)$.

The impact of accounting for spatial heat transport in consumption discount rate, disregarding any impacts on the local consumption growth rate, is determined by

$$\frac{\partial}{\partial \hat{r}}\left(\frac{d\hat{D}(x, t)}{dt}\right) = \dot{T}_0'(t) + \dot{T}_2'(t) P_2(x). \tag{7.146}$$

In the proof of proposition 3 in the Appendix, the derivatives $\dot{T}_0'$ and $\dot{T}_2'(t)$ are explicitly calculated as

$$T_0'(t) = \lambda \int_{s=0}^t e^{B(s-t)} E'(s) ds \tag{7.147}$$

$$T_2'(t) = -6\hat{r} D T_{b2} \int_{s=0}^t e^{(B+6D)(s-t)} T_0(s) ds. \tag{7.148}$$

Thus, if $E'(s) > 0$ for all s, accounting for spatial heat transport, that is increasing $\hat{r}$ from $\hat{r} = 0$ to $\hat{r} > 0$, will tend to reduce local consumption discount rates.

If we consider the case where arbitrage will force local rates $r_t(x)$ to a global equilibrium rate r_t, the spatially average consumption rate of interest will be defined as

$$r_t = \rho - \frac{d}{dt} \ln\left\{ \int_{x=0}^{x=1} [C(x, t)^{-\gamma} e^{-\phi(x)\hat{D}(x, t)}] dx \right\}. \tag{7.149}$$

7. CONCLUDING REMARKS AND SUGGESTIONS FOR FUTURE RESEARCH

This chapter is, to our knowledge, the first chapter in climate economics to consider the combination of spatial heat transport and polar amplification. We simplified the problem by stratifying the Earth into latitude belts and assuming as in North et al. (1981) and Wu and North (2007) that the two hemispheres were symmetric so that solutions to the climate dynamics could be expanded into an infinite series of even numbered Legendre polynomials.

In order to obtain analytical tractability of the climate dynamics across latitude belts and to solve the economic infinite horizon welfare economics problem, we introduced some approximations to the climate dynamics and some specializations to specific utility functions.

First we follow recent research suggesting that global mean warming is linearly proportional to cumulative carbon emissions (e.g. Matthews at al., 2009, Pierrehumbert,

2012–2013; 2014). Second, we truncated the Legendre polynomial expansion of the climate dynamics to a small number of modes. Third, we built upon work by Alexeev et al. (2005), Langen and Alexeev (2007), and Alexeev and Jackson (2012) to motivate our specification of the heat transport function across latitudes as a function of global average temperature. This specification imparts a nonlinearity which we approximated by series expansion around the case of no polar amplification where heat transport is linear.

In this chapter, we use logarithmic utility and exponential specification of climate damages as a function of temperature, except in Section 6 where we use a more general utility function. We analyzed spillover effects from higher latitudes onto lower latitudes because of amplification of warming on the higher latitudes. Our main contributions are the following.

First, we showed that it is possible to build climate economic models that include the very real climatic phenomena of heat transport and high latitude amplification of warming (i.e. "polar amplification") and still maintain analytical tractability. Since analytical tractability is essential for understanding the output of more complicated and realistic models, we view this as an important – maybe the most important – contribution of this line of research. It is interesting to note the importance of the work by North and others in showing how models with spatial transport in climate dynamics can be made analytically tractable by use of the "right" mathematics, e.g., bases of even number Legendre polynomials and spherical harmonics. This kind of work is used heavily to understand the computational output of much more complicated and realistic climate models. We consider our work as initiating a similar line of research for the joint modeling of coupled climate dynamics and economic dynamics. We believe that our finding regarding the link between heat transfer, polar amplification, optimal fuel taxes, and permits' markets illustrates the importance of directing future research in climate change economics towards addressing the impact of spatial energy transport across the globe.

Second, we showed that the optimal tax function, i.e., the marginal social cost of emissions, depended upon the distribution not only of welfare weights but also population across latitudes, the distribution of marginal damages across latitudes and cross latitude interactions of marginal damages, along with Nature dynamics. These dynamics are reflected in the decomposition of the temperature field into modes via the expansion of the climate dynamics into a series of even numbered Legendre polynomials. The formulas we obtained are quite interpretable and comparative dynamics can be quite easily done on their components.

Third, we derived and compared optimal solutions under (i) no heat transport, (ii) heat transport but no polar amplification, and (iii) both heat transport and polar amplification.

Fourth, we compared the solution for optimal taxes under the standard assumption of compensatory transfers, so that a unit of emissions is taxed the same no matter which latitude belt emitted it, with the solution for optimal taxes in which there are no compensatory transfers at all. In this latter case the poorer latitudes are taxed less per unit emissions than richer latitudes. While this is obvious for the direction of the tax, we give a formula that shows both how the interaction of the climate system with the economic system feeds into a formula for the optimal tax per unit emissions, and the way in which optimal taxes are differentiated across locations. We also discuss the possibility that an increase in the heat transfer towards the Poles may increase or reduce fossil fuels taxes. This is an important observation because it provides a direct link between spatial heat transport in climate dynamics, which are usually disregarded in IAMs at the analytically tractable level in the simplicity hierarchy, and optimal economic policy. In the context of policy analysis, we analyze the spatial transport impacts on the equilibrium trading price of emissions permits in the Chichilnisky/Sheeran market and we calculate the equilibrium level of wealth of each latitude under this allocation scheme.

Fifth, by using a more general utility function, we showed that an increase in the coefficient of relative risk aversion from the value of one will reduce the social price of the climate externality. The more general utility function also allowed us to characterize the impact of spatial heat/moisture transfer on the discount rate which is appropriate for discounting future consumption costs and benefits.

Future research could move in different directions. The most important is that extensive computational work should be done to locate sufficient conditions for spatial heat transport and polar amplification to quantitatively matter significantly for welfare economics at different locations on the planet. We believe the ideal would be to conduct computational work like that of Cai et al. (2015) to assess the quantitative importance of taking into account heat transport. In addition, it would be valuable to extend the results in this chapter to two-dimensional space where heat transport occurs across both latitude and longitude. Brock et al. (2013) did this for the case of linear heat transport but did not include polar amplification. Another area of future research would be to extend our current chapter and the Desmet and Rossi-Hansberg (2015) paper, which addresses migration responses to climate change, to include the impact of heat and moisture transport across the globe. This research could build on the work of Desmet and Rossi-Hansberg (2010; 2015), and Boucekkine et al. (2009; 2013). Moreover, we have ignored the linkage between the dynamics of the carbon cycle and temperature dynamics in order to focus sharply on the additional impact of spatial heat transport on the temperature dynamics. Future research is needed to model the interaction of spatial heat transport with the carbon cycle and land use changes.

We conclude this chapter by hoping that our analytical results have helped make the case that a serious dynamic climate science phenomenon like spatial heat transport can be included in analytically tractable simple climate economics models. We believe that the results in this chapter suggest that spatial heat transfer and polar amplification could have a potentially important impact on climate change policy.

REFERENCES

Alexeev, V.A. and C.H. Jackson, 2013, Polar amplification: is atmospheric heat transport important?, *Climate Dynamics*, **41**(2): 533–47. DOI 10.1007/s00382-012-1601-z.

Alexeev, V.A., P.L. Langen and J.R. Bates, 2005, Polar amplification of surface warming on an aquaplanet, in "ghost forcing" experiments without sea ice feedbacks, *Climate Dynamics*, **24**(7–8): 655–66. DOI 10.1007/s00382-005-0018-3.

Arrow, K.J., M.L. Cropper, C. Gollier, B. Groom, G.M. Heal, R.G. Newell, W.D. Nordhaus, R.S. Pindyck, W.A. Pizer, P.R. Portney, T. Sterner, R.S.J. Tol and M.L. Weitzman, 2014, Should governments use a declining discount rate in project analysis?, *Review of Environmental Economics and Policy*, **8**(2): 145–63. doi:10.1093/reep/reu008.

Boucekkine, R., C. Camacho and G. Fabbri, 2013, Spatial dynamics and convergence: the spatial AK model, *Journal of Economic Theory*, **148**(6): 2719–36.

Boucekkine, R., C. Camacho and B. Zou, 2009, Bridging the gap between growth theory and the new economic geography: the spatial Ramsey model, *Macroeconomic Dynamics*, **13**(1): 20–45.

Brock, W. and A. Xepapadeas, 2017, Climate change policy under polar amplification, *European Economic Review*, **99**: 93–112. https://doi.org/10.1016/j.euroecorev.2017.06.008.

Brock, W., G. Engström and A. Xepapadeas, 2014, Spatial climate-economic models in the design of optimal climate policies across locations', *European Economic Review*, **69**: 78–103.

Brock, W.A., G.G. Engström, D. Grass and A. Xepapadeas, 2013, Energy balance climate models and general equilibrium optimal mitigation policies, *Journal of Economic Dynamics and Control*, **37**(12): 2371–96.

Burgess, R., D. Deschenes, D. Donaldson, and M. Greenstone, 2014, The unequal effects of weather and climate change: evidence from mortality in India, Cambridge, USA: Massachusetts Institute of Technology, Department of Economics, available at: http://www.lse.ac.uk/economics/people/facultyPersonalPages/facultyFiles/RobinBurgess/UnequalEffectsOfWeather.

Cai, Y., K.L. Judd and T.S. Lontzek, 2015, The social cost of carbon with economic and climate risks, Working Papers 1504.06909, arXiv.org, revised Apr 2015.

Castruccio, S., D.J. McInerney, M.L. Stein, F. Liu Crouch, R.L. Jacob and E.J. Moyer, 2014, Statistical emulation of climate model projections based on precomputed GCM runs, *Journal of Climate*, **27**(5): 1829–44.

Chichilnisky, G., 2015, The need for sustainable development and a carbon market: avoiding extinction, in Bernard, L. and W. Semmler (eds), *The Oxford Handbook of the Macroeconomics of Global Warming*, Oxford, UK: Oxford University Press.

Chichilnisky, G. and K. Sheeran, 2009, *Saving Kyoto*, London, UK: New Holland.

Dasgupta, P., 2008, Discounting climate change, *Journal of Risk and Uncertainty*, **37**: 141–69.

Dell, M., B. Jones and B. Olken, 2012, Temperature shocks and economic growth: evidence from the last half century, *American Economic Journal: Macroeconomics*, **4**(3): 66–95.

Desmet, K. and E. Rossi-Hansberg, 2010, On spatial dynamics, *Journal of Regional Science*, **50**(1): 53–63.

Desmet, K. and E. Rossi-Hansberg, 2015, On the spatial economic impact of global warming, *Journal of Urban Economics*, **88**: 16–27.

Dietz, S. and N. Stern, 2015, Endogenous growth, convexity of damage and climate risk: how Nordhaus' framework supports deep cuts in carbon emissions, *The Economic Journal*, **125**(583): 547–620. ISSN 1468-0297.

Francis, J. and S. Vavrus, 2014, Evidence for a wavier jet stream in response to rapid Arctic warming, *Environmental Research Letters*, **10**: 1–12.

Gollier, C., 2007, The consumption-based determinants of the term structure of discount rates, *Mathematical Financial Economics*, **1**: 81–101.

Hasselmann, K., S. Hasselmann, R. Giering, V. Ocana and H. V. Storch, 1997, Sensitivity study of optimal CO_2 emission paths using a simplified Structural Integrated Assessment Model (SIAM), *Climatic Change*, **7**(2): 345–86.

Hof, A., 2015, Economics: welfare impacts of climate change, *Nature Climate Change*, **5**: 99–100. doi: 10.1038/nclimate2506.

Kummu, M. and O. Varis, 2011, The world by latitudes: a global analysis of human population, development level and environment across the North–South axis over the past half century, *Applied Geography*, **31**(2): 495–507.

Langen, P.L. and V.A. Alexeev, 2007, Polar amplification as a preferred response in an idealized aquaplanet GCM, *Climate Dynamics*, **29**(2–3), 305–17. DOI 10.1007/s00382-006-0221-x.

Leduc, M., H. Damon Matthews and R. de Elia, 2016, Regional estimates of the transient climate response to cumulative CO_2 emissions, *Nature Climate Change, Letters*, **6**, 474–8. DOI: 10.1038/NCLIMATE2913.

Lenton, T., H. Held, E. Kriegler, J. Hall, W. Lucht, S. Rahmstorf and H.J. Schellnhuber, 2008, Tipping elements in the Earth's climate system, *PNAS*, **105**(6): 1786–93.

MacDougall, A.H. and P. Friedlingstein, 2015, The origin and limits of the near proportionality between climate warming and cumulative CO_2 emissions, *American Meteorological Society*, **28**: 4217–30. DOI: 10.1175/JCLI-D-14-00036.1.

Matthews, H.D., S. Solomon and R. Pierrehumbert, 2012, Cumulative carbon as a policy framework for achieving climate stabilization, *Philosophical Transactions of the Royal Society*, **370**: 4365–79.

Matthews, H.D., N.P. Gillett, P.A. Stott and K. Zickfield, 2009, The proportionality of global warming to cumulative carbon emissions, *Nature*, **459**: 829–33.

Mendelsohn, R., A. Dinar and L. Williams, 2006, The distributional impact of climate change on rich and poor countries, *Environment and Development Economics*, **11**(2): 159.

Moore, F.C. and D.B. Diaz, 2015, Temperature impacts on economic growth warrant stringent mitigation policy, *Nature Climate Change, Letters*, **5**: 127–31. DOI: 10.1038/NCLIMATE2481.

Moyer, E.J., M.D. Woolley, M.J. Glotter, N.J. Matteson and D.A. Weisbach, 2014, Climate impacts on economic growth as drivers of uncertainty in the social cost of carbon, *Journal of Legal Studies*, **43**(2): 401–25. doi 10.1086/678140.

North, G.R., 1975a, Analytical solution to a simple climate model with diffusive heat transport, *Journal of the Atmospheric Sciences*, **32**: 1301–07.

North, G.R., 1975b, Theory of energy-balance climate models, *Journal of the Atmospheric Sciences*, **32**: 2033–43.

North, G., R. Cahalan and J. Coakely, 1981, Energy balance climate models, *Reviews of Geophysics and Space Physics*, **19**(1): 91–121.
Pierrehumbert, R.T., 2012–13, Cumulative carbon and just allocation of the global carbon commons, *Chicago Journal of International Law*, **13**(2): 527–48.
Pierrehumbert, R.T., 2014, Short-lived climate pollution, *Annual Review of Earth and Planetary Sciences*, **42**: 341–79.
Schmalensee, R. and R.N. Stavins, 2015, Lessons learned for three decades of experience with cap-and-trade, FEEM Working Paper, 107.2015.
Weitzman, M., 2014, Can negotiating a uniform carbon price help to internalize the global warming externality?, *Journal of the Association of Environmental and Resource Economists*, **1**(1–2): 29–49
Wu, W. and G.R. North, 2007, Thermal decay modes of a 2-D energy balance climate model, *Tellus A*, **59**(5): 618–26.

APPENDIX

Proof of Proposition 3

Recall the optimality condition for the optimal emission path

$$E(x, t) = \frac{\alpha v(x) L(x)}{\lambda_R(x, t) + \xi(t)}, \quad \xi = -\lambda\lambda_{T_0}^*. \tag{7.150}$$

Combining this condition with the constraint of finite fossil fuel reserves in each location, we obtain

$$\alpha v(x) L(x) \int_{s=0}^{\infty} \left[\frac{1}{\lambda_R(x, 0) e^{\rho s} + \xi} \right] ds = R_0(x). \tag{7.151}$$

We evaluate the last integral and solve for $\lambda_R(x, 0)$ to obtain:

$$\lambda_R(x, 0) = \xi \Big/ \left\{ \exp\left[\frac{\rho \xi R_0(x)}{\alpha v(x) L(x)} \right] - 1 \right\}, \tag{7.152}$$

since

$$\int_{s=0}^{\infty} \left[\frac{1}{\lambda_R(x, 0) e^{\rho s} + \xi} \right] ds = \frac{1}{\rho \xi} \left[\ln\left(\frac{\lambda_R(x, 0) + \xi}{\lambda_R(x, 0)} \right) \right]. \tag{7.153}$$

Differentiating (7.150) and (7.151) with respect to $\hat{r}$, we obtain

$$E'(x, t) = -\alpha v(x) L(x) \frac{\lambda'_R(x, 0) e^{\rho s} + \xi'}{[\lambda_R(x, 0) e^{\rho s} + \xi]^2} \tag{7.154}$$

$$-\alpha v(x) L(x) \int_{s=0}^{\infty} \frac{\lambda'_R(x, 0) e^{\rho s} + \xi'}{[\lambda_R(x, 0) e^{\rho s} + \xi]^2} ds = 0. \tag{7.155}$$

Multiplying the nominator and denominator of the integral in (7.155) by $e^{-\rho s}$ and solving for $\lambda'_R(x, 0)$, we obtain:

$$\frac{\partial \lambda_R(x, 0)}{\partial \hat{r}} = \tag{7.156}$$

$$\frac{-\int_{s=0}^{\infty} \xi'\{1/[\lambda_R(x, 0) e^{\rho s} + \xi]^2\} ds}{\int_{s=0}^{\infty} \{e^{\rho s}/[\lambda_R(x, 0) e^{\rho s} + \xi]^2\} ds} = \tag{7.157}$$

$$\frac{-\xi' \int_{s=0}^{\infty} \{1/[\lambda_R(x, 0) e^{\rho s} + \xi]^2\} ds}{\int_{s=0}^{\infty} \{e^{\rho s}/[\lambda_R(x, 0) e^{\rho s} + \xi]^2\} ds} = \tag{7.158}$$

$$-\xi'\{\rho\lambda_R(x, 0)[\lambda_R(x, 0)+\xi]\}\int_{s=0}^{\infty}\left\{\frac{1}{[\lambda_R(x, 0)e^{\rho s}+\xi]^2}\right\}ds, \quad (7.159)$$

since

$$\int_{s=0}^{\infty}\frac{1}{[\lambda_R(x, 0)e^{\rho s}+\xi]^2}ds = \frac{1}{\rho\lambda_R(x, 0)[\lambda_R(x, 0)+\xi]}. \quad (7.160)$$

It follows from (7.159) that $\frac{\partial \lambda_R(x, 0)}{\partial \hat{r}}$ and $\xi' = \lambda\frac{\partial \lambda_{T_0}^*}{\partial \hat{r}}$ have opposite signs.

Proof of Proposition 4

The impact of heat transport on the optimal temperature paths requires the computation of the derivative of $T(x, t)$ with respect to $\hat{r}$ which, using the two-mode approach, is defined as:

$$T'(x, t) = T = T_0'(t) + T_2'(t)P_2(x). \quad (7.161)$$

Recall that ' mean $\partial/\partial\hat{r}$. Differentiating the optimality conditions for the state variables we obtain:

$$\dot{T}_0'(t) = -BT_0'(t) + \lambda E'(t), \ T_0'(0) = 0 \quad (7.162)$$

$$\dot{T}_2'(t) = -(B + 6D)T_2' - 6\hat{r}DT_0(t)T_{b2}, \ T_2'(0) = 0 \quad (7.163)$$

$$E'(x, t) = \frac{-av(x)L(x)(\lambda_R'(x, 0)e^{\rho t} - \lambda\lambda_{T_0}'(t))}{(\lambda_R(x, 0)e^{\rho t} - \lambda\lambda_{T_0}(t))^2}. \quad (7.164)$$

We evaluate (7.162) at $T_{b2} < 0$, the solution of is:

$$T_0'(t) = e^{-Bt}\left(T_0'(0) + \lambda\int_{s=0}^{t}e^{Bs}E'(s)ds\right) = \lambda\int_{s=0}^{t}e^{B(s-t)}E'(s)ds \quad (7.165)$$

while the solution of (7.163), evaluated at $T_{b2} < 0$, is

$$T_2'(t) = e^{-(B+6D)t}\left(T_2'(0) - 6\hat{r}DT_{b2}\int_{s=0}^{t}e^{(B+6D)s}T_0(s)ds\right) \quad (7.166)$$

$$= -6\hat{r}DT_{b2}\int_{s=0}^{t}e^{(B+6D)(s-t)}T_0(s)ds > 0. \quad (7.167)$$

Assume that the fossil fuel reserves are infinite so that $\lambda_R(x, 0) = 0$ for all t. The derivative $E'(s)|_{s=0} = [avL\lambda\lambda_{T_0}'(t)]/(\lambda\lambda_{T_0}(t))^2$ could be either positive or negative, depending on the sign of the derivative of the social price of the externality λ_{T_0}', which is given in Proposition 2. Assume that socioeconomic and natural factors are such that $\lambda_{T_0}'(t) > 0$, $(\lambda_{T_0}(t) < 0)$, then $E'(s) > 0$ at all locations x and at all dates s when reserves are infinite. In this case

$T'_0(t) > 0$. Recall that $T_0(t)$ is global average temperature at date t. Hence we should expect global average temperature to increase when more fossil fuels are used. Solving (165) and using $T'_0(t) > 0$, $T_{b2} < 0$, we obtain $T'_2(t) > 0$ for all $t > 0$, or

$$T'_2(t) = e^{-(B+6D)t}\left(T'_2(0) - 6D\int_{s=0}^{t} e^{(B+6D)s}T_0(s)T_{b2}(s)\,ds\right)$$

$$= -\int_{s=0}^{t} e^{(B+6D)(s-t)}T_0(s)T_{b2}(s)\,ds > 0. \tag{7.168}$$

Then from the derivative (7.161) we obtain:

$$T'(x,t) = T'_0(t) + T'_2(t)P_2(x) = T'_0(t) + T'_2(t)\left[\frac{1}{2}(3x^2 - 1)\right] \tag{7.169}$$

$$T'(0,t) = T'_0(t) + T'_2(t)P_2(0) = T'_0(t) - T'_2(t)\left(\frac{1}{2}\right) \tag{7.170}$$

$$T'(1,t) = T'_0(t) + T'_2(t)P_2(1) = T'_0(t) + T'_2(t) > 0, \tag{7.171}$$

i.e., temperature may fall or even rise at the Equator and rises at the North Pole. Hence we obtain PA when $\hat{r}$ increases from $\hat{r} = 0$ in the case where reserves are infinite at all locations.

If $E'(s) < 0$, the signs of inequalities are reversed. That is, $T'_0(t) < 0$, $T'_2(t) < 0$ and

$$T'(1,t) = T'_0(t) + T'_2(t)P_2(1) = T'_0(t) + T'_2(t) < 0. \tag{7.172}$$

In this case temperature may fall or even rise at the Equator and fall at the North Pole.

Proof of Proposition 5

We differentiate the dynamical system (7.126)–(7.127) with respect to γ, using the utility function $U(C) = \frac{C^{1-\gamma}}{1-\gamma}$, to obtain at $\gamma = 1$:

$$\dot{\lambda}_{T_0} = (\rho + B)\lambda_{T_0} + \langle vL, \phi U'(\tilde{C})\tilde{C}\rangle + \lambda_{T_2}6\hat{r}DT_{b2} \tag{7.173}$$

$$\dot{\lambda}_{T_2} = (\rho + B + 6D)\lambda_{T_2} + \langle vL, \phi P_2 U'(\tilde{C})\tilde{C}\rangle$$

$$\dot{\lambda}_R(t) = \rho\lambda_R(t). \tag{7.174}$$

$$\frac{\partial\dot{\lambda}_{T_0}}{\partial\gamma} = (\rho + B)\frac{\partial\lambda_{T_0}}{\partial\gamma} + \langle vL, -\phi\ln\tilde{C}\rangle + 6\hat{r}DT_{b2}\frac{\partial\lambda_{T_2}}{\partial\gamma} \tag{7.175}$$

$$\frac{\partial\dot{\lambda}_{T_2}}{\partial\gamma} = (\rho + B + 6D)\frac{\partial\lambda_{T_2}}{\partial\gamma} + \langle vL, -\phi P_2\ln\tilde{C}\rangle \tag{7.176}$$

$$\frac{\partial \dot{\lambda}_R(t)}{\partial \gamma} = \rho \frac{\partial \lambda_R}{\partial \gamma}. \tag{7.177}$$

The quantity $\ln \tilde{C}$ can be computed at $(\hat{r}, \gamma) = (0, 1)$ as

$$\ln \tilde{C}(x, t; 0, 1) = \ln y(x, t) + \alpha \ln E(x, t; 0, 1) - \tag{7.178}$$

$$\phi(x)[T_0(t; 0, 1) + T_{b0}(t) + (T_2(t; 0, 1) + T_{b2}(t))P_2(x)]. \tag{7.179}$$

It is natural to put $T_{b0}(t) = \overline{T}_{b0}$, $T_{b2}(t) = \overline{T}_{b2}$ at steady-state values for all t because the climate system without humans would plausibly be at the steady state. Making the no-serious-poverty assumption at any location at any time, so that $\ln \tilde{C}(x, t) > 0$ for all (x, t), we can compute the forward solution for $\frac{\partial \lambda_{T_0}}{\partial \gamma}$ at $\hat{r} = 0$ from (7.715). Thus, we have for the forward solution and the steady-state value of $\frac{\partial \lambda_{T_0}(v; 0, 1)}{\partial \gamma}$:

$$\frac{\partial \lambda_{T_0}(v; 0, 1)}{\partial \gamma} = \int_{s=v}^{\infty} e^{-(\rho + B)(s - v)} \langle vL, \phi \ln \tilde{C} \rangle(s) ds > 0 \tag{7.180}$$

$$\frac{\partial \overline{\lambda}_{T_0}(0, 1)}{\partial \gamma} = \overline{\lambda}_{T_0 \gamma} = \frac{\langle vL, \phi \ln \tilde{C} \rangle}{(\rho + B)} > 0. \tag{7.181}$$

If $\hat{r} > 0$, with constant baseline temperatures the steady-state values for $\frac{\partial \lambda_{T_0}}{\partial \gamma}, \frac{\partial \lambda_{T_2}}{\partial \gamma}$ are:

$$\frac{\partial \overline{\lambda}_{T_2}}{\partial \gamma} = \frac{-\langle vL, -\phi P_2 \ln \tilde{C} \rangle}{(\rho + B + 6D)} > 0 \tag{7.182}$$

$$\frac{\partial \overline{\lambda}_{T_0}}{\partial \gamma} = \frac{-\langle vL, -\phi \ln \tilde{C} \rangle}{(\rho + B)} - \frac{6\hat{r} D T_{b2} \langle vL, \phi \ln \tilde{C} \rangle}{(\rho + B)(\rho + B + 6D)} > 0 \tag{7.183}$$

Assume that $\gamma > 1$ then $U'(\tilde{C})\tilde{C} = \tilde{C}^{-\gamma}\tilde{C}$, and $\frac{\partial(\tilde{C}^{-\gamma}\tilde{C})}{\partial \gamma} = -\tilde{C}^{1-\gamma} \ln \tilde{C}$. If $\ln \tilde{C}(x, t) > 0$ for all (x, t) than the same result holds.

8. Progressive adaptation strategies in European coastal cities: a response to flood-risk under uncertainty*

Luis M. Abadie, Elisa Sainz de Murieta, Ibon Galarraga and Anil Markandya

1. INTRODUCTION

Global damages associated with coastal flooding are expected to increase significantly in the future due to the rise of the global sea level but also to the increase in the number and value of assets at risk from coastal flooding worldwide (Wong et al., 2014). For a global sea level between 25 and 123 cm higher than today, the associated annual costs could reach 0.3–9.3% of global GDP (Hinkel et al., 2014). Observe that the worst-case estimate of this range represents almost half of the global costs of climate change calculated by Stern (2007). The impacts of sea-level rise (SLR) in 83 developing countries was carried out by Dasgupta et al. (2009), whose results show a loss of global GDP ranging from 1.3 per cent to 6.05% for sea-level rise of 1 m and 5 m respectively.

In the last 10 years, the United States has suffered two of the most catastrophic coastal extreme events of their recent history. In 2005, the city of New Orleans was hit by hurricane Katrina and the damages were estimated at 142 billion US dollars (2010), the highest costs ever recorded for a coastal extreme event. The death of almost 1800 people should be added to this number (Nicholls and Kebede, 2012). In 2012, another coastal extreme event hit the East Coast of the US. Only in New York City, damages were estimated in 19 billion US dollars, in addition to the 43 human lives lost and thousands of people affected (Steffen et al., 2014). An assessment of the risks of coastal extremes to the coast of California calculated that a 1.4 m increase in sea level by the end of the century could cause damage costs of more than 100 billion US dollars and the population at risk would be close to half a million people (Heberger et al., 2011).

In Australia, Steffen et al. (2014) reviewed the literature on the costs of sea-level rise using three different approaches. The first was based on the value of the infrastructure at risk of coastal flooding, which in Australia reaches 226 billion dollars (2008 prices) for a sea-level rise of 1.1 m. This rise in sea level is a high-end projection for 2100. In the second approach, the observed economic damages of previous coastal flood events were used.

---

* The authors acknowledge funding from the European Union's Seventh Framework Programme for research, technological development and demonstration under grant agreement no 603906, Project: ECONADAPT. Luis M. Abadie and Ibon Galarraga are grateful for financial support received from the Basque Government for support via project GIC12/177-IT-399-13. Elisa Sainz de Murieta acknowledges the funding by the Basque Government (Postdoctoral Fellowship grant no. POS_2018_2_0027). Anil Markandya and Luis M. Abadie are also grateful for the financial support, from MINECO via project ECO2015-68023-C2-1-R.

Damages for the period 1967–99 reached 28.6 billion dollars (2008), considering events of major flooding, tropical cyclones and severe storms. The third approach reviewed deals with the future costs of coastal flooding. One of the examples for Australia assessed the impacts of future coastal flooding in Gosford. The findings show that for a sea-level rise of 0.4 m by 2100, damage costs valued in 13.5 million dollars (adjusted to 2010 prices) would be generated (Lin et al., 2014).

In Europe, Brown et al. (2011) found that, in the absence of mitigation or adaptation, coastal flooding together with other impacts linked to sea-level rise, such as coastal erosion, would produce annual damage costs of 11 billion euros (2005 prices) by 2050. These costs, measured for the middle-of-the-road emission scenario A1B, would increase up to 25 billion euros (2005) by 2080. In addition to direct costs, other impacts were also included in this estimate, such as salinization of soil and water, land loss and the cost of moving. At the country level, these authors found great differences among European countries. The Netherlands would experience the greatest economic damages followed by France, the UK, Germany and Belgium. For these countries, annual damage costs vary between 0.3 and 12 million euros per kilometre of the coast. Following the estimates for Belgium closely, but well below the top four, Spain would be the sixth country in the ranking with expected annual damage costs of around 1.2 billion euros (2005) by 2080.

At city level, flooding in main coastal megacities caused, in 2005, global losses of around 6 billion US dollars, but this estimate could rise by 2050 to 52 billion US dollars annually only due to socio-economic development. If climate change and subsidence are added and no adaptation is implemented, losses could reach 1 trillion US dollars per year (Hallegatte et al., 2013).

Uncertainty is not a new challenge and different methods for taking it into account have been developed, for example, in financial economics to deal with risk. However, in the context of climate change, uncertainty has gained a new dimension and addressing it is critical for coastal cities that will need to cope with sea-level rise and more frequent and intense extreme events (Markandya, 2014). Still, incorporating uncertainty requires the application of alternative economic decision support tools (Watkiss et al., 2015). In this chapter, we develop a risk measure-based approach to assess the potential impacts of relative sea-level rise and extreme events in 19 European coastal cities. Risk measure approaches are economic tools that have long been used in financial economics to effectively account for uncertainty in many variables, such as prices, by adding stochastic behaviour to deterministic formulations (Abadie and Chamorro, 2013; Abadie and Galarraga, 2015). Abadie et al. (2017) have shown that risk measure-based approaches are particularly appropriate to assess low-probability, high-damage events, as is the case with coastal extremes.

This chapter is organized as follows: Section 2 describes the sea-level rise stochastic model, how risk measures can be applied to a context of rising sea levels and the way in which SLR and socio-economic development has been integrated. Section 3 presents the estimated damage costs for each city under different scenarios by the Interngovernmental Panel on Climate Change (IPCC) and for different time slots. Section 4 examines the case of Glasgow and illustrates an application of real options analysis for this city. Finally, section 5 provides some concluding remarks.

2. METHODS

2.1 A Stochastic Approach to Modelling Global Sea-Level Rise

The IPCC's 5th Assessment Report (AR5) (2013) provides a range of estimates of global SLR for every decade from 2007 until the end of the century, according to different representative concentration pathways (RCPs) (Table 8.1). Using these data, a continuous stochastic Geometric Brownian Motion (GBM) model is calibrated to model the probability distribution of relative SLR in each moment in time. The proposed model is then calibrated with local sea-level rise data (LSLR) for each European coastal city. The derived distributions form the basis for the valuation of different actions, as we show later.

Table 8.1 Global mean sea-level rise, measured in metres, compared to the period 1986–2005 at 1st January[1]

Year	RCP2.6			RCP4.5			RCP6.0			RCP8.5		
	Median	Low perc.	High perc.	Median	Low perc.	High perc.	Median	Low perc.	High perc.	Median	Low perc.	High perc.
2007	0.03	0.02	0.04	0.03	0.02	0.04	0.03	0.02	0.04	0.03	0.02	0.04
2010	0.04	0.03	0.05	0.04	0.03	0.05	0.04	0.03	0.05	0.04	0.03	0.05
2020	0.08	0.06	0.10	0.08	0.06	0.10	0.08	0.06	0.10	0.08	0.06	0.11
2030	0.13	0.09	0.16	0.13	0.09	0.16	0.12	0.09	0.16	0.13	0.10	0.17
2040	0.17	0.13	0.22	0.17	0.13	0.22	0.17	0.12	0.21	0.19	0.14	0.24
2050	0.22	0.16	0.28	0.23	0.17	0.29	0.22	0.16	0.28	0.25	0.19	0.32
2060	0.26	0.18	0.35	0.28	0.21	0.37	0.27	0.19	0.35	0.33	0.24	0.42
2070	0.31	0.21	0.41	0.35	0.25	0.45	0.33	0.24	0.43	0.42	0.31	0.54
2080	0.35	0.24	0.48	0.41	0.28	0.54	0.40	0.28	0.53	0.51	0.37	0.67
2090	0.40	0.26	0.54	0.47	0.32	0.62	0.47	0.33	0.63	0.62	0.45	0.81
2100	0.44	0.28	0.61	0.53	0.36	0.71	0.55	0.38	0.73	0.74	0.53	0.98

Source: Data from IPCC (2013), Table AII.7.7.

We use the simplest geometric Brownian motion stochastic process (GBM), defined as follows:

$$dV_t = \alpha_V V_t \, dt + \sigma_V V_t dW_t^V \quad (8.1)$$

Where V_t represents the value of SLR at time t, plus an initial value V_0. This current value grows at rate α_V; the term σ_V is the instantaneous volatility, and dW_t^V denotes the increment to a standard Wiener process. The initial value V_0 is used to ensure a better fit of the modelization of the IPCC scenarios. One of the characteristics of this model is that it does not generate negative values, so $V_t > 0$ at all times.[2] For this reason, the initial value V_0

---

[1] Note that the IPCC describes these ranges as 'likely', that is with a probability between 66% and 100%. We assume 95% in this subsection.

[2] One technical detail of the model is that the initial value V_0 must be different from zero not to get $dV_t = 0$. This is solved by adjusting the scale in the model at the initial moment. Of course, the results account for this change in scale.

must be different from zero, otherwise we would always obtain $dV_t = 0$. For this reason, V_0 has been placed below current sea level and its value has been calculated optimally. The effective sea-level rise at time t, S_t is then estimated as the difference between V_t and V_0.

At a time t, this distribution process generates a log-normal distribution where the first moment is defined as follows:

$$E(V_t) = V_0 e^{a_V t} \tag{8.2}$$

The variance is expressed as

$$Var(V_t) = V_0^2 e^{2a_V t}(e^{\sigma_V^2 t} - 1) \tag{8.3}$$

If we define $X_t = \ln V_t$ then X_t still follows a normal distribution, with the following moments:

$$E(X_t) = X_0 + \left(a_V - \frac{\sigma_V^2}{2}\right)t \tag{8.4}$$

$$Var(X_t) = \sigma_V^2 t \tag{8.5}$$

$$\ln(V_t) = \phi\left[\ln(V_0) + \left(\mu - \frac{\sigma_V^2}{2}\right)t, \sigma_V^2 t\right] \tag{8.6}$$

The GBM distribution model enables the estimation of risk corresponding to any value of α, but in this study, we will focus on the 5% worst cases and therefore $1-\alpha = 95\%$.

The model has been calibrated using the SLR data for 2100 from Table 8.1 (median and upper percentile) and the results of the calibration process are shown next in Table 8.2.

The values of a_V and V_0 are estimated using Equation (8.4), so the best fit is obtained by minimizing the sum on the square of the differences between the theoretical value in Equation (8.4) and the values in Table 8.1. The value of σ_V is calculated by adjusting a log-normal distribution to the top percentile. Figure 8.1 shows the degree of fit at the median value, which gets smaller the closer we get to 2100.

The expected mean values for all the RCPs over time is summarized in Figure 8.2, where

Table 8.2 *Estimated values obtained in the calibration process*

Scenario	Calibration parameters (V)			Outcome of the calibration process (S_t)		
	a_V	σ_V	V_0	Mean 2100	Median 2100	Upper percentile (95%) 2100
RCP2.6	0.0077	0.0120	0.3685	0.4455	0.4400	0.6100
RCP4.5	0.0088	0.0115	0.3685	0.5356	0.5300	0.7100
RCP6.0	0.0090	0.0113	0.3685	0.5554	0.5500	0.7300
RCP8.5	0.0111	0.0124	0.3685	0.7478	0.7400	0.9800

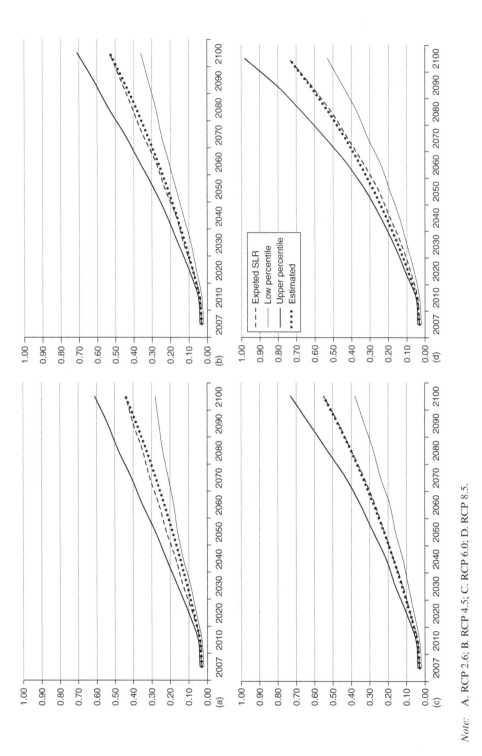

Note: A. RCP 2.6; B. RCP 4.5; C. RCP 6.0; D. RCP 8.5.

Figure 8.1 Median values, low and high percentiles for each SLR scenario obtained after the calibration of the GBM model

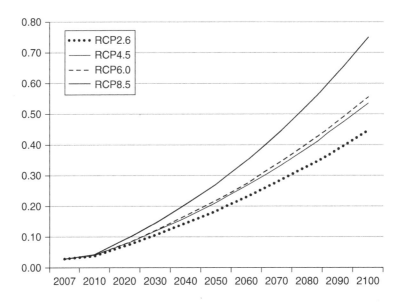

Figure 8.2 Expected SLR over time for the four IPCC RCP scenarios (mean values)

we also include the average of the four RCPs. It can be observed that RCP 2.6 provides the most optimistic SLR at any time, while RCP 8.5 presents the most pessimistic values. RCP 4.5 and 6.0 lie very close to each other showing a middle-of-the-road path. The figure reveals that the further in time, the greater the differences among RCPs and the largest the range of potential SLR. This is also confirmed by the probability distribution obtained for each RCP in 2050 and 2100.

Analysing the probability distribution of the latest IPCC sea-level rise projections is not new. For example, of Grinsted et al. (2015) that present probability distributions of local sea-level rise (LSLR) for several cities in Northern Europe under the most pessimistic RCP 8.5. Kopp et al. (2014) estimate LSLR for many locations and three RCP emission scenarios, RCP 2.6, 4.5 and 8.5. Jevrejeva et al. (2014) used probability distributions as well to determine the potential upper limit of sea-level rise by 2100. Hinkel et al. (2014) also considered probability distributions of three RCPs when assessing the damages and adaptation costs globally. The probability distribution of SLR is an input to our model, that will enable obtaining the distribution of damages in each coastal city, which is our final objective.

Figure 8.4 shows the probability distribution of SLR scenario RCP 8.5 over time. Both the expected value and the risks increase rapidly as SLR increases. Volatility also increases with SLR, and therefore with time.

2.2 Estimating Local Sea-Level Rise (LSLR)

The regional distribution of sea-level rise can differ significantly from the global signal due to different regional or local factors (Church et al., 2013). These are determinants related to: (i) absolute changes in sea level due to, for example, thermal expansion or ice

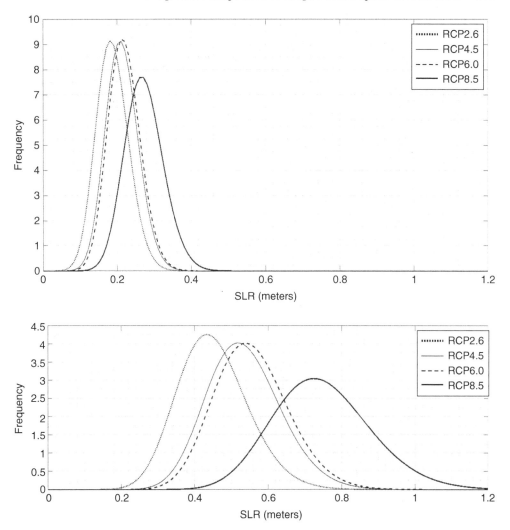

Note: Upper figure shows the distribution in 2050 and the lower figure in 2100.

Figure 8.3 Probability distribution of IPCC RCPs

melting; (ii) static equilibrium effects;[3] or (iii) vertical land movements caused by glacial isostatic adjustments or other local processes, such as groundwater depletion, sediment compaction or tectonism (Perrette et al., 2013; Abadie et al., 2016).

The general methodology for global sea-level rise described in section 2.1 is now applied for the assessed European cities using the probability percentiles of the relative sea-level

---

[3] Defined as 'perturbations in the Earth's gravitational field and crustal height associated with the redistribution of mass between the cryosphere and the ocean' (Kopp et al., 2014: 383).

174 *Handbook on the economics of climate change*

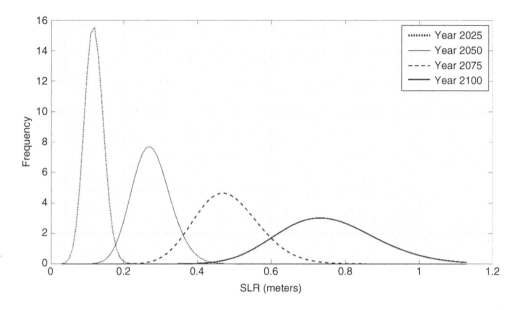

Figure 8.4 *Probability distribution for scenario RCP 8.5 over time*

rise projections obtained by Kopp et al. (2014). Now, we have 19 European cities and three sea-level rise scenarios.[4] The locally calibrated parameters are presented in Table 8.3.

2.3 Risk Measures and Sea-Level Rise

The central distribution of the IPCC's latest SLR projections (66–100% probabilities) does not provide all the information that needs to be considered from a coastal risk management perspective, which requires information on the 'upper-tail end' of the probability distribution, i.e. the worst-case scenario (Hinkel et al., 2014: 188). Risk measures are very useful analytical tools in situations of uncertainty and have been often used in economics to account for uncertainties of prices or other variables in investment projects (Abadie and Chamorro, 2013). In this section, we describe how these tools can be used to deal with uncertainties derived from climate risks, more precisely from the risk of sea-level rise and extreme coastal events. Risk measures enable analysing, precisely, the upper-tail end of the distribution. Note that risk measures can be used to determine the upper-tail end of both the SLR and the damage distribution. This is a rather innovative approach rarely used in the literature.

While the well-known derivative pricing models are based on risk-neutral measurements so that cash-flows can be discounted at the risk-neutral interest rate by altering the drift, risk measurement methods are based on real stochastic models. In this chapter, we

---

[4] RCP 8.5, RCP 4.5 and RCP 2.6. The scenario RCP 6.0 has not been calculated by Kopp et al. (2014) because these authors believe that it is very similar to RCP 4.5.

Table 8.3 Estimated values obtained in local sea-level rise calibration process

City	RCP2.6		RCP4.5		RCP8.5	
	a_V	σ_V	a_V	σ_V	a_V	σ_V
1. Istanbul	0.0119	0.0523	0.0123	0.0505	0.0141	0.0468
2. Odessa	0.0128	0.0559	0.0134	0.0575	0.0151	0.0522
3. Izmir	0.0092	0.0382	0.0101	0.0223	0.0121	0.0237
4. St. Petersburg	0.0029	0.0302	0.0073	0.0424	0.0075	0.0280
5. Rotterdam	0.0039	0.0161	0.0049	0.0143	0.0066	0.0159
6. Glasgow	0.0029	0.0252	0.0044	0.0222	0.0064	0.0211
7. Dublin	0.0043	0.0204	0.0055	0.0179	0.0070	0.0187
8. Lisbon	0.0068	0.0235	0.0079	0.0214	0.0099	0.0209
9. Marseille	0.0076	0.0402	0.0085	0.0217	0.0107	0.0224
10. London	0.0021	0.0108	0.0028	0.0097	0.0039	0.0121
11. Barcelona	0.0075	0.0375	0.0083	0.0226	0.0103	0.0226
12. Hamburg	0.0031	0.0119	0.0040	0.0113	0.0052	0.0126
13. Porto	0.0064	0.0246	0.0078	0.0224	0.0096	0.0217
14. Copenhagen	0.0037	0.0228	0.0055	0.0201	0.0077	0.0241
15. Amsterdam	0.0028	0.0141	0.0038	0.0126	0.0052	0.0140
16. Naples	0.0084	0.0405	0.0093	0.0225	0.0114	0.0228
17. Athens	0.0003	0.0340	0.0308	0.1047	0.0388	0.0695
18. Stockholm	0.0015	0.0384	0.0282	0.1667	0.0386	0.0832
19. Helsinki	0.0041	0.0901	0.0092	0.0580	0.0143	0.0513

use a real stochastic diffusion process to model medium- or long-term risks, such as SLR, by the year 2100.

There are mainly two risk measures that can be used for this purpose: Value-at-Risk (VaR) and Expected Shortfall (ES). Value-at-Risk (VaR) is the most standard measurement of risk, well recognized by international financial regulatory bodies such as the Basel Committee on Banking Supervision (Bank for International Settlements). The $VaR(\alpha)$ of an investment portfolio at the confidence level α is the value at which the probability of obtaining higher values is exactly $1 - \alpha$.

$$P[x > VaR(\alpha)] = 1 - \alpha \qquad (8.7)$$

That is, $VaR(\alpha)$ is the value for which the probability of greater losses is $1 - \alpha$. In our case, the $VaR(\alpha)$ is the level of damage caused by SLR for which the probability of a higher level of damage is $1 - \alpha$. This risk analysis provides a distribution of losses, where the relevant value is located in the right tail of the distribution.

In other words, the $VaR(\alpha)$ of damage caused by SLR represents the maximum losses that could occur with a given confidence level for a given time frame, usually a confidence level α of 90, 95 or 99%. Note that $VaR(\alpha)$ refers to a given time frame T, for example, 1 day, 1 week, 1 month, 1 year or more than 1 year. In this chapter, we are considering from the baseline to 2100, which is an appropriate time frame for analysing SLR related issues.

While VaR tends to work well when losses are normally distributed, it performs less well when the distribution has a fat tail. It is therefore important to find a way of dealing with fat tails. This is achieved by the use of a Geometric Brownian Motion (GBM) distribution process to generate log-normal distributions for each time frame in a continuous fashion. With the calculated parameters our log-normal distribution does not show significant fat tails.

Despite its widespread use, VaR has poor mathematical properties for optimization applications because it is non-convex[5] and not sub-additive, thus it is not a coherent risk measure (Artzner et al., 1999), i.e. does not meet the four conditions of monotonicity, sub-additivity, positive homogeneity and translation invariance.[6] This justifies the use of the so-called Expected Shortfall (ES). This is a coherent risk measurement meeting the above-mentioned conditions. For the case of the damage considered here, the ES is the expected value when the damage is greater than $VaR(x\%)$. In this sense, ES is a better risk measure and in practice makes many large-scale calculations possible showing numerical efficiency and stability of such calculations (Rockafellar and Uryasev, 2002). When comparing both risk measures, the tail risk of VaR can cause serious problems in certain cases, in which expected shortfall can serve more aptly in its place (Yamai and Yoshiba, 2005).

The Expected Shortfall (ES)[7] is the expected loss when VaR loss is exceeded, so as a risk measurement it provides more information on expected losses in less favourable situations than VaR.

For a function with a continuous distribution of losses, the $ES(\alpha)$ is defined as follows:

$$E[x|x > VaR(\alpha)] = ES(\alpha) \tag{8.8}$$

$$ES(\alpha) = \frac{1}{\alpha} \int_{VaR(\alpha)}^{+\infty} x f_x(x) dx \tag{8.9}$$

where:

$$\alpha = \int_{VaR(\alpha)}^{+\infty} f_x(x) dx \tag{8.10}$$

With a stochastic model, it is possible to generate a large number of scenarios using Monte Carlo simulation methods and to calculate both $ES(\alpha)$ and $VaR(\alpha)$ fairly accurately. This is the approach followed in this chapter: we obtain measurements of the risk of experiencing damages caused by SLR for different scenarios and time frames with and without optimal adaptation measures. We then compare both situations and apply Real

---

[5] Subadditivity and positive homogeneity ensure that the risk measure is a convex function (Artzner et al., 1999).

[6] It is considered that a coherent risk measure R(D), in which D represents the damages, must meet the following four conditions: monotonicity, sub-additivity, positive homogeneity and translation invariance:
    a) monotonous: $R(D) \leq 0$.
    b) sub-additive: $R(D1+D2) \leq R(D1)+R(D2)$.
    c) positive homogeneous: $R(kD) = kR(D)$.
    d) translation invariant: $R(D+l) = R(D)-l$.

[7] Also known as Expected Tail Loss (ETL), Mean Excess Loss and Conditional Value-at-Risk (CVaR).

Options analysis to choose the optimal moment in time to undertake the investment in adaptation.

2.4 Assessing Damage Costs and Adaptation Needs in Coastal Cities

A study by Hallegatte et al. (2013) estimated the annual average damage costs due to the combined effect of sea-level rise and extreme events in 136 major mega-cities. The authors defined a damage function in which flood losses depend on the following variables: global sea-level rise (S), the level of protection existing in coastal cities (P), subsidence (SUB),[8] extreme events (E), the socio-economic scenario (SE), and the defence failure model and of the characteristics of the defences (DF).

$$D = f(S, P, SUB, E, SE, DF) \qquad (8.11)$$

These data has been an input to our study, but while Hallegatte et al. (2013) follow a deterministic approach to SLR, we model it in a stochastic way to account for uncertainty. The modelling is done following RCP 2.6, RCP 4.5 and RCP 8.5 regionalized by Kopp et al. (2014). In addition, we use the risk measures defined earlier, VaR and ES, to develop a risk-based assessment of the potential damages. These two innovations are the major contributions of the method presented in this chapter.

The damage function
The damage function is based in two main components: first, local sea-level rise in each city, which is estimated using the stochastic GBM model as described in Section 2.2. The three LSLR scenarios are distinguished via the index i, so that at a time T the local SLR in scenario RCP 2.6 and city k is $S^{k,1,T}$, assuming the stochastic behaviour defined as laid out. The subsidence of each city has not been considered in our damage function as it has been accounted for, together with other local determinants, in the SLR regionalization process developed by Kopp et al. (2014).

The second component of the damage function is represented by the socio-economic development of each city in the future, as defined by Hallegatte et al. (2013).[9] Accordingly, damages also vary with time T in a deterministic way, due to the effect of the socio-economic scenario. Note that the function does not include extreme events as a variable. However, the original values coming from Hallegatte et al. (2013) and used in this chapter incorporate the probability of extreme events.

In areas where coastal defences exist, we assume that these fail to provide any protection once they are overcome by flooding. This implies that the damage function for each local SLR scenario i and for each city k at each time T is defined as follows:

$$D^{k,i,t} = f(S^{k,i,t}, T) \qquad (8.12)$$

---

[8] Defined as the sum of subsidence and land-uplift.
[9] Hallegatte et al. (2013) develop three possible socio-economic scenarios: a constant scenario, a scenario with no city limit (every city within a country grows at the same rate) and a scenario with city limit assuming that no city will exceed 35 million inhabitants. In this study, we have used the data corresponding to the latter.

At time T the function presents the following form:

$$D^{k,i,t} = f^1(S^{k,i,t}, +f^2(T) \tag{8.13}$$

where $f^1(S^{k,i,T})$ represents the impact of sea-level rise in city k at time T, including subsidence, while $f^2(T)$ shows the socio-economic impacts in the absence of SLR.

This way, following the data from Hallegatte et al. (2013), we calibrate a continuous damage function for each city using discontinuous data. While both factors, sea-level rise and socio-economic development contribute to increasing risk with time, the main factor was found to be $S^{k,i,T}$.

For illustrative purposes, we detail next the functions f^1 and f^2 estimated for the city of Glasgow.

Figure 8.5 presents two sections of the damage function for this city, i.e. damages versus LSLR and time respectively, while the full representation of the damage-function is the surface area shown in Figure 8.6.

Assessing adaptation needs

Future adaptation needs can be assessed assuming an increase in the level of protection defences to level A^k, which can be expressed as follows:

$$D^{k,i,T} = f^j(S^{k,i,T} - A^k, T) \tag{8.14}$$

In other words, this approach assumes that the rise in sea level is A^k or lower, but would not exceed future defences.

However, adaptation needs A^k can be assessed from other different perspectives, for example, maintaining the risk level or the likelihood of defences being overcome. But there is an optimum decision-making time and the possibility of an upgrade, i.e. of raising the level of defences subsequently if the initial adaptation is based on a more solid foundation which can be built on. This is an example of real options analysis (RO) as it is an option for the future, the application of which would entail a cost at the time of the initial adaptation.

2.5 Risk Calculation with Monte Carlo Simulation

For a GBM, it is possible to find a discretization algorithm which is both exact and simple, that is, the differential equation can be integrated exactly; the result is as follows:

$$S_t = S_0 e^{(\alpha - \frac{\sigma^2}{2})t + \sigma \int_0^t dW} \tag{8.15}$$

Now, over a time step Δt we have:

$$S_{t+\Delta t} = S_t + \Delta S = S_t e^{(\alpha - \frac{\sigma^2}{2})\Delta t + \sigma \sqrt{\Delta t} \varepsilon} \tag{8.16}$$

Where ΔS denotes the change in S over Δt, and ε stands for a random sample from an $N(0,1)$ distribution. This can be seen in Abadie and Chamorro (2013) using the Monte Carlo Method together with stochastic diffusion models.

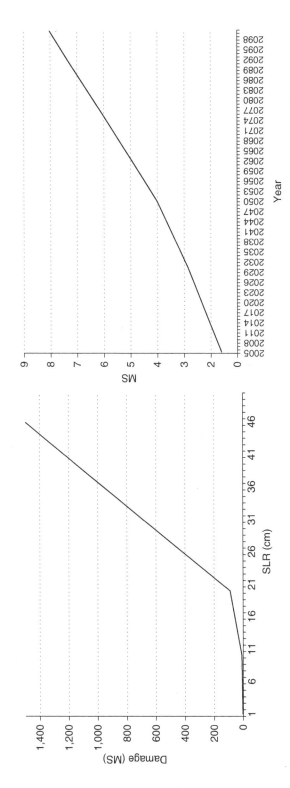

Figure 8.5 Damage function for Glasgow

Note: The figure in the left shows damage as a function of SLR. The image in the right presents damage as a function of time.

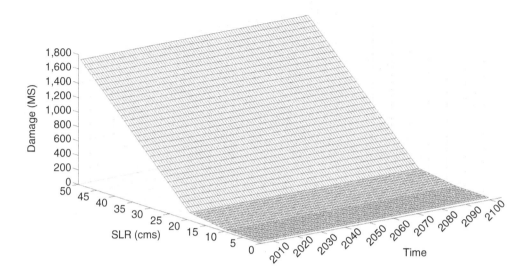

Figure 8.6   Damage as a function of both SLR and time in Glasgow

Note that this is an exact expression. Therefore, Δt need not be small. Indeed, if there is just one risk value which depends only on the terminal value of the asset then the latter can be simulated in a great leap using a time step of length T. However, there remains a minor error that can arise from using a finite number of random numbers.

In this study, we use 1,000,000 Monte Carlo simulations for each case, i.e. the combination of the climate-induced SLR scenario (RCP 2.6, RCP 4.5, RCP 8.5), time (2020–2100 with five-year intervals), which is linked to the socio-economic development and city (19 European cities). The number of simulations enables us to approximate almost exactly the theoretical distribution of $SLR^{k,\,i,T}$ at time T. This can be checked by finding the mean and standard deviation of the value simulated and comparing them with the theoretical values. Using the 95% percentile we have 50,000 simulated values for the most unfavourable situations, which enables us to obtain highly accurate values of VaR (95%) and ES (95%) using Equation (8.9). Any value of α could be used, but as in this case we consider the 5% worst cases, α=95%.

Building the global scenario: a combination of IPCC's SLR scenarios
Additionally to the approach explained so far, we decide to explore the damages of a combination of the three IPCC sea-level rise scenarios. This fourth scenario (Global Scenario, henceforth) is obtained by weighting a 25% of RCP 2.6, 50% of RCP 4.5 and 25% of RCP 8.5. For this case, a million Monte Carlo simulations have been carried out for the years 2025, 2050, 2075 and 2100: 250,000 for the scenarios RCP 2.6 and RCP 8.5 and 500,000 for the RCP 4.5 scenario. The result obtained for the average damage is, logically, the weighted average for the scenarios. However, the figures for VaR (95%) and ES (95%) are expected to be higher than the average because the figures simulated for RCP 8.5 predominate in the calculation. ES (95%) is determined in this case by the average of the 50,000 worst iterations.

3. DAMAGE COSTS IN MAIN EUROPEAN COASTAL CITIES

Using the methodology explained above we can estimate mean annual losses in 2050 for the main 19 coastal European cities. These numbers are shown in Table 8.4 below. In aggregate, these values range from US$5,154 million to US$6,792 million, depending on the RCP considered in each case. Compared to the losses obtained by Hallegatte et al. (2013), our results show higher annual average costs: losses are more than twice in RCP 2.6 than following a deterministic approach and more than triple under RCP 8.5. Note that we are using regionalized climate projections and not global sea-level rise. The ranking of greatest damages also varies depending on the approach used, but the top five cities are the same Table 8.4. This difference is mainly due to the use of the regionalized SLR scenarios from Kopp et al. (2014) that present greater SLR than the 20 cm assumed by Hallegatte et al. (2013) in 2050.

By 2100, the ranking of cities remains similar in 2050 but annual average costs increase dramatically (Table 8.5). Under the most optimistic scenario (RCP 2.6) the total damage costs could reach US$24,140 million per year, while under RCP 8.5 losses increase up

Table 8.4 European cities ranked by annual average losses (AAL) in 2050

City	AAL damage costs (US$ million)			Optimistic sea-level rise from Hallegatte et al. (2013)	
	RCP2.6	RCP4.5	RCP8.5	Urban Agglomeration	SLR 20 cm
1 Istanbul	1,382	1,312	1,400	1 Istanbul	327
2 Rotterdam	718	851	1,219	2 Izmir	314
3 Izmir	730	668	915	3 Odessa	280
4 Odessa	872	816	836	4 Rotterdam	256
5 St. Petersburg	200	738	577	5 St. Petersburg	187
6 Lisbon	235	267	354	6 Lisbon	100
7 Dublin	155	180	272	7 Glasgow	95
8 Hamburg	140	164	221	8 Dublin	85
9 Glasgow	125	146	218	9 London	65
10 London	118	136	190	10 Hamburg	61
11 Amsterdam	86	101	137	11 Amsterdam	56
12 Barcelona	106	99	124	12 Marseille	43
13 Marseille	101	76	101	13 Barcelona	37
14 Porto	59	63	87	14 Copenhagen	36
15 Copenhagen	44	53	81	15 Naples	35
16 Naples	43	39	52	16 Porto	34
17 Helsinki	6	5	8	17 Athens	18
18 Stockholm	16	1	1	18 Stockholm	14
19 Athens	17	0	0	19 Helsinki	11
TOTAL	5,154	5,716	6,792	TOTAL	2,054

Table 8.5 European cities ranked by annual average losses (AAL) in 2100

City	AAL damage costs (US$ million) by IPCC RCPs		
	RCP2.6	RCP4.5	RCP8.5
1 Istanbul	7,318	7,654	9,806
2 Odessa	5,023	5,219	6,577
3 Izmir	3,472	4,059	5,746
4 Rotterdam	2,666	3,637	5,511
5 St. Petersburg	788	3,266	3,274
6 Lisbon	1,033	1,306	1,891
7 Glasgow	516	844	1,558
8 Dublin	710	1,010	1,504
9 Marseille	464	538	810
10 Barcelona	471	561	797
11 Hamburg	423	550	775
12 London	356	478	703
13 Amsterdam	254	350	523
14 Porto	244	317	460
15 Copenhagen	140	223	368
16 Naples	175	204	290
17 Stockholm	27	38	91
18 Athens	43	30	86
19 Helsinki	17	24	53
TOTAL	24,140	30,307	40,824

to US$40,824 million annually. Thus, damages by the end of the 21st century would be between 4.5 and more than 6 times higher compared to 2050.

This traditional approach of estimating damage costs does not account for uncertainty, which should play a major role when assessing climate change impacts. The risk measure approach we propose can effectively incorporate uncertainty into the assessment, through the two risk indicators VaR (95%) and ES (95%).

In the risk calculation process, 1,000,000 simulated LSLR values were obtained for each RCP and time T, and each produces a specific damage cost. Our damage function enables us to calculate the 95% percentile VaR (95%), i.e. the value of the loss corresponding to the damage function in the 95% percentile. The mean expected loss of the 5% worst cases is the expected shortfall ES (95%). As we have obtained 50,000 values for these most adverse cases, the value of ES (95%) is highly accurate.

The 95% percentile is used in this case to represent low-probability, high-damage impacts often discussed in climate change economics literature (Weitzman, 2007, 2009; Nordhaus, 2011; Weitzman, 2013). The importance of these low-probability events lies in the enormous magnitude of their potential damage (Pindyck, 2011). Results of the VaR (95%) and ES (95%) by 2050 are presented in Table 8.6, ranked according to RCP 8.5. Observe that the first four cities of this ranking, those most affected by flooding, remains the same, although overall damages are much higher when assessing the worst 5% cases compared to average annual values.

Damage costs due to low probability events by 2100 are more than twice the mean

Table 8.6 European cities risk ranked by Expected Shortfall (ES) in 2050, under RCP8.5

City	RCP2.6		RCP4.5		RCP8.5	
	VaR (95%)	ES (95%)	VaR (95%)	ES (95%)	VaR (95%)	ES (95%)
1 Istanbul	5,809	8,007	5,596	7,719	5,716	7,748
2 Odessa	3,800	5,350	3,792	5,450	3,749	5,281
3 Rotterdam	2,299	2,846	2,394	2,892	3,151	3,758
4 St. Petersburg	1,158	1,827	3,418	4,730	2,326	3,062
5 Izmir	2,746	3,622	1,902	2,341	2,429	2,949
6 Glasgow	699	1,048	745	1,060	956	1,278
7 Lisbon	730	914	756	927	901	1,081
8 Dublin	630	834	652	835	855	1,060
9 London	351	429	361	432	497	593
10 Hamburg	345	408	372	435	474	549
11 Barcelona	439	589	313	392	367	453
12 Marseille	446	614	263	339	331	418
13 Amsterdam	245	300	257	306	329	388
14 Copenhagen	149	195	162	204	240	297
15 Porto	183	232	186	232	227	275
16 Naples	159	211	105	129	129	156
17 Helsinki	21	44	14	28	25	38
18 Stockholm	131	238	2	16	1	2
19 Athens	125	221	1	2	1	1
TOTAL	20,466	27,930	21,291	28,469	22,704	29,387

annual losses. The worst-case damages from 2050 to 2100 increase significantly, reaching values three to four times greater (Table 8.7).

A key output of this model is that it enables a continuous distribution of risk in time. That is, it allows identification of which moment the risk threshold is expected to be exceeded, and therefore by when adaptation would need to be implemented. The deadline for taking action is calculated by finding the precise time in which ES (95%) is expected to overcome the maximum acceptable damage with the current information,[10] in this case defined as 0.1%, 0.5% and 1% of each city's GDP in 2020 (see Table 8.8).

Based on these results, we then define several risk aversion thresholds based on each city's GDP. In situations with a very high risk aversion threshold (0.1% of GDP), adaptation measures would have to begin immediately to be available by 2020–25 for the top three cities in the ranking (Odessa, Izmir and Rotterdam) under RCP 2.6 (Table 8.8). When selecting a risk aversion equivalent to 1% of GDP there is more time for adaptation, which could be postponed a few years, although results vary significantly depending on the city and the scenario selected.

---

[10] The arrival of new information can alter this perception.

Table 8.7 European cities risk ranked by Expected Shortfall (ES) in 2100, under RCP8.5

City	RCP2.6		RCP4.5		RCP8.5	
	VaR (95%)	ES (95%)	VaR (95%)	ES (95%)	VaR (95%)	ES (95%)
1 Istanbul	22,003	29,675	22,418	29,945	25,895	33,649
2 Odessa	15,556	21,322	16,593	22,960	18,707	24,994
3 Izmir	9,255	11,774	7,569	8,789	10,300	11,912
4 Rotterdam	5,987	7,064	6,914	7,937	9,803	11,170
5 St. Petersburg	3,531	4,881	10,645	14,143	8,135	9,986
6 Glasgow	2,089	2,775	2,604	3,264	3,685	4,424
7 Lisbon	2,256	2,691	2,539	2,963	3,340	3,834
8 Dublin	1,896	2,316	2,205	2,601	2,944	3,421
9 Marseille	1,446	1,897	1,101	1,296	1,524	1,773
10 Barcelona	1,374	1,769	1,134	1,334	1,488	1,729
11 London	803	941	912	1,040	1,310	1,493
12 Hamburg	812	930	952	1,072	1,283	1,438
13 Amsterdam	568	668	662	757	922	1,046
14 Porto	556	670	642	755	834	963
15 Copenhagen	385	477	488	579	769	913
16 Stockholm	210	389	164	637	418	692
17 Naples	494	638	388	452	517	596
18 Athens	287	459	170	391	366	564
19 Helsinki	68	129	86	127	154	207
TOTAL	69,577	91,467	78,187	101,043	92,396	114,805

3.1 Results under the Global Scenario

As explained in Section 2.4, one additional scenario has been defined. We refer to this as the Global Scenario and it has been estimated as the combination of the three previous scenarios (IPCC RCPs), assigning different weights to each: 25% to RCP 2.6, 50% to RCP 4.5 and 25% to RCP 8.5. We estimated this for all 19 European cities in four years: 2025, 2050, 2075 and 2100. The results show that the top five most vulnerable cities are always the same, but the rest of the ranking varies with time (Table 8.9 to Table 8.12).

By 2025 Istanbul enters the top five, followed by St. Petersburg, Rotterdam, Odessa and Izmir (Table 8.9). By mid-century the top five is the same (Table 8.10). From 2025 to 2050 risk, measured in terms of ES (95%), increases between three and more than six times. By 2075 Rotterdam swaps places with Izmir, and both are then followed by Glasgow and Lisbon. The ranking does not change in 2100. Damages increase significantly: in 2075 average damages range from US$11 to US$4,005 million; by 2100 damage costs reach US$47–8,102 million.

4. A CLOSER LOOK AT THE CITY SCALE: THE CASE OF GLASGOW

In the case of Glasgow the annual average losses (AAL), the VaR and the ES remain at moderate levels in the first part of the century but increase exponentially in the second

Table 8.8 Year in which cities should start adaptation based on RCP scenarios and different threshold of acceptable risk

City	0.1% damage			0.5% damage			1% damage		
	2.6	4.5	8.5	2.6	4.5	8.5	2.6	4.5	8.5
1 Amsterdam	2015	2015	2015	2045	2045	2060	2085	2075	2060
2 Athens	2030	2080	2075	2100	2100	2100	2100	2100	2100
3 Barcelona	2025	2030	2030	2060	2075	2100	2100	2100	2100
4 Dublin	2015	2020	2020	2030	2030	2035	2040	2040	2035
5 Glasgow	2015	2015	2020	2025	2025	2035	2030	2035	2035
6 Hamburg	2015	2015	2015	2045	2045	2060	2085	2075	2060
7 Helsinki	2055	2065	2055	2100	2100	2100	2100	2100	2100
8 Istanbul	2015	2015	2015	2020	2020	2030	2025	2025	2030
9 Izmir	2010	2015	2015	2015	2025	2025	2020	2030	2025
10 Copenhagen	2020	2025	2020	2065	2060	2070	2100	2095	2070
11 Lisbon	2015	2020	2020	2030	2030	2045	2045	2045	2045
12 London	2040	2045	2035	2100	2100	2100	2100	2100	2100
13 Marseille	2015	2025	2025	2030	2045	2055	2045	2065	2055
14 Naples	2030	2040	2040	2085	2100	2100	2100	2100	2100
15 Odessa	2010	2010	2015	2010	2010	2015	2015	2015	2015
16 Porto	2020	2025	2025	2045	2045	2060	2070	2070	2060
17 Rotterdam	2010	2010	2010	2015	2015	2020	2020	2020	2020
18 St. Petersburg	2015	2010	2015	2025	2020	2030	2035	2020	2030
19 Stockholm	2015	2060	2070	2090	2085	2100	2100	2100	2100

part for all RCPs and the Global Scenario, reaching levels that are probably unacceptable to any decision-maker. In the case of scenario RCP 8.5, the average of the 5% of worst losses, i.e. the ES (95%) in the city would entail a figure of US$4.4 billion by 2100. When looking at adaptation needs, for a very high risk aversion (0.1% of GDP), adaptation measures would have to begin immediately to be available by 2020–25. For a risk aversion equivalent to 1% of GDP, adaptation measures would need to be available by 2030–35, so in this case, adaptation action could be postponed for a few years (Table 8.13).

Considering the Global Scenario, in Glasgow, the VaR (95%) would exceed US$2.9 billion and ES (95%) US$3.6 billion in 2100 (Table 8.14). This last figure is markedly greater than the weighted average for the three scenarios, which works out to US$3,432M.

We can learn from these results that risk is largely determined by the extreme scenarios, at least at the outset. But this is just an initial and limited view of risk. As time goes on, the perception of risk changes as new information (be it favourable or unfavourable) becomes available. Depending on the degree of risk aversion, the construction of adaptation infrastructures should immediately begin or the decision to start building should be postponed. If the decision is postponed now, it will be made in the future based on more available information, and thus more accurate estimates of the expected trend and volatility could be obtained, and also that of the likelihood of each scenario. This may reveal that a particular scenario (be it one of the four posited or an alternative scenario) is highly likely, enabling it to be assumed thereafter that there is only one plausible scenario.

Table 8.9 *European cities ranked by ES (95%) under the Global Scenario in 2025*

City	Mean	Standard Deviation	VaR (95%)	ES (95%)
1 Istanbul	131	345	531	1,363
2 St. Petersburg	76	226	326	879
3 Rotterdam	125	184	480	772
4 Odessa	82	204	248	743
5 Izmir	76	126	252	463
6 Lisbon	41	42	93	179
7 Hamburg	42	35	117	155
8 London	35	35	107	155
9 Glasgow	19	42	67	144
10 Dublin	25	35	70	134
11 Amsterdam	26	25	76	106
12 Barcelona	13	21	33	80
13 Marseille	11	19	30	64
14 Copenhagen	14	13	33	56
15 Porto	12	10	28	45
16 Stockholm	2	14	3	38
17 Naples	6	9	23	37
18 Athens	1	10	2	22
19 Helsinki	2	2	4	7

Table 8.10 *European cities ranked by ES (95%) under the Global Scenario in 2050*

City	Mean	Standard Deviation	VaR (95%)	ES (95%)
1 Istanbul	1,350	2,058	5,678	7,780
2 Odessa	834	1,394	3,782	5,368
3 St. Petersburg	563	1,022	2,717	3,944
4 Rotterdam	909	869	2,614	3,196
5 Izmir	745	762	2,246	2,906
6 Glasgow	158	287	799	1,124
7 Lisbon	281	258	795	973
8 Dublin	197	242	711	910
9 London	145	129	401	492
10 Hamburg	172	122	400	473
11 Barcelona	107	124	355	466
12 Marseille	88	116	322	449
13 Amsterdam	106	87	277	334
14 Porto	68	64	198	246
15 Copenhagen	58	60	184	237
16 Naples	43	42	124	162
17 Stockholm	5	32	8	95
18 Athens	4	28	8	86
19 Helsinki	6	8	19	35

Table 8.11 European cities ranked by ES (95%) under the Global Scenario in 2075

City	Mean	Standard Deviation	VaR (95%)	ES (95%)
1 Istanbul	4,005	4,524	12,907	17,116
2 Odessa	2,618	3,259	9,083	12,386
3 St. Petersburg	1,446	2,018	5,509	7,584
4 Izmir	2,234	1,569	5,087	6,221
5 Rotterdam	2,236	1,551	5,073	6,024
6 Glasgow	481	627	1,788	2,296
7 Lisbon	755	502	1,678	1,985
8 Dublin	565	491	1,502	1,820
9 Marseille	291	252	759	962
10 Barcelona	313	250	773	955
11 London	311	218	713	852
12 Hamburg	359	198	715	829
13 Amsterdam	227	150	502	592
14 Porto	182	128	421	502
15 Copenhagen	136	118	365	457
16 Naples	117	83	267	331
17 Stockholm	13	74	62	214
18 Athens	11	49	55	179
19 Helsinki	15	21	57	86

Table 8.12 European cities ranked by ES (95%) under the Global Scenario in 2100

| Urban Agglomeration | AAL damage costs (US$ million) | | | |
	Mean	Standard Deviation	VaR (95%)	ES (95%)
1 Istanbul	8,102	7,909	23,308	30,925
2 Odessa	5,505	5,932	16,936	23,129
3 Sankt Peterburg (St. Petersburg)	2,647	3,198	8,907	12,074
4 Izmir	4,333	2,544	8,906	10,750
5 Rotterdam	3,862	2,255	7,952	9,378
6 Glasgow	940	1,005	2,930	3,667
7 Lisboa (Lisbon)	1,384	777	2,799	3,293
8 Dublin	1,058	744	2,428	2,900
9 Marseille-Aix-en-Provence	587	409	1,325	1,636
10 Barcelona	597	394	1,307	1,589
11 London	504	302	1,051	1,245
12 Hamburg	574	276	1,068	1,233
13 Amsterdam	369	213	754	884
14 Porto	334	200	700	828
15 København (Copenhagen)	238	182	582	725
16 Stockholm	48	221	274	613
17 Napoli (Naples)	218	133	456	555
18 Athínai (Athens)	47	124	270	470
19 Helsinki	84	29	137	153

188 *Handbook on the economics of climate change*

Table 8.13 Summary of Glasgow risk estimates for different times and IPCC RCPs

Summary Results for Glasgow		Year	RCP2.6	RCP4.5	RCP8.5
Annual average losses (AAL)		2050	125	146	218
		2100	516	844	1,558
Value at Risk VAR (95%)		2050	699	745	956
		2100	2,089	2,604	3,685
Expected Shortfall ES (95%)		2050	1,048	1,060	1,278
		2100	2,775	3,264	4,424
Year to start adaptation	Damage 0.1% GDP		2015	2015	2020
	Damage 0.5% GDP		2025	2025	2035
	Damage 1% GDP		2030	2035	2035

Table 8.14 Glasgow risk estimates under the Global Scenario in 2100

Year	Mean	Standard Deviation	VaR (95%)	ES (95%)
2025	19	42	67	144
2050	158	287	799	1,124
2075	481	627	1,788	2,296
2100	940	1,005	2,930	3,667

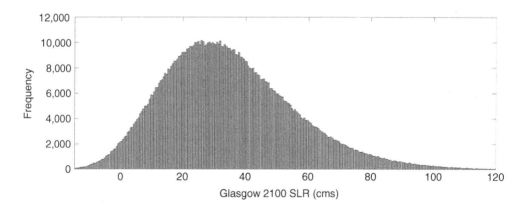

Figure 8.7 Glasgow 2100 SLR distribution under the Global Scenario

The distribution of SLR under the Global Scenario (Figure 8.7) combined with the damage that depends on SLR and the level of socio-economic development (Figure 8.8, left and right respectively) gives the full distribution of damage (Figure 8.9) under the Global Scenario. Figure 8.10 represents the whole calculation process.

The average damage, the VaR and the ES expected for Glasgow in future years based on the information currently available for the Global Scenario are presented in Figure 8.11. The year when adaptation is expected to begin, depending on the degree of risk aversion measured in terms of percentage of GDP is shown in Figure 8.12. For example, if the risk aversion level in terms of ES (95%) in Glasgow is defined as 0.5% of GDP, adaptation

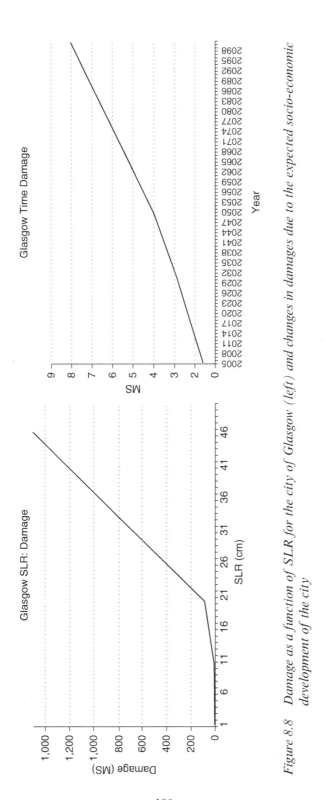

Figure 8.8 Damage as a function of SLR for the city of Glasgow (left) and changes in damages due to the expected socio-economic development of the city

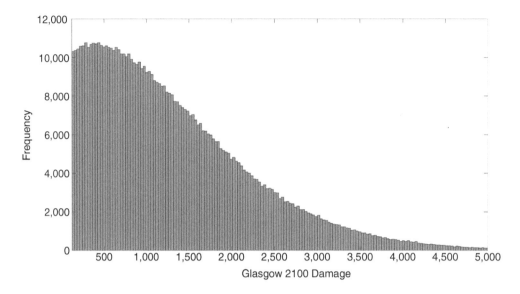

Figure 8.9   Glasgow damage distribution in 2100 under the Global Scenario

should be initiated already in 2030. However, for lower risk aversion levels, i.e. 5% of GDP, adaptation could wait until 2080. In other words, if the year shown is a long way off it would be preferable to wait and obtain further information.

Considering the results so far several decisions can be taken. The first option would be to construct an adaptation infrastructure to cover the risk for the year 2100. The problem with this decision lies in the fact that it entails investing before it is strictly necessary, that the structure may prove inflexible, i.e. it may turn out to be too big or too small. It must be borne in mind that risks are calculated today with the information currently available, but that as time passes and further information exists risk levels may change. It is, therefore, important that protection measures are flexible enough to cope with risks in the most effective way possible.

The second option would be to wait until the optimum time to build an infrastructure to cover risks up to 2100 is reached. Table 8.15 shows the expected optimum time by which the relevant investments should be completed based on the information currently available, in line with the risk aversion levels measured in terms of the maximum expected shortfall ES (95%) accepted as a percentage of GDP in 2020. The figures obtained from Glasgow are those shown in Table 8.15, but it must be remembered that they are based on current information and are therefore merely expected figures. If the level of risk tolerance is very low then construction should already have started: if damages equal 0.1% of GDP, there would already be an adaptation shortfall. Taking the damage of 0.5% of GDP, from our current standpoint in 2015 and given the time required to build adaptation infrastructures (e.g. 10 years), construction would need to be planned to take place shortly. In the case of a risk tolerance of 1% of GDP, it would be advisable to wait for further information until 2025 in line with the IPCC scenario considered most credible.

The third option would be to build the necessary structure in line with the medium-term risk at each given time in such a way that the structure is flexible so that the levels of

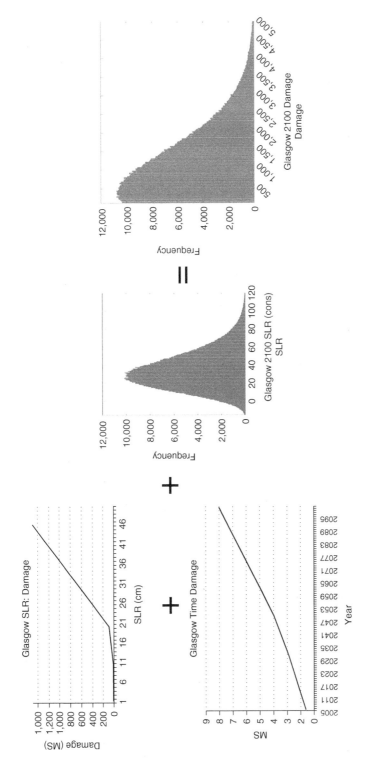

Figure 8.10   Schematic representation of the damage calculation process for Glasgow, by 2100 under the Global Scenario

192 *Handbook on the economics of climate change*

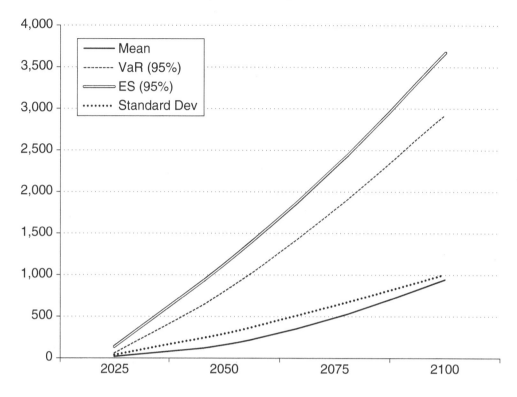

Figure 8.11 Glasgow risk values under the Global Scenario

protection can be increased in the future. To that end, the initial structure must be built with the option of enlarging it in mind. This usually entails greater initial spending, e.g. on setting aside more space or building stronger foundations. This outlay can be seen as the premium on a real option that may be exercised on one or more occasions in the future as risks increase. This avoids spending too much too early in terms of initial outlay in cases in which subsequent enlargements prove unnecessary. We work below with this setup, in which there are a number of real options that will be exercised if necessary to keep the level of risk under control and below a level that depends on the risk aversion of policymakers. Risk aversion levels may differ from one country to another, or even at the city level.

We now develop a case of real options based on the risks for the city of Glasgow. First, we take the Global Scenario and assume that the true scenario will be revealed by mid-2037. We take 2015 as the current year and consider periods of 12.5 years as shown in Figure 8.13, even though, exceptionally, the first period is 10 years long (2015–25). We assume a period of 12.5 years for the construction of infrastructures. Depending on the level of risk aversion, the decisions shown in Table 8.15 are made (from the viewpoint of 2015 and in line with the risks expected to exist in 2025).

The only case in which the first option would be exercised immediately is when risk aversion is limited to 0.1% of GDP. There is an optimum size for adaptation, but a standard minimum size would need to be selected (e.g. two or three meters, which would be sufficient). It will also be possible in the future to exercise another adaptation option

Progressive adaptation strategies in European coastal cities 193

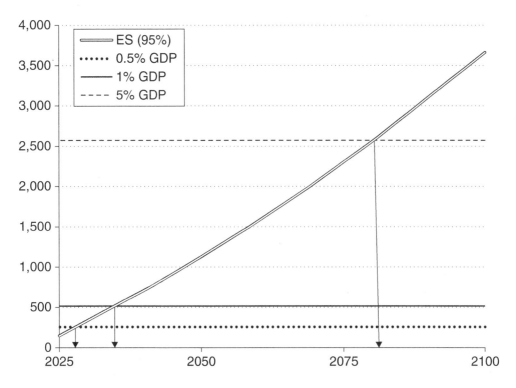

Figure 8.12 Glasgow ES (95%) risk value under the Global Scenario

Table 8.15 Glasgow investment decision under the Global Scenario at present time (2015)

Risk Aversion	Decision	Risk and GDP
0.1% GDP	Invest Immediately	(ES (95%) 2025 = US$144) > (0.1% GDP = US$51)
0.5% GDP	Wait	(ES (95%) 2025 = US$144) < (0.5% GDP = US$257)
1.0% GDP	Wait	(ES (95%) 2025 = US$144) < (1.0% GDP = US$515)
2.0% GDP	Wait	(ES (95%) 2025 = US$144) < (2.0% GDP = US$1,029)
5.0% GDP	Wait	(ES (95%) 2025 = US$144) < (5.0% GDP = US$2,573)

Figure 8.13 Time steps

if it proves necessary to build an extra level of defences to keep the risk below the limit set.

For the other risk levels, the option would not be exercised at the outset and a decision would be put off. Below, we consider a limit of 1% of GDP, i.e. an investment criteria in which risk, measured by ES (95%) for the coming period, must not exceed 1% of GDP.

194 *Handbook on the economics of climate change*

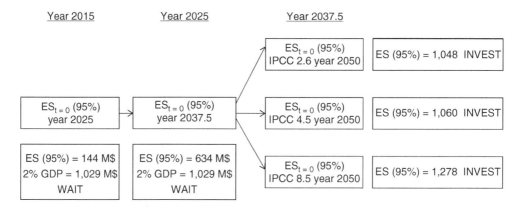

Figure 8.14   Schematic decision-making process with current information

The logical course of action would be to wait until 2025, see how far sea level has actually risen by that time and then calculate the new ES (95%) risk level with the overall scenario of, for example, one million simulations for each RCP by 2037.5. To that end, the trend and volatility figures for these scenarios would be used but starting from the SLR figure for 2025, though it may also be necessary to use new values if more information becomes available and more precise figures can be calculated for both parameters of the stochastic model. Once again, if calculating the risks for 2037.5 gives values for ES (95%) exceeding 1.0% of the city's GDP then investment will be made, and if not the decision will again be postponed. It must be noted that in this case it was decided not to invest in 2015. The process could thus continue through successive periods, taking into account possible previous investments in adaptation that might render new investment unnecessary or might influence the scale of new investments in adaptation required. Figure 8.14 shows this process under the assumption that all expected values are met and all variables behave as per the information available at the outset (in 2015).

According to the decision tree built with the information currently available and presented in Figure 8.15 for risk aversion 2% GDP, if the forecasts are met there would be no investment either at the present time or in 2025. However, the investment required for adaptation in Glasgow will need to have been made by 2050, given that by that time the expected risk levels in all scenarios exceed the levels acceptable under a risk aversion of 2% of GDP. Indeed, the investments will need to have been made by sometime between 2035 and 2045, depending on the scenario considered (see Table 8.15). If the risk aversion level is 5% of GDP (equivalent to US$2.573M) then adaptation can be postponed, but the relevant measures must be available after 2075. However, the time required to build infrastructures must be taken into account when calculating when construction should begin.

Figure 8.15 shows various possible behaviour patterns with risk aversion measured by 2% of GDP. With the information available at the outset (t=0), i.e. in 2015, the decision is to wait, given that the risk level in 2025 is below the limit. Risk calculations in 2025 with the information available at t=10 could result in the same expected value as initially (US$634M) or a higher or indeed lower figure. For the sake of illustration, figures of

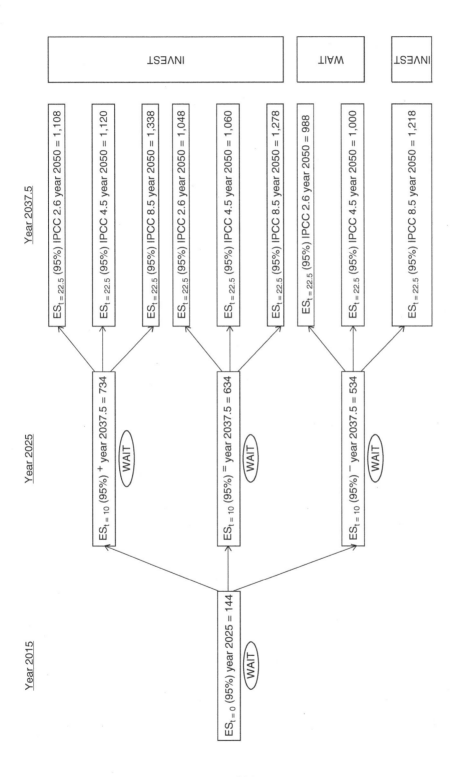

Figure 8.15 Decision with current and future information

+US$100 million and –US$100 million of the amount envisaged are considered. If the risk remains the same or has increased then it is time to begin investing so that those risks are covered in 2037.5. In the less likely event that the risk has decreased, then we again wait until further information becomes available.

It is assumed that when 2037.5 arrives we will know which RCP materializes (trend + volatility), but SLR may be higher or lower than expected, so different risk values appear for the same scenario. There are cases where the real option is exercised at the second possible time, and it needs to be determined whether it is necessary to exercise a second option and reinforce the flexible defences. In none of the cases, the real option is exercised in the first time (2025), but by 2037 it would be exercised in several cases.

Real options for adaptation are therefore exercised (or not) at different times, and to exercise a second real option (and successive real options) it is necessary to assess the size of the defences already constructed in case the additional adaptations and changes in size need to be made.

5. CONCLUSIONS

This chapter presents a novel method to model SLR in a non-deterministic way. This is done with a stochastic diffusion model, the so-called GBM. The use of this model allows us to estimate two risk measures, VaR and ES that, while relatively well-known in financial economics, have not been used often to address issues such as climate impacts even though they can be suitable measures to analyse the risk of SLR, as shown in this chapter. The VaR will enable the understanding of the threshold of average damages that leads to the worst 5% of the cases. The ES instead shows what the average damage will be for those 5% worst cases. Dealing with these low probability, high impact situations is a very relevant issue in climate change economics. We have illustrated the use of the methodology to assess the risk of SLR in 19 major European coastal cities using regionalized SLR data from Kopp et al. (2014) and comparing the results with values from Hallegatte et al. (2013) that use a deterministic approach, which does not account for uncertainty in the way described here. The values estimated show that, when uncertainty is present, VaR and ES figures are more appropriate than the usual average values.

The method is then used to show how a certain acceptable risk level can be determined above which no city is willing to go. This threshold allows estimation not only of how much adaptation is needed but also when adaptation should start. This information could be very relevant complementing other data used for decision-making processes.

We have finally focused on the case of one city, Glasgow, to show how the methodology presented here can be combined to undertake a Real Options Analysis. This is a very sophisticated economic assessment tool that allows for an adequate assessment of investment decisions. That is, allows comparison of the outcomes of postponing a decision until more information on the impacts of SLR (or other climate impacts) becomes available with other possibilities such as building a flexible defence that will allow a greater degree of protection in the future.

Our approach could incorporate new data from city-specific models to obtain high-resolution damages in each city. Nevertheless, we believe that our results can be useful as a guide of what could be expected in each city in the long run under each RCP scenario and also under the Global Scenario that combines the three RCPs.

Climate change requires new analytical instruments that allow for positive treatment of the various uncertainties associated. The treatment presented in this chapter is one of the methodologies that can, in addition, be combined with ROA. The latter is one of the robust decision-making tools that economists are claiming will guide decisions in climate change policy.

REFERENCES

Abadie, L.M. and J.M. Chamorro (2013), *Investment in Energy Assets under Uncertainty: Numerical Methods in Theory and Practice*, 1st edn, London: Springer.

Abadie, L.M. and I. Galarraga (2015), 'Managing energy price risk', in U. Chandra Sharma, R. Prasad and S. Sivakumar (eds), *Compendium of Energy Science and Technology*, Vol. 12, New Delhi: Studium Press LLC, p. 827.

Abadie, L.M., I. Galarraga and E. Sainz de Murieta (2016), 'Climate risk assessment under uncertainty: an application to main European coastal cities', *Frontiers in Marine Science*, **3**, 1–13.

Abadie, L.M., I. Galarraga and E. Sainz de Murieta (2017), 'Understanding risks in the light of uncertainty: low-probability, high-impact coastal events in cities', *Environmental Research Letters*, **12**(1), 014017.

Artzner, P., F. Delbaen, J.M. Eber and D. Heath (1999), 'Coherent measures of risk', *Mathematical Finance*, **9**(3), 203–28.

Brown, S., R.J. Nicholls, A.T. Vafeidis, J. Hinkel and P. Watkiss (2011), 'The impacts and economic costs of sea-level rise in Europe and the costs and benefits of adaptation', Final Report, Volume 1: Europe, The ClimateCost Project, Stockholm Environment Institute, Sweden.

Church, J.A., P.U. Clark, A. Cazenave, J.M. Gregory, S. Jevrejeva, A. Levermann, M.A. Merrifield, G.A. Milne, R.S. Nerem, P.D. Nunn, A.J. Payne, T. Pfeffer, D. Stammer and A.S. Unnikrishnan (2013), 'Sea level change', in T.F. Stocker, D. Qin, G.K. Plattner, M. Tignor, S.K. Allen, J. Boschung, A. Nauels, Y. Xia, V. Bex, P.M. Midgley (eds), *Climate Change 2013: The Physical Science Basis. Contribution of Working Group I to the Fifth Assessment Report of the Intergovernmental Panel on Climate Change*, Cambridge: Cambridge University Press, pp. 1137–216.

Dasgupta, S., B. Laplante, C. Meisner, D. Wheeler and J. Yan (2009), 'The impact of sea level rise on developing countries: a comparative analysis', *Climatic Change*, **93**, 379–88.

Grinsted, A., S. Jevrejeva, R. Riva and D. Dahl-Jensen (2015), 'Sea level rise projections for northern Europe under RCP8.5', *Climate Research*, **64**(1), 15–23.

Hallegatte, S., C. Green, R.J. Nicholls and J. Corfee-Morlot (2013), 'Future flood losses in major coastal cities', *Nature Climate Change*, **3**, 802–6.

Heberger, M., H. Cooley, P. Herrera, P.H. Gleick and E. Moore (2011), 'Potential impacts of increased coastal flooding in California due to sea-level rise', *Climatic Change*, **109**, 229–49.

Hinkel, J., D. Lincke, A.T. Vafeidis, M. Perrette, R.J. Nicholls, R.S.J. Tol, B. Marzeion, X. Fettweis, C. Ionescu and A. Levermann (2014), 'Coastal flood damage and adaptation costs under 21st century sea-level rise', *Proceedings of the National Academy of Sciences*, **111**(9), 3292–7.

IPCC (2013), 'Annex II: climate system scenario tables', in M. Prather, G. Flato, P. Friedlingstein, C. Jones, J.F. Lamarque, H. Liao and P. Rasch (eds), *Climate Change 2013: The Physical Science Basis. Contribution of Working Group I to the Fifth Assessment Report of the Intergovernmental Panel on Climate Change*, Cambridge and New York: Cambridge University Press.

Jevrejeva, S., A. Grinsted and J.C. Moore (2014), 'Upper limit for sea level projections by 2100', *Environmental Research Letters*, **9**(10), 104008.

Kopp, R.E., R.M. Horton, C.M. Little, J.X. Mitrovica, M. Oppenheimer, D.J. Rasmussen, B.H. Strauss and C. Tebaldi (2014), 'Probabilistic 21st and 22nd century sea-level projections at a global network of tide-gauge sites', *Earth's Future*, **2**(8), 383–406.

Lin, B.B., B.Y. Khoo, M. Inman, C.H. Wang, S. Tapsuwan and X. Wang (2014), 'Assessing inundation damage and timing of adaptation: sea level rise and the complexities of land use in coastal communities', *Mitigation and Adaptation Strategies for Global Change*, **5**, 551–68.

Markandya, A. (2014), 'Incorporating climate change into adaptation programmes and project appraisal: strategies for uncertainty', in A. Markandya, I. Galarraga, E. Sainz de Murieta (eds), *Routledge Handbook of the Economics of Climate Change Adaptation*, New York: Routledge International Handbooks, pp. 97–119.

Nicholls, R.J. and A.S. Kebede (2012), 'Indirect impacts of coastal climate change and sea-level rise: the UK example', *Climate Policy*, **12**, S28–S52.

Nordhaus, W.D. (2011), 'The economics of tail events with an application to climate change', *Review of Environmental Economics Policy*, **5**(2), 240–57.

Perrette, M., F. Landerer, R. Riva, K. Frieler and M. Meinshausen (2013), 'A scaling approach to project regional sea level rise and its uncertainties', *Earth System Dynamics*, **4**, 11–29.

Pindyck, R.S. (2011), 'Fat tails, thin tails, and climate change policy', *Review of Environmental Economics Policy*, **5**(2), 258–74.

Rockafellar, R.T. and S. Uryasev (2002), 'Conditional value-at-risk for general loss distributions', *Journal of Banking & Finance*, **26**(7), 1443–71.

Steffen, W., J. Hunter and L. Hughes (2014), 'Counting the costs: climate change and coastal flooding', Climate Council of Australia Ltd, Australia.

Stern, N. (2007), *The Economics of Climate Change: The Stern Review*, Cambridge: Cambridge University Press.

Watkiss, P., A. Hunt, W. Blyth and J. Dyszynski (2015), 'The use of new economic decision support tools for adaptation assessment: a review of methods and applications, towards guidance on applicability', *Climatic Change*, **132**(3), 401–16.

Weitzman, M.L. (2007), 'A review of the *Stern Review on the Economics of Climate Change*', *Journal of Economic Literature*, **45**(3), 703–24.

Weitzman, M.L. (2009), 'On modeling and interpreting the economics of catastrophic climate change', *Review of Economics and Statistics*, **91**(1), 1–19.

Weitzman, M.L. (2013), 'A precautionary tale of uncertain tail fattening', *Environmental and Resource Economics*, **55**(2), 159–73.

Wong, P.P., I.J. Losada, J.P. Gattuso, J. Hinkel, A. Khattabi, K.L. McInnes, Y. Saito and A. Sallenger (2014), 'Coastal systems and low-lying areas', in Field, C.B., V.R. Barros, D.J. Dokken, K.J. Mach, M.D. Mastrandrea, T.E. Bilir, M. Chatterjee, K.L Ebi, Y.O. Estrada, R.C. Genova, B. Girma, E.S. Kissel, A.N. Levy, S. MacCracken, P.R. Mastrandea and L.L. White (eds), *Climate Change 2014: Impacts, Adaptation, and Vulnerability. Part A: Global and Sectoral Aspects. Contribution of Working Group II to the Fifth Assessment Report of the Intergovernmental Panel on Climate Change*, Cambridge: Cambridge University Press, pp. 361–409.

Yamai, Y. and T. Yoshiba (2005), 'Value-at-risk versus expected shortfall: a practical perspective', *Journal of Banking & Finance, Risk Measurement*, **29**(4), 997–1015.

9. Economic growth and the social cost of carbon: additive versus multiplicative damages*
Armon Rezai, Frederick van der Ploeg and Cees Withagen

1. INTRODUCTION

Integrated assessment models of climate change aim to integrate economics and climate science and to assess the impact the economy has on the climate and vice versa. This analysis is a crucial step in the design of optimal policies to fight the potentially negative effects of climate change on economic well-being. There are various ways in which the relationship between economics and climate can be and is addressed. Our main aim is to investigate the implications of different types of modeling damage and production technology in a simple but well-calibrated framework of aggregate economic growth. We are particularly interested in the different kinds of modeling of the potential damage inflicted upon the economy and the substitutability of energy in production.

Global warming stems from the negative externality associated with the emission of CO_2 from the use of fossil fuels. This can be corrected by pricing carbon emissions at the social cost of carbon either via a carbon tax or a market for carbon emission permits. In the absence of distortions in raising public funds and other second-best issues, the optimal social cost of carbon corresponds to the present value of all future global warming damages from burning an extra unit of fossil fuel.[1] It matters *how* global warming damages are specified. Higher global temperatures lead to respiratory and other diseases and thus induce lower levels of health, productivity and aggregate output. Global warming also destroys productivity of agriculture and reduces aggregate output in that way. Both of these channels justify a specification with damages proportional to output with the proportion increasing in temperature. This is customary in applied integrated economic assessment models and equivalent to the assumption of unit-elasticity of substitution between output and damage. Nordhaus (2008) and Stern (2007) are prominent examples

---

* The first author is grateful for support from the OeNB Anniversary Fund grant on Green Growth (no. 14373). The last two authors are grateful for support from the ERC Advanced Grant "Political Economy of Green Paradoxes" (FP7-IDEAS-ERC Grant No. 269788). This chapter together with Rezai and van der Ploeg (2016, 2017) forms part of a larger research project on the optimal price of carbon and the specification of climate damage.

[1] Determining the social cost of carbon is less straightforward in a world with exhaustible fossil fuel, increasing efficiency of carbon-free alternatives, gradual and abrupt transitions from fossil fuel to renewables, fossil extraction cost, and endogenous growth and structural change. Here, one also needs to consider the interaction between carbon pricing and the market prices of fossil and renewable energy. These prices depend on expectations about future prices of fossil energy, back-stop technology and the degree of learning by doing (e.g., Rezai and van der Ploeg, 2017).

and this approach is also chosen in Tol (2002) and Golosov et al. (2014).[2] But higher global temperatures also lead to rising sea levels and destroy part of the capital stock, in which case damages might be proportional to the aggregate capital stock rather than to aggregate output. Global warming also leads to destruction of natural habitats, e.g., the coral reefs, and to less biodiversity. In that case, one might suppose that global warming damages are specified in final goods units and increases in temperature, not proportional to output or GDP (Stern, 2013). We refer to this as *additive* in contrast to the more usual case of *multiplicative* global warming damages. The additive damage specification implies an infinite elasticity of substitution between global warming damages and aggregate output. It also implies that with a positive rate of economic growth and with additive rather than with multiplicative damages, marginal damages do not rise with world GDP as the economy grows.

Our first objective is to analyze the effects of these seemingly innocuous differences in specification for both the level and the time profile of the social cost of carbon and the optimal carbon tax. Of course, there are other ways of specifying global warming damages (e.g. Stern, 2013, pp. 846–50). For example, one might relax the assumption that the proportion of what is left from production after global warming damages and the level of total factor productivity (the state of technical progress) appear separable in the production function. More importantly, the rate of technical progress rather than the level of output may suffer from global warming. This should capture that the assumption of exogenous technical progress is simply not realistic given the scale of disruption to output that might result from global warming. In fact, global warming may also adversely affect production factors such as the stock of capital, infrastructure or land directly. We will focus on the difference between multiplicative and additive global warming damages and abstract from these other ways of specifying damages in a first approximation. We will also abstract from capturing global warming damages directly in the social welfare function or, more implicitly, via a ceiling as is often done in the theoretical literature on climate policy.[3] Global warming damages have been carefully split up into its multiplicative production and its additive utility components (e.g., Barrage, 2020). As is well known, our multiplicative formulation is equivalent to additive linear damages in utility. A more

---

[2] Tol (2002) is a noteworthy exception in explicitly considering effects on ecosystems, vector borne diseases and heat and cold stress rather than subsuming them under production losses.

[3] First, a branch of the literature assumes catastrophic changes once the carbon concentration reaches or passes a deterministic or stochastic threshold because decay of atmospheric stops and climate change is drastic or irreversible (e.g., Tsur and Zemel (1996) and (1998)). A problem with this approach is that usually nothing goes wrong until the ceiling is reached, whereas one would expect damages to arise already for temperatures not too far from the ceiling (see also Dullieux et al. (2011) and Chakravorty et al. (2006)). Second, damages can appear as an externality in the social welfare function (e.g., van der Ploeg and Withagen (1991, 2012a, 2012b, 2014) and John and Pecchenino (1994)). Utility from consumption can be strongly separable from the damages from climate change or not. Bretschger and Smulders (2007) argue that in the latter case with Cobb–Douglas, production-balanced growth is feasible (assuming away exhaustibility of non-renewables). For the case of additive separability, Stokey (1998) shows that the growth process of the economy comes to an end, if more and more output is devoted to abatement. Weitzman (2009) considers additive and multiplicative damages in the social welfare function as well. He shows that the differences in optimal outcomes are considerable.

complex model would also have to account for the differential impact of climate change on factors of production and agents in the economy (e.g. Karp and Rezai, 2018).

Our second objective is to see how the optimal climate policy under multiplicative and additive specifications of global warming affects the amount of fossil fuel to be left in the crust of the earth and the timing at which the world switches to a carbon-free economy. We allow for scarcity of fossil fuel and stock-dependent extraction costs, so that costs of extraction rise as fewer reserves remain and reserves need not be fully depleted. This contrasts with many integrated assessment models in which fossil fuel reserves are typically abundant and extraction costs do not increase as fossil fuel reserves are depleted. Furthermore, in such models the elasticity of substitution between energy and a capital-labor composite is typically set to zero[4] whilst we allow for substitution possibilities. Fossil fuel demand at any point of time in such models does not depend on expectations about the price of the future renewable backstops and consequently the transition times simply occur when the price of fossil fuel, inclusive of the carbon tax, reaches the price of the renewable. In contrast, we will determine the levels and the time profiles of the social cost of carbon and the market prices of fossil and renewable energy within the context of a fully calibrated integrated assessment model of climate change and Ramsey growth with exhaustible fossil fuel, transition to carbon-free renewable energy sources, stock-dependent extraction costs, and technical progress in the production of renewable energy.

The price of fossil fuel contains two forward-looking components: the scarcity rent of fossil fuel (the present discounted value of all future increases in extraction costs resulting from extracting an extra unit of fossil fuel) and the social cost of carbon (the present discounted value of all future marginal global warming damages). This complicates the calculation of the transition times, since expectations about future developments such as technological progress in using the renewable matter. We study not only the social cost of carbon and market prices of all energy sources but also the optimal transition times for abandoning fossil fuel altogether as well as the amount of untapped fossil fuel. We derive our results based on a calibrated and much richer version of the analytical growth and climate model put forward in van der Ploeg and Withagen (2014) and take into account recent empirical findings by Hassler et al. (2011) on the substitutability of energy in production.

Our third objective is to demonstrate that there is a hump-shaped relationship between the optimal carbon tax and world GDP. In contrast, Golosov et al. (2014) offer a tractable Ramsey growth model which generates the result that an optimal carbon tax is proportional to GDP.[5] Their result depends on bold assumptions: logarithmic utility, Cobb–Douglas production, 100% depreciation of capital in each period, zero fossil fuel extraction costs, and multiplicative production damages captured by a negative exponential function. We find that their result is not robust in a general integrated assessment model of climate change and Ramsey growth with exactly the same carbon cycle, especially if the coefficient of intergenerational inequality aversion differs from unity.

---

[4] For example, the seminal study of Nordhaus (2008) assumes energy demand is exogenously decreasing.

[5] This formula is used already by others too (e.g., Hassler and Krusell, 2012; Gerlagh and Liski, 2018). Copeland and Taylor (1994) prose a similar framework.

Our results demonstrate that, if damages are not proportional to aggregate production output and the economy is along a development path, the social cost of carbon and the optimal carbon tax are smaller than with multiplicative damages, as damages can more easily be compensated for by higher output and damages do not increase with a growing economy. Consequently, the economy switches later from fossil fuel to the carbon-free backstop and leaves less oil in situ. If intergenerational inequality aversion is weaker (i.e., the elasticity of intertemporal substitution is larger), we show that the optimal carbon tax is still smaller with additive damages, but that the effect is less substantial. We find that with an elasticity of intertemporal substitution of 0.5 the social cost of carbon for additive damages from global warming is about half that for multiplicative damages from global warming.

In contrast, Weitzman (2009) finds for the same elasticity of intertemporal substitution, 0.5, that the optimal willingness to forsake current consumption to avoid future global warming is seven times larger with additive damages and a growth rate of 2% per annum. This effect disappears in a stagnant economy. Our results are not comparable as Weitzman (2009) has a partial equilibrium model whilst we derive our results within a fully specified general equilibrium model. This has the advantage that we can allow for growth and development from an initial capital stock that is below the steady state and for the exhaustibility of fossil fuel whilst Weitzman (2009) assumes an exogenous growth path. Furthermore, he deals with damages in utility and we focus on damages in production. However, Weitzman's insights do not survive once we look at *production* damages instead of *utility* damages in a fully fledged integrated climate assessment model.

The outline of this chapter is as follows. Section 2 sets out our general equilibrium model of climate change and Ramsey economic growth with additive and multiplicative global warming damages, exogenous population growth and labor productivity growth. Production combines energy with a capital-labor composite. Energy and the composite are imperfect substitutes in production. The two sources of energy, however, carbon-free energy and exhaustible fossil fuel, are perfect substitutes. The cost of carbon-free energy is exogenous and benefits from exogenous technical progress. The extraction cost of fossil fuel increases as fewer reserves are left in the crust of the earth. Our carbon cycle allows for permanent and transient components of atmospheric carbon and abstracts from positive feedback loops. We provide intuition for the different effects of additive and multiplicative damage and derive closed-form solutions for the social cost of carbon under simplifying assumptions. We show that the implications of different specifications of damage hinge on the growth prospects of the economy. In a growing economy, environmental policy is less ambitious with multiplicative than with additive damages. Section 3 uses a calibrated version of the model of section 2 and presents the simulations paying particular attention to the level and time profile of the optimal carbon tax as well as to how this tax affects the moments in time that the economy switches from fossil fuel to the carbon-free renewable and shows how the tax and transition times depend on whether global warming damages are additive or multiplicative and on how elastic energy use is. We present details for the price dynamics of fossil and renewable energy and also investigate the sensitivity of our core results with respect to the elasticity of substitution in production, the discount rate and the elasticity of intertemporal substitution. Section 4 concludes.

2. AN INTEGRATED ASSESSMENT MODEL OF RAMSEY GROWTH AND ENERGY TRANSITIONS

Using a simple Ramsey growth model, we derive the social cost of carbon for two different ways of modeling damages, additive and multiplicative. The social welfare function is utilitarian, with instantaneous per capita utility U depending on per capita consumption C_t/L_t, where L_t is total population, possibly non-constant over time, but exogenous. With ρ the constant rate of time preference social welfare is given by

$$\sum_{t=0}^{\infty}(1+\rho)^{-t}L_t U_t(C_t/L_t) = \sum_{t=0}^{\infty}(1+\rho)^{-t}L_t\left[\frac{(C_t/L_t)^{1-1/\eta}-1}{1-1/\eta}\right]. \tag{9.1}$$

U is of CES form and h is the elasticity of intertemporal substitution. The ethics of climate policy depend on how much weight is given on the welfare of future generations (and thus on how small r is) and on how small intergenerational inequality aversion is or how easy it is to substitute current for future consumption per head (how low $1/h$ is). Ceteris paribus, climate policy is most ambitious if society has a low rate of time preference and little inequality aversion (low r, high h).

Optimal climate policy takes place under a number of constraints in the form of a set of difference equations governing the global economy. First, output at time t, $Z(K_t, L_t, F_t + R_t)$, is produced using three inputs, man-made capital K_t, labor, L_t, and energy. We model two types of energy: fossil fuels like oil, natural gas and coal, F_t, and renewables, R_t, such as solar and wind energy. The aggregate general production function $Z(K_t, L_t, F_t + R_t)$ allows for imperfect factor substitution and perfect substitution between the two types of energy. Renewable energy is infinitely elastically supplied at potentially exogenously decreasing cost, b_t. Fossil fuel extraction cost at time t is $G(S_t)F_t$, with S_t the existing stock of fossil fuel reserves at the start of period t. Extraction becomes costlier as the less accessible fields have to be explored, $G' < 0$. We also allow for technical progress in aggregate and renewable energy production and for an exogenous profile for the time path of population growth. Mean temperature or the concentration of atmospheric carbon creates a convex combination of (ξ) multiplicative and ($1 - \xi$) additive climate damages (with $0 < \xi < 1$). This specification allows us to contrast multiplicative global warming damages in production with additive damages in production.

What is left of production after covering the cost of resource use and climate damage is allocated to consumption C_t, investments in man-made capital $K_{t+1} - K_t$, depreciation d K_t with a constant rate of depreciation d:

$$K_{t+1} = (1-\delta)K_t + (1-\xi D(T_t))Z(K_t, L_t, F_t + R_t)$$
$$+ (1-\xi)D(T_t)Z_0 - G(S_t)F_t - b_tR_t - C_t, \tag{9.2}$$

where damages, $D(T_t)$, increase with temperature and the initial stock of capital K_0 is given.

With $\xi = 1$, there are purely multiplicative damages and with $\xi = 0$ purely additive damages which are proportional to $Z_0 = Z(K_0, L_0, F_0 + R_0)$. The development of the finite fossil fuel stock follows from:

$$S_{t+1} = S_t - F_{1t}, \quad \sum_{t=0}^{\infty} F_t \leq S_0, \quad (9.3)$$

where initial reserves S_0 are given. We follow Golosov et al. (2014) in modeling a three-stock carbon cycle with carbon as fossil fuel reserves in the crust of the earth S, and a permanent component E_1 and a transient component E_2 of the stock of carbon in the atmosphere.[6] Equations (9.4), and (9.5) show the dynamics of the permanent and transient component of the stock of atmospheric carbon with j_L the fraction of emissions that stays up permanently in the atmosphere, j the speed at which the temporary component of the atmospheric stock of carbon decays, and j_0 a coefficient to calibrate how much of carbon is returned to the surface of the oceans and earth within a decade.

$$E_{1t} = E_{1t-1} + \varphi_L F_t, \quad (9.4)$$

$$E_{2t} = \varphi E_{2t-1} + \varphi_0 (1 - \varphi_L) F_t, \quad (9.5)$$

Ignoring the lags between stocks of atmospheric carbon and global warming discussed by Rezai and van der Ploeg (2016), van der Bijgaart et al. (2016), and Gerlagh and Liski (2018), we define global mean temperature, T_t, as the deviation from the pre-industrial temperature in degrees Celcius. The equilibrium climate sensitivity, ω, defines the rise in global mean temperature following a doubling of the total stock of carbon in the atmosphere, $E_t = E_{1t} + E_{2t}$. The usual formulation for radiative forcing capturing this relationship is:

$$T_t = \omega \ln\left(\frac{E_t}{596.4}\right) / \ln(2) \quad (9.6)$$

where 596.4 GtC is the IPCC figure for the pre-industrial stock of atmospheric carbon.[7] Using (9.6) we can thus write damages just as well as a function of the total stock of carbon in the atmosphere, $D(E_t)$.

---

[6] We focus on the effects of fossil fuel use on global warming in a detailed calibrated model of growth and climate change but following Golosov et al. (2014) and based on Archer (2005) and Archer et al. (2009) we adopt a tractable model of the carbon cycle which is linear and allows for decay of only part of the stock of atmospheric carbon. This model of the carbon cycle abstracts from a delay between the carbon concentration and global warming and the dynamics of multiple carbon reservoirs (e.g., Gerlagh and Liski, 2018). Abstracting from such a lag biases the estimate of the social cost of carbon and the carbon tax upwards. A more realistic model of the carbon cycle should also model the dynamics of the stocks of carbon in the upper and lower parts of the ocean and the time-varying coefficients originally put forward in the path-breaking paper of Bolin and Erikkson (1958). We also capture catastrophic losses at high levels of atmospheric carbon but abstract from positive feedback effects and the uncertain climate catastrophes that can occur in climate and growth models once temperature exceeds certain thresholds (e.g., Lemoine and Traeger, 2014; van der Ploeg and de Zeeuw, 2018).

[7] We abstract from a lag between temperature and atmospheric carbon stock, but Rezai and van der Ploeg (2016) discuss how the analysis is modified with such a lag. The implications of such a lag on optimal climate policy will be discussed within the context of our calibrated model in section 3.3 below.

The Lagrangian for the social planner's problem (maximize (9.1) subject to (9.2)–(9.6)) of our model of Ramsey growth and climate change reads as follows:

$$L \equiv \sum_{t=0}^{\infty}(1+\rho)^{-t}[L_t U_t(C_t/L_t) - \mu_t(S_{t+1} - S_t + F_t) + v_{1,t}(E_{1,t+1} - E_{1,t} - \varphi_L F_t)$$

$$+ v_{2,t}\{E_{2,t+1} - (1-\varphi)E_{2,t} - \varphi_0(1-\varphi_L)F_t\}] - \sum_{t=0}^{\infty}(1+\rho)^{-t}\lambda_t[K_{t+1} - (1-\delta)K_t$$

$$- D(E_t)\{\xi Z(K_t, L_t, F_t + R_t) + (1-\xi)Z_0\} + G(S_t)F_t + b_t R_t + C_t],$$

where μ_t denotes the shadow value of in-situ fossil fuel, v_{1t} and v_{2t} the shadow disvalue of the permanent and transient stocks of atmospheric carbon, and λ_t the shadow value of man-made capital. Necessary conditions for a social optimum are:

$$U'(C_t/L_t) = (C_t/L_t)^{-1/\eta} = \lambda_t, \quad (9.7a)$$

$$(1-\xi D_t)Z_{F_t+R_t} \leq G(S_t) + [\mu_t + \varphi_L v_{1t} + \varphi_0(1-\varphi_L)v_{2t}]/\lambda_t, \quad F_t \geq 0, \text{ c.s.}, \quad (9.7b)$$

$$(1-\xi D_t)Z_{F_t+R_t} \leq b_t, \quad R_t \geq 0, \text{ c.s.}, \quad (9.7c)$$

$$(1-\delta + (1-\xi D_{t+1})Z_{K_{t+1}})\lambda_{t+1} = (1+\rho)\lambda_t, \quad (9.7d)$$

$$\mu_{t+1} = (1+\rho)\mu_t + G'(S_{t+1})F_{t+1}\lambda_{t+1}, \quad (9.7e)$$

$$v_{1t+1} = (1+\rho)v_{1t} - D'(E_{t+1})\{\xi Z_{t+1} + (1-\xi)Z_0\}\lambda_{t+1}, \quad (9.7f)$$

$$(1-\varphi)v_{2t+1} = (1+\rho)v_{2t} - D'(E_{t+1})\{\xi Z_{t+1} + (1-\xi)Z_0\}\lambda_{t+1}. \quad (9.7g)$$

Equations (9.7a) and (9.7d) give the Euler equation for the growth in consumption per capita as an increasing function of the return on capital and decreasing function of the rate of time preference:

$$\frac{C_{t+1}/L_{t+1}}{C_t/L_t} = \left(\frac{1+r_{t+1}}{1+\rho}\right)^{\eta}, \quad r_{t+1} \equiv (1-\xi D_{t+1})Z_{K_{t+1}} - \delta, \quad (9.8)$$

The effect of the return on capital (r_{t+1}) on per-capita consumption growth now depends on the elasticity of intertemporal substitution (h) and the effect is stronger if the EIS is high (or intergenerational inequality aversion ($1/h$) is low).

Equation (9.7b) implies that, if fossil fuel is used, its marginal product should, again, equal its marginal extraction cost (which now equals $G(S_t)$) plus its scarcity rent (defined as $s_t \equiv \mu_t/\lambda_t$) plus the social cost of carbon ($\tau_t \equiv [\varphi_L v_{1t} + \varphi_0(1-\varphi_L)v_{2t}]/\lambda_t$). The scarcity rent and the social cost of carbon are defined in units of final goods (not utility units). If the marginal product of fossil fuel is below the total marginal cost (extraction cost *plus* scarcity rent *plus* social cost of carbon), it is not used. Equation (9.7c) states that, if the renewable is used, its marginal product must equal its marginal cost, b_t. We get:

$$(1-\xi D_t)Z_{F_t+R_t} \leq G(S_t) + s_t + \tau_t, \quad F_t \geq 0, \text{ c.s.}, \quad (9.9a)$$

$$(1 - \xi D_t)Z_{F_t + R_t} \leq b_t, \quad R_t \geq 0, \quad \text{c.s.} \tag{9.9b}$$

The dynamics of the scarcity rent follows from (9.7e) and (9.7d) and yields the Hotelling rule:

$$s_{t+1} = (1 + r_{t+1})s_t + G'(S_{t+1})F_{t+1} \text{ or } s_t = -\sum_{\varsigma=0}^{\infty}[G'(S_{t+1+\varsigma})F_{t+1+\varsigma}\Delta_{t+\varsigma}], \tag{9.10}$$

where the compound discount factors are $\Delta_{t+\varsigma} \equiv \prod_{\varsigma'=0}^{\varsigma}(1 + r_{t+1+\varsigma'})^{-1}, \varsigma \geq 0$. Hence, the scarcity rent of keeping an extra unit of fossil fuel unexploited must equal the present discounted value of all future reductions in fossil fuel extraction costs.

Finally, using (9.7f), (9.7g) and (9.7d), the *SCC*, defined as the present discounted value of all future marginal global warming damages from burning an additional unit of fossil fuel, equals:

$$\tau_t = -\sum_{\varsigma=0}^{\infty}[\{\varphi_L + \varphi_0(1 - \varphi_L)(1 - \varphi)^\varsigma\}\Delta_{t+\varsigma}D'(E_{t+1+\varsigma})\{\xi Z_{t+1+\varsigma} + (1-\xi)Z_0\}]. \tag{9.11}$$

It takes into account that one unit of carbon released from burning fossil fuel affects the economy in two ways: the first part remains in the atmosphere for ever and the second part gradually decays over time at a rate corresponding to roughly 1/300 per year.

2.1 Additive Global Warming Damages

The expression for the social cost of carbon (9.11) already allows us to make a first comparison between additive and multiplicative damages on two levels:

First, suppose we propose two feasible paths to the economy, which at each instant of time are identical with regard to investments, renewable inputs and fossil fuel use. They yield the same capital stock, resource stock and CO_2 concentration. The differences between the resulting consumption paths are given by $C_t^A - C_t^M = D(E_t)(Z_t - Z_0)$, where superscripts refer to additive and multiplicative damage. Hence, additive damages allow higher consumption levels (and the social cost of carbon is lower) if production is above Z_0, i.e. if the rate of economic growth is positive.

Second, consider the conditions from profit maximization: $(1 - D(E_t^M))Z_{F_t}^M = G(S_t^M) + s_t^M + \tau_t^M$ and $Z_{F_t}^A = G(S_t^A) + s_t^A + \tau_t^A$. The equations imply that the marginal product of fossil fuel equals the direct extraction cost plus the social cost of carbon plus the scarcity rent. The first equation holds for the multiplicative case, the latter for the additive case. If two economies would follow identical paths then the extraction costs are identical, the marginal potential products of fossil fuels are identical, as well as the scarcity rents. It then follows that this would require higher carbon taxes in the multiplicative economy. Hence, since the taxes represent the *SCC*, the *SCC* is higher in the multiplicative economy.

2.2 The Social Cost of Carbon Under Additive Global Warming Damages

A tractable model of the optimal carbon tax has been put forward by Golosov et al. (2014) based on a decadal Ramsey growth model. Relying on logarithmic utility, Cobb–Douglas production function for capital, labor and energy, 100% depreciation each period (and thus has a coarse calibration grid), exponential damages, and labor-only energy production costs, they show that the *social cost of carbon (SCC)* is proportional to the optimal level of GDP and independent of technology. Following the exposition of Rezai and van der Ploeg (2014), the optimal carbon tax of Golosov et al. (2014) can be generalized to different specifications of damage. Under the assumptions of Golosov et al. (2014), our general carbon tax of the Ramsey model, equation (9.11), simplifies to

$$\tau_t = \left[\left(\frac{1+\rho}{\rho-n}\right)\varphi_L + \left(\frac{1+\rho}{1+\rho-\varphi(1+n)}\right)\varphi_0(1-\varphi_L)\right]D'(E_t)\{\xi Z_t + (1-\xi)Z_0\}. \quad (9.11')$$

Comparing equations for different specifications of damage (i.e. different values of ξ), we see that

$$[\tau_t]^{additive} < [\tau_t]^{multiplicative} \Leftrightarrow Z_0 < Z_t.$$

The social cost of carbon is, thus, higher under multiplicative damages if the rate of economic growth (in potential output) is positive (i.e. $Z_t > Z_0$, $t > 0$). The reason is, of course, that positive economic growth implies that marginal damages are growing under multiplicative but not under additive global warming damages and consequently the optimal social cost of carbon as a fraction of GDP is higher under multiplicative than under additive damages. Since the elasticity of GDP with respect to the carbon tax is likely to be less than one, we conclude that the social cost of carbon will be higher under additive than under multiplicative damages as damages no longer grow with GDP.

2.3 Policy Scenarios

The missing market for carbon permits is the only externality in our model and the social optimum can be implemented in the market economy with a specific carbon tax t_t which is set to the social cost of carbon (9.11). Under "laissez-faire" the climate externality remains uncorrected, i.e. $t_t = 0$.

In principle, three regimes can occur in our fully specified Ramsey model: a regime with only fossil fuel use, a regime with only use of renewable energy and a regime with simultaneous use of fossil fuel and renewable energy. In our numerical IAM, outlined in the next section, it is optimal to start with an initial phase with only fossil fuel use since initially renewable energy is not competitive. This holds for additive as well as multiplicative damages. After some time, renewable energy is phased in and an intermediate phase with simultaneous use of fossil fuel and renewable energy commences. After some more time fossil fuel is phased out and the final carbon-free era starts. Since fossil fuel extraction costs become infinitely large as reserves are exhausted, fossil fuel reserves will not be fully exhausted and thus some fossil fuel will be left untapped in the crust of the earth at the end of the intermediate phase. From that moment on, the in-situ stock of fossil fuel will remain unchanged, but the carbon in the atmosphere will gradually decay leaving ultimately only the permanent component of the carbon stock.[8]

During the initial phase, fossil fuel demand follows from (9.7b), holding with equality. Setting its marginal product, $(1 - \xi D_t)Z_{F_t}$, to the sum of extraction cost,

---

[8] For our calibration it is never optimal to use oil again, despite the decrease in the damages.

scarcity rent and carbon tax, $G(S_t) + s_t + \tau_t$. We have strict inequality in the second part of (9.7c). During the intermediate phase fossil fuel and renewable demand follow from $(1 - \xi D(E_t))Z_{F_t+R_t} = G(S_t) + s_t + \tau_t = b_t$. Since fossil fuel and renewable energy are perfect substitutes, simultaneous use of the two energy types in a competitive economy beyond one period of time, requires a renewable subsidy or a carbon tax.[9] Simultaneous can thus not be ruled out in an optimum. During the final phase, we have $(1 - \xi D(E_t))Z_R(L_t, K_t, R_t, t) = b_t$, which gives renewable use as an increasing function of capital and a decreasing function of global mean temperature or the concentration of carbon in the atmosphere.

The time profile of the carbon tax is crucial in determining whether it is optimal to return to simultaneous use after a carbon-free era. The carbon tax suggested by Golosov et al. (2014) to be rephrased in (9.11′) sets it proportional to output and in equilibrium this will be proportional to optimal output. We find that this rule is a poor approximation to the optimal carbon tax in our model of economic growth and climate change with exogenous growth in technical progress and population. We find that it is optimal for the carbon tax to be hump-shaped, since it must fall in the carbon-free era as the temporary component of atmospheric carbon dissipates. If the market price of fossil energy falls below the market price of renewables in this transition, (partial) re-switching to fossil fuel is optimal.

One of our key objectives is to study the optimal timing of transitions from introducing renewable energy alongside fossil fuel and from phasing out fossil fuel altogether because in most of the prevailing integrated assessment models these transitions are exogenous. We are interested in how the timing of these transitions is affected by different assumptions on the climate-economy and energy-output relationships, for example, by how much do optimal carbon taxes bring forward the carbon-free era when damages are additive or multiplicative damage and elasticity of substitution between energy and capital is high or low. The stock of fossil fuel to leave untapped in the earth at the end of the intermediate phase follows from the condition that the economy is indifferent between fossil fuel and the renewable and that the scarcity rent has vanished:

$$G(S_t) + \tau_t < b_t, 0 \leq t < t_{CF}, \text{ and } G(S_t) + \tau_t \geq b_t, S_t = S_{t_{CF}}, \forall t \geq t_{CF}. \quad (9.12)$$

where t_{CF} is the time at which the economy for the first time relies on using only the renewable (carbon free). The amount of fossil fuel to be left in situ increases in the renewable subsidy and the carbon tax.

3. POLICY SIMULATION AND OPTIMIZATION

In our numerical simulations time runs from 2010 till 2600 and is measured in decades, $t = 1, 2, \ldots, 60$, so period 1 corresponds to 2010–20, period 2 to 2020–30, etc. The final

---

[9] Technological progress lowers the market price of renewable energy. Since both energy sources are perfect substitutes, simultaneous use would imply that it is optimal to sell energy at a lower price in the future rather than meeting full demand at a higher price today. This cannot be the case under positive discounting.

Table 9.1 Policy scenarios for the setting of the global carbon tax

	First best τ_t	Zero ("Laissez faire")	Proportional to GDP (9.11')
multiplicative damages ($\xi = 1$)	———	– – – –	- - - - - -
additive damages ($\xi = 0$)	———	- - - -	- - - - - -

time period is $t = 60$ or 2600–10, but we highlight the transitional dynamics in the earlier parts of the simulation. The functional form and calibration of the carbon cycle, temperature module and global warming damages are discussed in more detail in the appendix. The functional forms and benchmark parameter values for the economic part of our IAM of growth and climate change are also discussed in the appendix. On the whole, our benchmark parameter values assume relatively low damages, low fossil fuel extraction cost and a high cost for renewable energy. This biases our model toward fossil fuel use.

We report full results for three simulations with multiplicative damages ($\xi = 1$) and another three simulations with additive damages ($\xi = 0$). We consider three scenarios for each type of damages. First, the first-best optimum. Second, the "laissez faire" outcome (not taking damages into account and setting the carbon tax to zero). Third, a scenario based on a carbon tax set according to a proportional rule, as proposed by Golosov et al. (2014), (9.11'), where solving the rule together with the equations of the model ensures that the carbon tax reacts to optimal GDP. The reported simulations use an elasticity of substitution between energy and the capital-labor aggregate equal to $\vartheta = 0.5$. Table 9.1 illustrates the six simulations and the coding that is used to distinguish them in the simulation figures. We also analyze the sensitivity of the social cost of carbon with respect to the elasticities of intertemporal and factor substitution and the social rate of discount. All our simulations of optimal policies allowed in principle for simultaneous use of the two types of energy, but we found that it rarely occurred in our simulations and if it occurred never more than for one period of a decade.

3.1 Climate Policy is More Ambitious Under Multiplicative than Under Additive Damage

We start with the *first-best* outcomes and first compare additive and multiplicative damages. These correspond to, respectively, the solid light and dark lines in Figure 9.1. The first, second and third panels show aggregate consumption, total net output and the aggregate capital stock. Over the entire period of time under consideration output net of damage is monotonically increasing. Moreover, net output is almost the same in both situations. The same holds for the capital stock and consumption. Surprisingly, this implies that the total welfare is hardly affected by whether the function capturing global warming damages is additive or multiplicative. Still, there are essential differences in how this is achieved under these two types of specification for damages. These differences concern mostly the use of fossil fuel and the timing of the transition to renewable energy. The economy's endowments and technological change allow the economy to grow. So,

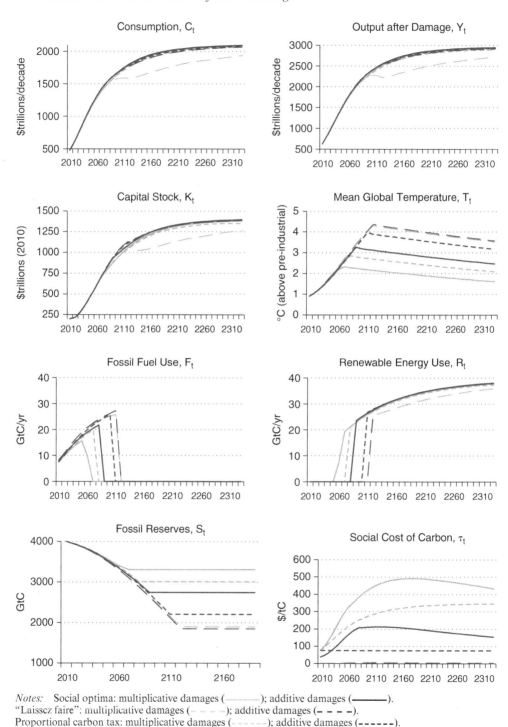

Figure 9.1 Simulations with CES production technology ($\vartheta = 0.5$)

Table 9.2 *Transition times and carbon budget*

		Fossil fuel Only	Simultaneous use	Renewable Only	Carbon used	max. T
multip.	First best	2010–2050	2060	2070 –	690 GtC	2.3°C
	"Laissez faire"	2010–2110	x	2120 –	2100 GtC	4.3°C
	Proportional tax	2010–2070	x	2080 –	990 GtC	2.8°C
additiv.	First best	2010–2050	x	2090 –	1250 GtC	3.2°C
	"Laissez faire"	2010–2110	x	2120 –	2150 GtC	4.3°C
	Proportional tax	2010–2100	x	2110 –	1800 GtC	3.9°C

if the economy with multiplicative global warming damages would use the same rate of fossil fuel, damages in terms of loss of production would be much higher over time. Therefore, the economy with multiplicative damages uses less fossil fuel, leaves more fossil fuel unexploited and makes the transition to renewables at an earlier stage. Despite the effects on consumption, capital and output being small, the differences in timing of energy use and how much fossil fuel is stranded are considerable – see also Table 9.2. With multiplicative damages 700 GtC are burnt, and the transition to renewables takes place as soon as 2050. For the additive case, much less fossil fuel is left in situ, i.e. 1250 GtC are burnt, and renewables are phased in much later, i.e., during 2090. This difference in fossil fuel consumption leads to higher temperature trajectories under additive damages than under multiplicative damages, but despite the higher damages the effects on consumption, output and capital and thus on welfare is not large. This leads us to conclude that climate change, if addressed through optimal policy, can be avoided at relatively low costs. Depending on the nature of climate damages, the costs of inaction are potentially large.

Under multiplicative damages, temperature slightly overshoots the 2°C warming limit, peaking at 2.3°C, whereas with additive damages temperature peaks at 3.2°C above pre-industrial levels. The imposition of the carbon tax ensures that the transition to the carbon-free era is speeded up. In both cases the optimal path for the social cost of carbon and global carbon tax is inverted U-shaped, which results from the fact that CO_2 emissions first rise and then come to an end. The location of the two curves is, of course, different. In the multiplicative case the social cost of carbon starts at a level of 75$/tC and reaches a maximum of 490$/tC in the year 2180. With additive damages the carbon tax starts with half that under multiplicative damages, 37.5$/tC, and the maximum is reached at 210$/tC in the year 2120.

Weitzman (2010) finds that with additive global warming damages (in instantaneous welfare) the willingness to sacrifice current consumption to avoid future global warming is seven times higher than with multiplicative damages. In contrast, we find that additive damages lead to half the social cost of carbon at each point of time compared to multiplicative damages. However, this stark difference should serve as a reminder that the additive utility damages used in Weitzman and the additive production damages used in our model are not comparable. Weitzman's number applies to the willingness to pay to avoid any change in temperature along an exogenous consumption trajectory, whereas the social cost of carbon of our simulations reports the willingness to pay to avoid a marginal increase in atmospheric carbon along the optimal path. Our numerical results

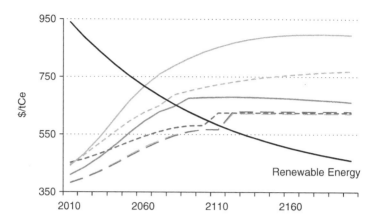

Notes: Social optima: multiplicative damages (———); additive damages (———).
"Laissez faire": multiplicative damages (– – – –); additive damages (– – – –).
Proportional carbon tax: multiplicative damages (- - - - -); additive damages (- - - - - -).

Figure 9.2 The market prices of energy

are in line with the analytical findings in section 2: the social cost of carbon is higher under multiplicative damages provided that the rate of economic growth (which changes over time and is endogenous) is positive. Within a fully specified integrated climate assessment model additive damages lead to a less ambitious climate policy. The nature of climate damage is irrelevant for laissez-faire fossil fuel use and the climate dynamics it implies.

3.2 Time Paths for the Market Price of Fossil Fuel and the Renewable in the Various Scenarios

The imposition of a carbon tax increases the too low (relative to the first-best) market price of fossil energy. The prices of all energy sources under the different scenarios are depicted in Figure 9.2. The energy price in the first-best outcome is the shadow price of fossil fuel, which consists of the marginal extraction cost, the Hotelling rent (the present discounted sum of all extraction cost savings due to a higher fossil fuel stock) and the social cost of carbon.

The shadow price of fossil fuel increases initially because all three components of the social cost increase. Since the social cost of carbon is higher under multiplicative damages, the carbon tax is higher initially and rises faster than under additive damages. This leads to a higher market price of fossil energy and induces lower fossil fuel consumption and higher in situ stocks as discussed above. Once the market price of fossil energy exceeds the market cost of renewable energy, renewable energy takes over. From then on, the marginal extraction cost of fossil fuel is constant and the Hotelling rent is zero. The reason is that some fossil fuel is left in the crust of the earth, but no extraction takes place. However, the social cost of carbon continues to rise for some time, because decay of atmospheric carbon is limited and consumption is increasing, yielding smaller marginal utility of consumption and thus a higher social cost of carbon, expressed in the numeraire (see equation (9.11)). However, after some point of time decay of atmospheric

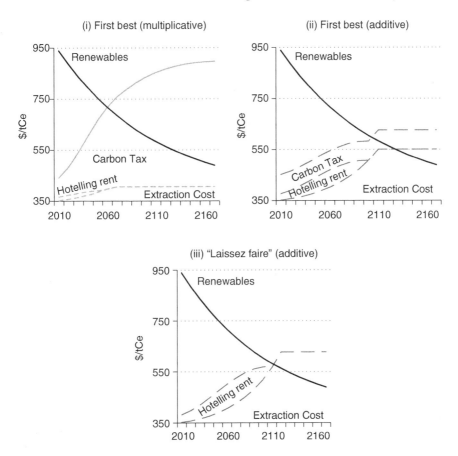

Figure 9.3 Decomposition of the market price of fossil relative to renewable energy

carbon dominates the decrease in marginal utility of consumption and the social cost of carbon starts to fall.[10] Under "laissez faire" the *SCC* is not imposed on the market price of fuel which consequently is lower. We are not finding any Green Paradox effects in our simulations (cf. van der Ploeg and Withagen, 2012a). Figure 9.2 also illustrates the sub-optimality of the proportionality rule in (9.11′) compared to the fully numerically optimized model. With additive damages the proportional carbon tax increases the market price of fossil energy beyond its optimal level initially but fails rise rapidly enough to curb carbon emissions sufficiently. Figure 9.3 presents a decomposition of the market price of fossil energy for selected scenarios.

---

[10] In theory, the falling social cost of carbon can decrease the market price of fossil energy sufficiently to make fossil fuel competitive again. As Figure 9.2 indicates, the permanent stock of atmospheric carbon and extraction costs are too high and the time horizon too short to make re-switching optimal.

214 *Handbook on the economics of climate change*

Under multiplicative damages the *SCC* is large relative to the other cost components of the market price of fossil energy, namely the scarcity rent and extraction cost. The carbon tax rises rapidly and induces a transition to renewable energy mid of the current century. Under additive damages the *SCC* is significantly lower, allowing for more extraction of fossil fuel. Higher extraction increases the value of the in situ resource and the scarcity rent.

The dynamics of the "laissez faire" market price of fossil energy are essentially identical under additive and multiplicative damages as pointed out in section 2 and can be seen in Figure 9.2. As the carbon tax is set to zero in both cases, almost the same amount of fossil fuel is used for approximately the same period of time, until 2120.[11] Again, higher cumulative extraction increases the Hotelling rent. At the end of the fossil era slightly more fossil fuel is left in situ in the economy with multiplicative damages than in the one with additive damages (see Figure 9.2). But, as is to be expected, damages to production are much higher in the multiplicative case, and therefore consumption will be lower. This becomes particularly manifest toward the end of this century. An interesting feature of the simulations is that with multiplicative damages capital is decreasing for several decades immediately after the economy stops using fossil fuel. We also see that, in spite of higher input of fossil fuel, net output decreases over a short period of time, preceding the transition to renewable energy. This indicates that in the fossil fuel phase capital is over-accumulated, which is then corrected in the phase where renewable energy is used. We also observe that much more fossil fuel is used in the absence of a carbon tax and that the transition to renewable energy takes place much later.

3.3 No Policy Leads to Overinvestment and Too Little Use of the Renewable Energy

Although not our primary focus, it is interesting to see how the market outcome differs from the first-best outcome. Output, consumption and capital accumulation take place at very similar levels for the first-best and "laissez-faire" outcomes under additive damages. The reason for this is mainly that in a growing economy net output is not much affected by temperature changes if affordable mitigation is available (as in the first-best scenarios) or damages are low (as in the additive BAU scenario). Under multiplicative damages of global warming the impacts of the climate externality are large enough to drastically change accumulation paths. Table 9.3 summarizes total welfare relative to multiplicative first-best. "Laissez-faire" yields a welfare loss of 17% of initial GDP relative to first-best under multiplicative damage. If damages are additive, welfare falls by a mere 3% due to zero tax on carbon.

Sinn (2007) and Stern (2010) point out that "laissez-faire" leads to an inefficient allocation of resources, because economic decision makers do not recognize the deleterious effects of greenhouse gas (GHG) emissions. Private and social cost calculations diverge; agents overvalue the returns to conventional capital stock and undervalue the investments

---

[11] In the "laissez faire" scenario the social cost of carbon does not vanish completely, which is due to the numerical implementation of the program where each individual agent is aware of the fact that she is responsible for less than 1% of total emissions and thus for some damage (see appendix for details on "laissez faire").

Table 9.3 Welfare gains and losses in % of initial GDP (relative to multiplicative first-best)

	multiplicative	additive
First-best	0%	5%
"Laissez faire"	–17%	2%

in green energy sources. Imperfect price signals (λ, μ, ν) induce excess fossil fuel extraction and capital accumulation, leading to high climate damages over the time horizon. The inefficiency of "laissez faire" manifests itself in low consumption to allow accumulation in early periods of the program leading to low consumption in the future due to high climate damages.[12]

Comparing welfare across different specification in Table 9.3, scenarios under additive global warming damages yield higher welfare than multiplicative damages even if no carbon tax is imposed. The reason is, again, that positive economic growth implies that marginal damages are growing under multiplicative but not under additive global warming damages and rising production costs of fossil energy drive the transition to renewable energy rather than mounting environmental damage. The higher social cost of carbon under multiplicative damages brings forward the carbon-free era but also increases the cost of the energy transition (market prices of energy increase by at most 10%) and lowers consumption (by at most 3%). This implies that welfare decreases by 5% of today's GDP under multiplicative damages under first-best relative to the outcome under additive damages.

We conclude from Table 9.3 that the climate problem is potentially large if not addressed by optimal policy. Under "laissez faire" the nature of global warming damages matters greatly. In the social optimum the problem of climate change can be managed at relatively modest cost. Interestingly, whether damages are additive or multiplicative leads to small differences in welfare under the optimal carbon tax but to large differences in welfare under "laissez faire".

3.4 The Optimal Carbon Tax is Not Proportional to Aggregate Consumption or World GDP

To examine whether the linear formula for the optimal carbon tax is really proportional to global GDP as suggested by Golosov et al. (2014) and demonstrated in equation (9.11'), we examine whether it holds up in a more general integrated assessment model of Ramsey growth and climate change. Figure 9.4 therefore plots the ratio of the optimal carbon tax to both world GDP and aggregate consumption; the short-dashed lines in Figure 9.1 provide further details. For sake of comparison we also use equation (9.11') to plot a similar simple formula for the optimal carbon tax when global warming damages are additive.

We immediately observe that the optimal carbon taxes (solid lines) are not well described by a constant proportion of world GDP or aggregate consumption. The general

---

[12] Rezai et al. (2012) discuss this mechanism in more detail and demonstrate the relevance of this inefficiency for the debate on the (opportunity) cost of climate change in a simpler model of Leontief production technology and unlimited stocks of oil.

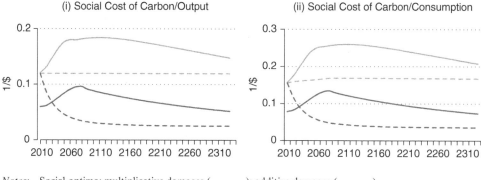

Notes: Social optima: multiplicative damages (———); additive damages (———).
Proportional carbon tax: multiplicative damages (– – –); additive damages (- - - -).

Figure 9.4 The social cost of carbon as ratio of aggregate world GDP and consumption

pattern is that during the initial phases of fossil fuel use the social cost of carbon rises as a proportion of world GDP as more carbon emissions push up marginal damages of global warming whilst during the carbon-free phases the social cost of carbon falls as a proportion of world GDP as a significant part of the stock of carbon in the atmosphere is gradually returned to the surface of the oceans and the earth.

Further sensitivity analysis shows that setting the carbon tax to a constant proportion of world GDP is a poor approximation to the optimal carbon tax under multiplicative or additive climate damages, regardless of whether the elasticity of factor substitution is zero (Leontief) or one (Cobb–Douglas), the elasticity of intertemporal substitution is high or low, and the pure rate of time preference is high or low. These simple formulas for the optimal carbon tax are, in fact, non-optimal and induce more fossil fuel to be burnt (about 50% more relative to the first-best, see also Table 9.2) and more severe climate change. They are an especially poor description of the optimal carbon tax under additive global warming damages, since the carbon tax then falls during the fossil fuel era as a result of economic growth whilst the optimal carbon tax should be increasing during that period (see Figure 9.4 for details). In general, the social cost of carbon under multiplicative damage is about 2 to 2½ times its value under additive damages. Weitzman (2009) uses a partial equilibrium model and finds that the willingness to pay, in terms of giving up present consumption, for reducing future temperature is seven times higher in the additive case compared to the multiplicative case along a suboptimal trajectory. In our fully fledged integrated climate assessment model we consistently find the opposite result.

4. CONCLUSIONS

How global warming damages are modeled and calibrated matters for the social cost of carbon and climate policy. We find that the climate policy is less ambitious, energy use higher, the stock of fossil fuel left in situ lower, global mean temperature higher and the optimal carbon tax lower with additive damages provided that the rate of economic growth (in potential output) is positive. This ranking is independent of how tough society finds it to substitute present for future consumption, the social rate of discount, and the elasticity of substitution of energy in production. Interestingly, the time paths for global consumption, capital and GDP are not much affected by whether damages are additive or multiplicative despite the temperature path being higher under additive damages. Stern (2013) criticizes the current generation of IAMs for focusing on too limited a set of functional and parametric relationships. Our analysis is a first step in broadening the scope, admittedly, leaving ample room for further improvement. Our analysis, however, also leads us to conclude that climate change, if addressed through optimal policy, can be avoided at relatively low costs. Depending on the nature of climate damages, the costs of inaction are potentially large even if many additional reasons for concern Stern highlights are not taken into account.

Our integrated assessment model also indicates that a higher elasticity of intertemporal substitution and a lower social rate of discount lead to a higher optimal carbon tax and a quicker phasing in of renewables and more fossil fuel left in the crust of the earth, less so under additive than multiplicative global warming damages. A higher elasticity of intertemporal substitution corresponds to a lower coefficient of intergenerational inequality aversion. Since society is more concerned with fighting global warming than with avoiding big differences in consumption of different generations, the carbon tax is borne much more by earlier generations than by later generations, both in the additive and in the multiplicative case.

Golosov et al. (2014) have offered a tractable fully consistent general equilibrium model of climate change and Ramsey growth which yields the optimal carbon tax a simple proportion of world GDP. However, they employ unrealistically low damages at higher temperatures and need to make some very bold assumptions to ensure that both aggregate consumption and the carbon tax are a fixed proportion of world GDP. Our model of Ramsey growth and climate change has more realistic global warming damages, capital-intensive extraction costs, CES instead of Cobb–Douglas utility and production functions, and realistic time profiles for the evolution of population and technology. We find that the optimal carbon tax is a hump-shaped function with the carbon tax falling in the carbon-free era as temporary component of the atmospheric stock of carbon fades away. Our findings of a hump-shaped carbon tax and a lower carbon tax are robust to changes in key parameters of the model.

Finally, our analysis also pays careful attention to how fast and how much fossil fuel should be abandoned and how quickly and how much renewable energy should be phased in. Future developments in the productive capacity of the economy are important determinants of the social cost of carbon and the optimal carbon tax. Future prices of clean and dirty sources of energy and their necessity in the general production process heavily influence relative prices and the allocation of resources today. We have examined the effects of variations in the substitutability between energy and conventional capital through a CES production function with a fixed rate of technological progress. Recent contributions by Acemoglu et al. (2012) and Mattauch et al. (2015) highlight the importance of learning and lock-in effects by making the rate of technical progress as endogenous. It is possible to use the empirical estimates of the determinants of growth rates in total factor and energy productivities given in Hassler et al. (2011) in our model of economic growth and climate change. This will allow much more substitution possibilities between energy and the capital-labor aggregate in the long run than in the short run. The logic of directed technical change suggests that it is more important to have substantial R&D subsidies for green technology to kick-start green innovation and fight global warming.

REFERENCES

Acemoglu, D., P. Aghion, L. Bursztyn and D. Hemous (2012), 'The environment and directed technical change', *American Economic Review*, **102**(1), 131–66.
Archer, D. (2005), 'The fate of fossil fuel CO_2 in geologic time', *Journal of Geophysical Research*, **110**, C09S05.
Archer, D., M. Eby, V. Brovkin, A. Ridgwell, L. Cao, U. Mikolajewicz, K. Caldeira, K. Matsumoto, G. Munhoven, A. Montenegro and K. Tokos (2009), 'Atmospheric lifetime of fossil fuel carbon dioxide', *Annual Review of Earth and Planetary Sciences*, **37**, 117–34.
Barrage, L. (2020), 'Optimal dynamic carbon taxes in a climate-economy model with distortionary fiscal policy', *Review of Economic Studies*, **87**(1), 1–39. https://doi.org/10.1093/restud/rdz055.
Bijgaart, I. van den, R. Gerlag and M. Liski (2016), 'A simple formula for the social cost of carbon', *Journal of Environmental Economics and Management*, **77**, 75–94.
Bolin, B. and B. Eriksson (1958), 'Changes in the carbon dioxide content of the atmosphere and sea due to fossil fuel combustion', in B. Bolin (ed.), *The Atmosphere and the Sea in Motion: Scientific Contributions to the Rossby Memorial Volume*, New York: Rockefeller Institute Press, pp. 130–42.
Bretschger, L. and S. Smulders (2007), 'Sustainable resource use and economic dynamics', *Environmental and Resource Economics*, **36**, 1–13.
Chakravorty, U., B. Magne and M. Moreaux (2006), 'A Hotelling model with a ceiling on the stock of pollution', *Journal of Economic Dynamics and Control*, **30**, 2875–904.
Copeland, B. and M.S. Taylor (1994), 'North–South trade and the environment', *Quarterly Journal of Economics*, **109**, 755–87.
Dullieux, R., L. Ragot and K. Schubert (2011), 'Carbon tax and OPEC's rents under a ceiling constraint', *Scandinavian Journal of Economics*, **113**, 798–824.
Gerlagh, R. and M. Liski (2018), 'Consistent climate policies', *Journal of the European Economic Association*, **16**(1), 1–44.
Golosov, M., J. Hassler, P. Krusell and A. Tsyvinski (2014), 'Optimal taxes on fossil fuel in general equilibrium', *Econometrica*, **82**(1), 48–88.
Hassler, J. and P. Krusell (2012), 'Economics and climate change: integrated assessment in a multi-region world', *Journal of the European Economic Association*, **10**(5), 974–1000.
Hassler, J., P. Krusell and C. Olovsson (2011), 'Energy-saving technical change', mimeo, Institute of International Economic Studies, Stockholm University.
IEA (2008), 'World energy outlook 2008', accessed April 1, 2018 at http://www.iea.org/textbase/nppdf/free/2008/weo2008.pdf.
John, A. and R. Pechenino (1994), 'An overlapping generations model of growth and the environment', *Economic Journal*, **104**, 1393–410.
Karp, L. and A. Rezai (2018), 'Climate policy and asset prices', mimeo.
Lemoine, D. and C.P. Traeger (2014), 'Watch your step: optimal policy in a tipping climate', *American Economic Journal: Economic Policy*, **6**, 1–31.
Mattauch, L., F. Creutzig and O. Edenhofer (2015), 'Avoiding carbon lock-in: policy options for advancing structural change', *Economic Modelling*, **50**, 49–63.
Nordhaus, W. (2008), *A Question of Balance: Economic Models of Climate Change*, New Haven, CT: Yale University Press.
Ploeg, F. van der and C. Withagen (1991), 'Pollution control and the Ramsey problem', *Environmental and Resource Economics*, **1**, 215–36.
Ploeg, F. van der and C. Withagen (2012a), 'Is there really a Green Paradox?', *Journal of Environmental Economics and Management*, **64**(3), 342–63.
Ploeg, F. van der and C. Withagen (2012b), 'Too much coal, too little oil', *Journal of Public Economics*, **96**, 62–77.
Ploeg, F. van der and C. Withagen (2014), 'Growth, renewables and the optimal carbon tax', *International Economic Review*, **55**, 283–311.
Ploeg, F. van der and A. de Zeeuw (2017), 'Climate tipping and economic growth: precautionary capital and the price of carbon', *Journal of the European Economic Association*, **16**(5), 1577–617.
Rezai, A. (2011), 'The opportunity cost of climate policy: a question of reference', *Scandinavian Journal of Economics*, **113**, 885–903.
Rezai, A. and F. van der Ploeg (2014), 'Robustness of a simple rule for the social cost of carbon', *Economics Letters*, **132**, 48–55.

Rezai, A. and F. van der Ploeg (2016), 'Intergenerational inequality aversion, growth and the role of damages: Occam's rule for the global carbon tax', *Journal of the Association of Environmental and Resource Economics*, **3**, 493–522.
Rezai, A. and F. van der Ploeg (2017), 'Abandoning fossil fuel: how fast and how much?', *Manchester School*, **85**, e16-e44.
Rezai, A., D. Foley and L. Taylor (2012), 'Global warming and economic externalities', *Economic Theory*, **49**, 329–51.
Sinn, H.W. (2007), 'Pareto optimality in the extraction of fossil fuels and the greenhouse effect: a note', NBER Working Paper No. 13453.
Stern, N. (2007), *The Economics of Climate Change: The Stern Review*, Cambridge: Cambridge University Press.
Stern, N. (2010), 'Imperfections in the economics of public policy, imperfections in markets, and climate change', *Journal of the European Economic Association*, **8**, 253–88.
Stern, N. (2013), 'The structure of economic modeling of the potential impacts of climate change: grafting gross underestimation of risk onto already narrow science models', *Journal of Economic Literature*, **51**(3), 838–59.
Stokey, N. (1998), 'Are there limits to growth?', *International Economic Review*, **39**, 1–31.
Tol, R. (2002), 'Estimates of the damage costs of climate change', *Environmental and Resource Economics*, **21**, 135–60.
Tsur, Y. and A. Zemel (1996), 'Accounting for global warming risks: resource management under event uncertainty', *Journal of Economic Dynamics and Control*, **20**, 1289–305.
Tsur, Y. and A. Zemel (1998), 'Pollution control in an uncertain environment', *Journal of Economic Dynamics & Control*, **22**, 967–75.
Weitzman, M. (2009), 'Additive damages, fat tailed climate dynamics and uncertain discounting', *Economics: The Open-Access, Open-Assessment E-Journal*, **3**, 2009–39.
Weitzman, M. (2010), 'What is the "damage function" for global warming – and what difference does it make?', *Climate Change Economics*, **1**, 57–69.

APPENDIX: FUNCTIONAL FORMS, CALIBRATION AND COMPUTATIONAL IMPLEMENTATION

Preferences

In the simulations we will use an iso-elastic utility function $U(C/L) = \frac{(C/L)^{1-1/\eta}-1}{1-1/\eta}$, where the elasticity of intertemporal substitution is $EIS = -\frac{U'}{U''C} = -\frac{(C/L)^{-1/\eta}}{(-1/\eta)(C/L)^{-1/\eta}} = \eta$. We set the elasticity of intertemporal substitution to $\eta = ½$ and thus intergenerational inequality aversion to 2. The pure rate of time preference ρ is set to 10% per decade which corresponds to 0.96% per year.

Cost of Energy

We employ an extraction technology of the form $G(S) = \gamma_1(S_0/S)^{\gamma_2}$, where γ_1 and γ_2 are positive constants. This specification implies that reserves will not be fully be extracted; some fossil fuel remains untapped in the crust of the earth. Extraction costs are calibrated to give an initial share of energy in GDP between 5%–7% depending on the policy scenario. This translates to fossil production costs of $350/tC ($35/barrel of oil), where we take one barrel of oil to be equivalent to 1/10 ton of carbon. This gives approximately $G(S_0) = \gamma_1 = 0.75$. The IEA (2008) long-term cost curve for oil extraction gives a doubling to quadrupling of the extraction cost of oil if another 1000 GtC are extracted. Since we are considering all carbon-based energy sources (not only oil) which are more abundant and cheaper to extract, we assume a more doubling but less than quadrupling of production costs if a total 3000 GtC is extracted. With $S_0 = 4000$ GtC,[13] this gives $\gamma_2 = 0.75$.[14] In general, we assume very low extraction costs and a high initial stock of reserves.

The unit cost of renewable energy is calibrated to the percentage of GDP necessary to generate all energy demand from renewables. Under a Leontief technology, with $\vartheta \to 0$, energy demand is σZ_t with Z_t potential, pre-damage output and σ the carbon intensity of output. The cost of generating all energy carbon free is $\sigma Z_t b_t / Z_t = \sigma b_t$. Nordhaus (2008) assumes that it costs 5.6% of GDP to achieve this. We take double this number $\sigma b_1 = 0.12$ (i.e. we assume 12%) or, with $\sigma = 0.62$ as derived below, $b_1 = 2$. In the future, this cost falls to current prices of fossil energy (with energy amounting to about 5% of GDP), that is, b_t approaches 0.8. We assume that exogenous technical progress lowers the unit cost at a falling rate starting at a reduction of 1% per year. Specifically, $b_t = 0.8 + 1.2e^{-0.1t}$. This calibration is done for a Leontief technology. We assume that for a more general technology the same parameter values can be applied. Our calibration assumes that renewable energy is initially very expensive and falls to current levels only in the very long run. This, together with the assumption about fossil energy, biases the model against rapid de-carbonization.

---

[13] Stocks of carbon-based energy sources are notoriously hard to estimate. IPCC (2007) assumes in its A2-scenario that 7000 GtCO$_2$ (with 3.66 tCO2 per tC this equals 1912 GtC) will be burnt with a rising trend this century alone. We roughly double this number to get our estimate of 4000 GtC for initial fossil fuel reserves. Nordhaus (2008) assumes an upper limit for carbon-based fuel of 6000 GtC in the DICE-07.

[14] Since and $G(1000)/G(4000) = (4000/1000)^{\gamma_2} = 4^{\gamma_2}$ and $4^{0.75} = 2.8$.

Initial Capital Stock and Depreciation Rate

The initial capital stock is set to 200 (US$ trillion), which is taken from Rezai et al. (2012). We set d to be 0.5 per decade, which corresponds to a yearly depreciation rate of 6.7%.

Population Growth and Labor-Augmenting Technical Progress

Population in 2010 (L_1) is 6.5 billion people. Following Nordhaus (2008) and UN projections population growth is given by $L_t = 8.6 - 2.1e^{-0.35t}$. Population growth starts at 1% per year and falls below 1% per decade within six decades and flattens out at 8.6 billion people. Without loss of generality the efficiency of labor $A_t^L = 3 - 2e^{-0.2t}$ starts out with $A_1^L = 1$ and an initial Harrod-neutral rate of technical progress of 2% per year. The efficiency of labor stabilizes at three times its current level.

Global Production and Global Warming Damages

Output before damages is $Z_t = [(1-\beta)(AK_t^a(A_t^L L_t)^{1-a})^{1-1/\vartheta} + \beta(\frac{F_t + R_t}{\sigma})^{1-1/\vartheta}]^{\frac{1}{1-1/\vartheta}}$, $\vartheta \geq 0, 0 < a < 1$ and $0 < \beta < 1$. This is a constant-returns-to scale CES production function in energy and a capital-labor composite with ϑ the elasticity of substitution, b the share the parameter for energy, and σ the carbon intensity of output. The capital-labor composite is defined by a constant-returns-to-scale Cobb–Douglas function with a the share of capital, A total factor productivity and A_t^L the efficiency of labor. The two types of energy are perfect substitutes in production. Damages are calibrated so that they give the same level of global warming damages for the initial levels of output and mean temperature. It is convenient to rewrite production before damages as $Z_t = Z_0[(1-\beta)(\frac{AK_t^a(A_t^L L_t)^{1-a}}{Z_0})^{1-1/\vartheta} + \beta(\frac{F_t + R_t}{\sigma Z_0})^{1-1/\vartheta}]^{\frac{1}{1-1/\vartheta}}$.

We set the share of capital to $a = 0.35$, the energy share parameter to $b = 0.05$, and the elasticity of factor substitution to $\vartheta = 0.5$. World GDP in 2010 is 63 $trillion. The energy intensity of output s is calibrated to current energy use. In the Leontief case energy demand (only fossil fuel initially) is $F_0 = \sigma D_0 Z_0$. With carbon input equal to 8.36GtC in 2010, we obtain $\sigma = (8.36/2.13)/63 = 0.062$. Finally, given $A_1^L = 1$ we can back out $A = 34.67$.

Climate Dynamics and Damage

Following Golosov et al. (2014), E_t^P denotes the stock of carbon (GtC) that stays thousands of years in the atmosphere and E_t^T the stock of atmospheric carbon (GtC) that decays at rate $\varphi = 0.0228$, this carbon cycle supposes that 20% of carbon emissions stay up "forever" in the atmosphere and the remainder has a mean life of about 300 years, so $\varphi_L = 0.2$. The parameter $\varphi_0 = 0.393$ is calibrated so that about half of the carbon impulse is removed after 30 years. We set current atmospheric carbon at $E_0^P = 103$ GtC and $E_0^T = 699$ GtC. It is commonly assumed that an increase of atmospheric carbon to double its pre-industrial level, leads to a temperature increase of 3°C, so $w = 3$.

Nordhaus (2008) supposes that with global warming of 2.5°C damages are 1.7% of world GDP and uses this for purposes of his DICE-07 model to calibrate the following function for the fraction of output that is left after damages from global warming: $D(T_t) = \frac{1}{1 + 0.00284 T_t^2} = \frac{1}{1 + (T_t/18.8)^2}$.

Weitzman (2010) argues that global warming damages rise more rapidly at higher levels of mean global temperature than suggested by Nordhaus (2008). With output damages equal to 50% of world GDP at 6°C and 99% at 12.5°C, Ackerman and Stanton (2012) calibrate what is left of output after global warming damages as: $D(T_t) = \frac{1}{1 + (T_t/20.2)^2 + (T_t/6.08)^{6.76}}$. The extra term in the denominator captures potentially catastrophic losses at high temperatures.

Computational Implementation

The transversality condition for the model is $\lim_{t \to \infty} e^{-\rho t}(\lambda_t K_t + \mu_t S_t + \eta_{1t} E_t^P + \eta_{2t} E_t^T) = 0$. In our simulations we solve the model for finite time and use the turnpike property to approximate the infinite-horizon problem. All equilibrium paths approach the steady state quickly such that the turnpike property renders terminal conditions essentially unimportant. We allow for continuation stocks to reduce the impact of the terminal condition on the transitions paths in the early periods of the program. We use the computer program GAMS and its optimization solver CONOPT3 to solve the model numerically. The social planner optimum in which the externality is taken into account, fit the program structure readily. To solve for the "laissez faire" equilibrium paths, we adopt the iterative approach discussed in detail in Rezai (2011). Briefly, to approximate the externality scenario, the aggregate economy is fragmented into N dynasties. Each dynasty has $1/N$th of the initial endowments and chooses consumption, investment and energy use in order to maximize the discounted total utility of per capita consumption. The dynasties understand the contribution of their own emissions to the climate change, but take carbon emissions of others as given. The climate dynamics are affected by the decisions of all dynasties. This constitutes the market failure.

It might seem easier to simply assume that there is one dynasty that ignores the externality but this would not be a rational expectation equilibrium. The externality problem is not an optimization but an equilibrium problem. The CONOPT3 solver of GAMS is powerful in solving maximization problems and it is more efficient to adopt an iterative routine in which a planner of a fragmented economy solves an optimization problem representatively than to attempt solving the equilibrium conditions directly. Given our specifications, the computation of the equilibrium problem takes less than one minute. To introduce this approximate externality, we make the following adjustments to the initial stocks $K(0) = K_0/N$, $S(0) = S_0/N$ and $L(0) = L_0/N$. All production and cost functions are homogeneous of degree 1 and therefore invariant to N. The introduction of the pollution externality only requires a modification of the transition equation of atmospheric carbon to include emissions regarded as exogenous by each dynasty:

$$E_{t+1}^P = E_t^P + \varphi_L(F_t + Exg_t) \text{ and}$$

$$E_{t+1}^T = (1 - \varphi)E_t^T + \varphi(1 - \varphi_L)(F_t + Exg_t)$$

In the "laissez faire" scenario dynasties essentially play a dynamic non-cooperative game, which leads to a Nash equilibrium in which each agent forecasts the paths of emissions correctly and all agents take the same decisions as all dynasties are identical. Equilibrium requires $Exg_t = (N - 1)F_t$. Under "laissez faire" the planner only adjusts her controls to take into account the effects of her own decisions (i.e. $1/N$th of the climate externality). If $N = 1$ the externality is internalized and we obtain the social optimum. As $N \to \infty$, we obtain the "laissez faire" outcome characterized in section 2.

Following Rezai (2011), the numerical routine starts by assuming a time path of emissions exogenous to the dynasty's optimization, Exg_t, at an informed guess. GAMS solves for the representative dynasty's welfare-maximizing investment, consumption, and energy use choices conditional on this level of exogenous emissions. $(N - 1)$ times the dynasty's emission trajectory implied by these choices, F_t, defines the time profile of exogenous emissions in the next iteration. The same applies for the knowledge trajectory. The routine is repeated and Exg_t are updated until the difference in the time profiles between iterations meets a pre-defined stopping criterion. In the reported results iterations stop if the deviation $|(N - 1)F_t/Exg_t - 1|$ each time period is at most 0.001%.

We set $N = 400$ to account for the fact that in the present world economy, the externality in the market of GHG emissions is already internalized to a very small extent through the imposition of carbon taxes or tradable emission permits and non-market *regulation* (e.g. through the Kyoto Protocol or the establishment of the European Union Emission Trading Scheme). In our "laissez faire" simulations, the dynastic planner takes into account less than 0.25% of global emissions.

10. Optimal global climate policy and regional carbon prices*
Mark Budolfson and Francis Dennig

INTRODUCTION

It is often stated that optimal global climate policy requires global harmonization of marginal abatement costs – i.e., a single carbon price throughout the world. Chichilnisky and Heal (1994) have shown quite generally that this is only the case if distributional issues are ignored, or if lump-sum transfers are made between countries. Else, a policy in which different regions face different carbon prices may be superior to one with a single global carbon price from a welfare point of view. Still, most integrated assessment models (IAMs) assume away distributional issues and report a single optimal carbon price.[1] We calculate utilitarian-optimal carbon prices under zero cross-regional lump-sum transfers in the multi-region IAM NICE. The result is optimal global climate policy with different regional carbon prices in which the poorest regions face initially low prices, while the richest regions face very high prices from the outset. This entails significant welfare gains over the standard single price optima commonly reported, which, as we argue briefly in conclusion, can be improved upon still by allowing international trading in the corresponding emissions allocations. If implemented in a way that makes trading competitive, such a scheme would result in a globally harmonized carbon price. Such a result would constitute an efficient use of carbon resources in a way that addresses the distributional issues internal to the climate problem.

NICE is based on William Nordhaus' multi-regional model RICE but includes representation of sub-regional inequalities based on World Bank income distribution data (World Bank, 2014). This allows us to show not only the effect on optimal prices of allowing differential regional prices while otherwise holding fixed the assumptions of RICE, but also the effect of differential pricing on optimal policy given a variety of

---

* We thank Charles Beitz, Navroz Dubash, Marc Fleurbaey, Ewan Kingston, Armon Rezai, Itai Sher, Dean Spears, Lucas Stanczyk, Gerard Vong, and an anonymous reviewer for helpful comments. FD thanks the United Kingdom Economic and Social Research Council Grant ES/I903887/1 for funding. Both authors thank the University Center for Human Values at Princeton University, as well as Bert Kerstetter for financial support of the Climate Futures Initiative.

[1] The most prominent models are DICE/RICE (Nordhaus and Sztorc, 2013), FUND (Tol, 1996), and PAGE (Hope, 2006). These are either globally aggregated models with no regional heterogeneity (as DICE and PAGE) or regional models run with Negishi weights or a constraint requiring all regions to have the same carbon price (as RICE). Exceptions are Anthoff (2009), which uses FUND, computes an optimum with regionally different carbon prices, as well as Hassler and Krusell (2012) who have a modification of RICE in which there is trade in fossil fuels but no other trade, and report that the optimum would only impose carbon taxes on oil producing countries.

alternative assumptions about the distribution, within regions, of both mitigation cost and damages by income quintile.[2]

We find that the effect on optimal policy of allowing different regional prices can be large, even for the relatively low value of 1 for the elasticity of marginal utility, the parameter which determines the intensity of concern for inequality in cost–benefit climate models such as NICE.[3] The optimal regional prices span the whole range of prices that are found using globally aggregated models and that are currently debated as optimal global prices. The richest regions have carbon prices greater than those prescribed with low discounting parameters in the Stern review (Stern, 2006), and the poorest regions have even lower prices than found optimal by studies with very high discounting parameters (e.g. Nordhaus, 2007). This is robust to a large range of combinations in the other relevant parameter values. The welfare gains and change in global mitigation effort from allowing different regional prices depends on model parameters, in particular, the two income elasticities which determine how mitigation costs and climate damages are distributed across income groups within regions. For example, for elasticity values in the middle of our reported range we find that the overall mitigation effort (measured by total global emissions) is comparable in the harmonized and differential price optima, but the welfare gain from allowing poorer regions to mitigate less is still substantial: over 1 per cent of perpetual equally distributed consumption. Regardless of the parameter values, the welfare gain from considering optimal differential prices is *always* positive since the removal of the harmonization constraint cannot result in a welfare loss.

In the discussion section, we argue that the differential prices optimum is a natural focal point for climate policy, as it gives proper weight to common but differentiated responsibilities and provides a reference for judging the relative adequacy of national commitments (NDCs) in the emerging post-Paris 'bottom-up' international climate regime. Because the differential prices optimum can be used to calculate the welfare-optimal shares of emissions, these shares are then a natural welfare-based focal point for judging the adequacy of shares of a given level of global emissions reductions. Once such commitments are established and deemed adequate, international emissions trading can provide further gains still, as in any situation with differential prices the same emissions level can be achieved in a Pareto improving way by allowing a region with a higher price to pay a region with a lower price for a portion in the latter's emission share – e.g., in an emissions trading scheme that allocates permits according to the emission shares in the differential prices optimum.[4] The resulting gain over the standard harmonized price optimum would be twofold: first from allowing different regional levels of mitigation cost

---

[2] NICE does not model health co-benefits of CO_2 mitigation, which, as shown in Scovronick et al. (2019) and Boyce (Chapter 1 in this volume), would result in higher carbon prices overall.

[3] For greater values of this elasticity, the spread in regional prices becomes greater. The value 1 is at the lower end of the range of primary disagreement over this parameter in the literature.

[4] The result reached after trading will have a single international marginal abatement cost, but it will depend on the initial allocation, and will, in general, not be the same as the harmonized price optimum. In fact, Chichilnisky et al. (2000) show that for some initial allocations the result of emissions trading may not even be Pareto efficient. Still, such a result would be superior to the harmonized price optimum.

burden, and second from the efficiency gain due to trading. The latter is accompanied by some degree of international transfers, but rather than being lump-sum transfers stemming from cosmopolitan redistributive aims they stem from the logic of common but differentiated responsibilities internal to the climate policy challenge.

MODELING: OPTIMAL CLIMATE POLICY WITH DIFFERENTIAL REGIONAL PRICES IN NICE

Very few papers produce an optimal global response with different price paths for different regions – Anthoff (2009) computes optimal differential prices in the FUND model and also provides an overview of the relevant economic theory. Here, we compute the utilitarian-optimal carbon prices in the multi-region integrated assessment model NICE and compare the prices and welfare levels to the harmonized-price constrained optimum as well as to the Negishi-weighted and globally aggregated models.[5] NICE is based on William Nordhaus' multi-region model RICE, but includes sub-regional inequalities represented by aggregating World Bank data on the distribution of income within nations to regional income distributions. These regional income distributions are treated as a proxy for the distribution of consumption prior to both mitigation cost and damage from climate change. The post-mitigation cost and damage consumptions then depend on the way in which these two impacts correlate with consumption, as measured by the impact elasticity of consumption.

Following RICE 2010 (Nordhaus, 2010), on which NICE is based, and most of the literature, NICE evaluates public policy with a discounted and separable constant elasticity social welfare function. In general, we don't use Negishi weights (though we report the Negishi-weighted optimum for comparison to our results),[6] but only population weights:

$$W(c_{ijt}) = \sum_{ijt} \frac{L_{ijt}}{(1+\rho)^t} \frac{c_{ijt}^{1-\eta}}{1-\eta} \qquad (10.1)$$

---

[5] In all our optima we adopt a constraint against direct international transfers, and we specify an exogenous savings rate of 25.8%, which can be interpreted as the optimal savings rate of private savers with a time-separable and discounted objective with a logarithmic utility function (Dennig et al., 2015; Golosov et al., 2014). There are two alternative treatments of savings in the climate-economy modeling literature: one approach assumes that economic agents endogenously look forward to climate damages and policies and optimally adjust their planned savings (a leading example is in Nordhaus' original versions of DICE and RICE), and another that assumes that savings do not so respond to climate policy optimization (leading examples are FUND and PAGE; in a DICE/RICE framework, see Dennig et al., 2015). Although both approaches are defensible, we prefer and use the second approach, because we find it more realistic to assume that society has a fixed appetite for savings that is essentially insensitive to climate change and climate policy decisions.

[6] More precisely, we run a 'globally aggregated' version of the model in which all individuals in all regions are assumed to consume the global average consumption. For the logarithmic utility which we use ($\eta = 1$) the Negishi-weighted optimal policy is identical to the globally aggregated optimum. In this sense, a Negishi-weighted regional model gets a single global carbon price by ignoring distributional issues. See Anthoff (2009) for this result.

Here W denotes social welfare, L population, c per capita consumption, ρ the pure rate of time preference, and η the elasticity of marginal utility. The subscripts i, j, and t are region, quintile, and time indices respectively.

The main equation embodying the economic trade-off in the RICE model, inherited by NICE is:

$$Y_{it} = \left(\frac{1 - \Lambda_{it}}{1 + D_{it}}\right) Q_{it} \tag{10.2}$$

Here Y denotes (net) economic output post-mitigation cost and climate damage, Q denotes pre-cost and damage (gross) output, and Λ and D are mitigation cost and damage respectively. Thus, mitigation comes at a cost that subtracts from output, as do climate damages, which increase as temperature rises relative to preindustrial levels. Temperature increase is a function of the stock of emissions in the atmosphere, which can be controlled (at a cost) by past abatement. As RICE is a regional model, each of these variables is specified by region. Gross output Q is a Cobb–Douglas function with exogenous regional and time varying *total factor productivity*, which is computed as a residual for 2005 and projected forward with empirical growth estimates as well as a modest convergence assumption. The regional damage functions $D_i(T_t)$ are quadratic functions of global mean temperature above pre-industrial levels with coefficients that vary by region. The abatement cost $\Lambda_{it}(\mu_i)$ is a convex function of the regional mitigation rate μ_i, with regional coefficients that reflect current carbon intensities and are projected into the future with modest convergence assumptions analogous to those for TFP. We denote by carbon price the marginal cost of mitigating a ton of carbon.[7]

What is specific to NICE is the representation of sub-regional heterogeneity by attributing regional output to population quintiles by income. We use a fixed savings rate, equal in every region and period, denoted by s. Regional average per-capita consumption is

$$\bar{c}_{it} = \frac{1 - s}{L_{it}} Y_{it} \tag{10.3}$$

In NICE, pre-mitigation cost, pre-climate damage, per-capita consumption of quintile j is given by

$$c_{ijt}^{pre} = 5\bar{c}_{it}\left(\frac{1 + D_{it}}{1 - \Lambda_{it}}\right) q_{ij} \tag{10.4}$$

where q_{ij} is the income share of quintile j in region i.[8]

Post-mitigation cost and post-damage average per capita consumption (of quintile j in region i at time t) is given by

$$c_{ijt} = \frac{5\bar{c}_{it}}{1 - \Lambda_{it}}((1 + D_{it})q_{ij} - (1 - \Lambda_{it})D_{it}d_{ij} - (1 + D_{it})\Lambda_{it}e_{ij}) \tag{10.5}$$

---

[7] A detailed description of the RICE model, as well as the DICE model, on which it is based, can be found in Nordhaus (2010) and Nordhaus and Sztorc (2013).

[8] These regional quintile shares are computed by aggregation of the national quintile shares provided in the World Bank Development Indicators (World Bank, 2014).

where e_{ij} is the share of mitigation cost and d_{ij} is the share of damages of quintile j in region i. These quintile shares of mitigation cost and damage are computed for different values of elasticity parameters ξ and ω such that[9]

$$d_{ij} = k_{i\xi}q_{ij}^{\xi}; \ e_{ij} = k_{i\omega}q_{ij}^{\omega}. \tag{10.6}$$

This implies a constant elasticity relationship for the quintile mitigation cost and damage shares as a function of income. By modifying the parameters ξ and ω, we are thus able to vary the distribution across quintiles of mitigation cost and climate damages.

To illustrate the meaning of ξ (and ω), consider an 'economy' comprised of two (equally populous) consumption groups A and B, with A consuming USD 4,000, and B USD 40,000 a year. If this 'economy' suffers 5% damage from climate change, they jointly lose USD 2,200. If $\xi = 1$, A loses USD 200 and B loses USD 2,000. If $\xi = 0$, both A and B lose USD 1,100. If $\xi = -1$, A loses USD 2,000 and B loses USD 200. B goes from losing 5% to 2.75% to 0.5%, while A goes from losing 5% to 27.5% to 50% of pre-damage consumption. (Similar remarks apply to ω.)

The distribution of damages, and thus the value of ξ, depends on where and how the climate changes and modifies the ecosystem at a sub-regional level, on how vulnerable the populations are given the organization of the economy and the infrastructure set-up, and on policy response.[10] The value of ξ has not received much scrutiny so far in the empirical literature, perhaps partly due to the fact that the importance of this parameter had until recently not been demonstrated. However, many studies argue that the poor will disproportionately suffer from climate change (Hallegatte et al., 2016; Oppenheimer et al., 2014; Mendelsohn et al., 2006; Leichenko and O'Brien, 2008; Cutter et al., 2003; Kates, 2000), meaning that ξ is likely to be less than 1, and might even be negative (in particular in the case of health and mortality impacts). We consider that a relevant range for ξ in the present investigation, is from -1 to $+1$.

The distribution of mitigation cost, and thus the value of ω, is even more dependent on policy decisions. Several studies (Krey, 2014; Daioglou et al., 2012; Riahi et al., 2012; Bacon et al., 2010) analyze the share of energy in household expenditures and conclude that an increase in energy prices will hit the poor more than proportionally in the absence of compensatory measures, at least in developed nations.[11] This suggests a value of ω less than 1 (but greater than 0) for a carbon tax alone with no compensatory measures. Several other studies (Cullenward et al., 2014; Williams et al., 2014; Sterner, 2012; Metcalf, 2009) agree with the studies just cited, but also conclude that if an increase in energy prices is combined with compensatory measures it need not disproportionately hit the

---

[9] For equation (10.6), the parameter values $k_{i\xi}$ and $k_{i\omega}$ are chosen such that $\Sigma_j d_{ij} = 1$ and $\Sigma_j e_{ij} = 1$ respectively. This ensures that only the distribution, and not total amounts of cost and damage, are modulated by the elasticity parameters.

[10] Note that the measurement of damages itself has both empirical and ethical dimensions: valuing losses to different parts of the income distribution in the wake of climate change depends both on relatively objective data on property damage, capital losses, etc., and the more ethically challenging questions regarding valuation of loss of life, health, and livelihood.

[11] The papers in Sterner (2012) suggest that even without compensatory measures, a carbon tax in developing nations might not be regressive.

poor, and could even make all but the highest quintile net beneficiaries – for example, if the compensatory measures involve equal per capita redistribution of the revenues from a carbon tax.[12] In light of this, we consider that a relevant range for ω is from 0 to 2, the latter value being obtained when the cost is borne more heavily by the rich.

RESULTS: WELFARE GAINS FROM REPRESENTING INEQUALITIES AND ALLOWING DIFFERENT REGIONAL PRICES

In previous work (Dennig et al., 2015) we show with our co-authors that the value of ξ is of great importance to climate policy. For example, if damages are distributed inversely proportionally to income, optimal mitigation effort under the discounting and inequality aversion assumptions of Nordhaus (2010) is equivalent to optimal mitigation in the more aggregated RICE model under the much lower discounting and inequality aversion assumptions of the Stern Review (Stern, 2006).

Here we stress that allowing different carbon prices in different regions of the world is another important way in which a utilitarian improvement can be achieved by being sensitive to the interests of the poor, especially when combined with careful consideration of the sub-regional distribution of damages and mitigation cost. As an indication of the importance of these factors, especially the magnitude of the effect on the optimum that allowing differential prices can have, consider the following (Figure 10.1), which compares the range of optimal prices under the harmonized-price constraint (left-most panel, Figure 10.1a, showing optima in NICE under a wide range of different mitigation cost and damage distribution assumptions) to the optimum with differential regional prices (middle graph, Figure 10.1b, showing the wide range of different regional prices given discounting assumptions $\rho=2\%$ and $\eta=1$, and proportional sub-regional mitigation cost and equal absolute sub-regional damage distributions).[13] Figure 10.1c shows the temperature paths relative to pre-industrial levels for the policies in Figure 10.1a and Figure 10.1b.

As the comparison between the carbon prices in Figure 10.1a and 10.1b demonstrates, the effect on optimal policy of allowing differential regional carbon prices can be large; Figure 10.1b shows the regional carbon prices that emerge from the assumptions behind the middle (ω=1 and $\xi=0$) line in Figure 10.1a when the harmonized-price constraint is removed. This comparison shows that even holding fixed the other assumptions of RICE including discount rates but merely allowing differential regional prices leads to optimal carbon prices in several rich regions that are higher than they would be in a globally

---

[12] As a consequence, progressive compensatory measures can also arguably improve the political feasibility of carbon taxes, at least as measured by percentage of voters who are net beneficiaries of the policy.

[13] We use a relatively high value $\rho=2\%$, and a relatively low value $\eta=1$ throughout. For comparison, Stern (2006) used $\rho=0.1\%$ and $\eta=1$, and Nordhaus and Sztorc (2013) use $\rho=1.5\%$ and $\eta=1.45$. Increasing ρ reduces all prices. Increasing η spreads the prices in the differential price optimum (Figure 10.1b) even more, and makes the welfare effects reported in Figure 10.2 slightly greater. Using $\eta=1$, which is at the lower end of the primary range of disagreement, underestimates the difference between the regional price optimum and the harmonized price optimum.

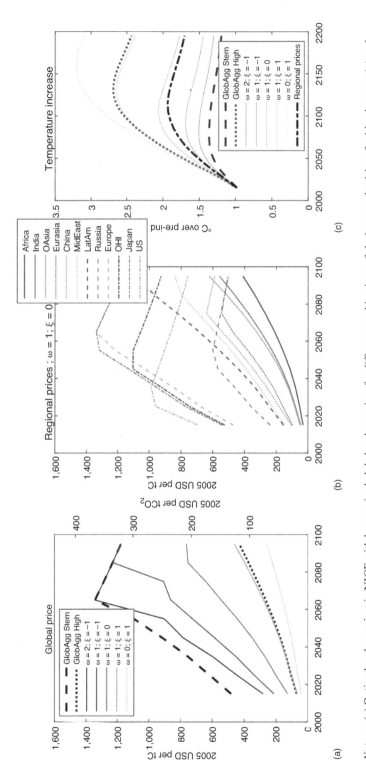

Figure 10.1 Optimal policy

Notes: (a) Optimal carbon prices in NICE with harmonized global carbon prices for different combinations of the income elasticity of mitigation cost (ω) and damage (ξ), all for the discounting assumptions ρ=2% and η=1: the prices increase with increasing ω and with decreasing ξ. Also shown are the optimal policies when all regions (and quintiles therein) are assigned the average global consumption (essentially our implementation of DICE, or RICE with Negishi weights) with Stern's discounting assumptions (Global-Stern, with ρ=0.1% and η=1), as well as alternative higher discounting assumptions (Glob Agg, with ρ=2% and η =1). (b) Optimal carbon prices in NICE allowing differential regional carbon prices given higher discounting assumptions (ρ=2% and η=1) and with proportional distribution of mitigation cost and equal absolute distribution of damages within regions (i.e. ω=1 and ξ=0). (c) Temperature relative to preindustrial levels in the optima reported in (a) and (b), where 'Regional Prices' refers to the global temperature path under the optimum in (b). Note that the temperature paths for ω=1; ξ=0 and Regional Prices are for runs that involve the same assumptions about discounting as well as mitigation cost and damage distribution (ρ=2%, η=1, ω=1, and ξ=0), but differ only in that the former but not the latter impose harmonization as a constraint.

aggregated model with the discounting assumptions of the Stern Review, and in the poorest region are significantly lower than they would be in a globally aggregated model with equivalent discounting assumptions ($\rho=2\%$, $\eta=1$). The effect on optimal policy of imposing a harmonized price can be quite large, and so the question of whether to allow different carbon prices across regions or insist on a globally harmonized carbon price is not merely a theoretical curiosity.

The main regional difference driving the heterogeneity in prices is the difference in TFP and capital stocks, since these determine the vastly different consumption levels. Greater consumption levels imply lower disutility from abatement cost, thus resulting in higher marginal abatement costs (carbon prices) in richer regions. This is the effect described in Chichilnisky and Heal (1994). In our model the aggregate disutility to the region also depends on the sub-regional income distribution, along with the distribution of costs across the sub-regional income quintiles. When mitigation cost is distributed more regressively than in proportion to income ($\omega < 1$) a given amount of mitigation cost will result in greater aggregate disutility in a more unequal region, thus reducing the optimal carbon price in that region relative to what it would be if there was no inequality in that region. A similar argument gives the converse when mitigation cost is distributed more progressively than in proportion to income ($\omega > 1$). In this way, different average regional consumption levels *and* the distribution of consumption have a first order effect on the carbon prices at the optimum. The mitigation cost functions Λ_{it} are also different across regions, which implies that at a given price two regions would bear different costs, leading to a (second order) level effect on the optimal carbon prices.

To compare the welfare effects of the different policies we measure the welfare gains over business-as-usual (BAU) welfare levels. In our model the BAU runs are simply the model runs with zero carbon prices.[14] The welfare loss from using a policy that ignores distribution altogether (the 'Global' policies) or one that considers the distributional impacts, but is constrained to a globally harmonized price depends on the distribution of costs and damages (ω and ξ) and can be quite large.[15] In Figure 10.2, we show the gain in welfare over business-as-usual, as a percentage of BAU consumption, from implementing the 'Global' policy (GlobAgg High path in Figure 10.1a), the harmonized-price optimum, and the differential-price optimum, *for different values of* ω *and* ξ.[16]

Assuming that both sub-regional mitigation-cost and damage are proportional ($\omega=1$ and $\xi=1$) the gain over BAU of the Global and harmonized-price optimum is 0.3% of consumption equivalent welfare (note that these two policies are almost identical in Figure 10.1a). The gain over and above that by allowing differential prices is another 0.2%. So, allowing for differential prices almost doubles the welfare gain over BAU relative to the Global policy, if we assume that sub-regional costs and damages are proportional

---

[14] Since we use a fixed savings rate rather than one which maximizes the overall welfare level, our notion of BAU does not contain an amount of mitigation-by-savings-rate, as described in Rezai et al. (2012).

[15] It also depends on the discounting parameters. Figure 10.2 reports the results for $\rho=2\%$ and $\eta=1$. If η is greater, then the welfare gains (from allowing differential prices) are larger.

[16] The quantity plotted in Figure 10.2 is the gain in welfare over BAU as a proportion of BAU welfare, measured in consumption units. If W_{BAU} is BAU welfare, as computed by (10.1), and W_P is the welfare at for some policy P, then Figure 10.2 plots $(W_P^{\frac{1}{1-\eta}} - W_{BAU}^{\frac{1}{1-\eta}})/W_{BAU}^{\frac{1}{1-\eta}}$.

to consumption.[17] If we assume instead (see Mendelsohn et al., 2006; Hallegatte et al., 2016), that damage is distributed equally across income groups ($\xi=0$) while mitigation cost is still proportional, then the Global policy already provides a large proportion of the possible gain. However, the gains are almost two orders of magnitude greater, and the differential-price optimum yields an additional 1% welfare gain over BAU relative to the harmonized-price optimum, and additional 2% over the Global policy. These are large utilitarian welfare gains that are left on the table by focusing only on harmonized-price optimal or Global optimal policies.

We have run Figures 10.1 and 10.2 corresponding to alternative values of ρ and η. Since the effect of ρ is to discount the future, regardless of the spatial distribution of outcomes, increasing ρ simply reduces the price paths. Since the effect of increasing η is to increase the sensitivity to distribution, the effect is that for higher values the results corresponding to Figure 10.1b have more spread out carbon prices across the regions. We also find that the welfare effects in Figure 10.2 are slightly greater for larger values of η. Additionally, we look at versions of Figure 10.1b for different values of ξ and ω. Lower values of ξ lead to higher carbon prices overall. Higher values of ω also lead to higher prices. As discussed above, alongside the other determinants of the structure of the differential price optimum, ω affects the relative magnitude of carbon prices for regions with different degrees of inequality. For example, India, which has relatively low levels of inequality has significantly higher carbon prices than Africa when $\omega = 0$, however, when $\omega = 2$ India and Africa have almost identical carbon prices. Overall this effect is small when compared with more modest changes in the value of η, even for the sizeable change from $\omega = 0$ to $\omega = 2$. Still, the *overall* effect of changing the elasticity parameters on the carbon prices is high, and as shown in Figure 10.2, the welfare implications of getting the policy wrong depend importantly in those elasticities.

DISCUSSION: DIFFERENTIAL PRICES AND INTERNATIONAL CLIMATE POLICY

The first principle of the United Nations Framework Convention on Climate Change (UNFCCC) states that 'The Parties should protect the climate system for the benefit of present and future generations of humankind, on the basis of equity and in accordance with their common but differentiated responsibilities and respective capacities. Accordingly, the developed country Parties should take the lead in combating climate change and the adverse effects thereof' (United Nations, 1992). In context, this implies that respect for equity and common but differentiated responsibilities (CBDR) are part of the objective of the UNFCCC as agreed by the parties to that convention, where those

---

[17] If we assume $\eta = 1$, when costs are distributed proportionally ($\omega=1$ and $\xi=1$), the regional distribution does not affect the optimal policy. This is because with such a (logarithmic) utility function, the marginal utility of any income group is proportional to one over its consumption. In this case the marginal utility of a unit of damage (or mitigation cost) to the average consumer is the same as the same unit of damage, distributed proportionally over all income groups. That is to say, proportional cost and damage distributions lead to the same policies as models that aggregate at the regional level.

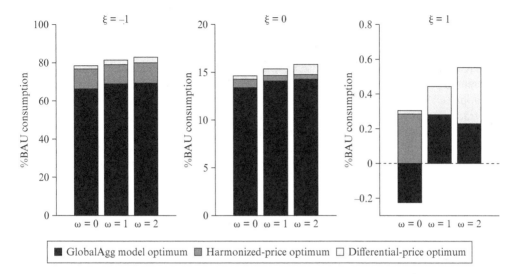

Figure 10.2 Welfare gain versus business-as-usual

Note: All three bars in all three panels plot the same three stacked quantities: the welfare gain over business-as-usual of implementing the 'Global' policy, the harmonized-price optimum, and the differential-price optimum. The 'Global' policy consists of the same carbon price path for all nine bars, whereas the harmonized-price and differential-price optima are computed optimally for the corresponding (ω, ξ) pair. All outcomes assume $\rho=2\%$, $\eta=1$. Notice that if $\omega=0$ and $\xi=1$, using the policy recommended by the globally aggregated model results in a *loss* relative to BAU. This is because at that particular distribution of costs and damages the 'Glob Agg-High' optimum *over-mitigates*. At such a high carbon price the loss to the mitigators is greater than the gain in avoided damage. This is visible in Fig 10.1a, as the price path for 'Glob Agg-High' is greater than for '$\omega=0$; $\xi=1$'.

values of equity and CBDR are meant to be weighty values that should not be traded off lightly in pursuit of the concurrent goals of protecting current and future generations with climate policy.

In light of this objective, a natural focal point for optimal policy in the absence of large international transfers is the welfare maximizing differential regional prices optimum explained and computed in previous sections. In contrast, imposing a globally harmonized carbon price in the absence of large transfers would result in a sub-optimal outcome by the lights of the UNFCCC, since it requires developing nations to make welfare sacrifices in the pursuit of further cost minimization that are larger than the welfare gains elsewhere that result from those further moves toward cost minimization.

Because the differential prices optimum is a natural focal point for understanding optimal forward-looking[18] CBDR without increased transfers from rich nations to poor, it can

---

[18] Here and in what follows, our modeling and discussion focuses only on 'forward-looking' considerations (namely, future welfare consequences), and so sets aside 'backward-looking' considerations such as historical responsibility that are relevant to optimal policy according to many normative frameworks, including some interpretations of CBDR. In setting aside backward-looking considerations, we do not intend to take a stand on whether they are actually relevant to optimal policy – our thought is merely that (a) it is much more controversial whether they should

also serve as a 'CBDR baseline' to judge whether alternative policies are improvements for all nations over this constrained baseline: insofar as rich nations prefer to move toward harmonized prices they must at least then compensate for any welfare loss relative to this CBDR baseline that might otherwise be implied for developing nations.[19] In this way, the differential prices optimum can be used as a baseline to evaluate whether particular alternative approaches that combine harmonization with progressive instruments (e.g. a global cap and trade system with a progressive allotment of permits) would also satisfy the UNFCCC objective, because the differential prices optima, again, reflect the utilitarian weighting of the interests of developing nations: one possible interpretation of the CBDR terms articulated in the convention, unlike welfare weightings that ignore distributional issues.

For example, Pareto improving transfers, whereby rich nations pay off poorer nations in order to emit more, are a much touted mechanism for rich nations to meet their obligations while assisting poorer nations financially for their (additional) mitigation efforts. The clean development mechanism of the Kyoto Protocol is just such a mechanism. It is based on the fact that once obligations in different regions have been established, differences in marginal abatement costs may be quite large, and mutual gains may be achieved by a region with lower cost using less than what is required by its obligation, and thereby allowing the richer nation to pay it in order to emit more. The result is a mutual gain whereby the same global emissions level is still achieved. In the context of Table 10.1, this would allow the US to persist in emitting substantial amounts of carbon in 2035, but it would have to pay India or a country in Africa for the privilege to do so.[20] Notice that for such a mechanism to work, mitigation obligations must be previously established. We claim that the differentiated price optimum could serve such a purpose and is a natural focal point given the objective stated by the UNFCCC.

Of course, the addition of such a trading scheme would imply some regional transfers. However, these transfers would leave regional differences in wealth largely undisturbed, as they are not the massive transfers suggested by general cosmopolitan redistributive aims, but merely the transfers required by the utilitarian objective once it is decided that cosmopolitan general redistribution will not happen, which must then be taken into account by subsequent policy decisions that have distributional consequences. Furthermore, if the

---

be included in the climate policy objective, (b) they are not uncontroversially recognized as relevant by the UNFCCC objective, and (c) if they are included in the objective this would tend to move optimal policy even further in the direction we are arguing, so proponents of backward-looking considerations can simply add a further adjustment on top of our calculation of the forward-looking considerations. (Similar remarks apply to other justice-based considerations, such as giving extra weight (or even lexical priority) to meeting the urgent needs of the global poor, etc.) So, we set aside backward-looking considerations here for ease of exposition, and focus only on forward-looking considerations based on current and future income levels, which are at least a large part of the objective from all normative perspectives.

[19] However, it is important to stress that even if richer nations make these transfers, this might be insufficient in the real world to protect the poor relative to how they would have fared under the differential prices optimum, as transfers from rich nations to poor nations are unlikely to be distributed as intended across income groups, and to a first approximation might predictably only benefit the richest quintile in some developing nations.

[20] These two regions would have the lowest marginal abatement cost at the proposed optimum, as can be seen from Figure 10.1b.

Table 10.1   Optimal regional emission shares in 2035

Region (%)	$\eta=1; \omega=1; \xi=0$		$\eta=2; \omega=1; \xi=0$		$\eta=1; \omega=0; \xi=0$		$\eta=1; \omega=1; \xi=1$		Population Shares 2035
	HP	DP	HP	DP	HP	DP	HP	DP	
Africa	4	6	4	12	4	6	4	5	16
India	9	15	9	23	9	12	9	12	18
Other n-OECD Asia	9	15	10	25	9	13	9	12	16
Non-Russia Eurasia	2	2	1	3	2	3	2	2	1.9
China	22	29	20	18	23	28	24	28	17
Middle East	10	12	10	9	10	12	10	11	7.8
Latin America	7	8	7	9	7	8	7	7	8.4
Russia	3	2	2	0	3	2	3	3	1.5
OECD Europe	11	6	11	0	10	8	10	8	6.3
OHI	6	2	6	0	5	2	5	4	1.8
Japan	4	2	4	0	3	2	3	2	1.4
USA	15	0	15	0	15	4	15	6	4.4

Notes: Optimal regional emission shares and world totals for industrial carbon emissions in 2035. Each pair of columns contrasts the emission shares under the harmonized and differential price optima. The first column assumes logarithmic utility and the mitigation and damage cost distributions from the middle panel in Figure 10.1. This is our reference scenario. Each other column pair changes one of those three parameters. The pure rate of time preferences is fixed at $\rho=2$ in all columns. Notice that increasing η has a large effect on spreading the distribution of optimal emission shares, to the point where the richest four regions emit zero by 2035. Decreasing the mitigation cost elasticity and increasing the damage elasticity have similar effects: they increase the overall global emissions allowing the richer regions a greater share of that increase relative to our reference scenario.

resulting international market in permits is fully competitive, this would result in a globally harmonized carbon price. However, it will be a distinct outcome from the constrained harmonized optimum, as it will have lower emissions, and less of the emission burden will be on the poorest nations. It will be cost efficient, like the constrained harmonized optimum, but a welfare improvement over the differentiated price optimum, which is the baseline from which the cost savings would be made. Unfortunately, our model is not equipped to compute the distribution of emissions and global carbon price that would emerge from such a trading scheme. A global general equilibrium model would be necessary for that.

In the absence of such a trading scheme and the equilibration of a global carbon price, differential prices cause a competitiveness differential that could lead to relocation of energy intensive industries. The large literature on 'carbon leakage' looks at this issue and policy proposals to counteract the effect. The broad conclusion of this literature is that there are two channels for leakage – competitiveness differences due to carbon price differences, and fossil fuel price level reductions due to decreased global demand. The consensus is that the second, price level, effect is the dominating one.[21] The corollary of this insight is that border tax adjustments (BTAs), which correct for competitiveness differences, therefore do not help avoid most of the leakage. However, these models presume unilateral emissions reductions in some regions, and equilibration through increases of

---

[21] See Monjon and Quirion (2013), Lockwood and Whalley (2010), Aldy and Pizer (2009), and Felder and Rutherford (1993) for examples.

demand by other regions that impose no cap at all. This is the source of the price level effect. A differential price optimum requires different effort in different regions, but does impose a cap on all regions, meaning that the level effect is shut down as a channel for leakage, leaving only the competitiveness channel. We know of no global general equilibrium model that evaluates the competitiveness effect for a global policy with differential prices and the BTAs required to shut down leakage. Such a complementary analysis would be important to flesh out this global policy proposal.

Finally, the differential prices optimum can also be used to judge national commitments in the emerging post-Paris international climate regime, in which the international community has pivoted to a bottom-up approach and international cooperation via nationally determined contributions (NDCs). In this regime, there is a fundamental need to judge the adequacy of national commitments relative to a context in which it is common knowledge that they do not collectively add up to a globally optimal level of mitigation. Because the differential prices optimum can be used to calculate the welfare-optimal shares of emissions, these shares are then also a natural focal point for judging the adequacy of shares of a given level of global emissions reductions, even if that global level is itself suboptimal. These shares can serve as a focal point for the negotiated relative contributions of nations as they gradually deepen their commitments in coming decades.

Table 10.1 shows the difference between the harmonized price optimum and the differentiated price optimum in terms of shares of industrial CO_2 emissions in 2035. In our reference scenario the harmonized price optimum emits 6.5 Gigatons of carbon while the differentiated price optimum emits 5.1 Gigatons of carbon.[22] As reductions from the 11.7 Gigatons in business-as-usual, the total mitigation effort is not too different in the two optima. However, the distribution is radically different. For example, in the harmonized price optimum, the US and Europe would continue to cause 15% and 11% of total emissions while India and Other (non OECD) Asia would only cause 9% of emissions each. In the differentiated price optimum, the US would be expected to have (net) zero per cent of emissions and Europe 6%, while the two developing regions would be emitting 15% of global emissions each.

CONCLUSION

In sum, a differential prices optimum is generally welfare superior to the harmonized global prices optimum produced by standard IAMs. While it is often stated that optimal global climate policy requires global harmonization of marginal abatement costs, this is only the case if distributional issues are ignored, or if lump-sum transfers are made between countries.

As our results indicate, the welfare gain and change in global mitigation effort from allowing different regional prices depends on model parameters, in particular, the two income elasticities which determine how mitigation costs and climate damages are distributed across income groups within regions. We find that the effect on optimal policy of allowing different regional prices can be large, even for the relatively low value of 1 for the elasticity of marginal utility.

---

[22] These are just the industrial carbon emissions that are endogenous in the model.

A differential prices optimum is also a natural focal point for climate policy that gives proper weight to common but differentiated responsibilities and respective capacities, and for judging the relative adequacy of national commitments in the emerging 'bottom-up', NDC-focused international climate regime. The resulting gain in welfare is the main argument for grounding policy analysis on the differential prices optimum rather than the welfare inferior harmonized price optimum that ignores distributional issues, as the welfare gain is driven only by the logic of common but differentiated responsibilities and different capacities internal to the climate policy challenge, rather than by general redistributive aims. Once such commitments are established and deemed adequate, international emissions trading can provide further gains still, as in any situation with differential prices the same emissions level can be achieved in a Pareto improving way by allowing a region with a higher price to pay a region with a lower price for a share in the latter's emission share – i.e. an emissions trading scheme that allocates permits in accord with the emission shares in the differential prices optimum.

In general, when national and subnational inequalities are properly represented it is especially problematic to insist on ignoring the negative welfare effects of imposing harmonized abatement costs as a modeling constraint. Improved representation of these inequalities and recognition of their relevance to climate policy (as in NICE with differential regional prices) should become new best practices in climate economy IAMs.

REFERENCES

Aldy, J. and W. Pizer (2009), 'The US competitiveness impacts of domestic greenhouse gas mitigation policies', Pew Center on Global Climate Change.

Anthoff, D. (2009), 'Optimal global dynamic carbon taxation', Working Paper 278, ESRI Working Paper Series.

Bacon, R., S. Bhattacharya and M. Kojima (2010), 'Expenditure of low-income households on energy: evidence from Africa and Asia', Extractive Industries for Development Series #16, World Bank, accessed 17 October 2016 at https://openknowledge.worldbank.org/bitstream/handle/10986/18284/549290NWP0eifd10box34943 1B01PUBLIC1.pdf?sequence=1.

Burniaux, J.M., J. Chateau and R. Duval (2012), 'Is there a case for carbon-based border tax adjustment? An applied general equilibrium analysis', *Applied Economics*, **45**, 2231–40.

Chichilnisky, G. and G. Heal (1994), 'Who should abate carbon emissions? An international viewpoint', *Economics Letters*, **44**(4), 443–9.

Chichilnisky, G., G. Heal and D. Starrett (2000), 'Equity and efficiency in environmental markets: global trade in CO_2 emissions', in Chichilnisky, G. and G. Heal (eds), *Environmental Markets: Equity and Efficiency*, New York: Columbia University Press, pp. 46–67.

Cullenward, D., J. Wilkerson, M. Wara and J. Weyant (2016), 'Dynamically estimating the distributional impacts of U.S. climate policy with NEMS: a case study of the Climate Protection Act of 2013', *Energy Economics*, **55**, 303–18.

Cutter, S., B. Boruff and W.L. Shirley (2003), 'Social vulnerability to environmental hazards', *Social Science Quarterly*, **84**(2), 242–61.

Daioglou, V., B. Van Ruijven and D. Van Vuuren (2012), 'Model projections for household energy use in developing countries', *Energy*, **37**(1), 601–15.

Dennig, F., M. Budolfson, M. Fleurbaey, A. Siebert and R. Socolow (2015), 'Inequality, climate impacts on the future poor, and carbon prices', *Proceedings of the National Academy of Sciences*, **112**(52), 15827–32.

Felder, S. and T. Rutherford (2001), 'Unilateral CO_2 reductions and carbon leakage: the consequences of international trade in oil and basic materials', *The Economics of International Trade and the Environment*, **25**(2), 217–29.

Golosov, M., J. Hassler, P. Krusell and A. Tsyvinski (2014), 'Optimal taxes on fossil fuel in general equilibrium', *Econometrica*, **82**(1), 41–88.

Hallegatte, S., M. Bangalore, L. Bonzanigo, M. Fay, T. Kane, U. Narloch, J. Rozenberg, D. Treguer and A.

Vogt-Schilb (2016), *Shock Waves: Managing the Impacts of Climate Change on Poverty*, Washington, DC: World Bank.

Hassler, J. and P. Krusell (2012), 'Economics and climate change: integrated assessment in a multi-region world', *Journal of the European Economic Association*, **10**(5), 974–1000.

Hope, C. (2006), 'The marginal impact of CO_2 from PAGE2002: an integrated assessment model incorporating the IPCC's five reasons for concern', *Integrated Assessment Journal*, **6**, 19–56.

Interagency Working Group on Social Cost of Carbon, United States Government (2000), 'Technical support document: technical update of the social cost of carbon for regulatory impact analysis – under executive order 12866', US Government.

Kates, R.W. (2000), 'Cautionary tales: adaptation and the global poor', *Societal Adaptation to Climate Variability and Change*, **45**(1), 5–17.

Krey, V. (2014), 'Global energy climate scenarios and models: a review', *WIREs Energy and Environment*, **3**, 363–83.

Leichenko, R. and K. O'Brien (2008), *Environmental Change and Globalization: Double Exposures*, Oxford: Oxford University Press.

Lockwood, B. and J. Whalley (2010), 'Carbon-motivated border tax adjustments: old wine in green bottles?', *The World Economy*, **33**(6), 810–19.

Mendelsohn, R., A. Dinar and L. Williams (2006), 'The distributional impact of climate change on rich and poor countries', *Environment and Development Economics*, **11**(2), 159–78.

Metcalf, G. (2009), 'Designing a carbon tax to reduce US greenhouse gas emissions', *Review of Environmental Economics and Policy*, **3**(1), 63–83.

Monjon, S. and P. Quirion (2011), 'Addressing leakage in the EU ETS: border adjustment or output-based allocation?', *Ecological Economics*, **70**(11), 1957–71.

Moore, F. and D. Diaz (2015), 'Temperature impacts on economic growth warrant stringent mitigation policy', *Nature Climate Change*, **5**(2), 127–31.

Nordhaus, W. (2007), 'A review of the *Stern Review on the Economics of Climate Change*', *Journal of Economic Literature*, **45**, 686–702.

Nordhaus, W. (2010), 'Economic aspects of global warming in a post-Copenhagen environment', *Proceedings of the National Academy of Sciences*, **107**(26), 11721–6.

Nordhaus, W. and P. Sztorc (2013), 'DICE 2013R: introduction and user's manual', accessed 17 October 2016 at http://www.econ.yale.edu/~nordhaus/homepage/documents/DICE_Manual_103113r2.pdf.

Nordhaus, W. and Z. Yang (1996), 'A regional dynamic general-equilibrium model of alternative climate-change strategies', *The American Economic Review*, **86**(4), 741–65.

Oppenheimer, M., M. Campos, R. Warren, J. Birkmann, G. Luber, B. O'Neill and K. Takahashi (2014), 'Emergent risks and key vulnerabilities', in *Climate Change 2014: Impacts, Adaptation, and Vulnerability*, Contribution of Working Group II to the Fifth Assessment Report of the Intergovernmental Panel on Climate Change, Cambridge: Cambridge University Press.

Rezai, A., D. Foley and L. Taylor (2012), 'Global warming and economic externalities', *Economic Theory*, **49**(2), 329–51.

Riahi, K., F. Dentener, D. Gielen, A. Grubler, J. Jewell, Z. Klimont, V. Krey, D. McCollum, S. Pachauri, S. Rao, B. van Ruijven, D. van Vuuren and C. Wilson (2012), 'Energy pathways for sustainable development', in *Global Energy Assessment – Toward a Sustainable Future*, Cambridge: Cambridge University Press, pp. 1203–306.

Scovronick, N., M. Budolfson, F. Dennig, F. Errickson, M. Fleurbaey, W. Peng, R. Socolow, D. Spears and F. Wagner (2019), 'The impact of human health co-benefits on evaluations of global climate policy', *Nature Communications*, **10**, 2095, https://doi.org/10.1038/s41467-019-09499-x.

Sandmo, A. (2006), 'Global public economics: public goods and externalities', *Économie Publique*, **18–19**, 3–21.

Stern, N. (2006), *The Economics of Climate Change: The Stern Review*, Cambridge: Cambridge University Press.

Sterner, T. (ed.) (2012), *Fuel Taxes and the Poor: The Distributional Effects of Gasoline Taxation and their Implications for Climate Policy*, Abingdon, UK: Routledge.

Tol, R. (1996), 'The Climate Framework for Uncertainty, Negotiation and Distribution', in Miller, K. and R. Parkin (eds), *An Institute on the Economics of the Climate Resource*, Boulder, CO: University Corporation for Atmospheric Research, pp. 471–96.

United Nations (1992), 'Framework Convention on Climate Change', accessed 17 October 2016 at https://unfccc.int/resource/docs/convkp/conveng.pdf.

Williams, R., H. Gordon, D. Burtraw, J. Carbone and R. Morgenstern (2014), 'The initial incidence of a carbon tax across income croups', Resources for the Future Working Paper Working Paper.

World Bank (2014), 'World Development Indicators', Table 2.9: Distribution of income or consumption, Washington, DC: World Bank, accessed 17 October 2016 at wdi.worldbank.org/table/2.9.

11. Tipping and reference points in climate change games
Alessandro Tavoni and Doruk İriş

1. INTRODUCTION

We live in a world characterized by discontinuities, where thresholds for abrupt and irreversible change are omnipresent, both in economic and ecological dynamics. Such thresholds, often referred to as tipping points, trigger nonlinear responses on the part of individuals or ecosystems.

Climate change is a prominent example of the pervasiveness of tipping points, since they appear both in the strategic decision to embark in costly mitigation (Heal and Kunreuther, 2012) and in the Earth's climate system (Lenton et al., 2008).

In this chapter we will focus on "behavioral tipping points", to distinguish them from the ecological ones. As it will become apparent, though, the two are closely linked, since planetary boundaries define "the safe operating space for humanity with respect to the Earth system and are associated with the planet's biophysical subsystems or processes" (Rockström et al., 2009). Hence, to discuss strategies one has to account for the underlying physical processes and how they are perceived (Tavoni and Levin, 2014).

Whether a country or a subnational actor (a city, an NGO or a firm) decides to invest in a clean technology, or more broadly in actions aimed at reducing greenhouse gas emissions, depends on its expectations with regards to the actions of others. This is particularly salient in the context of a public good such as climate change mitigation. Depending on the choice of the underlying parameters, the climate change game is generally characterized either by coordination (selecting the mutually preferable outcome in a chicken game), or a unique inefficient outcome resulting from widespread defection in a prisoner's dilemma game (Barrett, 2016). In the latter class of games, which comprises linear public goods games for plausible parameters capturing the temptation to defect (not contributing to the mitigation good), the worst-case scenario is for an actor to take costly action while the others refuse to do so, the so-called "sucker's payoff". Arguably, this may have been the case for the European Union in climate negotiations up to COP 20 in Lima, with unilateral commitments by the EU routinely unmatched by other large economies. COP 21 in Paris was perhaps the first Conference of the Parties to mark a greater willingness to show leadership by other large powers, such as China and the United States (although the election of Trump as president casts a long shadow over the prospects of the Paris agreement). A possible interpretation, in keeping with the above arguments, is that enough action at various scales had accumulated in the years leading to the Paris summit that even less committed countries showed an increased willingness to act.[1]

---

[1] Indeed, while the Nationally Determined Contributions agreed upon in Paris are insufficient to meet the target of "holding the increase in the global average temperature to well below 2°C

These emerging trends are potentially game-changing, provided that enough actors lead the way by taking action early on. Once a tipping point for sufficient investments in low carbon technologies has been reached, and constituencies with stakes in the nascent markets have formed, standard economic forces will sustain the transition to a carbon-neutral economy.

We will review some of the recent literature that provides clues about when such reinforcing dynamics take place. In doing so, we will come across related concepts, such as diffusion and feedback. Importantly, given the wide scientific uncertainties surrounding the location of the thresholds, we will discuss the role of expectations and argue that reference points are crucial for supporting cooperation. Intuitively, under uncertainty asymmetries about views on the expected losses from climate change are as important as differences in objective (but elusive) vulnerabilities.

2. REGIME SHIFTS AND CATASTROPHIC CLIMATE CHANGE

A key concept relating to tipping points, or regime shifts as they are otherwise referred to, is that of irreversibility: "Ecological regime shifts are large, sudden changes in ecosystems [...]. They entail changes in the internal dynamics and feedbacks of an ecosystem that often prevent it from returning to a previous regime" (Biggs et al., 2009). In other words, once a threshold has been passed, the system will enter a new basin of attraction, resulting in a sudden and persistent shift to a new stable fixed point (fixed point 2 in Figure 11.1).

A related concept in the social sciences is that of diffusion, which Rogers (1995) defines as "the process by which an innovation is communicated through certain channels over time among the members of a social system". The link here is the idea that the actions of others can reinforce one's own choices. Different terms have been coined for this feedback mechanism, depending on the disciplinary focus, such as bandwagon effect in fashion-oriented behavior (Leibenstein, 1950), adoption thresholds (Granovetter, 1978), entrapment (Dixit, 2003), global cascades (Watts, 2002) and tipping (Gladwell, 2000), to mention a few.

Early models of diffusion focus on the societal adoption rate (of a technology or behavior), whose dynamics are governed by the overall ratio of adopters to non-adopters at a given point in time (Bass, 1969; Young, 2009). The main insight from the Bass model, also known as S-shaped diffusion curve, is that diffusion follows a nonlinear trend, with a fast acceleration in the initial phase of adoption and a subsequent saturation (Figure 11.2).

Of course, the channels through which innovations diffuse are less mechanistic than those depicted in Figure 11.2. An important driver of the speed of diffusion, whether of an opinion, a fad or a technology, is the topology of the network on which the agents are embedded, since it will spread through society according to dynamics that depend on the patterns of social connections (Currarini et al., 2015).

---

above pre-industrial levels and to pursue efforts to limit the temperature increase to 1.5°C above preindustrial levels", these commitments are viewed as genuine, in the sense that they exceed what states would have done in the absence of the agreement (Averchenkova and Bassi, 2016).

Tipping and reference points in climate change games 241

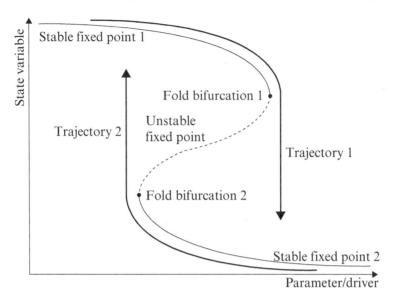

Note: Fold bifurcations occur when a stable fixed point (solid lines) collides with an unstable fixed point (dashed line), and can lead to regime shifts.

Source: Lade et al. (2013).

Figure 11.1 Regime shifts

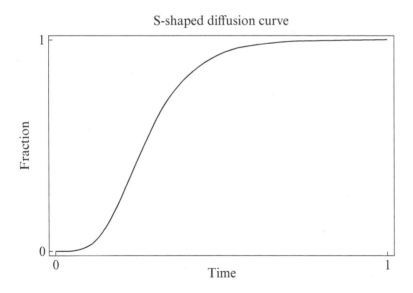

Note: The fraction of adopters is plotted against time.

Figure 11.2 An example of S-shaped diffusion curve

What are the implications of the nonlinearities identified above in ecological and social processes for the economic analysis of climate change? We hope to contribute to answering this question by reviewing some of the recent literature focusing on the avoidance of a threshold for dangerous climate change.

Perhaps the most influential paper on the economics of catastrophic climate change is due to Weitzman (2009), who focused on high-impact, low-probability catastrophes and catalyzed attention on the so-called (bad) fat tail of the probability density function (PDF) of what might happen in the absence of serious mitigation policy.[2] His dismal theorem points to the "potentially unlimited exposure to catastrophic impacts" in the presence of "deep structural uncertainty in the science coupled with an economic inability to evaluate meaningfully the catastrophic losses from disastrous temperature changes" (Weitzman, 2009). The implications for both theorists and practitioners are big: if we are serious about accounting for unlikely, yet possible future climate disasters, we cannot rely on the standard cost–benefit analysis calculus dealing with uncertainty in the form of a known PDF with thin tails. Consequently, Weitzman advocates a "generalized precautionary principle" that accounts for the potential catastrophic costs of inaction.

A not-so-dismal account is found in Heal and Kunreuther (2012), which instead investigates the implications of the existence of a tipping point in the adoption of climate policies by the international community. The authors offer illustrative evidence on the role of early adopters (in the abstract representation of Figure 11.2, those who are located at the bottom of the diffusion curve) in triggering a global shift away from the use of damaging pollutants, towards greener alternatives. They cite two pieces of evidence. One concerns the adoption of unleaded gasoline in replacement of leaded gasoline: the unilateral adoption by the United States meant that the subsequent adoption's costs for other countries was limited to modifying refinery capacity, since motor industries exporting to the United States had to transition to lead-free fuel immediately after the move. Thanks to these reduced costs for the followers, the new technology spread quickly worldwide. The second example refers to phasing out chlorofluorocarbons (CFCs), a remarkable achievement of the Montreal Protocol on Substances that Deplete the Ozone Layer. In this case, the U.S. decision to sign the Montreal Protocol hinged on a technological innovation by Du Pont, the world's largest producer of CFCs, allowing the company to gain from elimination of CFCs. As in the previous example, strategic complementarity led most countries to phase out ozone-depleting chemicals.

3. KEY FEATURES OF DANGEROUS CLIMATE CHANGE

Avoiding disastrous climate change requires overcoming several barriers to collective action. As other public goods situations, such as protecting the global climate from *gradual* climate change, *dangerous* climate change requires widespread cooperation in the face of individual incentives to free-ride on the efforts of others. Moreover, it is characterized by sudden transitions to harmful states (tipping points) and irreducible uncertainty

---

[2] A fat-tailed distribution assigns a higher probability to rare events in the extreme tails than does a thin-tailed one.

on the location of the threshold and on the consequences of crossing it (Dannenberg and Tavoni, 2016). A complicating factor is that the ability to contribute to the solution and benefits from doing so vary greatly among parties. Together, these characteristics compound the difficulty of securing sufficient mitigation effort, with potentially disastrous consequences.

The Framework Convention on Climate Change has warned that climate change may be "abrupt and catastrophic", rather than linear and smooth, once greenhouse gas concentrations in the atmosphere exceed a certain threshold. This threshold is commonly identified as the concentration level that would translate in a 2°C average temperature increase above preindustrial levels. More recently, the Paris Agreement raised the ambition to limiting the increase to 1.5°C. The scientific literature confirms the possibility of dangerous thresholds but also shows that there is large uncertainty about their location (Rockström et al., 2009). This uncertainty is compounded by the political and technical uncertainty arising from translating such thresholds into the necessary mitigation measures required to avert disaster (on top of the uncertainty about the economic value of the catastrophic losses from failure to avoid dangerous climate change). In addition, international climate negotiations that aim to reduce global greenhouse gas emissions are strongly influenced by a conflict between rich and poor countries. Consequently, the Kyoto Protocol addressed this North–South equity issue by recognizing the industrialized nations' special responsibilities through the principle of "common but differentiated responsibilities". However, major polluters and great powers, such as China and the U.S., have until recently proven reluctant to bind themselves to internationally agreed ambitious emission reductions. While progress has been made in Paris, the move to a bottom-up architecture where states unilaterally announce their proposed targets confirms the burden-sharing difficulties.

How can this deadlock be broken? As real-world data on alternative mechanisms is not available, we must rely on theory and experimental economics. Here, we focus on several simple and recent applications of both methods. Although both theory and experiments rely on simplified settings, and in the case of experiments on convenience samples, these methods have the advantage of allowing one to isolate in a controlled manner the effect of relevant features (such as inequality and uncertainty) on cooperation. The next two sections tackle, in turn, two classes of games that have recently been modified to accommodate some of the features we have discussed thus far in the context of the avoidance of disastrous climate change: public goods games and coalition formation games.

4. THRESHOLD PUBLIC GOODS GAMES

In order to account for the threat of disastrous losses from abrupt and irreversible climate change, a recent literature has resorted to studying public goods games, both theoretically and in the laboratory. These can accommodate the above features by introducing a discontinuity, in the form of a threshold for dangerous climate change. Threshold public goods games (TPGs), unlike linear ones, feature multiple Nash equilibria. Restricting attention to symmetric play, we have two equilibria: a Pareto-superior one where players shoulder the same burden and avoid catastrophe *just*, alongside the unique (Pareto-inferior) equilibrium prescribing defection. The latter characterizes the quintessential cooperation problem of under-provision (or lack of provision) of public goods. Thus, the presence of a discontinuity in the form of a threshold beyond which losses shoot up appears to facilitate the problem by effectively serving as a coordination device (Barrett, 2013).

In addition to work that focuses on equilibrium behavior as in traditional game theory, variants of the above TPGs have also been employed in evolutionary game theory models (EGT) to investigate off-equilibrium dynamics. The premise behind the replicator dynamics, the most commonly employed imitation dynamics in EGT, is that agents imitate the behavior of the other players in the population whenever these appear to be more successful. While the implied myopic behavior represents a departure from traditional game theory, which models optimization through instantaneous best responses of all players to one another, long-run behavior in evolutionary games often coincides with static Nash equilibria (Weibull, 1997).

Vasconcelos et al. (2014) investigate the collective action problem posed by disastrous climate change using EGT to assess disaster avoidance success in TPGs. Among other features, they focus on two that are relevant for our discussion: inequality and polycentric governance (Ostrom, 2009). To tackle the former, they include an asymmetric distribution of wealth that mimics the existing inequalities among nations. The authors find that the combination of homophily (high likelihood of imitating the strategies of agents similar to oneself) and inequality can lead to the collapse of cooperation, due to a drop in the poor countries' contributions. To assess the effectiveness of local sanctioning institutions in coping with dangerous climate change, Vanconcelos and colleagues distinguish between a global setup, in which the entire population constitutes the sole negotiating group, and a local setup, in which individuals interact in several smaller groups. They find that local institutions are more successful at promoting the emergence of widespread cooperation, particularly under low risk perception (when the perceived probability of disaster under non-provision of the public good is small). The reason is that cooperation can only thrive if at least some cooperators in the population do better than the defectors, and this is more likely to happen in the local setup because it allows for more heterogeneity in behavior.[3]

In sum, the above findings cast an optimistic light on the prospects for climate negotiations, especially in the aftermath of the Paris Agreement, which was hailed as an unprecedented success and relied heavily on unilateral pledges, known as Intended Nationally Determined Contributions (INDCs). However, as noted above, uncertainty about the location of the threshold complicates matters, by bringing us back to a gradual losses setup, as the sharp discontinuity at the threshold is replaced by a smooth probability density function of expected losses. Experiments confirm the detrimental effect of threshold uncertainty; see Barrett and Dannenberg (2012) and Dannenberg et al. (2014). A crucial question is then to what extent can climate treaties coordinate actions in spite of structural uncertainty. To address it, we turn to the literature on the economics of international environmental agreements, before returning in greater detail to the experimental evidence.

---

[3] Note however that the local institutions are assumed to be independent of each other. Hence, further research is needed to clarify if the results also hold in a more realistic setting with spillovers across regions.

5. COALITION FORMATION GAMES

Building on the work of d'Aspremont et al. (1983), international environmental agreements have long been modeled in game theory through coalition formation games (Hoel, 1992; Carraro and Siniscalco, 1993; Barrett, 2016). The premise is that the equilibrium number of signatories to a self-enforcing international agreement follows from the conditions of internal and external stability, which respectively guarantee that no signatory is better off leaving the coalition, and that there is no incentive for a non-signatory to join the coalition. These conditions are required since treaties such as the Kyoto Protocol (let alone the Paris Agreement, which lacks legal force) cannot be enforced by external institutions and must therefore rely on incentives to overcome the compliance issue.

The accepted insight from the theory of international environmental agreements is that self-enforcing treaties fail to deliver, especially when cooperation is most needed (i.e. when the potential gains from cooperation are large due to high mitigation costs and high benefits from mitigation). Scott Barrett summarizes how the introduction of a known tipping point changes this calculus:

> The standard model of a self-enforcing international environmental agreement predicts that collective action in reducing greenhouse gas emissions will be grossly inadequate. When this model is modified to incorporate a certain threshold with catastrophic damages, treaties can become highly effective. If the benefits of avoiding the threshold are high relative to the costs, the prospect of catastrophe transforms treaties into coordination devices. (Barrett, 2013)

However, as discussed above, uncertainty is detrimental to the prospects of disaster avoidance:

> While the uncertain prospect of approaching catastrophes may commend substantially greater abatement in the full cooperative outcome, it may make little difference to non-cooperative behavior or to the ability of a climate treaty to sustain substantial cuts in global emissions. (Barrett, 2013)

What else can we then rely upon, to design an effective treaty? Is there a way out of the grim results found in most of the models reviewed so far? The game-theoretic literature on International Environmental Agreements (IEAs) has identified several mechanisms that have the potential to mitigate the issue of shallowness of the mitigation efforts (or small stable coalition size, both of which translate into unambitious treaties and increased threat of catastrophe).[4] These range from expanding the strategy space via side payments and issue linkage (Barrett, 2005), to introducing minimum participation rules and heterogeneity (Weikard et al., 2014) and imposing trade sanctions on non-participants (Nordhaus, 2015).

Preferences also matter for the outcome of IEAs. The literature has traditionally modeled negotiators as rational agents with standard preferences. A recent strand has instead explored the implications of departing from these assumptions. Examples include introducing preferences for equity (Lange and Vogt, 2003), for reciprocity (Nyborg, 2015), reference dependence (İriş and Tavoni, 2018), and appetite for campaign contributions by

---

[4] For a recent review, see de Zeeuw (2015).

policy makers subject to lobbying pressure (Marchiori et al., 2017).[5] We review some of the non-standard preferences literature (employing other regarding and reference-dependent preferences) in Section 7, where we restrict attention to coalition formation and public goods games (Bosetti et al., 2017).

Broadly speaking, all of the above modifications of the standard model ease the collective action problem, to some extent. The question remains, however, about which representation of the climate negotiations is more realistic. For instance, some improving mechanisms, such as imposing trade sanctions on those outside the "climate club", while theoretically desirable may prove difficult to implement due to the threat of retaliations and potential escalation to trade wars. Equally important, we lack empirical evidence about the relative effectiveness of different schemes, or about the preferences of the negotiators. We tackle these issues in the next section.

6. CLIMATE CHANGE EXPERIMENTS

A productive way to gather some relevant insights is to conduct controlled experiments with subjects who are assigned to different treatments aimed at capturing relevant features of the game. An obvious advantage is that one doesn't need to wait decades to assess how a given IEA, such as the one agreed in Paris in December 2015, has affected global emissions (in addition to the fact that one can easily compare treatments to the baseline "untreated" status quo in an experiment, while in reality the counterfactual is not easy to identify). The price to pay in order to have such control (internal validity) is that one has to greatly simplify the problem in the laboratory, possibly at a cost in terms of external validity.

In this section we briefly review some of the recent experimental literature that uses TPGs as a metaphor for studying cooperation on the avoidance of dangerous climate change. In these experiments small groups (of four to ten players) have the option to contribute part of their endowment to avoid a collective loss (in one or several successive rounds). The aggregate contributions are then evaluated to see how they compare to the investment required to avoid the tipping point, and a large fraction of the remaining endowment is lost if the target has not been met.[6]

Milinski et al. (2008) test the role of risk perception on the success rate in disaster avoidance by manipulating the probability of losing one's savings (p) if the group fails to invest the target sum (€120) by the end of ten successive rounds. Thus, we have a TPG with three treatments, corresponding to either 10%, 50% or 90% probability of loss when missing the provision threshold. They find that the ensuing share of groups who managed to avoid the loss are respectively 0%, 10% and 50%. One should not be surprised that none of the groups averted disaster when p=10%, since the expected loss is so small that it is collectively rational *not* to provide the public good (i.e., there is only mild climate change, and no collective action problem). It is instead noteworthy that even at very high expected costs from miscoordination from the Pareto-superior equilibrium of collecting €120 when

---

[5] For a paper focusing on reciprocal strategies in IEAs, see Ochea and de Zeeuw (2015).
[6] For a more comprehensive review of dangerous climate change games, see Dannenberg and Tavoni (2016).

p=90%, half of the groups fail the target (while still contributing approximately €113). One likely explanation for such spectacular coordination failure is that the subjects in this experiment were not allowed to signal their intentions or communicate in any way. As mentioned above, TPGs with certain threshold location is a coordination game, which naturally begs for communication opportunities.

Even in the face of inequality in the ability to contribute to the public good of loss avoidance, a clear coordination target proves a powerful mechanism to select the "good" equilibrium, provided that the communication channels are in place to facilitate redistribution. For instance, in a similar setting of repeated contributions to a TPG with p=50%, Tavoni et al. (2011) find that the majority of groups are able to avoid disaster when they have an opportunity to signal future contribution intentions via pledges. Namely, when communication was possible 60% of the groups with unequally wealthy participants (three "Poor" and three "Rich" countries) coordinated on disaster avoidance, compared to only 20% of successful groups when pledging was not an option. This is noteworthy, given the non-binding nature of pledges; yet, one should not be surprised that for a coordination game to be played well, communication is indispensable.[7]

Dannenberg et al. (2014) revisit this TPG, with groups of six players facing dangerous climate change with p=90%, by introducing uncertainty on the location of the tipping point. In two treatments they test the effect of risk, i.e., known (uniform) distribution over a range of potential thresholds, and ambiguity, i.e., lack of knowledge even on the distribution. They find that contributions become more erratic under threshold uncertainty, particularly so under ambiguity. Early commitment in the risk treatment, demonstrated by a willingness to invest in the public good early on and fulfill the pledges, helps groups to reduce the negative effect of uncertainty. This result resonates with the one by Tavoni et al. (2011) that the negative effect of wealth inequality is mitigated by leadership and communication.

In a one-shot TPG without sequential decision-making, Barrett and Dannenberg (2012) test a simplified version of the model developed in Barrett (2013), to assess the effect of uncertainty on two variables: the extent of the damages from dangerous climate change, and the location of the threshold. They find that while the first is unimportant for the prospects of catastrophe avoidance, when there is significant uncertainty on the location of the tipping point, the game reverts to a prisoner's dilemma and cooperation drops, with no group succeeding in avoiding the large loss.

Recently, scholars have begun to introduce political economy features, such as delegation, into catastrophe avoidance games mimicking dangerous climate change. The reason is simple: since the goal is to shed light on the ability of stylized climate agreements to improve upon non-cooperative behavior, one should aim to capture realistic aspects of their decision-making process. An intuitively relevant one is delegation: do elected delegates, representing the subgroup (country) to which they belong, by deciding how much to contribute to the public account on their behalf, behave more

---

[7] It appears that the positive effect of communication is indeed stronger than the negative effect of inequality. In the symmetric wealth treatments where endowments were equal across the six players, success rate in disaster avoidance went up less markedly when communication was allowed, from 50% to 70%. This is also to be expected, given that coordination is a much easier task under symmetric payoffs.

or less cooperatively than when acting independently? In other words, are the findings from the experiments mentioned above robust to delegation? İriş et al. (2019) compare the baseline case of four-subjects' groups where each "country" independently decides how much to contribute to the public good (again in the face of uncertainty), with two treatments where the countries are no longer singletons, but are themselves composed of a three-player constituency responsible to elect a delegate to represent them in the negotiations. Hence, while the number of countries is still four, in these two treatments the group is made of 12 subjects, given that in each country there is one delegate deciding for herself as well as for the two unelected candidates. The only difference between the two delegation treatments is that in one the delegate decides on contributions in the TPG without being exposed to public scrutiny from the constituency, while in the other the two non-delegates in each country are in the same room as their delegate and send their non-binding preferred contribution suggestions to the delegate. İriş et al. (2019) find that delegation without public pressure does not affect contributions much, relative to the baseline. However, even if messages are payoff-irrelevant in the experiment, public pressure has a significant negative effect on the delegates' contributions. This happens since the majority of delegates were elected because they signaled a low propensity to contribute in the practice phase, and, once elected, delegates focused on the lower of the two contributions preferred by their teammates, thus behaving more selfishly than in the treatments without public pressure. This finding echoes the one of Milinski et al. (2016), who use a similar setup with six three-player "countries" voting twice whether to confirm or vote out the incumbent representative. They find that selfish representatives are preferentially elected, and that once in power, they indeed contribute less to the public good than their fair share.

To summarize, in this section we have established the following: (i) the higher the (perceived) probability of disaster, the more likely it is that the catastrophic tipping point is avoided. (ii) Communication, in the form of pledges for achieving the target (e.g., INDCs), also increases the prospects for success, given that they facilitate the coordination problem; (iii) this is particularly needed in the presence of threshold uncertainty, although when uncertainty is too large failure is widespread. (iv) Delegates tend to focus on the least ambitious suggestion when confronted with public pressure, and act in a more self-interested manner than individuals.

In the next section we turn to the insights gathered in recent theoretical work on tipping points and reference dependence in related climate change games.

7. NON-STANDARD PREFERENCES

During the last decades, abundant evidence from both the laboratory and the field demonstrate that people exhibit discontinuities in their preferences as well. They often not only care about the outcome, but also about how it stands relative to a reference level. Economists and other social scientists argue that salient reference levels can relate to the status quo (Tversky and Kahneman, 1991), or alternatively emerge from social comparisons (Fehr and Schmidt, 1999; Rabin, 1993; Shafir et al., 1997), or can be based on goals and aspirations (Heath et al., 1999).

In this section, we focus on two widely used non-standard preferences: i) reference-dependent preferences; and ii) other-regarding preferences (also referred to as social preferences). In particular, we review their applications in public goods and coalition formation games.[8]

7.1 Fairness and Other-Regarding Preferences

Standard economic models assume that economic agents are self-interested, meaning that they care only about their payoffs and are not concerned with others' payoffs. However, theoretical and empirical studies have shown that people have strong preferences for equity, even in market settings (Kahneman et al., 1986; Fehr et al., 1993; Fehr et al., 1997).

The literature classifies two groups of models for other-regarding preferences. The first group contains models in which people reciprocate based on the actions and perceived intentions of other players. The second group contains models in which people care about the distributions of payoffs.

If another player's action, or the outcome it leads to, is more positive (or negative) than the one that is considered fair—the reference level—then reciprocal players perceive it as a kind (or unkind) action. In the context of climate change, reciprocal players would be willing to undertake costly mitigation effort (e.g., abatement) insofar as others do too. Thus, reciprocal players do not only care about the outcomes of their actions, but also about the actions or intentions of other players (Rabin, 1993; Dufwenberg and Kirchsteiger, 2004; Falk and Fischbacher, 2006; Cox et al., 2007; Segal and Sobel, 2007).

Hadjiyiannis et al. (2012) is the first paper that incorporates reciprocity preferences into self-enforcing IEAs, in which reciprocal countries play an infinitely repeated prisoner's dilemma game. The authors compare the depth of the cooperation in the case of full participation in two games: one with reciprocal countries and the other with self-interested countries. The authors find that reciprocal countries that have moderate expectations with respect to others' national abatement strategies can support a greater degree of environmental cooperation than self-interested ones. They define reciprocal countries as those where abatement standards are considered fair if they are within certain boundaries. For moderate expectations, reciprocal countries perceive others' cooperative abatement levels as kind and non-cooperative (Nash) abatements as unkind, leading to more cooperation. However, when only very high abatement standards are deemed fair (i.e., fair abatement levels are higher than the most cooperative abatement levels countries can sustain), then reciprocity could have a detrimental effect on international environmental cooperation as countries perceive even the cooperative abatements as unkind. Their model therefore provides a novel perspective on the role of expectations in environmental negotiations. Consistent with their findings, failure in Copenhagen (COP15 summit), owing to the very high expectations, has been noted by some experts.[9] Consequently, leaders emphasized the

---

[8] DellaVigna (2009) reviews the empirical evidence of such non-standard preferences as well as other concepts in non-standard beliefs and non-standard decision making. Camerer et al. (2003) and Kahneman and Tversky (2000) collect early theoretical and applied studies pioneered behavioral economics. Camerer's (2003) and Kagel and Roth (Volume 1, 1995; Volume 2, forthcoming) are the three handbooks in experimental games.

[9] "Can Copenhagen still be saved?" *The Economist*, November 17, 2009.

importance of moderating expectations before and during the Paris conference (COP21) in December 2015.[10]

Nyborg (2015) studies the impact of reciprocal preferences on stable IEAs by using coalition formation games. To keep the model tractable, she focuses on a discrete strategy space, so that countries can either abate or pollute. Unlike Hadjiyiannis et al. (2012), she follows more closely Rabin (1993) in both incorporating reciprocity into the utility function and defining equitable payoff. She shows that no country participates in the coalition in the game with only self-interested countries. However, if some countries exhibit reciprocity, three stable coalition sizes become feasible: no participation, a minority coalition, and a majority or even full participation. The minority coalition improves upon zero participation, but only weakly. Moreover, for the majority or full participation coalition to be stable, most of the countries should exhibit strong reciprocal preferences.

A different agreement literature also studies the impact of reciprocity in which signatory countries *bind* themselves to share the cost of public good investment (Jang et al., 2016). Their model of reciprocity follows Dufwenberg and Kirchsteiger (2004), which extends Rabin's (1993) reciprocity model for a one-shot game to a sequential game with updating beliefs. Dufwenberg and Patel (2017) additionally investigate the network effects for discrete public goods in this literature.

On the other hand, players with distributive concerns care about relative payoffs, that is, both about own payoff and how it compares with other players' payoffs (their reference level). A significant part of the literature focuses on models in which people exhibit self-centered inequality-aversion (IA): individuals care about their payoff but are also willing to reduce the differences between their payoff and those of the others. If their payoff is lower than the others' (disadvantageous inequality), they envy better-off players and accordingly incur in a disutility. If their payoff is higher than the others' (advantageous inequality), they feel guilt or compassion. Inequality-averse players are assumed to dislike disadvantageous inequality more than advantageous inequality. As a result, they are willing to sacrifice some of their payoffs to reduce either type of inequality, but with a stronger willingness to reduce disadvantageous inequality (Ochs and Roth, 1989; Loewenstein et al., 1989; Bolton, 1991; Kirchsteiger, 1994; Fehr et al., 1998; Fehr and Schmidt, 1999; Bolton and Ockenfels, 2000). Charness and Rabin (2002) introduce an alternative model of distribution concerns in which players maximize social surplus and are also inclined to help other players who receive lower payoffs relative to others.

Lange and Vogt (2003) is the first paper that incorporates IA to IEAs. They study IA in three different games to analyze international environmental cooperation: (i) a symmetric prisoner's dilemma game with binary choice of cooperation or defection; (ii) a symmetric emission game with continuous strategy space; and (iii) a coalition formation game with discrete abatement choice where the second stage is a simultaneous move game (cf. Carraro and Siniscalco, 1993; Hoel, 1992). The authors follow Bolton and Ockenfels' (2000) IA model, with country i's utility defined by:

$$a_i u(y_i) + b_i r(\sigma_i), \tag{11.1}$$

---

[10] "Paris Deal Would Herald an Important First Step on Climate Change", *The New York Times*, November 29, 2015.

where $a_i, b_i \geq 0$. The first term y_i captures the standard utility from payoff y_i. The second term represents fairness utility based on the relative share $\sigma_i = y_i/\Sigma_j y_j$. Furthermore, $u(\cdot)$ is differentiable, strictly increasing, and concave; and $r(\cdot)$ is differentiable, concave, and has its maximum at equal share $\sigma_i = 1/N$, where N is the number of countries. The authors find that, depending on its distribution among players, IA can lead to an equilibrium in which majority or all countries cooperate in game (i), since when others cooperate a country values its payoffs to be closer to the equal share, compared to their absolute payoffs. However, in game (ii) with continuous strategies, a country with a possibility to increase its absolute payoff would do so if it receives less than the equal share. Since all countries have such incentives, IA has no effect at all: countries act as if they were maximizing their own payoff in equilibrium. Finally, in game (iii) with coalition formation, countries' strong preference for the equal share leads efficiency gains and even the grand coalition can be stable. In particular, fringe countries outside the coalition incentivize self-interested countries to join and reward the coalition's higher abatement efforts by their own efforts. However, this incentive diminishes as the countries with strong preference for equality join the coalition.

Lange (2006) further analyzes different notions of equity when countries are either industrialized or developing, such as IA with respect to the differences in per capita emissions, and IA with respect to differences in abatement targets. In the model, only industrialized countries take on emission reductions, while developing countries could participate if industrialized countries finance them. He shows that IA concerns on the differences in per capita emissions lead to higher emission reductions in industrialized countries, but no qualitative impact on the incentives to cooperate. On the other hand, IA concerns on the differences between countries' abatement targets and the average of abatement by the other industrialized countries lead (i) developing countries to both participate and abate more, and (ii) higher coalition sizes and even grand coalition if countries are sufficiently IA.

Grüning and Peters (2010) also study the role of fairness in a coalition formation game. In addition to the usual welfare, which consists of benefit minus cost of abating, they subtract the variance in the environmental policy (abatement levels) of all countries. This incentivizes countries to abate similarly and at much higher levels, both within and outside an IEA and, thus, increases the coalition size.

Kolstad (2013) adapts Charness and Rabin's (2002) social preferences to study linear public goods and coalition formation games in which he allows countries to differ in their sizes. More specifically, in his model there are N countries and each country i chooses its abatement level g_i (contribution to public good), or equivalently its emission level $x_i = w_i - g_i$, where w_i is the highest possible emission level. Country i's welfare function has the following form:

$$u_i(x_i, G) = \lambda_i \pi_i(x_i, G) + \delta_i \min_{j \neq i} \pi_j + \varepsilon_i \sum_j \pi_j. \tag{11.2}$$

The first term is the self-interested welfare of country i, with $G = \Sigma_i g_i$ and a scaling factor λ_i measuring its relative importance for country i. The second term captures the level of country i's care for equity, particularly by the country with the lowest self-interested welfare, scaled by δ_i. Finally, the third term with its scaling factor ε_i captures the preference for efficiency. The author shows that this formulation increases individual countries' willingness to contribute to the public good. However, it decreases the coalition size. Additionally, heterogeneity in countries' sizes destabilizes coalitions.

Bucholz and Sandler (2016) study the role of other-regarding preferences in public good games with Stackelberg-type leader-follower relations. The standard literature, which is reviewed by the authors, finds that the leaders' unilateral actions to contribute to a public good trigger a reduction in the followers' efforts. However, the authors show that a general other-regarding preference, which captures reciprocity and inequality-aversion, could eliminate this crowding-out effect by leaders' contributions influencing the followers' beliefs positively.

Bucholz and Sandler (2016), and the references cited therein, mention comments by negotiators' highlighting that the IEAs have to be fair and equitable in sharing the burden, as well as emphasizing their willingness to contribute more if others also do so (reciprocity). Lange et al. (2007) and Dannenberg et al. (2010) find empirical support for equity principals and preferences by using data from people involved in international environmental policy. However, Lange et al. (2010) find that equity arguments are often used for self-serving purposes.[11] Therefore, further empirical research is needed to understand the roles of equity concerns and reciprocity on IEAs.

In sum, to the extent that reciprocity and inequity-aversion indeed play a relevant role in IEAs, then they tend to facilitate sustaining an effective agreement and increase the coalition size. However, reciprocity and inequity-aversion can also have detrimental effects, when countries have very high expectations from each other, or when they have different levels of equity-concerns.

7.2 Loss-Aversion and Reference-Dependent Preferences

Kahneman and Tversky's (1979) prospect theory consists of four key elements: i) reference-dependence—the perception of outcomes as gains or losses relative to a reference level; ii) loss-aversion—the tendency towards avoiding losses rather than acquiring gains; iii) diminishing sensitivity—the higher sensitivity to changes around the reference level than to changes away from it; and iv) probability weighting—the subjective overweighting of low probabilities and underweighting of high probabilities. Most of the follow-up literature employs a simplified version of the theory, and utilizes only reference-dependence and loss-aversion (DellaVigna, 2009).

İriş and Tavoni (2018) study the role of loss-aversion and the threat of catastrophic damages, which they call jointly "environmental threshold concerns", on international environmental agreements. They aim to understand whether a threshold for dangerous climate change serves as an effective coordination device for countries to overcome the global free-riding problem. Loss-averse countries decide their abatement level under the threat of either high environmental damages (loss domain), or low damages (gain domain). They find that such concerns can cause the size of the coalition to increase when countries display identical environmental threshold concerns. When countries have heterogeneous threshold concerns, countries with high threshold concerns tend

---

[11] For more about equity principals and burden sharing rules in international environmental policy, see Cazorla and Toman (2001), Ringius et al. (2002), and Najama et al. (2003).

Table 11.1   Fourfold pattern of risk attitude by Markowitz (1952)

	Small Probability	High Probability
Gains	Risk-seeking	Risk-aversion
Losses	Risk-aversion	Risk-seeking

to join the coalitions, and the coalition size may diminish depending on the level of asymmetry.

İriş (2016) studies loss-aversion and political parties' economic target concerns in an infinitely repeated emission game. Despite optimization's common use in the literature, he argues that political parties often fail to optimize their countries' overall wellbeing since they have additional incentives to be elected. As economic issues influence voters' decisions more than the environmental issues, political parties aim to additionally satisfy their economic targets. He finds that stronger economic target concerns deter the most cooperative emission levels that countries could jointly sustain. Asymmetry either in economic target concerns or in technology levels hinders sustainability. However, when both asymmetries are in effect, they may cancel each other out. The paper concludes that such adverse incentives require efforts at the citizen level to sustain sound international environmental agreements.

Levy (1996) analyzes the impact of prospect theory for international conflict and for bargaining. Following prospect theory that also explains the fourfold pattern of risk attitude (Markowitz, 1952; see Table 11.1), he argues that political parties of adversarial states behave differently when they bargain over losses or gains. Moreover, leaders would be less willing to compromise and more willing to risk large losses to eliminate small losses (diminishing sensitivity).

Similar to Levy (1996), Gsottbauer and van den Bergh (2013) employ prospect theory to investigate how framing by leaders involved in climate agreements affects voters' risk perceptions and preferences for climate policy, in a self-serving way that justifies their standpoint.[12] For instance, the Bush Administration and Al Gore frame climate agreements differently to support their arguments. Specifically, the Bush Administration emphasizes that a strict climate agreement would bring a certain and substantial economic cost, while climate change and its consequences are uncertain. Thus, it frames the climate agreements as a loss, leading to risk-seeking behavior for uncertain losses, and advocates for climate inaction. On the other hand, Al Gore's documentary "An Inconvenient Truth" emphasizes that climate inaction would lead to a high environmental cost and damages, while the economic cost of an agreement is uncertain but should be moderate. Thus, it frames no agreement as a loss, thus justifying support for an effective climate agreement.[13]

Loss-aversion and reference-dependent preferences provide mixed results in terms of their impacts on international environmental cooperation, depending on the framing of the narrative and how the reference levels are determined. Similarly, to almost all the papers employing reference-dependent preferences, the ones reviewed here assume the

---

[12] Their general objective is to examine the impact of bounded rationality and other-regarding preferences from the perspective of voters who elect the negotiators involved with international climate negotiations.

[13] See the paper for examples of positive frames used by Nordhaus (1992) and Stern (2007).

reference levels to be exogenous. Endogenizing reference levels remains an open question in this literature.

Recently, however, Kőszegi and Rabin (2006) have developed a new model of reference-dependent preferences in which the reference levels are determined endogenously. Specifically, the authors assume that one's reference point is the person's rational expectations about the outcomes, which are determined in a personal equilibrium. The personal equilibrium requires the expectations to be consistent with the optimal behavior, given expectations. Kőszegi and Rabin (2006) and its extensions could provide an interesting agenda for further research on the role of expectations in IEAs.

8. DISCUSSION

We have reviewed both theoretical and experimental papers featuring tipping points and reference dependence, with the aim of extending our understanding of their potentially game-changing impacts on climate change cooperation. To this end, we have examined the role of thresholds and reference levels in public goods and coalition formation games, since they capture important features of dangerous climate change and its impacts on human behavior.

While we have started by highlighting the role of technology in creating strategic complementarities, a key message of this review is that the same holds for the psychological mechanisms described later in the chapter. This is the case because reciprocity leads to strategic complementarity as well (i.e., countries and other actors tend to cooperate conditionally on observable effort by their peers), leading again to a coordination game. The same can be said, to a lesser extent, about inequity aversion, which rationalizes infinitely many equilibria, such that coordination equilibria can be supported (Isaksen et al., 2016). It therefore appears natural to study social and ecological tipping points in a unified framework.

A further reason for integration is that ecological and behavioral tipping points are highly linked. The existence of ecological tipping points associated with abrupt and catastrophic, rather than gradual, climatic change has important behavioral repercussions in terms of the incentives to cooperate on mitigation efforts. Namely, when the threshold for acceptable emission levels, i.e., the amount of emissions that is compatible with gradual change, can be identified with a good degree of confidence, the climate change negotiations can be modeled as a coordination game. Hence, the problem becomes much easier to tackle than in the absence of a tipping point. Intuitively, this is so because the threshold provides an anchor for individuals to coordinate efforts upon. However, uncertainty on the location of the threshold removes the anchor for coordination, and if enough uncertainty surrounds the tipping point the game reverts back to a prisoner's dilemma. This is bad news, since the only self-interested equilibrium in this class of games is defection by all: free-riding incentives lock us into inaction.

An important question is thus which is the best approximation of the real negotiations. One may find reasons for optimism from the fact that negotiators at the COP21 Summit in Paris agreed to take steps towards limiting average global warming to "well below 2°C". Candidate thresholds are thus the symbolic 1.5°C and 2°C targets. However, much uncertainty still plagues the problem of translating such goals into the required actions at the national and subnational level. This is especially important given that the

Paris Agreement lacks legal force and relies instead on pledged nationally determined contributions, which even if fully implemented will not suffice for achieving even the 2°C target. Thus, increased ambition will be needed, in spite of the incentives to delay action.

Non-standard preferences, such as other-regarding and reference-dependent preferences, also have game-changing features that could explain important phenomena regarding climate change. Among the implications that arise from the literature that we have reviewed here are: whether strong leadership in mitigation efforts induces cooperation by others; why countries' high expectations about others' abatement efforts could have detrimental effects; and why developing countries have been relatively reluctant to exert even limited abatement efforts.

Of course, caution must be used when extrapolating from the games reviewed here to real-world issues such as collective action on climate change mitigation. The problem faced by negotiators to international environmental agreements has many more layers of complexity that will make the matter of coordination more difficult. Moreover, implementing agreements, such as the one negotiated in Paris and recently entered into force, that rely on nationally determined pledges, is likely to be further hindered by myopic policymaking. However, introducing realistic behavioral and ecological features, such as reference dependence and tipping points, into mainstream economic modeling appears to be an important step in the right direction.

REFERENCES

Averchenkova, A. and S. Bassi (2016), 'Beyond the targets: assessing the political credibility of pledges for the Paris Agreement', Grantham Research Institute On Climate Change and The Environment, accessed 1 April 2014 at www.lse.ac.uk/GranthamInstitute/publication/beyond-the_targets/.

Barrett, S. (1992), 'International environmental agreements as games', in R. Pethig (ed.), *Conflict and Cooperation in Managing Environmental Resources*, Berlin: Springer-Verlag, pp. 11–37.

Barrett, S. (2005), 'The theory of international environmental agreements', in K.G. Maeler and J. Vincent (eds), *Handbook of Environmental Economics*, Vol. 3, Amsterdam: Elsevier, pp. 1457–516.

Barrett, S. (2013), 'Climate treaties and approaching catastrophes', *Journal of Environmental Economics and Management*, **66**(2), 235–50.

Barrett, S. (2016), 'Coordination vs. voluntarism and enforcement in sustaining international environmental cooperation', *Proceedings of the National Academy of Sciences*, **113**(51), 14515–22.

Barrett, S. and A. Dannenberg (2012), 'Climate negotiations under scientific uncertainty', *Proceedings of the National Academy of Sciences*, **109**(43), 17372–6.

Barrett, S., T.M. Lenton, A. Millner, A. Tavoni, S. Carpenter, J.M. Anderies, C. Folke et al. (2014), 'Climate engineering reconsidered', *Nature Climate Change*, **4**(7), 527.

Bass, F.M. (1969), 'A new product growth for model consumer durables', *Management Science*, **15**(5), 215–27.

Benartzi, S. and R.H. Thaler (1995), 'Myopic loss aversion and the equity premium puzzle', *The Quarterly Journal of Economics*, **110**(1), 73–92.

Biggs, R., S.R. Carpenter and W.A. Brock (2009), 'Turning back from the brink: detecting an impending regime shift in time to avert it', *Proceedings of the National Academy of Sciences*, **106**(3), 826–31.

Bolton, G. (1991), 'A comparative model of bargaining: theory and evidence', *American Economic Review*, **81**, 1096–136.

Bolton, G. and A. Ockenfels (2000), 'A theory of equity, reciprocity, and competition', *American Economic Review*, **90**(1), 166–93.

Bosetti, V., M. Heugues and A. Tavoni (2017), 'Luring others in: coalition formation games with threshold and spillover effects', *Oxford Economics Papers*, **69**(2), 410–31.

Bowman, D., D. Minehart and M. Rabin (1999), 'Loss aversion in a consumption–savings model', *Journal of Economic Behavior & Organization*, **38**(2), 155–78.

Buchholz, W. and T. Sandler (2016), 'Successful leadership in global public good provision: incorporating behavioral approaches', *Environmental and Resource Economics*, **67**(3), 591–607.

Camerer, C.F. (2003), *Behavioral Game Theory: Experiments in Strategic Interaction*, Princeton, NJ: Princeton University Press.
Camerer, C.F., G. Loewenstein and M. Rabin (eds) (2003), *Advances in Behavioral Economics*, Princeton, NJ: Princeton University Press.
Carraro, C. and D. Siniscalco (1993), 'Strategies for international protection of the environment', *Journal of Public Economics*, **52**, 309–28.
Cazorla, M.V. and M.A. Toman (2001), 'International equity and climate change policy', in M.A. Toman (ed.), *Climate Change Economics and Policy*, Resources for the Future, Washington, DC, pp. 235–47.
Charness, G. and M. Rabin (2002), 'Understanding social preferences with simple tests', *Quarterly Journal of Economics*, **117**, 817–69.
Ciccarone, G. and E. Marchetti (2013), 'Rational expectations and loss aversion: potential output and welfare implications', *Journal of Economic Behavior & Organization*, **86**, 24–36.
Cox, J.C., D. Friedman and S. Gjerstad (2007), 'A tractable model of reciprocity and fairness', *Games and Economic Behavior*, **59**(1), 17–45.
Currarini, S., C. Marchiori and A. Tavoni (2015), 'Network economics and the environment: insights and perspectives', *Environmental and Resource Economics*, **65**(1), 159–89.
Daido, K., K. Morita, T. Murooka and H. Ogawa (2013), 'Task assignment under agent loss aversion', *Economics Letters*, **121**(1), 35–8.
Dannenberg, A. and A. Tavoni (2016), 'Collective action in dangerous climate change games', in Anabela Botelho (ed.), *WSPC Reference of Natural Resources and Environmental Policy in the Era of Global Change. Vol. 4: Experimental Economics*, World Scientific, pp. 95–120.
Dannenberg, A., B. Sturm and C. Vogt (2010), 'Do equity preferences matter for climate negotiations? An experimental investigation', *Environmental and Resource Economics*, **47**(1), 91–109.
Dannenberg, A., G. Löschel, C. Paolacci, A. Reif and A. Tavoni (2014), 'On the provision of public goods with probabilistic and ambiguous thresholds', *Environmental and Resource Economics*, **61**(3), 365–83.
d'Aspremont, C., A. Jacquemin, J. Gabszewicz and J. Weymark (1983), 'On the stability of collusive price leadership', *Canadian Journal of Economics*, **16**(1), 17–25.
DellaVigna, S. (2009), 'Psychology and economics: evidence from the field', *Journal of Economic Literature*, **47**(2), 315–72.
de Zeeuw, A. (2015), 'International environmental agreements', *Annual Review of Resource Economics*, **7**(1), 151–68.
Diamantoudi, E. and E. Sartzetakis (2006), 'Stable international environmental agreements: an analytical approach', *Journal of Public Economic Theory*, **8**(2), 247–63.
Dixit, A. (2003), 'Clubs with entrapment', *American Economic Review*, **93**(5), 1824–9.
Dufwenberg, M. and G. Kirchsteiger (2004), 'A theory of sequential reciprocity', *Games and Economic Behavior*, **47**(2), 268–98.
Dufwenberg, M. and A. Patel (2017), 'Reciprocity networks and the participation problem', *Games and Economic Behavior*, **101**, 260–72.
Eisenkopf, G. and S. Teyssier (2013), 'Envy and loss aversion in tournaments', *Journal of Economic Psychology*, **34**, 240–55.
Falk, A. and U. Fischbacher (2006), 'A theory of reciprocity', *Games and Economic Behavior*, **54**, 293–315.
Fehr, E. and K. Schmidt (1999), 'A theory of fairness, competition, and cooperation', *The Quarterly Journal of Economics*, **114**(3), 817–68.
Fehr, E., G. Kirchsteiger and A. Riedl (1993), 'Does fairness prevent market clearing? An experimental investigation', *Quarterly Journal of Economics*, **108**(2), 437–59.
Fehr, E., S. Gächter and G. Kirchsteiger (1997), 'Reciprocity as a contract enforcement device: experimental evidence', *Econometrica*, **65**, 833–60.
Fehr, E., G. Kirchsteiger and A. Riedl (1998), 'Gift exchange and reciprocity in competitive experimental markets', *European Economic Review*, **42**, 1–34.
Finus, M. (2008), 'Game theoretic research on the design of international environmental agreements: insights, critical remarks, and future challenges', *International Review of Environmental and Resource Economics*, **2**(1), 29–67.
Freund, C. and Ç. Özden (2008), 'Trade policy and loss aversion', *The American Economic Review*, **98**(4), 1675–91.
Genesove, D. and C. Mayer (2001), 'Loss aversion and seller behavior: evidence from the housing market', *The Quarterly Journal of Economics*, **116**(4), 1233–60.
Gladwell, M. (2000), *The Tipping Point: How Little Things Make a Big Difference*, Boston, MA: Little, Brown.
Granovetter, M. (1978), 'Threshold models of collective behavior', *The American Journal of Sociology*, **83**(6), 1420–43.
Greene, D.L. (2011), 'Uncertainty, loss aversion, and markets for energy efficiency', *Energy Economics*, **33**(4), 608–16.

Grüning, C. and W. Peters (2010), 'Can justice and fairness enlarge the size of international environmental agreements?', *Games*, **1**, 137–58.
Gsottbauer, E. and J.C. van den Bergh (2013), 'Bounded rationality and social interaction in negotiating a climate agreement', *International Environmental Agreements: Politics, Law and Economics*, **13**(3), 225–49.
Hadjiyiannis, C., D. İriş and C. Tabakis (2012), 'International environmental cooperation under fairness and reciprocity', *The B.E. Journal of Economic Analysis & Policy*, **12**(1), 1–30.
Heal, G. and H. Kunreuther (2012), 'Managing catastrophic risk', Paper No. w18136, National Bureau of Economic Research.
Heath, C., R. Larrick and G. Wu (1999), 'Goals as reference points', *Cognitive Psychology*, **38**(1), 79–109.
Hoel, M. (1992), 'International environment conventions: the case of uniform reductions of emissions', *Environmental and Resource Economics*, **2**(2), 141–59.
IPCC (2013), 'Summary for policymakers, in Climate Change 2013: The Physical Science Basis', in T.F. Stocker et al. (eds), *Contribution of Working Group I to the Fifth Assessment Report of the Intergovernmental Panel on Climate Change*, Cambridge: Cambridge University Press.
İriş, D. (2016), 'Economic targets and loss-aversion in international environmental cooperation', *Journal of Economic Surveys*, **30**(3), 624–48.
İriş, D. and A. Tavoni (2018), 'Loss aversion in international environmental agreements', *Environmental and Resource Economics Review*, **27**(2), 363–97.
İriş, D., J. Lee and A. Tavoni (2019), 'Delegation and public pressure in a threshold public goods game', *Environmental and Resource Economics*, **74**(3), 1331–53.
Isaksen, E., A. Richter and K.A. Brekke (2016), 'When kindness generates unkindness. Why positive framing cannot solve the tragedy of the commons', presented at EAERE 2016.
Jang, D., A. Patel and M. Dufwenberg (2016), 'Reciprocity and agreements', unpublished manuscript.
Kagel, J.H. and A.E. Roth (eds) (1995), *The Handbook of Experimental Economics*, Princeton, NJ: Princeton University Press.
Kahneman, D. (2003), 'A psychological perspective on economics', *The American Economic Review*, **93**(2), 162–8.
Kahneman, D. and A. Tversky (1979), 'Prospect theory: an analysis of decision under risk', *Econometrica*, **47**(2), 263–91.
Kahneman, D. and A. Tversky (eds) (2000), *Choices, Values, and Frames*, New York: Cambridge University Press.
Kahneman, D., J.L. Knetsch and R.H. Thaler (1986), 'Fairness as a constraint on profit seeking: entitlements in the market', *American Economic Review*, **76**(4), 728–41.
Kahneman, D., J.L. Knetsch and R.H. Thaler (1991), 'Anomalies: the endowment effect, loss aversion, and status quo bias', *Journal of Economic Perspectives*, **5**(1), 193–206.
Kirchsteiger, G. (1994), 'The role of envy in ultimatum games', *Journal of Economic Behavior and Organization*, **25**, 373–89.
Köbberling, V. and P.P. Wakker (2005), 'An index of loss aversion', *Journal of Economic Theory*, **122**(1), 119–31.
Kolstad, C.K. (2013), 'International environmental agreements with other-regarding preferences', unpublished paper, Stanford University.
Kőszegi, B. and M. Rabin (2006), 'A model of reference-dependent preferences', *The Quarterly Journal of Economics*, **121**(4), 1133–65.
Lade, S., A. Tavoni, S. Levin and M. Schlüter (2013), 'Regime shifts in a social-ecological system', *Theoretical Ecology*, **6**, 359–72.
Lange, A. (2006), 'The impact of equity-preferences on the stability of international environmental agreements', *Environmental and Resource Economics*, **34**, 247–67.
Lange, A. and C. Vogt (2003), 'Cooperation in international environmental negotiations due to a preference for equity', *Journal of Public Economics*, **87**, 2049–67.
Lange, A., C. Vogt and A. Ziegler (2007), 'On the importance of equity in international climate policy: an empirical analysis', *Energy Economics*, **29**, 545–62.
Lange, A., A. Löschel, C. Vogt and A. Ziegler (2010), 'On the self-serving use of equity principles in international climate negotiations', *European Economics Review*, **54**, 359–75.
Leibenstein, H. (1950), 'Bandwagon, snob, and Veblen effects in the theory of consumers' demand', *Quarterly Journal of Economics*, **64**(2), 183–207.
Lenton, T.M. (2011), 'Early warning of climate tipping points', *Nature Climate Change*, **1**(4), 201–9.
Lenton, T., H. Held, E. Kriegler, J.W. Hall, W. Lucht, S. Rahmstorf and H.J. Schellnhuber (2008), 'Tipping elements in the Earth's climate system', *Proceedings of the National Academy of Sciences*, **105**(6), 1786–93.
Levy, J. (1996), 'Loss aversion, framing, and bargaining: the implications of prospect theory for international conflict', *International Political Science Review*, **17**(2), 179–95.
Loewenstein, G., L. Thompson and M. Bazerman (1989), 'Social utility and decision making in interpersonal contexts', *Journal of Personality and Social Psychology*, **57**, 426–41.

Marchiori, C., S. Dietz and A, Tavoni (2017), 'Domestic politics and the formation of international environmental agreements', *Journal of Environmental Economics and Management*, **81**, 115–31.
Markowitz, H. (1952), 'The utility of wealth', *Journal of Political Economy*, **60**, 151–8.
Milinski, M., R.D. Sommerfeld, H.J. Krambeck, F.A. Reed and J. Marotzke (2008), 'The collective-risk social dilemma and the prevention of simulated dangerous climate change', *Proceedings of the National Academy of Sciences*, **105**(7), 2291–4.
Milinski, M., C. Hilbe, D. Semmann, R. Sommerfeld and J. Marotzke (2016), 'Humans choose representatives who enforce cooperation in social dilemmas through extortion', *Nature Communications*, **7**, 10915.
Najama, A., S. Huq and Y. Sokona (2003), 'Climate negotiations beyond Kyoto: developing countries concerns and interests', *Climate Policy*, **3**, 221–31.
Nordhaus, W.D. (1992), 'The "DICE" model: background and structure of a dynamic integrated climate economy model of the economics of global warming', Cowles Foundation Discussion Paper No. 1009.
Nordhaus, W.D. (2008), *A Question of Balance: Weighing the Options on Global Warming Policies*, New Haven, CT and London: Yale University Press.
Nordhaus, W.D. (2015), 'Climate clubs: overcoming free-riding in international climate policy', *American Economic Review*, **105**(4), 1339–70.
Nyborg, K. (2015), 'Reciprocal climate negotiators', IZA Discussion Paper No. 8866.
Ochea M. and A. de Zeeuw (2015), 'Evolution of reciprocity in asymmetric international environmental negotiations', *Environmental and Resource Economics*, **62**(4), 837–54.
Ochs, J. and A.E. Roth (1989), 'An experimental study of sequential bargaining', *American Economic Review*, **79**, 355–84.
Ostrom, E. (2009), 'A polycentric approach for coping with climate change', World Bank Policy Research Working Paper 5095, Washington, DC: World Bank.
Rabin, M. (1993), 'Incorporating fairness into game theory and economics', *American Economic Review*, **83**(5), 1281–302.
Ringius, L., A. Torvanger and A. Underdal (2002), 'Burden sharing and fairness principles in international climate policy', *International Environmental Agreements: Politics, Law and Economics*, **2**, 1–22.
Rockström, J., W. Steffen, K. Noone, A. Persson, F.S. Chapin III, E.F. Lambin and B. Nykvist et al. (2009), 'A safe operating space for humanity', *Nature*, **461**, 472–5.
Rogers, E.M. (2003), *Diffusion of Innovations*, New York: Free Press.
Scheffer, M. and S.R. Carpenter (2003), 'Catastrophic regime shifts in ecosystems: linking theory to observation', *Trends in Ecology & Evolution*, **18**, 648–56.
Scheffer, M., S. Carpenter, J.A. Foley, C. Folke and B. Walker (2001), 'Catastrophic shifts in ecosystems', *Nature*, **413**, 591–6.
Segal, U. and J. Sobel (2007), 'Tit for tat: foundations of preferences for reciprocity in strategic settings', *Journal of Economic Theory*, **136**, 197–216.
Shafir, E., P. Diamond and A. Tversky (1997), 'Money illusion', *The Quarterly Journal of Economics*, **112**(2), 341–74.
Stern, N. (2007), *The Economics of Climate Change: The Stern Review*, Cambridge: Cambridge University Press.
Tavoni, A. and S. Levin (2014), 'Managing the climate commons at the nexus of ecology, behaviour and economics', *Nature Climate Change*, **4**, 1057–63.
Tavoni, A., A. Dannenberg, G. Kallis and A. Löschel (2011), 'Inequality, communication and the avoidance of disastrous climate change in a public goods game', *Proceedings of the National Academy of Sciences*, **108**(29), 11825–9.
Tovar, P. (2009), 'The effects of loss aversion on trade policy: theory and evidence', *Journal of International Economics*, **78**(1), 154–67.
Tversky, A. and D. Kahneman (1991), 'Loss aversion in riskless choice: a reference-dependent model', *The Quarterly Journal of Economics*, **106**(4), 1039–61.
Vasconcelos, V.V., F.C. Santos, J.M. Pacheco and S.A. Levin (2014), 'Climate policies under wealth inequality', *Proceedings of the National Academy of Sciences*, **111**(6), 2212–16.
Young, H.P. (2009), 'Innovation diffusion in heterogeneous populations: contagion, social influence and social learning', *American Economic Review*, **99**, 1899–924.
Watts, D.J. (2002), 'A simple model of global cascades on random networks', *Proceedings of the National Academy of Sciences*, **99**(9), 5766–71.
Weibull, J.W. (1997), *Evolutionary Game Theory*, Cambridge, MA: MIT Press.
Weikard, H., L. Wangler and A. Freytag (2014), 'Minimum participation rules with heterogeneous countries', *Environmental and Resource Economics*, **62**(4), 711–27.
Weitzman, Martin L. (2009), 'On modeling and interpreting the economics of catastrophic climate change', *Review of Economics and Statistics*, **91**(1), 1–19.

PART III

CLIMATE CHANGE AND SUSTAINABILITY

12. Climate change, Malthus and collapse
Norman Schofield

1. MALTHUS AND THE FALL OF ROME

In 1798, Thomas Malthus published *An Essay on the Principle of Population* in which he argued that population has a tendency to grow exponentially, while food resources tend to expand arithmetically. The conclusion was that there would always be competition for food; as Charles Darwin acknowledged, this argument provided the logical basis for Darwin's *The Origin of Species by Means of Natural Selection*, while Malthus saw it as a refutation of Condorcet's optimistic *Esquisse* of 1798. In this essay I shall consider the possibility that climate change provides reason to suppose that Malthus was correct, and that the optimistic arguments of Condorcet and what Israel (2010, 2014) has called the *Radical Enlightenment*, will prove to be wrong. Gregory Clark (2007a, b) has recently reconsidered what he calls the "Malthusian trap" and has argued that world population remained essentially flat from the time of the agricultural revolution about 10,000 BCE until about 1500 CE.[1] Of course there were local periods when population grew, and it is worth following Malthus and considering how this was done in China and the Roman Empire. Both agricultural societies were discussed by Malthus, and as he observed, China built an internal empire, based on intensive irrigation, from the Yellow and Yangtze rivers, with wheat and rice in the south; this empire was protected from northern barbarians by the Great Wall. By intense efforts enough food was provided for the population, as long as it was capped.[2] China paid little attention to expansion, though there was the voyage of Admiral Zheng He about 1405 CE to 1433 CE, which reached Madagascar. Neighboring kingdoms such as Indo China and Burma were treated as barbarian suzerainities.

In contrast, Rome established external dominions, starting with the conquest of Gaul by Julius Caesar in 55–54 BCE, followed by that of Brittania by Claudius in 47 CE. Rome built a trading empire importing the food it needed for its growing population, initially from North Africa after it defeated Carthage in the second Punic War in 201 BCE. After the defeat of Marc Anthony and Cleopatra, by Octavian in 30 BCE, Egypt was added to the empire. Egypt was particularly significant, as the annual flooding of the Nile meant that Egyptian productivity did not decline with time. Octavian added Hispania (Spain) in

---

[1] In fact, Wells (2011) presents evidence that the population of *Homo* may only have been about 1,000 before the exit from Africa about 45,000 years ago. The population grew after this "crash" to a few million hunter gatherers just before the dawn of agriculture. After the Agricultural Revolution circa 10,000 BCE, population first grew rapidly and then reached a Malthusian plateau, as noted by Clark. Wells observes that the anthropologists believe that the average health of agricultural populations was lower than in the earlier hunter-gatherer populations, suggesting that there was indeed a Malthusian barrier, imposed by agriculture itself.

[2] For example, Maddison estimates that the population of China was more or less stationary at 160 million until 1600 CE though it fell to about 138 million by 1700 CE.

19 BCE. The conquest of Greece and Hispania brought rich silver mines under Roman rule, which provided for the financial stability of the empire.[3]

Even so, once the boundary of the roman empire was delineated by Hadrian's Wall in Britannia (122 CE) and the conquest of Dacia by Trajan in 101 CE, we may infer that Rome found it increasingly difficult to maintain agricultural productivity for its population. It was efficient at maintaining transport infrastructure, such as roads and the navy, but it probably was not understood that it was also necessary to maintain the fertility of the soil. Rome's continuing need for agricultural land can be illustrated by the invasion of Caledonia by the Roman Emperor Septimius Severus in 208 CE. In addition, ownership of land became increasingly concentrated, and we may say the agrarian interest dominated, bringing about political and economic instability. Historians now consider that Malthusian population pressure, together with the consequent disease pandemics brought about the "Fall of Rome" (Harper, 2016). Moreover, climate change may have been the cause of massive nomadic population movements, first from Jutland and Germania to Poland, then the Crimea and the Balkans, and later to Northern Italy and Gaul. Once the Vandals conquered Hispania and North Africa between 400 CE and 420 CE, the Western Roman Empire could not survive, and indeed the last Roman legion left Britannia in 410 CE.

Before the Western Empire finally fell in 460 CE, the greatest Western Emperor, Constantine the Great, was acclaimed Emperor of the West in York in 306 CE, and founded the city of Constantinople in 323 CE in what had been Byzantium. This Eastern Roman or Byzantine Empire flourished initially until it was eventually taken in 1453 CE, by the Ottomans under Mehmed II. Under Justinian I, Byzantium had reconquered North Africa, Sicily and Northern Italy. Byzantium held on to its agricultural dominions in Egypt, Syria and the Levant, and could feed its population until these dominions and North Africa were taken by Arab armies about 642 CE. This brief sketch of the fall of Rome suggests that the global economy that has been created since World War II may be facing insurmountable dangers, just as Rome did again through population pressure, but made more severe by climate change.

1.1 Pax Brittanica and Americana

The fall of Contantinople led to attempts to sail west to China and in 1492 CE, Christopher Columbus landed on an island in the Bahamas, leading eventually to the conquest of Mexico and Peru by Spain. The flow of gold and silver provided Philip II of Spain with the resources to attempt to expand or maintain his empire in Europe. Kennedy (1987) provides a useful discussion of the contrast between the Spanish and British empires.

Philip continued his attempt to take England even after the expensive failure of the Armada in 1588 CE, and pushed Spain into what was essentially bankruptcy on many occasions. The Spanish elite seemingly had little interest in trade, although the Portugese,

---

[3] Whether the earlier civilizations of the Mediterranean faced Malthusian traps has not been fully investigated, but Abulafia (2011) gives an extensive account of the development of Mediterranean trade, from 1000 BCE until 600 CE, after the collapse of Rome. What is well known is that these civilizations did face the Malthusian trap of disease such as the plague of Justinian in 540 CE.

in contrast, searched for trade routes to China around the African continent. The English established Jamestown, Virginia in 1607 CE, but their most important colony in North America was probably Jamaica, taken from the Spanish in 1655 CE. Jamaica was ideal for growing sugar, and eventually Great Britain became involved in the slave trade, shipping millions of slaves from West Africa to the Caribbean to work in the sugar plantations. This slave system spread to the southern states of North America.

Aware that Great Britain would be involved in a long war with France and would need resources to finance this war, Robert Walpole, the first Prime Minister of Britain, engaged in a complex play to offset the debts that Britain had already accumulated, by incorporating these into the South Sea Company in 1711. The company was supposed to finance the debts through its profits from the slave trade, but after the "South Sea Bubble" broke in 1720, Walpole backed its shares with those of the Bank of England (founded in 1694, after the Glorious revolution of 1688). To finance this debt, Britain adopted a system of restricting the importation of food. Eventually, these were formalized as the Corn Laws. These raised the price of food, and thus of land and were supported by the agrarian interest, dominant in Parliament. For these reasons Parliament had an interest in increasing agricultural output. The so called "Agricultural Revolution" involved the scientific applications of new principles of husbandry, with fallowing of fields, fertilizing and more importantly a series of "enclosure acts" that took away common land and sold it to agrarian interests. The result was three fold. Britain had the food to supply its growing population, seemingly contradicting Malthus. The population that was removed from the land then either emigrated to parts of the growing empire or were absorbed in the new industrial cities. Because the poor were disadvantaged by this system it was crucial to restrict the franchise. Industrial interests were opposed to this system because high grain prices put upward pressure on wage rates. The Corn Laws were eventually repealed during the Irish "potato" famine of 1845–1852, while Robert Peel was Prime Minister (with the support of the Duke of Wellington, under the fear of political unrest). The restriction of the franchise was eventaully changed by the reform act of 1867. Before the act, only one million out of seven million could vote. This number doubled after the Act. It was opposed by the Liberal Gladstone, and supported by Benjamin Disraeli, Chancellor of the Exchequer. It may be argued that Disraeli achieved victory by calling for support of the Empire. So we now turn to the expansion of the British Empire, after Disraeli.

We may think of North America's role in British expansion as akin to the role of Egypt in the Roman Empire.

The flow of settlers to North America opened up an entire continent to the production of food. Although Hamilton and Jefferson were opposed over how the new US should be structured, Jefferson won the election in 1800, and the US focused on becoming an agricultual power, depending on Britain for its manufactures. To finance this flow of foodstuffs and raw materials, Britain required a trade surplus with India.

Originally set up by the East India Company (EIC), founded in 1600 CE, the EIC traded in cotton, silk, indigo, saltpetre, tea and opium. To pay for its purchase of tea from China, the EIC instigated the first Opium war (1839–1842 CE) forcing China to accept the importation of opium from India.

To facilitate the flow of goods from India and China, Britain participated in the building of the Suez canal, from the Mediterranean to the Red Sea, completed in 1869 CE. Under the influence of Disraeli, Britain took control of the canal in 1875 CE and then Egypt in 1882 CE. Suez meant that it was crucial for Britain to maintain control of its naval bases (Gibraltar, 1713 CE; Malta, 1814 CE; Cyprus, 1878 CE; Aden, 1839 CE; Singapore, 1819 CE). It was perhaps inevitable that Britain would take control of the Sudan in 1873 CE, under General Gordon. Gordon was killed in Khartoum by the army of the muslim Mahdi in 1885 CE.

The same period saw the explorations of David Livingstone from Zanzibar searching for the source of the Nile, through impenetrable jungle. By 1888 CE, the British had established a Protectorate in Kenya, a settler colony producing tea, coffee, sisal and wheat. British expansion in Africa continued with the Second Boer War between Britain and the Transvaal Republic and Orange Free State (1899–1902 CE). In 1895, the new East African settler colony of Rhodesia was named after Cecil Rhodes, who was instrumental in the attempt to construct a railway from the settler colony of South Africa to Cairo. By 1916, during the First World War, Britain took the German colony of Tanganyika, which became a league of Nations mandate in 1922. The great Depression of the 1920s stopped the construction of the Cape–Cairo railroad.

The Suez canal opened up a much shorter trade route to Asia, but required that Britain defend the route, by establishing, for example, the British Crown Colony of Aden, in what is now the Yemen. During the First World War, Lawrence of Arabia roused the Bedouin tribes to attack Aqaba, eventually leading to the taking of Jerusalem by General Allenby in 1917 CE. By 1922 CE, Britain had a mandate to govern Palestine, Transjordan and Syria. Under the secret Sykes-Picot Agreement, France was assigned the mandate for Lebanon and part of Syria. To ensure the protection of India from the advance of the Russians in Asia, Britain had been earlier embroiled in the First Afghan War of 1839–1842 CE. The EIC had already taken control of Bengal in the late 18th century. Just as the Roman Empire depended on control of the Mediterranean sea routes, so did the British Empire depend on the control exercised by the British navy over what we might call the African-Asian litoral.[4]

The dominions that Britain had established in Asia and its trade route through the Suez canal had created the first Global Economy, based on the gold standard and free trade. By 1900 CE, however, there was a move led by Joseph Chamberlain to transform this sytem by creating a system of Imperial Preference. As Belich (2009) has pointed out, the British Empire had begun to depend more on its settler colonies of Australia, New Zealand, South Africa, and Canada, than the US. By the First World War, the industrial development of the US had suggested that the US was more likely to be a future economic opponent rather than a partner to the British Empire. By combining the economic resources of the settler colonies, Britain could offset the economic costs of the War. This logic became even more pronounced during the Great Depression, and World War II, and was certainly on Churchill's mind during his negotiations with Roosevelt in 1940 CE. The transfer of British bases in Newfoundland, Bermuda, and the Bahamas to the United States in 1940, and the fall of Singapore to the Japanese in 1942 were all indications that Britain's naval dominance was about over. The final crisis came in 1956, when Britain with Israel and France attempted to regain control of the Suez canal from

---

[4] This is not to say, the Empire ensured sufficient food for all its population. There was the great famine of India in 1876–78, and the Bengal famine of 1943 CE. On these see Sen (1982).

Egypt's President Nasser, but were rebuffed by President Eisenhower, leading to the fall of Prime Minister, Anthony Eden. In the earlier negotiations over the Bretton Woods system immediately after the War, Keynes also attempted to persuade the Americans to create a global economic system that would maintain Britain's empire. In 1947 however, India, Pakistan and Bengal became independent, and the basis of the Empire was finally gone. Following the Japanese occupation of the Malayan states in 1941–45, Malaysia eventually gained independence. It was clear by this time that Britain could no longer afford to run an empire. Indeed, Roosevelt was opposed to the notion of empire, and as a *quid pro quo* for post war aid, Britain was forced to make the pound convertible, paying its substantial debts in gold. In 1960, Prime Minister Harold Macmillan spoke of the "wind of change" blowing through the African continent, while in 1968, Enoch Powell gave his "Rivers of Blood" speech, essentially decrying the loss of empire.

This brief history of empire has attempted to describe the similarities between the operation of the British and Roman empires. Both depended on the expansion of dominions, and the maintenance of trade. By these means, Rome had fed its growing population for a period of over 1000 years, from circa 753 BCE to 456 CE. In the nineteenth century, the public schools of Britain had trained generations of administrators in the Roman classics of government, such as the works of Cicero. Just as Romans believed in the virtue of Roman civilization, so did the British believe that their empire could bring civilization.[5] America may also exhibit a belief in its virtue. But America has not trained generations of imperial consuls. Because America is a great agricultural power, it must focus on ensuring the security of the necessary non-agricultural resources to supply its industrial machine. Nowadays, this means particularly oil and gas. America has constructed an extensive set of naval bases to support this global trade. As the *Limits to Growth* (Meadows, 1972) suggests, the growing world population may well have bought us to a Malthusian trap. The trap that Malthus saw in 1798 was avoided by the Agricultural and Industrial Revolutions and by the expansion of the British Empire to embrace new dominions. The avoidance of the Malthusian trap also depended on new sources of energy, particularly the extensive coal reserves in Britain. But as with agriculture, energy reserves exhibit decreasing returns with time. As reserves are depleted, the real cost of extraction increases. Since oil and gas provide the feedstock for fertilizer, the real price for fertilizer will also increase, thus increasing the price of food. Those parts of the world where political institutions are weak will then suffer from instability and perhaps civil war. We can see some evidence of this in Africa and the Middle East. *The World Watch Institute* has recently provided evidence that there is a growing mis-match between world food production and population. Figure 12.1 suggests that climate change will drastically affect our ability to feed the world population.

Britain was successful in building its empire because the spirit of what Armstrong (2011) calls *Logos* became dominant. *Logos* is a convenient term for the complex of scientific rationality that developed after the Enlightenment, in contrast to *Mythos*, pre-Enlightenment Religious belief. The Encyclical letter of Pope Francis, *Laudato Si*, exemplifies the contrast between *Logos* and *Mythos*. Francis blames the corporate capital-

---

[5] It is possible that Darwin's theory of evolution also played a role in the ideology underlying the Empire.

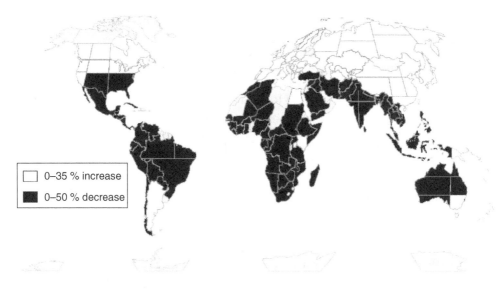

Source: Cline (2007).

Figure 12.1 The effect of climate change on agricultural production

ist system and modern technology for our unwillingness to deal with climate change and global poverty, and thus our unwillingness to care for our common home. He argues that returning to fundamental religious beliefs will provide meaning in our lives. In the rest of this chapter I shall add my own thoughts to those of Pope Francis.

2. *LOGOS* AND *MYTHOS*

Logos is the Enlightenment program to establish rationality as the basis for society, opposed to monarchy, religion and the church. Radical enlighteners included Thomas Jefferson, Thomas Paine and James Madison. They believed that society could be based on rational constitutional principles, leading to the "probability of a fit choice." Implicit in the Radical Enlightenment was the belief, originally postulated by Spinoza, that individuals could find moral bases for their choices without a need for a divine creator. An ancilliary belief was that the economy would also be rational and that the principles of the Radical Enlightenment would lead to material growth and the eradication of poverty and misery.[6] This enlightenment philosophy has recently had to face two troubling propositions. First are the results of social choice theory. These very abstract results suggest that no process of social choice can be rational Second, recent events suggest that the market models that we have used to guide our economic actions are deeply flawed. Opposed to

---

[6] See Pagden (2013) for an argument about the significance today of the enlightenment project, and a counter arguement by Gray (1995, 1997, 2000). The term the Radical Enlightenment was introduced by Israel (2010, 2014).

the Radical enlighteners, David Hume and Edmund Burke believed that people would need religion and nationalism to provide a moral compass to their lives. As Putnam and Campbell (2010) have noted, religion is as important as it has ever been in the US. Recent models of US Elections (Schofield and Gallego, 2011) show that religion is a key dimension of politics that divides voters one from another. A consequence of the Industrial Revolution, that followed on from the Radical Enlightenment, has been the unintended consequence of climate change. Since this is the most important policy dimension that the world economy currently faces, this chapter will consider aspects of *Logos* and *Mythos* to address the question whether we are likely to be able to make wise social choices to avoid future catastophe.

2.1 The Radical Enlightenment and Rationality

It was no accident that the most important cosmologist after Ptolemy of Alexandria was Nicolaus Copernicus (1473–1543), born only a decade before Martin Luther. Both attacked orthodoxy in different ways.[7] Copernicus formulated a scientifically based heliocentric cosmology that displaced the Earth from the center of the universe. His book, *De revolutionibus orbium coelestium* (*On the Revolutions of the Celestial Spheres*, 1543 CE), is often regarded as the starting point of the Scientific Revolution.

The ideas of Copernicus influenced many scholars; the physicist, mathematician, astronomer, and philosopher, Galileo Galilei (1564–1642); the mathematician and astronomer, Johannes Kepler (1571–1630).

Philosophiæ Naturalis Principia Mathematica (1687), by the physicist, mathematician, astronomer and natural philosopher, Isaac Newton (1642–1726) is considered to be the most influential book in the history of science. After Newton, a few scholars realized that the universe exhibits laws that can be precisely written down in mathematical form. Moreover, we have, for some mysterious reason, the capacity to conceive of exactly those mathematical forms that do indeed govern reality. This mysterious connection between mind and reality was the basis for Newton's philosophy. While celestial mechanics had been understood by Ptolemy to be the domain most readily governed by these forms, Newton's work suggested that *all* reality was governed by mathematics. The influence of Newton can perhaps be detected in the work of the philosopher, mathematician, and political scientist, Marie Jean Antoine Nicolas de Caritat, Marquis de Condorcet (1743–1794), known as Nicolas de Condorcet.[8] His work in formal social choice theory (Condorcet ([1785], 1994) was discussed in Schofield (2002) in connection with the arguments about democracy by Madison and Jefferson. The work on Moral Sentiment by the Scottish Enlightenment writers, Francis Hutcheson (1694–1746), David Hume (1711–1776), Adam Smith (1723–1790) and Adam Ferguson (1723–1816), also influenced Jefferson and Madison. Between Copernicus and Newton, the writings of Thomas Hobbes (1588–1679), René Descartes (1596–1650), John Locke (1632–1704), Baruch

---

[7] Weber (2013) speculated that there was a connection between the values of Protestantism and Capitalism. It may be that there are connections between the preference for scientific explanation and protestant belief about the relationship between God and humankind.

[8] See Hacking (1990) for the influence of Condorcet on the later work by Poisson and Laplace on *Social Mathematics*.

Spinoza (1632–1677), and Gottfried Leibniz (1646–1716) laid down foundations for the modern search for rationality in life. Hobbes was more clearly influenced by the scientific method, particularly that of Galileo, while Descartes, Locke, Spinoza, and Leibniz were all concerned in one way or another with the imperishability of the soul and the relationship between *Logos* and *Mythos*. The mathematician, Leibniz, in particular was concerned with an

> [E]xplanation of the relation between the soul and the body, a matter which has been regarded as inexplicable or else as miraculous.

Without the idea of a soul it would seem difficult to form a general scheme of ethics. Indeed, the progress of science and the increasing secularization of society have caused many to doubt that our civilization can provide meaning to life. Hawking and Mlodonow (2010) argue for a strong version of the universal mathematical principle of *logos*, which they call *model-dependent realism*, citing its origins in Pythagoras (580 BCE to 490 BCE), Euclid (383–323 BCE and Archimedes (287–212 BCE), and the recent developments in mathematical physics and cosmology.

Hawking and Mlodinow (2010) argue that it is only through a mathematical model that we can properly perceive reality. However, this mathematical principle faces two philosophical difficulties. One stems from the Godel 1931–Turing 1937 undecidability theorems. The first theorem asserts that mathematics cannot be both complete and consistent, so there are aspects of *Logos* that in principle cannot be verified. Turing's work, though it provides the basis for our computer technology also suggests that not all programs are computable. The second problem is associated with the notion of *chaos* or *catastrophe*. Even Newton believed God was necessary to ensure that the solar system did not collapse into chaos.

2.2 Chaos and the Prisoners' Dilemma

Since the early work of Garrett Hardin (1968) the "tragedy of the commons" has been recognized as a global prisoners' dilemma. In such a dilemma, no agent has a motivation to provide for the collective good. In the context of the possibility of climate change, the outcome is the continued emission of greenhouses gases like carbon dioxide into the atmosphere and the acidification of the oceans. There has developed an extensive literature on the n-person prisoners' dilemma in an attempt to solve the dilemma by considering mechanisms that would induce cooperation.[9]

The problem of cooperation has also provided a rich source of models of evolution, building on the early work by Trivers (1971) and Hamilton (1964, 1970). Nowak (2011) provides an overview of the recent developments. Indeed, the last twenty years has seen a growing literature on a game theoretic, or mathematical, analysis of the evolution of social norms to maintain cooperation in prisoners' dilemma-like situations. Gintis (2000), for example, provides evolutionary models of the cooperation through strong reciprocity and

---

[9] See for example, Hardin (1971, 1982), Taylor (1976, 1982), Axelrod and Hamilton (1981), Axelrod (1981, 1984), Kreps et al. (1982), Margolis (1982).

internalization of social norms.[10] The anthropological literature provides much evidence that, from about 500,000 years ago, the ancestors of *homo sapiens* engaged in cooperative behavior, particularly in hunting and caring for offspring and the elderly.[11] On this basis, we can infer that we probably do have very deeply ingrained normative mechanisms that were crucial, far back in time, for the maintenance of cooperation, and the fitness and thus survival of early hominids. These normative systems will surely have been modified over the long span of our evolution. However, these theories seem to suggest that we are very complicated mixtures of aggression and cooperation. Our hominid ancestors were very prone to murder each other.

Current work on climate change has focused on how we should treat the future. For example, Stern (2007, 2009), Collier (2010) and Chichilnisky (2009a,b) argue essentially for equal treatment of the present and the future.

The fundamental problem of climate change is that the underlying dynamic system is extremely complex, and displays many positive feedback mechanisms.[12] As the Godel-Turing theorems suggest, we can never be sure that such a complex system is not chaotic. In other words though, we may attempt to model the climate we can never be sure of the accuracy of our predictions.

Our society has recently passed through a period of economic disorder, where "black swan" events, low probability occurrences with high costs, have occurred with some regularity. Recent discussion of climate change has also emphasized so-called "fat-tailed climate events" again defined by high uncertainty and cost.[13]

2.3 International Cooperation and Uncertainty

Cooperation could in principle be attained by the action of a hegemonic leader such as the US as suggested by Kindleberger (1973) and Keohane and Nye (1977). Below, we give a brief exposition of the prisoners' dilemma and illustrate how hegemonic behavior could facilitate international cooperation. However, the analysis suggests that in the present economic climate, such hegemonic leadership by the US is constrained by the opposition of economic and political interests.

An earlier prophet of uncertainty was, of course, Keynes (1936) whose ideas on "speculative euphoria and crashes" would seem to be based on understanding the economy in terms of the qualitative aspects of its coalition dynamics.[14] An extensive literature has tried to draw inferences about market instability from recent economic events.

Similar uncertainty holds over political events. The fall of the Berlin Wall in 1989 was

---

[10] Strong reciprocity means the punishment of those who do not cooperate.

[11] Indeed, White et al. (2009) present evidence of a high degree of cooperation among very early hominids dating back about 4MYBP (million years before the present). The evidence includes anatomical data which allows for inferences about the behavioral characteristics of these early hominids.

[12] See the discussion in Schofield (2011). See also Nordhaus (2013) for an economic model of climate change.

[13] Weitzman (2009) and Chichilnisky (2010, 2014). See also Chichilnisky and Eisenberger (2010) on other catastophic events such as collision with an asteroid.

[14] See Minsky (1975, 1986) and Keynes's earlier work (1921).

not at all foreseen. Political scientists wrote about it in terms of "belief cascades"[15] as the coalition of protesting citizens grew apace. As the very recent democratic revolutions in the Middle East and North Africa suggest, these coalitional movements are extremely uncertain.[16] In particular, whether the autocrat remains in power or is forced into exile is as uncertain as anything Keynes discussed. Even when democracy is brought about, it is still uncertain whether it will persist.[17]

Section 3 introduces the Condorcet (1994 [1795]) Jury Theorem. This theorem suggests that majority rule can provide a way for a society to make a wise choice when the individuals have a common goal. Schofield (2002) has argued that Madison was aware of this theorem while writing Federalist X (Madison, 1999 [1787]) so it can be taken as perhaps the ultimate justification for democracy. However, models of belief aggregation that are derived from the Jury Theorem can lead to belief cascades that bifurcate the population. In addition, if the aggregation process takes place on a network, then centrally located agents, who have false beliefs, can dominate the process.[18]

Section 4 introduces the notion of "our common home" that I suggest could provide a teleology to guide us in making wise choices for the future. Section 5 concludes.

3. THE PRISONERS' DILEMMA, COOPERATION AND MORALITY

> For before constitution of Sovereign Power . . . all men had right to all things; which necessarily causeth Warre. (Hobbes, 2009 [1651]).

Kindleberger (1973) gave the first interpretation of the international economic system of states as a "Hobbesian" prisoners' dilemma, which could be solved by a leader, or "hegemon."

> A symmetric system with rules for counterbalancing, such as the gold standard is supposed to provide, may give way to a system with each participant seeking to maximize its short-term gain. . . . But a world of a few actors (countries) is not like [the competitive system envisaged by Adam Smith]. . . . In advancing its own economic good by a tariff, currency depreciation, or foreign exchange control, a country may worsen the welfare of its partners by more than its gain. Beggar-thy-neighbor tactics may lead to retaliation so that each country ends up in a worse position from having pursued its own gain . . .
> This is a typical non-zero sum game, in which any player undertaking to adopt a long range solution by itself will find other countries taking advantage of it . . .

In the 1970s, Robert Keohane and Joseph Nye (1977) made use of the idea of a hegemonic power in a context of "complex interdependence" of the kind envisaged by Kindleberger.

The essence of the theory of hegemony in international relations is that if there is

---

[15] Karklins and Petersen (1993); Lohmann (1994). See also Bikhchandani, Hirshleifer, and Welsh (1992).
[16] The response by the citizens of these countries to the demise of Osama bin Laden on May 2, 2011, is in large degree also unpredictable.
[17] See for example Carothers (2002) and Collier (2009).
[18] Golub and Jackson (2010).

a degree of inequality in the strengths of nation states then a hegemonic power may maintain cooperation in the context of an n-country prisoners' dilemma. Clearly, as the earlier discussion suggests, the British Empire in the 1800s is the role model for such a hegemon (Ferguson, 2002).

Hegemon theory suggests that international cooperation was maintained after World War II because of a dominant cooperative coalition. At the core of this cooperative coalition was the US; through its size it was able to generate collective goods for this community, first of all through the Marshall Plan and then in the context first of the post-World War II system of trade and economic cooperation, based on the Bretton Woods agreement and the Atlantic Alliance, or NATO. Over time, the US has found it costly to be the dominant core of the coalition In particular, as the relative size of the US economy has declined, it would seem that international cooperation has become more difficult to maintain. Indeed, the global recession of 2008–10 suggests that problems of debt and instability could induce "begger thy neighbor strategies," just like the 1930s.

The future utility benefits of adopting policies to ameliorate the possible climate changes in the future depend on the discount rates that we assign to the future. Obviously enough, different countries will in all likelihood adopt very different evaluations of the future. It is probable that developing countries like the BRICs (Brazil, Russia, India and China) will choose growth and development now rather than choosing consumption in the future. It is true however that China seems aware of the dangers in the future, and may prove to act as a hegemon in this context.

There have been many attempts to "solve" the prisoners' dilemma in a general fashion. For example, Binmore (2005) suggests that in the iterated nPD there are many equilibria with those that are *fair* standing out in some fashion. However, the criterion of "fairness" would seem to have little weight with regard to climate change. It is precisely the poor countries that will suffer from climate change, while the rapidly growing BRICs believe that they have a right to choose their own paths of development.

An extensive literature over the last few years has developed Adam Smith's ideas as expressed in the *Theory of Moral Sentiments* (1984 [1759]) to argue that human beings have an inate propensity to cooperate. This propensity may well have been the result of co-evolution of language and culture (Boyd and Richerson, 2005; Gintis, 2000).

Since language evolves very quickly (McWhorter, 2001; Deutcher, 2006), we might also expect moral values to change fairly rapidly, at least in the period during which language itself was evolving. In fact there is empirical evidence that cooperative behavior as well as notions of fairness vary significantly across different societies.[19] While there may be fundamental aspects of morality and "altruism," in particular, held in common across many societies, there is variation in how these are articulated. Gazzaniga (2006) suggests that moral values can be described in terms of various *modules*: reciprocity, suffering (or empathy), hierarchy, in-group and outgroup coalition, and purity/disgust. These modules can be combined in different ways with different emphases. An important aspect of cooperation is emphasized by Burkart, Hrdy and Van Schaik (2009), and Hrdy (2011), namely cooperation between man and woman to share the burden of child rearing.

---

[19] See Henrich et al. (2004, 2005), which reports on experiments in fifteen "small-scale societies," using the game theoretic tools of the "prisoners' dilemma," the "ultimatum game," etc.

It is generally considered that hunter-gatherer societies adopted egalitarian or "fair share" norms. The development of agriculture and then cities led to new norms of hierarchy and obedience, coupled with the predominence of military and religious elites (Schofield, 2011).

North (1990), North et al. (2009) and Acemoglu and Robinson (2006) focus on the transition from such oligarchic societies to open access societies whose institutions or "rules of the game," protect private property, and maintain the rule of law and political accountability, thus facilitating both cooperation and economic development. Acemoglu et al. (2009) argue, in their historical analyses about why "good" institutions form, that the evidence is in favor of "critical junctures."[20] For example, the "Glorious Revolution" in Britain in 1688 (North and Weingast, 1989), which prepared the way in a sense for the agricultural and industrial revolutions to follow (Mokyr, 2005, 2010; Mokyr and Nye, 2007) was the result of a sequence of historical contingencies that reduced the power of the elite to resist change. Recent work by Morris (2010), Fukuyama (2011), Ferguson (2011), and Acemoglu and Robinson (2008) has suggested that these fortuitous circumstances never occurred in China and the Middle East, and as a result these domains fell behind the West. Although many states have become democratic in the last few decades, oligarchic power is still entrenched in many parts of the world.[21]

At the international level, the institutions that do exist and that are designed to maintain cooperation, are relatively young. Whether they suceed in facilitating cooperation in such a difficult area as climate change is a matter of speculation. As we have suggested, international cooperation after World War II was only possible because of the overwhelming power of the US. In a world with oligarchies in power in Russia, China, and many countries in Africa, together with political disorder in almost all the oil producing countries in the Middle East, cooperation would appear unlikely.

In contrast to chaos, we now discuss the applicability of a theorem due to Condorcet. This theorem applies to a situation where people have a common goal but are uncertain about what choice to make. I shall argue below that we do indeed have a common goal, "the preservation of our common home." If people believe this, then we may have some trust in our ability to make a wise choice.

4. BELIEFS AND CONDORCET'S JURY THEOREM

The Jury Theorem formally only refers to a situation where there are just two alternatives $\{1,0\}$, and alternative 1 is the "true" option. Further, for every individual, i, it is the case that the probability that i picks the truth is ρ_{i1}, which exceeds the probability ρ_{i0}, that i does not pick the truth. We assume that $\rho_{i1} = \alpha > \frac{1}{2}$. To simplify the proof, we can assume that ρ_{i1} is the same for every individual, thus $\rho_{i1} = \alpha$ for all i. We use χ_i ($= 0$ or 1) to refer to the choice of individual i, and let $\chi = \Sigma_{i=1}^{n} \chi_i$ be the number of individuals who select the true option 1. We use Pr for the probability operator, and E for the expectation operator.

---

[20] See also Acemoglu and Robinson (2008).
[21] The popular protests in North Africa and the Middle East in 2011 were in opposition to oligarchic and autocratic power.

In the case that the electoral size, n, is odd, then a majority, m, is defined to be $m = \frac{n+1}{2}$. In the case n is even, the majority is $m = \frac{n}{2} + 1$. The probability that a majority chooses the true option is then

$$\alpha_{maj}^n = \Pr[\chi \geq m].$$

The theorem assumes that voter choice is *pairwise independent*, so that $\Pr(\chi = j)$ is simply given by the binomial expression $\binom{n}{j}\alpha^j(1-\alpha)^{n-j}$.

A version of the theorem can be proved in the case that the probabilities $\{\rho_{i1} = \alpha_i\}$ differ but satisfy the requirement that $\frac{1}{n}\Sigma_{i=1}^n \alpha_i > \frac{1}{2}$. Versions of the theorem are valid when voter choices are not pairwise independent (Ladha and Miller, 1996).

The Jury Theorem. If $1 > \alpha > \frac{1}{2}$, then $\alpha_{maj}^n \geq \alpha$, and $\alpha_{maj}^n \to 1$ as $n \to \infty$.

For both n being even or odd, as $n \to \infty$, the fraction of voters choosing option 1 approaches $\frac{1}{n}E(\chi) = \alpha > \frac{1}{2}$. Thus, in the limit, more than half the voters choose the true option. Hence the probability $\alpha_{maj}^n \to 1$ as $n \to \infty$.

Briefly, the Theorem suggest that jury, using majority rule will tend make a wise or true choice. Moreover, a very large jury, such as a democracy, will tend to make such a wise choice with certainty. Schofield (1972a, b) considered a model derived from the Jury Theorem where uncertain citizens were concerned to choose an ethical rule which would minimize their disappointment over the likely outcomes. He showed that majority rule was indeed optimal in this sense.

Models of belief aggregation extend the Jury Theorem by considering a situation where individuals receive signals, update their beliefs and make an aggregate choice on the basis of their posterior beliefs (Austen-Smith and Banks, 1996). Models of this kind can be used as the basis for analysing correlated beliefs and the creation of belief cascades (Easley and Kleinberg, 2010).

Schofield (2002) has argued that Condorcet's Jury Theorem provided the basis for Madison's argument in Federalist X (Madison, 1999 [1787]) that the judgments of citizens in the extended Republic would enhance the "probability of a fit choice." However, Schofield's discussion suggests that belief cascades can also fracture the society in two opposed factions, as in the lead up to the Civil War in 1860.[22]

There has been a very extensive literature recently on cascades but it is unclear from this literature whether cascades will be equilibrating or very volatile. In their formal analysis of cascades on a network of social connections, Golub and Jackson (2010) use the term *wise* if the process can attain the truth. In particular, they note that if one agent in the network is highly connected, then untrue beliefs of this agent can steer the crowd away from the truth. The recent economic disaster has led to research on market behavior to see if the notion of cascades can be used to explain why markets can become volatile or even irrational in some sense (Acemoglu et al., 2010; Schweitzer et al., 2009). Indeed, the literature that has developed in the last few years has dealt with the nature of herd instinct, the way markets respond to speculative behavior and the power law that characterizes

---

[22] Sunstein (2009, 2011) also notes that belief aggregation can lead to a situation where subgroups in the society come to hold very disparate opinions.

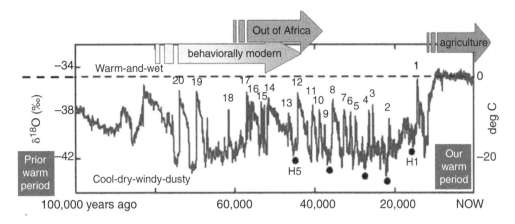

Source: Global-Fever.org.

Figure 12.2 Climate 100KYBP to now: chaos from 90KYBP to 10KYBP

market price movements.[23] The general idea is that the market can no longer be regarded as efficient. Indeed, as suggested by Ormerod (2001) the market may be fundamentally chaotic.

A more general remark concerns the role of climate change. Climate has exhibited chaotic or catastrophic behavior in the past.[24] There is good reason to believe that human evolution over the last million years can only be understood in terms of "bursts" of sudden transformations (Nowak, 2011) and that language and culture co-evolve through group or coalition selection (Cavalli-Sforza and Feldman, 1981). Calvin (2003) suggests that our braininess was cause and effect of the rapid exploration of the fitness landscape in response to climatic forcing. For example Figure 12.2 shows the rapid changes in temperature over the last 100,000 years. It was only in the last period of stable temperature, the "holocene," the last 10,000 years, that agriculture was possible, giving rise to the rapid increase in population and the Malthusisn traps discussed above.

Indeed, Figure 12.2 underplays the degree of volatility of Earth's climate. As Leakey and Lewin (1995) observe, there have been five major extinction events in Earth's history.[25] Some of these events included episodes of "snowball Earth" with the entire planet covered in ice. Leakey and Lewin (1995) argue that human activity is inducing a sixth extinction, dramatically reducing planetary biodiversity.

Stringer (2013) calls the theory of rapid evolution during a period of chaotic climate

---

[23] See, for example, Mandelbrot and Hudson (2004), Shiller (2003, 2005), Taleb (2007), Barbera (2009), Cassidy (2009), Fox (2009).

[24] Indeed as I understand the dynamical models, the chaotic episodes are due to the complex interactions of dynamical processes in the oceans, on the land, in weather, and in the heavens. These are very like interlinked *coalitions* of non-gradient vector fields.

[25] See Hays, Imbrie and Shackleton (1976) for a discussion of the work of Milutin Milankovitch, who postulated that global climate change on Earth was due to perturbations in Earth's orbit round the Sun.

change "the Social Brain hypothesis." The cave art of Chauvet in France, dating back about 36,000 years suggests that belief in the supernatural played an important part in human evolution.[26] Indeed, we might speculate that the part of our mind that enhances technological/mathematical development, *Logos*, and that part that facilitates social/religious belief, *Mythos*, are in conflict with each other.[27] It has also been suggested that *Logos* and indeed language, itself, has evolved in hominid culture in order to help us provide order to our chaotic environment. However, as recent elections in the US suggest, voters are very uncertain how to deal with climate change, and indeed are suspicious about the validity of *Logos*.

5. OUR COMMON HOME

If we accept that moral and religious beliefs are as important as rational calculations in determining the choices of society, then depending on models of preference aggregation will not suffice in helping us to make decisions over how to deal with climate change. Instead, I suggest a moral compass, derived from current inferences made about the nature of the evolution of intelligence on our planetary home. The anthropic principle reasons that the fundamental constants of nature are very precisely tuned so that the universe contains matter and that galaxies and stars live long enough to allow for the creation of carbon, oxygen, etc., all necessary for the evolution of life itself.[28] Gribbin (2011), goes further and points out that not only is the sun unusual in having the characterisics of a structurally stable system of planets,[29] but the Earth is fortunate in being protected by Jupiter from chaotic bombardment while the Moon also stabilizes our planet's orbit.[30] In essence, Gribbin gives good reasons to believe that our planet may well be the only planet in the galaxy that sustains intelligent life.[31] If this is true then we have a moral obligation to act as guardians of our common home. Parfit (2011) argues

> What matters most is that we rich people give up some of our luxuries, ceasing to overheat the Earth's atmosphere, and taking care of this planet in other ways, so that it continues to support intelligent life. If we are the only rational animals in the Universe, it matters even more whether

---

[26] It is interesting to note that Alfred Wallace (1898), who developed the theory of Natural Selection at the same time as Darwin, did not believe that the theory could provide an explanation for the development of mathematical abilities and moral beliefs in humankind.

[27] This is suggested by Kahneman (2011).

[28] As Smolin (2007) and Rees (2001) point out, the anthropic principle has been adopted because of the experimental evidence that the expansion of the universe is accelerating. Indeed, it has led to the hypothesis that there is an infinity of universes all with different laws. An alternative inference is the the principle of intelligent design. My own inference is that we require a teleology as proposed in the conclusion.

[29] The work by Poincare in the late 19th century focused on the structural stability of the solar system and, as a result, was the first to conceive the notion of chaos. Newton, himself, was worried about the stability of the solar system, and this is why he believed God was necessary to maintain divine order. In fact, the solar system has on occasion in the past been subject to cataclysms.

[30] The work by Poincare in the late 19th century focused on the structural stability of the solar system and was the first to conceive of the notion of chaos.

[31] See also Waltham (2014), Scharf (2014) and Wilson (2014).

we shall have descendants during the billions of years in which that would be possible. Some of our descendants might live lives and create worlds that, though failing to justify past suffering, would give us all, including those who suffered, reason to be glad that the Universe exists. (Parfit: 419)

6. CONCLUSION: COLLAPSE OF CIVILIZATION

Even if we believe that markets are well behaved, there is no reason to infer that markets are able to reflect the social costs of the externalities associated with production and consumption. Indeed, Gore (2007) argues that the globalized market place, what he calls *Earth Inc* has the power and inclination to maintain business as usual. If this is so, then climate change will undoubtedly have dramatic adverse effects, not least on the less developed countries of the world.[32] If this is so, then climate change may lead to instability and war. Collier et al. (2008) provides an estimate of the devastation that climate change may bring about in Africa. Taken together with Collier's (2009) analysis of the causes of civil war in less developed countries suggests what the long term consequences of climate change are likely to be.

In principle, we may be able to rely on a version of the Jury Theorem (Rae, 1969; Schofield, 1972a, b; Landemore, 2012), which asserts that majority rule provides an optimal procedure for making collective choices under uncertainty. However, for the operation of what Madison called a "fit choice" it will be necessary to overcome the entrenched power of capital. Although we now disregard Marx's attempt at constructing a teleology of economic and political development,[33] we are in need of a more complex over-arching and evolutionary theory of political economy that will go beyond the notion of equilibrium and might help us deal with the future.

Gamble (1993), in his discussion of how humans colonized our planet over many thousands of years, emphasizes that they not only used reason but were driven by *purpose*, what I here refer to as teleology, not in the sense of progress, but in the sense of safeguarding our heritage.[34]

---

[32] Zhang (2007) and Hsiang et al. (2013) have provided a quantative analysis of such adverse effects in the past. See also Parker (2013) for an historical account of the effect of climate change in early modern Europe, and Broodbank (2013) for the effects on the civilizations of the Mediterranean over a two-thousand-year period. There have also been suggestions that while other civilizations have arisen, possibly on nearby planets, we have seen no evidence say from radio messages, because advanced civilizations tend to destroy themselves.

[33] See Sperber (2013) for a discussion of the development of Marx's ideas, in the context of 19th century belief in the teleology of "progress" or the advance of civilization. The last hundred years has however, made it difficult to hold such beliefs.

[34] The empirical work by Piketty (2014) suggests that capital is becoming predominant, and may be threatening the ability of democratic capitalism to survive. Cooper (2014) also argues against the neo-classical equilibrium model in economics, and suggests that governments engage in investment that would benefit the less well off members of society. This could be done by expanding investment in new technologies to counter climate change. Vogel (2014) sketches the increasing ability of "Big Money" to dominate politics. It is an open question whether the bad times that Randers (2014) envisions in our future will weaken the power of big money by inducing a popular backlash.

In the aftermath of the Great Recession, many authors have argued that the institutions that served the west as it industrialized are no longer effective (Ferguson, 2012; Oreskes and Conway, 2014; Wolf, 2014; Fukuyama, 2014; Ackerman, 2011). An earlier argument by Tainter and Renfrew (1988) on the basis of a review of the anthropological and archeological literature on the collapse of complex societies is that all such societies develop increasing complex institutions, and that complexity itself induces increasing marginal cost. Without any doubt the institutions of capitalism have become more complex over time (Such complexity can be seen in the "Limits to Growth" models of Meadows et al. (1972). It is indeed plausible that Capitalism can collapse in a Malthusian catastrophe. (See also Diamond, 2011). The logic of this theory is that we face the collapse of the American hegemony, with the end of the period of cheap energy and resources. Jamieson (2014) looks for a new system of morality based on virtues appropriate to the Anthropocene. In my view, what may save us is a reinvigoration of a *Mythos*, based on the notion of care of our common home (see also Kauffman, 2016).

REFERENCES

Abulafia, D. (2011), *The Great Sea.* Oxford, Oxford University Press.
Acemoglu, D. and J. Robinson (2006), *Economic Origins of Dictatorship and Democracy.* Cambridge, Cambridge University Press.
Acemoglu, D. and J. Robinson (2008), 'Persistence of power, elites, and institutions', *American Economic Review* **98**, 267–293.
Acemoglu, D., S. Johnson, J. Robinson and P. Yared (2009), 'Reevaluating the modernization hypothesis', *Journal of Monetary Economics* **56**, 1043–1058.
Acemoglu, D., A. Ozdaglar and A.Tahbaz-Salehi (2010), 'Cascades in networks and aggregate volatility', NBER working paper no. 16516.
Ackerman, B. (2011), *Decline and Fall of the American Republic.* Boston, MA, Harvard University Press.
Armstrong, K. (2011), *The Battle for God.* New York, Random House.
Austen-Smith, D. and J. Banks (1996), 'Information aggregation, rationality, and the Condorcet jury theorem', *American Political Science Review* **90**, 34–45.
Axelrod, R. (1981), 'The emergence of cooperation among egoists', *American Political Science Review* **75**, 306–318.
Axelrod, R. (1984), *The Evolution of Cooperation.* New York, Basic.
Axelrod, R. and W.D. Hamilton (1981), 'The evolution of cooperation', *Science* **211**, 1390–1396.
Barbera, R. (2009), *The Cost of Capitalism: Understanding Market Mayhem.* New York, McGraw Hill.
Belich, J. (2009), *Replenishing the Earth: The Settler Revolution 1783–1939.* Oxford, Oxford University Press.
Bellah, R.N. (2011), *Religion in Human Evolution.* Cambridge, MA, Belknap Press.
Bikhchandani, S., D. Hirschleifer and I. Welsh (1992), 'A theory of fads, fashion, custom, and cultural change as information cascades', *Journal of Political Economy* **100**, 992–1026.
Binmore, K. (2005), *Natural Justice.* Oxford, Oxford University Press.
Bowles, S., J.K. Choi and A. Hopfensitz (2003), 'The co-evolution of individual behaviors and socal institutions', *Journal of Theoretical Biology* **223**, 135–147.
Boyd, J. and P.J. Richerson (2005), *The Origin and Evolution of Culture.* Oxford, Oxford University Press.
Broodbank, C. (2013), *The Making of the Middle Sea.* Oxford, Oxford University Press.
Burkart, J.M., S.B. Hrdy and C.P. van Schaik (2009), 'Cooperative breeding and human cognitive evolution', *Evolutionary Anthropology* **18**, 175–186.
Calvin, W.H. (2003), *The Ascent of Mind.* New York, Bantam.
Carothers, T. (2002), 'The end of the transition paradigm', *Journal of Democracy* **13**, 5–21.
Cassidy, J. (2009), *How Markets Fail: The Logic of Economic Calamities.* New York, Farrar, Strauss and Giroux.
Cavalli-Sforza, L. and M. Feldman (1981), *Cultural Transmission and Evolution.* Princeton, NJ, Princeton University Press.
Chichilnisky, G. (2009a), 'The topology of fear', *Journal of Mathematical Economics* **45**, 807–816.
Chichilnisky, G. (2009b), 'Avoiding extinction: equal treatment of the present and the future', Working Paper, Columbia University.

Chichilinsky, G. (2010), 'The foundations of statistics with black swans', *Mathematical Social Science* **59**, 184–192.
Chichilinsky, G. (2014), 'The topology of change: foundations of statistics with black swans', in: *Topics in Mathematical Economics: Essays in Honor of J. Marsden*, Fields Institute Communication Volume. Providence, RI, American Math Society.
Chichilinsky, G. and P. Eisenberger (2010), 'Asteroids: assessing catastrophic risks', *Journal of Probability and Statistics*, Article ID 954750.
Clark, G. (2007a), 'What made Britannia great? How much of the rise of Britain to world dominance by 1850 does the Industrial Revolution explain?', in: K. O'Rourke and A. Taylor (eds). *Comparative Economic History: Essays in Honor of Jeffrey Williamson*. Cambridge, MA, MIT Press.
Clark, G. (2007b), *A Farewell to Alms*. Princeton, NJ, Princeton University Press.
Cline, W. (2007), *Global Warming and Agriculture Impact Estimates by Country*. Washington, DC, Peterson Institute.
Collier, P. (2009), *Wars, Guns and Votes*. New York, Harper.
Collier, P. (2010), *The Plundered Planet*. Oxford, Oxford University Press.
Collier, P. G. Conway and T. Venables (2008), 'Climate change and Africa', Working Paper, Oxford University.
Condorcet, N. (1994 [1785]), 'Essai sur l'application de l'analyse à la probabilité des décisions rendues à la pluralité des voix', Paris, Imprimerie Royale. Translated in part in: I. McLean and F. Hewitt (eds). *Condorcet: Foundations of Social Choice and Political Theory*. Aldershot, UK, Edward Elgar Publishing.
Condorcet, N. (1798), *Esquisse d'un tableau historique des progres de l'esprit humaine*. Paris, Yves Gravier.
Cooper, G. (2014), *Money, Blood and Revolution*. Petersfield, UK, Harriman.
Corcos, A., J.-P. Eckmann, A. Malaspinas, Y. Malevergne and D. Sornette (2002), 'Imitation and contrarian behavior: hyperbolic bubbles, crashes and chaos', *Quantitative Finance* **2**, 264–281.
Darwin, C. (1859), *The Origin of Species by Means of Natural Selection*. London, John Murray.
Darwin, C. (1871), *The Descent of Man*. London, John Murray.
Deutscher, G. (2006), *The Unfolding of Language*. New York, Holt.
Diamond, J. (2011), *Collapse*. London, Penguin.
Easley, D. and Kleinberg, J. (2010), *Networks, Crowds and Markets*. Cambridge, Cambridge University Press.
Ferguson, N. (1997), 'Introduction', in: N. Ferguson (ed.). *Virtual History*. London, Picador.
Ferguson, N. (2002), *Empire: The Rise and Demise of the British World Order*. London, Penguin Books.
Ferguson, N. (2011), *Civilization*. London, Penguin.
Ferguson, N. (2012), *The Great Degeneration*. London, Penguin.
Fox, J. (2009), *The Myth of the Rational Market*. New York, Harper.
Fukuyama, F. (2011), *The Origins of Political Order*. New York, Farrar, Strauss and Giroux.
Fukuyama, F. (2014), *Political Order and Political Decay*. New York, Farrar, Strauss and Giroux.
Gamble, C. (1993), *Timewalkers: The Prehistory of Global Colonization*. London, Penguin Books.
Gazzaniga, M.S. (2006), *The Ethical Brain*. New York, Harper.
Gintis, H. (2000), 'Strong reciprocity and human sociality', *Journal of Theoretical Biology* **206**, 169–179.
Gleick, J. (1987), *Chaos: Making a New Science*. New York, Viking.
Gödel, K. (1931), 'Über formal unentscheidbare Sätze der Principia Mathematica und verwandter Systeme', *Monatshefte für Mathematik und Physik* **38**, 173–198. Translated as 'On formally undecidable propositions of *Principia Mathematica* and related systems', in Jean van Heijenoort (ed.). *Frege and Gödel: Two Fundamental Texts in Mathematical Logic*. Cambridge, MA, Harvard University Press.
Golub, B. and M. Jackson (2010), 'Naive learning in social networks and the wisdom of crowds', *American Economic Journal: Microeconomics* **2**, 112–149.
Gore, A. (2007), *The Assault on Reason*. London: Bloomsbury.
Gray, J. (1995), *Enlightenment's Wake*. London, Routledge.
Gray, J. (1997), *Endgames*. London, Blackwell.
Gray, J. (2000), *False Dawn*. London, New Press.
Gribbin, J. (2011), *Alone in the Universe*. New York, Wiley.
Hamilton, W. (1964), 'The genetical evolution of social behavior I and II', *Journal of Theoretical Biology* **7**, 1–52.
Hamilton, W. (1970), 'Selfish and spiteful behavior in an evolutionary model', *Nature* **228**, 1218–1220.
Hardin, G. (1968 [1973]), 'The tragedy of the commons', in: H.E. Daly (ed.). *Towards a Steady State Economy*. San Francisco, Freeman.
Hardin, R. (1971), 'Collective action as an agreeable n-prisoners' dilemma', *Behavioral Science* **16**, 472–481.
Hardin, R (1982), *Collective Action*. Baltimore, MD, Johns Hopkins University Press.
Hawking, S. and L. Mlodinow (2010), *The Grand Design: New Answers to the Ultimate Questions of Life*. New York: Bantam Books.
Harper, K. (2016), 'The enviromental fall of the Roman Empire', *Daedalus* Spring, 101–111.
Hays, J.D., J. Imbrie and N.J. Shackleton (1976), 'Variations in the Earth's orbit: pacemaker of the Ice Ages', *Science* **194**(4270), 1121–1132.

Henrich, J., R. Boyd, S. Bowles, C. Camerer, E. Fehr and H. Gintis (2004), *Foundations of Human Sociality: Economic Experiments and Ethnographic Evidence from Fifteen Small-Scale Societies*. Oxford, Oxford University Press.
Henrich, J. et al. (2005), '"Economic man" in cross-cultural perspective: behavioral experiments in 15 small-scale societies', *Behavioral Brain Science* **28**, 795–855.
Hobbes, T. (2009 [1651]), *Leviathan; or the Matter, Forme, and Power of a Common-wealth, Ecclesiastical and Civil*, J.C.A. Gaskin (ed.). Oxford, Oxford University Press.
Hrdy, S.B. (2011), *Mothers and Others: The Evolutionary Origins of Mutual Understanding*. Cambridge, MA, Harvard University Press.
Hsiang, S.M., M. Burke and E. Miguel (2013), 'Quantifying the influence of climate on human conflict', *Science Express*, **341**, 1235367.
Israel, J. (2002), *Radical Enlightenment*. Oxford, Oxford University Press.
Israel, J. (2006), *Enlightenment Contested*. Oxford, Oxford University Press.
Israel, J. (2010), *Revolution of the Mind*. Princeton, NJ, Princeton University Press.
Israel, J. (2014), *Revolutionary Ideas*. Princeton, NJ, Princeton University Press.
Jacques, M. (2009), *When China Rules the World*. London, Penguin.
Jamieson, D. (2014). *Reason In A Dark Time*. Oxford, Oxford University Press.
Kahneman, D. (2011) *Thinking Fast and Slow*. New York, Farrar Strauss and Giroux.
Karklins, R. and R. Petersen (1993), 'Decision calculus of protestors and regime change: Eastern Europe 1989', *Journal of Politics* **55**, 588–614.
Kauffman, S. (2016), *Humanity in a Creative Universe*. Oxford, Oxford University Press,
Kennedy, P. (1987), *The Rise and Fall of the Great Powers*. New York, Random House.
Keohane, R. and R. Nye (1977), *Power and Interdependence*. New York, Little Brown.
Keohane, R. (1984), *After Hegemony*. Princeton, NJ, Princeton University Press.
Keynes, J.M. (1921), *Treatise on Probability*. London, Macmillan.
Keynes, J.M. (1936), *The General Theory of Employment, Interest and Money*. London, Macmillan.
Kindleberger, C. (1973), *The World in Depression 1929–1939*. Berkeley, CA, University of California Press.
Kolbert, E. (2014), *The Sixth Extinction*. New York, Holt.
Kreps, D.M., P. Milgrom, J. Roberts and R. Wilson (1982), 'Rational cooperation in the finitely repeated prisoners' dilemma', *Journal of Economic Theory* **27**, 245–252.
Ladha, K. (1992), 'Condorcet's jury theorem, free speech and correlated votes', *American Journal of Political Science* **36**, 617–674.
Ladha, K. and G. Miller (1996), 'Political discourse, factions and the general will: correlated voting and Condorcet's jury theorem', in: N. Schofield (ed.). *Collective Decision Making*. Boston, MA, Kluwer.
Landemore, H. (2012), *Democratic Reason*. Princeton, NJ, Princeton University Press.
Laudato Si' (2015), Encyclical Letter of The Holy Father Francis 'On Care for our Common Home'. The Holy See, Rome.
Leakey, R. (1994), *The Origin of Humankind*. New York, Basic.
Leakey, R. and R. Lewin (1995), *The Sixth Extinction*. New York, Anchor.
Lohmann, S. (1994), 'The dynamics of information cascades', *World Politics* **47**, 42–101.
Maddison, A. (2007), *Contours of the World Economy 1–2030 AD: Essays in Macro-Economic History*. Oxford, Oxford University Press.
Madison, J. (1999 [1787]), 'Federalist X', in: J. Rakove (ed.). *Madison: Writings*. New York, Library Classics.
Malthus, T. (1798), *An Essay on the Principle of Population*. London, J. Johnson.
Mandelbrot, B. and R. Hudson (2004), *The (Mis)behavior of Markets*. New York, Perseus.
Margolis, H. (1982), *Selfishness, Altruism and Rationality*. Cambridge, Cambridge University Press.
McWhorter, J. (2001), *The Power of Babel*. New York, Holt.
Meadows, D., D.L. Meadows, J. Randers and W. Behrens (1972), *Limits to Growth*. New York, Signet.
Miller, G. and N. Schofield (2003), 'Activists and partisan realignment in the US', *American Political Science Review* **97**, 245–260.
Miller, G. and N. Schofield (2008), 'The transformation of the Republican and Democratic party coalitions in the United States', *Perspectives on Politics* **6**, 433–450.
Minsky, H. (1975), *John Maynard Keynes*. New York, Columbia University Press.
Minsky, H. (1986), *Stabilizing an Unstable Economy*. New Haven, Yale University Press.
Mokyr, J. (2005), 'The intellectual origins of modern economic growth', *Journal of Economic History* **65**, 285–351.
Mokyr, J. (2010), *The Enlightened Economy: An Economic History of Britain 1700–1850*. New Haven, CT, Yale University Press.
Mokyr, J. and V.C. Nye (2007), 'Distributional coalitions, the Industrial Revolution, and the origins of economic growth in Britain', *Southern Economic Journal* **74**, 50–70.
Morris, I. (2010) *Why the West Rules*. New York, Farrar, Strauss and Giroux.

Nagel, T. (2012), *Mind and Cosmos*. Oxford, Oxford University Press.
Nordhaus, W. (2013), *Climate Casino*. New Haven, CT, Yale University Press.
North, D.C. (1990), *Institutions, Institutional Change and Economic Performance*. Cambridge, Cambridge University Press.
North, D.C. and B.R. Weingast (1989), 'Constitutions and commitment: the evolution of institutions governing public choice in seventeenth-century England', *Journal of Economic History* **49**, 803–832.
North, D.C., B. Wallis and B.R. Weingast (2009), *Violence and Social Orders: A Conceptual Framework for Interpreting Recorded Human History*. Cambridge, Cambridge University Press.
Nowak, M. (2011), *Supercooperators*. New York, Free Press.
Oreskes, N. and E. Conway (2014), *The Collapse of Western Civilization*. New York, Columbia University Press.
Ormerod, P. (2001), *Butterfly Economics*. New York, Basic.
Pagden, A. (2013), *The Enlightenment*. New York, Random.
Parfit, D. (2011), *On What Matters*. Oxford, Oxford University Press.
Parker, G. (2013), *Global Crisis*. New Haven, CT, Yale University Press.
Penn, E. (2009), 'A model of far-sighted voting', *American Journal of Political Science* **53**, 36–54.
Piketty, T. (2014), *Capital*. Cambridge, MA, Harvard University Press.
Putnam, R.D. and D.E. Campbell (2010), *American Grace: How Religion Divides and Unites Us*. New York, Simon and Schuster.
Rae, D. (1969), 'Decision rules and individual values in constitutional choice', *American Political Science Review* **63**, 40–56.
Rees, M. (2001), *Our Cosmic Habitat*. Princeton, NJ, Princeton University Press.
Scharf, C. (2014), *The Copernicus Complex*. New York, Farrar, Strauss and Giroux.
Schofield, N. (1972a), 'Is majority rule special?', in: R.G. Niemi and H.F. Weisberg (eds). *Probability Models of Collective Decision-Making*. Columbus, OH, Charles E. Merrill Publishing Co.
Schofield, N. (1972b), 'Ethical decision rules for uncertain voters', *British Journal of Political Science* **2**, 193–207.
Schofield, N. (2002), 'Madison and the founding of the two party system in the US', in: S. Kernel (ed.). *James Madison: The Theory and Practise of Republican Government*. Stanford, CA, Stanford University Press, pp. 302–327.
Schofield, N. (2003), 'Evolution of the constitution', *British Journal of Political Science* **32**, 1–20.
Schofield, N. (2011), 'Is the political economy stable or chaotic?', *Czech Economic Review* **5**, 76–93.
Schofield, N. and M. Gallego (2011), *Leadership or Chaos*. Berlin, Springer.
Schweitzer, F., G. Fagiolo, D. Sornette, F. Vega-Redondo, A. Vespignani and D.R. White (2009). 'Economic networks: the new challenges', *Science* **325**, 422–425.
Sen, A. (1982), *Poverty and Famines*. Oxford, Clarendon Press.
Shiller, R. (2003), *The New Financial Order*. Princeton, NJ, Princeton University Press.
Shiller, R. (2005), *Irrational Exuberance*. Princeton, NJ, Princeton University Press.
Smale, S. (1966), 'Structurally stable systems are not dense', *American Journal of Mathematics* **88**, 491–496.
Smith, A. (1984 [1759]), *The Theory of Moral Sentiments*. Indianapolis, IN, Liberty Fund.
Smolin, L. (2007), *The Trouble with Physics*. New York, Houghton Mifflin.
Sperber, J. (2011), *Karl Marx: A Nineteenth-Century Life*. New York, Liveright.
Stern, N. (2007), *The Economics of Climate Change*. Cambridge, Cambridge University Press.
Stern, N. (2009), *The Global Deal*. New York, Public Affairs.
Stringer, C. (2013), *Lone Survivors*. New York, St. Martins.
Sunstein, C.R. (2009), *A Constitution of Many Minds*. Princeton, NJ, Princeton University Press.
Sunstein, C.R. (2011), *Going to Extremes*. Oxford, Oxford University Press.
Surowiecki, J. (2005), *The Wisdom of Crowds*. New York, Anchor.
Tainter, J.A. and C. Renfrew (1988), *The Collapse of Complex Societies*. Cambridge, Cambridge University Press.
Taleb, N.N. (2007), *The Black Swan*. New York, Random.
Taleb, N.N. and M. Blyth (2011), 'The black swan of Cairo', *Foreign Affairs* **90**(3), 33–39.
Taylor, M. (1976), *Anarchy and Cooperation*. London, Wiley.
Taylor, M. (1982), *Community, Anarchy and Liberty*. Cambridge, Cambridge University Press.
Tegmark, M. (2008), 'The mathematical universe', *Foundations of Physics* **38**, 101–150.
Tegmark, M. (2014), *Our Mathematical Universe*. New York, Random House.
Trivers, R. (1971), 'The evolution of reciprocal altruism', *Quarterly Review of Biology* **46**, 35–56.
Trivers, R. (1985), *Social Evolution*. Menlo Park, CA, Cummings.
Turing, A. (1937), 'On computable numbers with an application to the Entscheidungsproblem', *Proceedings of the London Mathematical Society* **42**, 230–265. Reprinted in Jack Copeland (ed.). *The Essential Turing*. Oxford, The Clarendon Press.
Vogel, K. (2014), *Big Money*. New York, Public Affairs.
Wallace, A.R. (1898), *Natural Selection*. New York, Classics US.

Waltham, D. (2014), *Lucky Planet*. New York, Basic.
Weber, M. (2013), *The Protestant Ethic and the Spirit of Capitalism*. London, Routledge.
Weitzman, M. (2009), 'Additive damages, fat-tailed climate dynamics, and uncertain discounting', *Economics* **3**, 1–22.
Wells, S. (2011), *Pandora's Seed*. New York, Random House.
White, T.D., B. Asfaw, Y. Beyene, Y. Haile-Selassie, C.O. Lovejoy, G. Suwa and G. WoldeGabriel (2009), '*Ardipithecus ramidus* and the paleobiology of early hominids', *Science* **326**, 65–86.
Wilson, E.O. (2014), *The Meaning of Human Existence*. New York, Norton.
Wolf, M. (2014), *The Shifts and the Shocks*. New York, Penguin.
Yanofsky, N.S. (2013), *The Outer Limits of Reason*. Cambridge, MA, MIT Press.
Zeeman, E.C. (1977), *Catastrophe Theory: Selected Papers, 1972–77*. New York, Addison Wesley.
Zhang, D.D. (2007), 'Global climate change, war, and population decline in recent human history', *Proceedings of the National Academy of Science* **104**(49), 19214–19219.

13. Greenhouse gas and cyclical growth*
Lance Taylor and Duncan Foley

How will accumulation of greenhouse gas (GHG) and economic growth affect each other over many decades? In what ways will output, income distribution, and employment respond to rising levels of atmospheric CO_2 concentration? In this chapter we address these questions, central to our times, by breaking from the mainstream consensus and assuming that economic activity is determined by aggregate demand in the "medium" run. The "long" run is set up as a steady state in which demand and supply effects commingle.

Standard models address the climate question strictly from the supply side, nearly always in a Ramsey optimal savings framework. This formulation raises several problems.

Given the havoc that climate scientists expect from global warming, the full employment assumptions built into supply side models strain credibility.

As will be seen, even with assumed full employment a Solow–Swan growth model linked with GHG accumulation generates complicated cyclical dynamics. Optimal growth models suppress cyclicality and use investment in "mitigation" of emissions to generate smooth trajectories of capital per capita and atmospheric GHG concentration toward a steady state. Using optimization to build such smooth behavior into a model's solutions is not necessarily wise. It elides dynamical complications and does not clarify how mitigation may fit into practical policy decisions.

We have argued elsewhere (Foley et al., 2013) that the smooth paths of state variables in optimizing models tend to be accompanied by strongly fluctuating values of the costates, i.e., the asset prices associated with capital and GHG concentration. Consequently, implicit interest rates vary strongly over time, with dynamics dependent on the detailed specification of a model. Because climate change is not a "small" perturbation to the economic system, the optimizing approach fails in its primary task of calculating a *constant* "appropriate" discount rate and social cost of carbon.

The key components of the demand-driven model are accumulation equations for atmospheric concentration of GHG (treated herein as CO_2 only) and capital per capita. Labor productivity also enters the specification as a third dynamic variable.

There is no aggregate production function with associated marginal conditions, so medium-run output has to be determined by effective demand. A specific formulation relating demand to the functional income distribution is provided. Two channels are considered for the effects of GHG concentration on the real economy

---

* Research supported by the Institute for New Economic Thinking (INET) under a grant to the Schwartz Center for Economic Policy Analysis at the New School. Thanks to Nelson Barbosa, Jonathan Cogliano, Gregor Semieniuk, Rishabh Kumar, Codrina Rada, and Armon Rezai for invaluable contributions.

– either a reduction in profits and therefore investment demand or capital stock destruction as induced by faster depreciation. "Damage functions" for the transmission of these effects are described informally in the text with details in an appendix.

In the central dynamic equations, higher capital per capita increases output which in turn increases the speed of CO_2 accumulation. On the other hand, higher atmospheric GHG concentration reduces output and growth of capital per capita. Hence we have a variation on "typical" predator-prey dynamics – CO_2 is the predator and capital per capita the prey. Numerical simulations suggest that there may be an upswing in capital per capita for around eight decades, followed by a crash *of output and capital only*. Contrary to familiar fox-and-rabbit models, the decay rate of CO_2 in the atmosphere is *very* slow (the "fox" is almost immortal). *Concentration remains high*, blocking any chance of economic recovery.

We follow the usual growth theory convention of setting up a model that converges to a steady state. In practice, the system *must* converge to a *stationary* state with constant capital stock, CO_2 concentration, productivity, etc. Otherwise, CO_2 accumulation would overwhelm the system. As in the optimal growth models investment in mitigation can offset the crash, and lead to a non-dismal stationary state. In numerical simulations the share of output required is more than half the total world defense spending and roughly double current worldwide energy consumption subsidies.

In the rest of this chapter, we describe medium-term adjustment and sketch the accumulation equations as well as dynamics of energy and labor productivity. Steady state behavior of the model is then considered, followed by discussion of transient dynamics toward the steady state in "business as usual" (BAU) and "mitigated" scenarios. Implications for employment and distribution are pointed out.

MACROECONOMIC RELATIONSHIPS

There are three dynamic variables: atmospheric CO_2 concentration in parts per million by volume or ppmv (G), capital stock per capita (κ), and the output/labor ratio ("labor productivity" ξ). The increase in G (or $\dot{G}$) is proportional to output (X) with the factor of proportionality reduced by outlays on mitigation (m) as a share of X. The increase in κ (or $\dot{\kappa}$) is driven by the investment/capital ratio ($g = I/K$) less depreciation (rate δ) and the population growth rate (n). In line with the literature on ecological economics the labor productivity growth rate ($\hat{\xi}$) depends on the growth rate of "energy intensity" or the energy/labor ratio ($e = E/L$).

As noted above, the long run must necessarily take the form of a stationary state with $\dot{G} = \dot{\kappa} = \hat{\xi} = \hat{e} = n = 0$. That is, G, κ, X, capital stock (K), employment (L) and population (N) must all be constant.

In the medium run X and L are determined by effective demand driven by g and m. "Capital utilization" $u = X/K$ *increases* with the profit share π via an increase in investment demand $g(\pi)$ so $u = u(\pi)$. In the usage of contemporary Keynesian growth theory (Taylor, 2004) demand for output is "profit-led." On the other hand, π is assumed to fall in response to tighter labor market as signaled by in an increase in the labor/population ratio $\lambda = L/N$.[1] Macro stability follows automatically from this profit-led/profit-squeeze combination.[2]

A key identity $\lambda = \kappa u/\xi$ ties the medium run together. Along with the investment function and macroeconomic balance it implies that higher κ increases λ, and reduces π and g. Lower g means that growth of κ slows, i.e., $d\dot\kappa/d\kappa < 0$ so growth in capital per capita is (locally) dynamically stable. Capital stock scales the system. In contrast to neoclassical supply-side models there is no aggregate production or cost function although the identity $\lambda = \kappa u/\xi$ and the assumption $\partial \pi/\partial \lambda < 0$ do apply.

Two variants of this specification are considered. In one, π is also squeezed by G as rising CO_2 concentration cuts into profits. The linkages could include direct losses of output such as crops, higher non-wage costs of production, and weather-induced shortening of supply chains which magnify mark-ups (Kemp-Benedict, 2014). The appendix gives details.

In the other variant, π decreases just with λ but higher GHG concentration raises the depreciation rate ("capital destruction") and slows growth of κ. Lower capital stock reduces the level of output directly.[3]

ACCUMULATION

Following the well-known "Kaya identity" from climate science (Waggoner and Ausubel, 2002), the accumulation equation for CO_2 is

$$\dot{G} = \chi E - \mu(m)X - \omega G = [(\chi/\varepsilon) - \mu(m)]X - \omega G \qquad (13.1)$$

with $\varepsilon = X/E$ as energy productivity. Subject to decreasing effectiveness through the function $\mu(m)$ higher mitigation reduces the factor of proportionality of $\dot{G}$ with respect to X. The parameter ω is small – atmospheric CO_2 dissipates very slowly.

There is a steady state at

$$G = \omega^{-1}[\chi(e/\xi) - \mu(m)]uN\kappa = \omega^{-1}[(\chi(e/\xi) - \mu(m)]X \qquad (13.2)$$

Note that steady state G is proportional to u, κ, and N (a Malthusian touch) or steady state X.

---

[1] Here we follow Marx, who in several passages in *Capital* sketched a theory of business cycles (formalized a century later by Goodwin, 1967) pivoting on shifts in the functional income distribution. At the bottom of a cycle, the real wage is held down by a large reserve army of un- or under-employed workers, and capitalists can accumulate freely. However, as output expands the reserve army is depleted and λ goes up. The real wage rises in response to a tighter labor market, forcing a profit squeeze. To keep the analysis to low dimensionality we omit detailed discussion of the dynamics of such a cycle.

[2] In a Solow–Swan or Ramsey model incorporating full employment, medium-term adjustment takes the form of shifts in investment in response to changes in saving. Higher GHG concentration cuts directly into output by shifting the aggregate production function downward.

[3] As noted above, capital stock K scales the system. In our simulations, when the negative effect of κ on π is taken into account the elasticity of X with respect to K is 0.75. The medium-run multiplier is 1.7.

284 *Handbook on the economics of climate change*

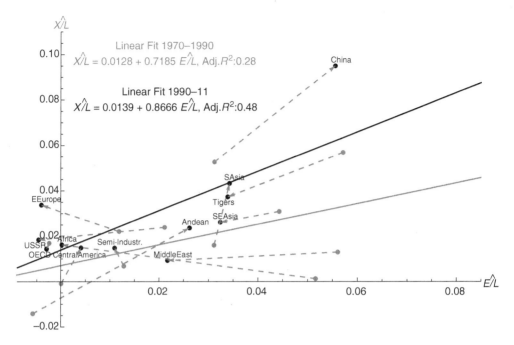

Figure 13.1 Evolution of average energy use per labor growth rate ($\hat{E/L}$) vs. labor productivity growth rate ($\hat{X/L}$) from 1971–90 (light gray, solid and dashed) to 1990–2011 (dark gray)

The other key accumulation equation is for capital per capita

$$\dot{\kappa} = \kappa(g - \delta - n). \tag{13.3}$$

There is a steady state when

$$g = \delta + n \tag{13.4}$$

and a stationary state when $n = 0$.

An ostinato theme in ecological economics is that labor productivity is closely tied to energy intensity – Figure 13.1 provides a recent illustration relating the growth rate of ξ to the growth rate of e. Hence we assume that producers choose a growth path for e that converges to steady state level, and labor productivity growth is determined as

$$\dot{\xi} = \xi T \hat{e} \tag{13.5}$$

with $T = 1.5$. Energy productivity for use in (13.1) is set by the identity $\varepsilon = \xi/e$. A value of $T > 1$ assures that c and e increase together, in line with much recent data.

STEADY STATES

For numerical illustration we assume constant steady state levels of population (initial level = 7 billion, final = 10), energy intensity $e = E/L$ (initial = 4 kilowatts per employed worker, final = 6) and labor productivity $\xi = X/L$ (initial = $20,000 per unit of labor, final = $35,000).

In the first medium-term variant, the steady state condition (13.4) with $n = 0$ means that $g = \delta$ determines π from investment demand. Then π sets u from macro balance. As described above we have $\pi = F(\lambda, G) = F(\kappa u/\xi, G)$ with negative partials from the profit squeezes. Hence, in steady state G and κ must trade off along a "nullcline" in the (κ, G) plane to hold π constant.

As shown in Figure 13.2, the slope of the nullcline for $\dot{G} = 0$ is sensitive to m so that mitigation can support a non-dismal steady state. With no mitigation, $\kappa < 20$ and $G = 759$ in a dismal BAU steady state. (Initial values are $\kappa = 28.57$ and $G = 400$.) The mitigated steady state $G = 486$ might correspond to 2.5°C of global warming over the pre-industrial baseline – more than the currently accepted "red line" of 2.0°. A steady state with $G = 759$ would mean 5 or 6 of warming,[4] a recipe for disaster.

If time trends for population and productivity were ignored, phase diagram aficionados would infer that from an initial position of $\kappa = 28.6$ and $G = 400$ the BAU solution trajectory would follow a counter-clockwise path toward the steady state. Both variables would rise until they hit the $\dot{\kappa} = 0$ nullcline. Thereafter, κ would start to fall and G to

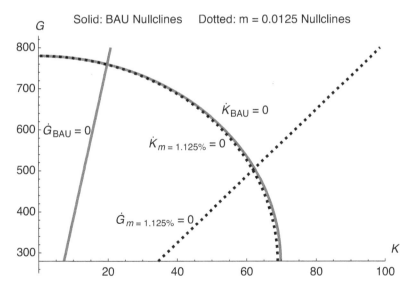

Figure 13.2 Nullclines for per capita capital stock (κ) and CO_2 concentration (G) when the profit share decreases with both G and κ

---

[4] The rule of thumb is that after an atmospheric lag of about ten years doubling pre-industrial CO_2 concentration of 280 ppmv should lead to a temperature increase of about 3°C. As of 2013 the increase was already about 0.8° and getting to 1.3° is well underway. The numbers in the text follow accordingly.

Table 13.1 Steady state Jacobians

BAU	k	G
$\dot{\kappa}$	−0.0570091	−0.0198071
$\dot{G}$	0.102817	−0.0163718
1.25% mitigation	k	G
$\dot{\kappa}$	−0.0418889	−0.00261627
$\dot{G}$	0.0217864	−0.00376279

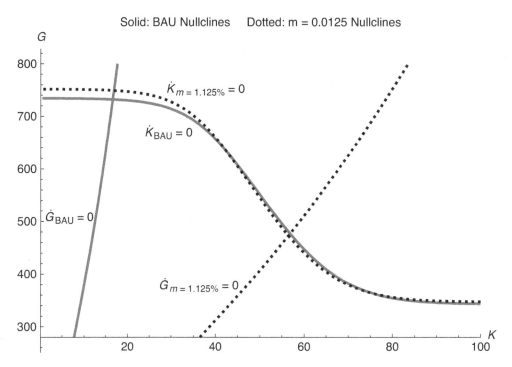

Figure 13.3 Nullclines for capital stock per capita (κ) and CO_2 concentration (G) when higher G increases capital depreciation

rise until they hit the $\dot{G} = 0$ nullcline and then both would spiral toward the steady state. In a mitigated solution, κ would rise with G initially falling slightly and then rising in a converging spiral. With time trends included both patterns are illustrated in simulations below.

Table 13.1 gives values for the Jacobian matrices for κ and G at the BAU and mitigated steady states. The values along the main diagonals show that convergence will be slow for κ and (especially) G. In the BAU solution the cross effects $\partial \dot{G}/\partial \kappa > 0$ and $\partial \dot{\kappa}/\partial G < 0$ are relatively strong; the magnitudes drop off in the mitigated solution. These results feed into the discussion below of dynamics away from the steady states.

In the second medium-run variant (Figure 13.3), there is no direct adverse effect of G on π, but higher CO_2 concentration raises the depreciation rate δ – there is more rapid

Table 13.2 *Initial and steady state values for BAU and mitigated paths for two model versions*

	Initial value	BAU	Mitigated
1) Profit share decreases with both			
G	400	759.4	486.2
κ	28.6	19.8	63.0
X/N	8.6	5.6	18.3
λ	0.429	0.153	0.5
2) Depreciation rate increases with			
G	400	698.6	464.7
κ	28.6	20.3	57.3
X/N	8.6	6.6	17.2
λ	0.429	0.181	0.468

destruction of capital stock. Now in steady state, higher G and δ must lead to higher π (investment is profit-led). But a higher π must be associated with a lower κ via lower λ. Again, we get a trade-off between G and κ. Mitigation can still support a high level steady state. No mitigation leads to low level stagnation. Similar results show up in, say, a Solow–Swan model in which $\delta = g = sf(\kappa, G)$ so δ determines κ from the *supply side*. Damages from GHG accumulation follow from an assumption that $\partial f/\partial G < 0$.

Table 13.2 gives a quick summary of steady state results for both variants. One can further show that higher steady state population strongly reduces κ and per capita output X/N under BAU; there is a relatively weak impact on G. The magnitudes reverse in mitigated solutions. Higher labor productivity (which also raises energy productivity) increases κ, G, and X/N, more strongly in mitigated solutions.

Transient Paths to Steady States

We set up simulations to track dynamics of κ, G, and ξ toward steady states. Growth trajectories are affected by assumed rates of increase of population, labor productivity, and energy intensity (modeled as logistic curves between initial and final levels). We first look at BAU growth when there is an adverse effect of CO_2 concentration on profitability (with similar results when higher concentration increases the depreciation rate). Figure 13.4 shows trajectories of variables of interest over a five-century time span.[5]

Along the lines discussed above, the model generates cyclical dynamics. Capital per capita and output rise for about eight decades, and then crash. Apart from energy and labor productivity levels which rise according to (13.5) the other economic variables follow a similar pattern. Output becomes constant near its initial level of $60 trillion so output per capita falls by around 35% at a final population level of 10 billion.

Atmospheric CO_2 concentration stabilizes at well over 700 ppmv, leading to a big temperature increase as noted above. There is no possibility for output per capita to recover. Two potential offsets are weak.

---

[5] The model's differential equations were simulated using *Mathematica 9*.

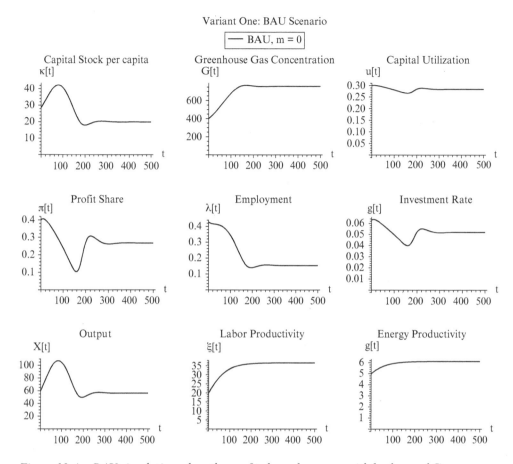

Figure 13.4 BAU simulation when the profit share decreases with both κ and G

First, there could be a reduction in the emissions/output ratio χ in (13.1) due to a shift in the mix of fossil fuels in use away from coal and oil toward natural gas. This change allows a modest reduction in CO_2 accumulation which permits better economic performance but the basic BAU oscillation persists.

The cycle also remains when there is slower growth of energy intensity, meaning that the growth rate of energy productivity will decrease as well. Then, GHG concentration should grow faster because the ratio χ/ε will fall less rapidly in (13.1). On the other hand, labor productivity growth will also drop, cutting into capital accumulation and output. The latter effect dominates. Economic performance deteriorates, marginally slowing the rise in greenhouse gas.

Next, we turn to growth with mitigation at initial cost of $160 per metric ton of carbon, or $44 per ton of CO_2 (in the mid-range of current estimates).[6] Outcomes with mitigation

---

[6] "Mitigation" comprises many different activities – reducing motor vehicle use, increasing energy efficiency of buildings, carbon sequestration, conservation tillage, ending deforestation, etc.

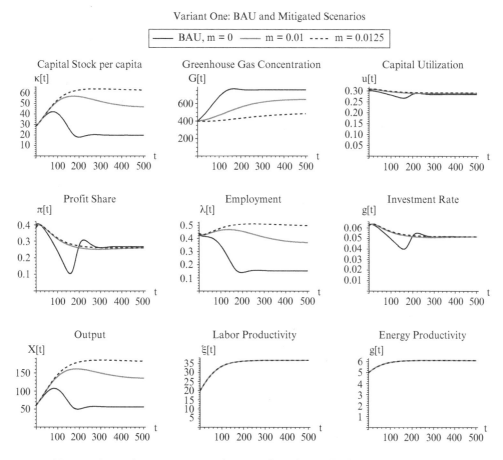

Figure 13.5 BAU and mitigation simulations when the profit share decreases with both κ and G

outlays of 1.0% and 1.25% of world output ($60 trillion initially) are illustrated in Figure 13.5. With 1.25% mitigation, CO_2 concentration can be stabilized. This outlay is around one-half of current level of defense spending and roughly twice the level of worldwide energy consumption subsidies.

As illustrated in Figure 13.6, the macro economy with mitigation follows a growth path of capital per capita to a stationary state that lies slightly below the one that the model generates in the absence of global warming. The BAU and 1.25% mitigation scenarios broadly correspond to the highest and lowest damage paths in the IPCC (2007).[7] One can show that "front-loading" mitigation leads to more favorable results (G converges to

---

As with reducing the parameter χ or the growth rate of energy intensity (see above) the effect of any single step toward mitigation will be "small," but in total they can have a big impact. See Pacala and Socolow (2004).

[7] At the time of writing, the fifth IPCC report on mitigation (IPCC 2014) is not yet published.

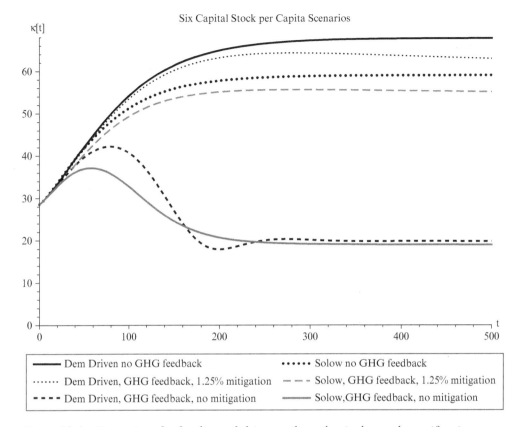

Figure 13.6   Dynamics of κ for demand-driven and neoclassical growth specifications

about 400). A "climate policy ramp" with low initial mitigation levels that gradually rise would be harmful.

The results in the demand-driven model are largely determined by convergence dynamics of κ and G to steady state levels. The same basic pattern appears under variant medium-run adjustments, e.g., higher CO_2 concentration reduces profitability *or* leads to capital destruction via faster depreciation *or* shifts down a neoclassical aggregate production function in a supply-driven full employment Solow–Swan growth model.

Figure 13.6 illustrates dynamics in a Solow–Swan scenario. The unconstrained solution for capital per capita lies below the demand-driven path, but under BAU both specifications converge to a similarly dismal stationary state. The full employment assumption partially stabilizes the dynamics but the cyclical rise and fall of κ going into the steady state persists. Finally, 1.25% mitigation returns the simulation close to its unconstrained variant.

IMPACTS ON LABOR

The demand-driven model can be used to explore implications of global warming for labor.

An initial observation is that at the macro level, the real wage is equal to the labor share of output (or one minus the profit share π) multiplied by labor productivity ξ. Approaching a "long run" stationary state, π will stabilize so that the real wage will increase along with productivity growth.[8] This standard result from growth accounting suggests that under BAU the economy will tend toward a high wage, low employment equilibrium. Indeed, the BAU steady state value of the employment/population ratio λ is 65% below its initial value due to high G, stagnating X and increases over time in labor productivity and population. In the mitigated solution, λ rises.

As in all demand-driven models away from steady state, employment is determined as a "lump of labor," or $L = X/\xi$. Alternatively, since $\pi = F(\lambda, G) = F(\kappa u/\xi, G)$, $L = \lambda N$ is a function of π, G, and N. The elasticity of λ with respect to ξ is -1 (higher productivity destroys jobs in proportion) in both BAU and mitigated steady states. It is about -0.8 along transient paths because X rises with higher ξ.

BOTTOM LINE

The intrinsic growth rate of capital per capita κ is low; the rate for atmospheric CO_2 concentration G is lower still. On the other hand, a higher level of κ (and output, employment, etc.) strongly stimulates growth of G. Ultimately, higher G slows the growth of κ and will make it turn negative. Moreover, because natural dissipation is very slow, once G reaches a high level there is no way for κ to recover. Under "business as usual" in our model's simulations there will be a climate crisis followed by economic stagnation. The time scale of the dynamics is such that the crisis could occur within a now young person's lifetime. These results carry through under differing specifications of medium-term macroeconomic adjustment and capital formation.

The conundrum for policy is that for several decades – half a human life span – the impending crisis could be masked by ongoing growth (see Figure 13.4). Effective mitigation of CO_2 emission could prevent the crisis and support a fairly high world level of economic activity with zero net new emissions and stable income per capita (assuming that population growth slows to zero or less). The sooner a mitigation effort gets underway at a level exceeding 1% of world GDP, the better. If serious mitigation is not implemented soon, prospects for the world economy are dismal.

---

[8] This sort of convergence only shows up over centuries. Over decades the real wage can be stagnant or fall, as in the USA since the 1980s.

REFERENCES

Ackerman, Frank and Elisabeth S. Stanton (2012), 'Climate Risks and Carbon Prices: Revising the Social Cost of Carbon', *Economics: The Open-Access, Open-Assessment E-Journal*, **6** (2012-10), 1–25.
Foley, Duncan K., Armon Rezai and Lance Taylor (2013), 'The Social Cost of Carbon Emissions: Seven Propositions', *Economics Letters*, **121**(1), 90–97.
Goodwin, Richard M. (1967), 'A Growth Cycle', in C.H. Feinstein (ed.), *Socialism, Capitalism, and Growth*, Cambridge: Cambridge University Press.
IPCC (2007), *Climate Change 2007: Mitigation of Climate Change. Contribution of Working Group III to the Fourth Assessment Report of the Intergovernmental Panel on Climate Change*, Cambridge and New York: Cambridge University Press.
IPCC (2014), 'Mitigation of Climate Change: Contribution of Working Group III to the Fifth Assessment Report of the Intergovernmental Panel on Climate Change', Cambridge and New York: Cambridge University Press.
Kemp-Benedict, Eric (2014), 'The Inverted Pyramid: A Post Keynesian View on the Economy–Environment Relationship', Bangkok: Stockholm Environmental Institute.
Pacala, Stephen W. and Robert M. Socolow (2004), 'Stabilization Wedges: Solving the Climate Problem for the Next 50 Years with Current Technologies', *Science*, **305**, 968–72.
Rezai, Armon, Duncan K. Foley and Lance Taylor (2012), 'Global Warming and Economic Externalities', *Economic Theory*, **49**, 329–51.
Taylor, L. (2009), *Reconstructing Macroeconomics: Structuralist Proposals and Critiques of the Mainstream*, Cambridge, MA: Harvard University Press.
Waggoner, Paul E. and Jesse H. Ausubel (2002), 'A Framework for Sustainability Science: A Renovated IPAT Identity', *PNAS*, **99**, 7860–65.

APPENDIX: FUNCTIONAL FORMS AND PARAMETERIZATION

Medium-term macroeconomic balance is determined by the equations

$$s(\pi) = s_w + (s_\pi - s_w)\pi, \quad (13A.1)$$

$$g(\pi) = g_0 + a\pi u, \quad (13A.2)$$

and

$$g(\pi) + h + mu - s(\pi)u - \tau u = 0 \quad (13A.3)$$

with h as the ratio of government spending to capital and τ as a tax rate.[9] In steady state (13A.1)–(13A.3) can be solved for π and u, given g; in the medium run for g and u, given π.

In the first medium-term adjustment scenario, we adopted the functional form

$$\pi(\lambda, G) = [\Phi Z(G)]^B \lambda^{-A} \quad (13A.4)$$

with

$$Z(G) = \left[1 - \left(\frac{G - 280}{\overline{G} - 280}\right)^{1/\eta}\right]^\eta. \quad (13A.5)$$

Here, $Z(G)$ is a concave decreasing damage function carried over from a supply-driven climate model constructed by Rezai et al. (2012) with 280 ppmv as pre-industrial CO_2 concentration and $\overline{G} = 780$ as a level at which extremely severe climate damage occurs. Figure 13A.1 illustrates the damage function for various values of the parameter η. After experimentation with sensitivities, in simulations we set $\eta = 0.5$.

The second scenario uses the equation

$$\delta = 0.016 + 0.00009G$$

which sets $\delta = 0.052$ at $G = 400$ and $\delta = 0.07$ at $G = 600$. The effect of labor market tightness (and capital per capita) on the profit share is given by the S-shaped relationship

$$\pi(\lambda) = 0.2 + 0.4/[1 + e^{\psi(\lambda - \lambda_0)}] \quad (13A.6)$$

which implies that π decreases from 0.6 to 0.2 as λ increases.

The functional form for decreasing effectiveness of mitigation in (13.1) is

$$\mu(m) = \psi \frac{1 - e^{-\phi m}}{\phi}. \quad (13A.7)$$

---

[9] One has to carry government spending and taxes in the accounting to fit the data and generate plausible multiplier values.

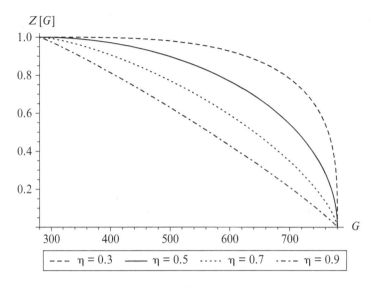

Figure 13A.1 Damage function for the profit share

For equations (13A.1)–(13A.3), world output or GDP is set at roughly $X = 60$ (trillion dollars). With $K = 200$, $u = 0.3$. With the share of government spending in output $H/X = 0.33$, $h = 0.099$. If the fiscal deficit is normally 3% of GDP, then $\tau = 0.3$. A plausible level of the world saving rate is $s(\pi) = 0.24$. If $m = 0$ initially, then the investment/capital ratio becomes $g = 0.063$. The profit share of output is roughly $\pi = 0.4$. If the saving rate from profits is $s_\pi = 0.28$ then the rate from wage income becomes $s_w = 0.2133$. With $\alpha = 0.25$, $g_0 = 0.033$ supports the investment function. Aggregate demand will be profit-led if $\alpha + s_w - s_\pi > 0$, a condition satisfied by these numbers.

For use in (13A.4)–(13A.5), we set world employment $L = 3$ (billion) and population $N = 7$ so that $\lambda = 0.42857$. With current output of 60, $\xi = 20$. The capital/population ratio is $\kappa = 200/7 = 28.5714$ (or $28,571 per capita). Together with $\eta = 0.5$ we set $A = 2$ and $B = 2$. The parameter Φ was "calibrated" to fit $\pi = 0.4$, taking the value of 0.2792.

In (13A.6) we set $\psi = 18.66$, broadly consistent with observed ranges of the profit share.

In (13.1), it is simplest to think of energy use in terms of terawatts of power (as opposed to exajoules of energy per year). The current world level is about 15 terawatts, of which 12 are provided by fossil fuels. Fossil fuel energy productivity becomes $\varepsilon = 60/12 = 5$. This energy use generates about seven gigatons of carbon emissions per year, corresponding to an increase in G of 3.37 ppmv. The observed increase is about 2 ppmv, so that $\hat{G} = 2/400 = 0.005$ with atmospheric dissipation of 1.37 ppmv. The dissipation coefficient becomes $\omega = 1.37/400 = 0.0034$.

Assuming that there is now no effective mitigation or $\mu(m) = 0$ the growth rate of G becomes

$$\hat{G} = (\chi/\varepsilon)\Gamma - \omega.$$

With emissions of 3.37 and fossil fuel energy use of 12, the ratio $\chi = 3.37/12 = 0.2808$ and $v = \chi/\varepsilon = 0.0562$. The balance equation for $\hat{G}$ works out to be

$$0.005 \approx (0.0562)(0.15) - 0.0034 = 0.0084 - 0.0034.$$

At present, G increases at a slower rate than κ.

In (13A.7) if 1% of output (or $0.6 trillion) is devoted to mitigation ($m = 0.01$) we have

$$\frac{1 - e^{-0.06}}{6} = 0.0097.$$

That is, the cost-effective outlay is $(0.6)(0.97) = 0.582$.

A fairly high estimate of the cost of removing one ton of carbon emissions is $160, or $44 per ton of CO_2 (roughly twice the level now being considered by the government in the USA, according to Ackerman and Stanton, 2012). To reduce atmospheric CO_2 concentration by 1.0 ppmv would require removal of 2.07 gigatons of carbon from emissions at a total cost of $(2.07)(\$0.16\text{ trillion}) = \$0.331$ trillion. Spending 1% of output would mitigate $0.582/0.331 = 1.7583$ ppmv. So, we get the change in emissions as $-\Delta G = 1.7583 = 0.582\psi$ or $\psi = 3.0211$.

In the Solow–Swan variant there is a neoclassical aggregate production function (Cobb–Douglas for simplicity)

$$X = Z(G)A(\xi\lambda N)^{0.6}K^{0.4}$$

with A as a calibration parameter. The employment/population ratio λ is no longer an adjusting variable but stays fixed at its base year level (a full employment assumption) and the damage function $Z(G)$ is the same as in the demand-driven version. If $v = X/N$ this equation becomes

$$v = Z(G)A(\xi\lambda)^{0.6}\kappa^{0.4}.$$

Replacing (13.3) the growth equation for κ is

$$\dot{\kappa} = (s + \tau - m)v - (h + n + \delta)\kappa$$

which can be solved using the base year values of λ and G along with the model's equations for $\dot{\xi}$ and $\dot{N}$.[10]

---

[10] Textbook presentations set $\kappa = K/\xi L$ but we avoid this specification to maintain comparability with the demand-driven model.

14. Growth and sustainability
Robin Hahnel

INTRODUCTION

What are we to make of statements like: "Anyone who believes that exponential growth can go on forever in a finite world is either a madman or an economist."[1] What are we to make of pleas to substitute the goal of a "steady state economy" for the traditional economic goal of increasing growth of production? What are we to make of the de-growth movement which argues that we must actually reduce output to make our economies environmentally sustainable? The key to clear thinking on these subjects is understanding the difference between environmental throughput and the economic "goods" we produce.

Ecological economists define *environmental throughput* as physical inputs from the natural environment (traditionally described as "raw materials") used in production processes such as iron ore and top soil; as well as physical outputs of production (usually thought of as waste or pollution) such as airborne particulate matter and greenhouse gases released back into natural environment where they are absorbed by what are now called natural "sinks." Throughput must be measured in some appropriate physical units such as tons of iron ore, cubic meters of top soil, pounds of particulate matter, and cubic tons of carbon dioxide.

But environmental throughput is not the same as the quantity of "goods" we produce. We produce potatoes, shoes, and steel, which we measure in appropriate units such as pounds of potatoes, pairs of shoes, and tons of steel.

If we are careful to interpret the above warnings to be referring to growth of environmental throughput they can be very insightful. On a planet where the quantity of nature available for throughput is finite, infinite growth of throughput is, indeed, impossible – if that is what Kenneth Boulding meant to say. Since it is increasingly apparent that many kinds of throughput have become so large that their continued growth is environmentally unstable, Herman Daly's call to strive to maintain *throughput* at a steady state, rather than seek to increase throughput, is sage advice – if that is what Daly meant to say. And since we know that for some parts of heterogeneous nature, such as the storage capacity for greenhouse gases in the atmosphere, maintaining throughput at present levels will prove disastrous, calling for de-growth for some kinds of *throughput* like carbon emissions is nothing more than simple sanity – assuming that is what those in the de-growth movement are calling for.

On the other hand, as the model in this chapter demonstrates, if these warnings are interpreted as referring to growth of the quantity of goods we produce, they do not follow logically and can be quite misleading. When we are careful to distinguish between

---

[1] Attributed to Kenneth Boulding in United States Congress, House of Representatives, *Energy Reorganization Act of 1973: Hearings, Ninety-Third Congress, First Session, on H.R. 11510*, p. 248.

environmental throughput and goods produced it becomes apparent that it is possible for the latter to grow while the former does not.[2]

After sorting out the "sense" from the "nonsense" in steady-state and de-growth economics, we return to the real issue: How can we continue to enjoy increases in the quantities of goods we produce without destroying the environment – not only in theory, but in a world where nature's bounty has been severely reduced, nature's many services are "heterogeneous" in meaningful ways, and some natural resources are non-reproducible?

THE SRAFFA MODEL

A model elaborated by followers of Piero Sraffa is very useful for our purposes. For rigorous proofs of all the propositions in this section see Hahnel (2017). Assume the square matrix of produced input coefficients, **A**, is non-negative, indecomposable, and productive. Assume the row vector of direct, hourly, labour input coefficients, **L**, is strictly positive. If we assume a uniform rate of profit in all industries, and assume that employers must pay only for produced inputs in advance, we can write the price and income equations for the economy as: $(1+r)\mathbf{pA} + w\mathbf{L} = \mathbf{p}$ where r is the uniform rate of profit, w is the hourly wage rate, and **p** is the row vector of relative prices for produced goods. It is well known that:

- For all permissible values of (r,w) there exists a vector of all positive relative prices for the income-price system: $(1+r)\mathbf{pA} + w\mathbf{L} = \mathbf{p}$.
- This vector is unique for all permissible values of (r,w), but changes, in general, when we change from one permissible combination of (r,w) to another.
- There is a negative relationship between the two distributive variables. In other words, when either r or w rises, the value of the other distributive variable must fall.
- If the column output vector for the economy, **x***, is an eigenvector of **A**, i.e. if we produce output in proportions equal to what Sraffa called his "standard commodity," then the negative relation between the wage and profit rates is linear, and can be written as: $1 = w + (1\backslash R)r$ – where R is the maximum value for r corresponding to a wage rate of zero, and for convenience the amount of labour applied, **Lx**, has been set equal to one unit, and the relative price vector has been normalized so the value of net output, $\mathbf{p}[\mathbf{I}-\mathbf{A}]\mathbf{x}$, is equal to one.

What if, in addition to homogeneous labour, a second homogeneous, primary, non-produced input from "nature" is needed for production? Assume h(j) is the number of units of "nature" needed directly to produce a unit of j, and **H** is a strictly positive row vector of nature input coefficients. In other words, h(j) and **H** are completely analogous to l(j) and **L**, with one exception. When l(j) hours of labour are used in production we do not think of the worker as being diminished by l(j) hours. We talk about disutility of labour, but we do not usually model this as a deletion from a fixed stock of labour hours

---

[2] See Hahnel (2013) for a more extensive analysis of common confusions in discussions about the relationship between economic growth and environmental sustainability.

the worker represents. No matter how much, or little, a person works today, the number of hours she can work tomorrow is the same. But in the case of an input from nature we will assume that when h(j) units of nature are used in production this does diminish nature by h(j) units. In truth what matters is the extent to which using h(j) units of nature in production causes nature to deteriorate in some way, which can be quite complicated and difficult to measure quantitatively. But for convenience we will think of the extent of deterioration as being representable by some number of units of nature that are "used up" in production, and therefore deleted from the existing "stock of nature."

Just as it does not matter what unit we use to measure inputs of homogeneous labour – we have used hours, but could just as well have used days, or years – it does not matter what unit we use to measure inputs from homogeneous nature used up in production. The problem is that while appropriate units come to mind whenever we talk about some particular part of nature such as oil, soil, old growth timber, airborne particulate matter, greenhouse gases, etc., there is no conventional unit of measurement that comes to mind when we think about nature as a homogeneous input in human production processes. When faced with a similar dilemma, others writing about "ecological footprints" use an "acre" as their unit of measurement. But since this can lead to confusion we use a "green" instead as our name for a unit of homogeneous nature.

Again if we assume that only produced inputs must be paid for in advance, i.e. that both wages (for hours of labour used), w, and rent (for greens of nature used up), m, can be paid for out of revenues at the end of the production period, we can write the price and income equations for the economy as: $(1+r)\mathbf{pA} + w\mathbf{L} + m\mathbf{H} = \mathbf{p}$. In this case it can be shown that:

- For all permissible values of (r,w,m) there exists a vector of all positive relative prices for the income-price system: $(1+r)\mathbf{pA} + w\mathbf{L} + m\mathbf{H} = \mathbf{p}$.
- This vector is unique for all permissible values of (r,w,m), but changes, in general, when we change from one permissible combination of (r,w,m) to another.
- There is a negative relationship between all three distributive variables.
- If the column output vector for the economy, $\mathbf{x}^*$, is an eigenvector of $\mathbf{A}$, i.e. if we produce output in proportions equal to Sraffa's "standard commodity," then the negative relation between r, w, and m is linear and can be written as: $1 = w\mathbf{Lx}^* + m\mathbf{Hx}^* + (1\backslash R)r$ where again for convenience $\mathbf{Lx}$ and $\mathbf{p[I-A]x}$ have each been set equal to one.

Moreover, this price-income system can easily be generalized to account for both heterogeneous labour and heterogeneous non-labour primary inputs from nature. In the Sraffa framework to account for heterogeneous labour we simply make our row vector of labour input coefficients, $\mathbf{L}$, into a matrix with as many rows as we have different kinds of labour, and we make our hourly wage rate, w, into a row vector, $\mathbf{w}$, of hourly wage rates for each category of labour. To account for all of the heterogeneous, primary inputs from nature needed for production we simply make our row vector of "nature" input coefficients, $\mathbf{H}$, into a matrix with as many different rows as we have different kinds of primary inputs from nature that are needed for production – each measured in its own appropriate physical units. And the row vector, $\mathbf{m}$, becomes a vector of the rental rates per unit of each category of nature used up in production. In which case:

- The negative linear relations among *all* of the distributive variables becomes: $1 = $ **wLx*** + **mHx*** + (1\R)r where again for convenience **Lx** and **p**[**I**−**A**]**x** have each been set equal to one.

In sum, this is how the Sraffa model explains the relations among distributive variables in capitalist economies, and how it explains relative price determination for any permissible values for the distributive variables, (**w**, **m**, r) and any given technologies, {**A**, **L**, **H**}. But what happens when capitalists discover new ways to produce things? For example, what happens when capitalists in industry j can choose between continuing to use [**a**(j), L(j), H(j)], or use [**a**(j)', L(j)', H(j)'] instead?

To know whether capitalists will replace an existing technique with a new technique we simply compare the cost of producing a unit of output using the old and new technologies at current prices, wages, and rents: If **pa**(j)'+ wL(j)' + mH(j)' < **pa**(j) + wL(j) + mH(j) profit maximizing capitalists in industry j will adopt the new technology, and otherwise they will not. Technological changes which lower production costs at current prices and values for distributive variables are often called *viable*.

TWO KEY THEOREMS

Technical change and labour productivity: It is well known that if we only wish to know if a technical change will raise or lower labour productivity, i.e. how it will affect labour productivity *qualitatively*, all we need to do is compare labour values in the economy before and after a technical change is introduced. The vector of labour values **V** = **L**[**I**−**A**]$^{-1}$ tells us how many hours of labour it takes, both directly and indirectly, to produce one unit of each good in the economy. Therefore, by comparing **V** = **L**[**I**−**A**]$^{-1}$ to **V'** = **L'**[**I**−**A'**]$^{-1}$ we can immediately tell if a technical change has increased or decreased labour productivity, and therefore the quantity of each good we will produce per hour we work. If **V'** ≤ **V** it now takes less labour to produce at least one good and no more labour to produce any good than it used to, so labour productivity has definitively increased, and if hours worked in each industry remain the same, we will produce more units of at least one good and no fewer units of any good. Conversely, if **V** ≤ **V'** labour productivity will decrease and output will either decrease or remain constant in every industry assuming hours worked in every industry remain constant. John Roemer (1981) called technological changes where **V'** ≤ **V** *progressive* and changes where **V** ≤ **V'** *retrogressive*.[3]

However, comparing labour values before and after any technical change does not tell us how much labour productivity has changed. The following result, proved as part iii of a theorem on dominant eigenvalues, profitability, and productivity in Hahnel (2016a),

---

[3] For a single technical change either (i) **V'** ≤ **V**, (ii) **V** ≤ **V'**, or (iii) **V** = **V'**. This is because in the industry where the change took place either v(j) fell, rose, or remained the same. If it fell and it enters into the production of another good, either directly or indirectly, the value of that good must also fall since there was no change in its input coefficients. If it does not enter into the production of another good, either directly or indirectly, the value of that good will stay the same. Similarly, if v(j) increases, the value of other goods must either increase or remain the same. Finally, if v(j) itself does not change, then no values will change.

explains how to calculate the size of changes in labour productivity in the economy as a whole stemming from any technological change.

Theorem on rate of change of labour productivity: For $\tilde{b}$ sufficiently large such that ß, the dominant eigenvalue of $[A+\tilde{b}L]$, is equal to one, the increase in overall labour productivity when $\{A, L\}$ is replaced by $\{A', L'\}$ is $\rho(l) = (1-ß')$ where ß' is the dominant eigenvalue of $[A'+\tilde{b}L']$.

Suppose, for example, we find that ß', the dominant eigenvalue for the new socio-technology matrix $[A'+\tilde{b}L']$, equals 0.97. In this case labour productivity has *increased* by $\rho(l) = (1-ß') = (1.00 - 0.97) = 0.03$, or 3 per cent. If people next year work the same number of hours as they did this year in every industry they will produce 3 per cent more of every good. Or people next year could consume exactly what they consumed this year while working 3 per cent fewer hours in every industry. On the other hand, suppose we find that ß', the dominant eigenvalue for the new socio-technology matrix $[A'+\tilde{b}L']$, equals 1.05. In this case labour productivity has *decreased* by $\rho(l) = (1-ß') = (1.0-1.05) = -0.05$, or negative 5 per cent, and people will either produce 5 per cent less of every good if they work the same number of hours in every industry or have to work 5 per cent more hours in every industry to produce the same amount of every good as before. For now, we will assume that hours worked in every industry do not change, i.e. that all increases in labour productivity are taken in the form of increases in production rather than as an increase in leisure.

Technical change, throughput, and throughput efficiency: At this point we need to distinguish between technical change that reduces the amount of labour used to produce goods, i.e. that increases labour productivity and the output of goods, and technical change that reduces the amount of nature "used up" producing goods, i.e. that increases throughput efficiency.

Just as the number of hours of labour used both directly and indirectly to make a unit of every good can be calculated as $V = L[I-A]^{-1}$, the number of greens of nature used up both directly and indirectly to make a unit of every good can be calculated as $N = H[I-A]^{-1}$. So, we can discover whether or not any technical change has increased or decreased throughput efficiency by comparing $N = H[I-A]^{-1}$ and $N' = H'[I-A']^{-1}$ If $N' \leq N$ throughput efficiency has increased. If $N \leq N'$ throughput efficiency has decreased.

However, the following result, proved in Hahnel (2016b), explains how to calculate the size of changes in throughput efficiency in the economy as a whole, stemming from any technological change in any industry.

Theorem on changes in throughput efficiency: For $\tilde{d}$ sufficiently large such that α, the dominant eigenvalue of $[A+\tilde{d}H]$, is equal to one, the size of the increase in throughput efficiency when $\{A, H\}$ is replaced by $\{A', H'\}$ is $\rho(n) = (1-α')$ where α' is the dominant eigenvalue of $[A'+\tilde{d}H']$.

So, if α' = 0.98 throughput efficiency will have increased by 2 per cent, while if α' = 1.04 throughput efficiency will have decreased by 4 per cent.

It is useful to distinguish between some different kinds of technical changes, and pause and review where we are: There are: (1) capital-using, labour-saving (CU-LS) technical changes: $A \leq A'$ and $L' \leq L$, (2) capital-saving, labour-using (CS-LU) technical changes: A'

≤ A and L ≤ L', (3) capital-using, nature-saving (CU-NS) technical changes: A ≤ A' and H' ≤ H, and (4) capital-saving, nature-using (CS-LU) technical changes: A' ≤ A and H ≤ H'.

1. We can only calculate a single measure of throughput efficiency, $\rho(n)$, when we have a single primary input from nature. While unfortunate, this is hardly surprising because we can only calculate a single measure of increases in labour productivity, $\rho(l)$, when labour is presumed to be homogeneous. Multiple primary inputs from nature render it impossible to calculate a single measure of increases in throughput efficiency – just as heterogeneous labour renders it impossible to calculate a single measure of increases in labour productivity.[4]
2. Just as we have to be careful not to confuse a reduction in a direct labour input coefficient, l(j), with a reduction in the total amount of labour used both directly and indirectly to make a unit of j, v(j); we must not confuse a reduction in a direct nature input coefficient, h(j), with a reduction in the total amount of nature used up both directly and indirectly to make a unit of j, n(j). It is possible that a capital-saving, nature-using (CS-NU) technical change might lower the total amount of nature used up to make commodities even though it increases the amount of direct nature used up. In other words, it is **N**, not **H** that we should care about, just as it is **V**, not **L** that matters.
3. Reductions in l(j)'s improve only labour productivity without affecting throughput efficiency, and reductions in h(j)'s improve only throughput efficiency without affecting labour productivity. On the other hand, any reduction in an a(ij) will improve both labour productivity and throughput efficiency. However – and this is very important – any capital using, labour-saving (CU-LS) technical change will necessarily reduce throughput efficiency, and any capital using, nature-saving (CU-NS) change will necessarily reduce labour productivity.[5]
4. Finally, it is worth considering what happens when capitalists choose technologies in a context where throughput from nature is under-priced. Assume in the extreme that the price of using nature is zero. In this case, there is no incentive for capitalists to choose pure nature-saving technologies (NS), much less capital-using, nature-saving technologies, CU-NS. Worse still, when capitalists discover viable CU-LS technologies they will adopt them without fail. But since they are CU, i.e. they use more of some a(ij)'s, they necessarily use more nature *indirectly* as well: Viable CU-LS changes will make **N'** > **N** and $\rho(n)$ negative.[6] There is every reason to believe that a great deal of technical change during the past few hundred years implemented by profit maximizing capitalists did just this. Certainly, in the case of carbon emissions

---

[4] Although we *can* calculate how much throughput efficiency increases for each component of heterogeneous nature individually, as explained below.
[5] Economic growth is often associated with equipping labour with more "capital" goods to work with. While pursuing growth via "capital deepening" is traditionally thought of as substituting more "capital" for less labour, what we have just discovered is that it is also, and perhaps more importantly, substituting more nature for less labour.
[6] The only circumstance under which a CU-LS technical change might not increase **N** and therefore make $\rho(n)$ negative is if it also just happened to be NS as well, i.e. if it was in fact a CU-LS-NS change.

where the price charged for carbon "throughput" has long been zero, there was no incentive to economize on carbon throughput, and whenever capitalists discovered and implemented viable CU-LS changes they necessarily increased carbon emissions indirectly and thereby decreased carbon throughput efficiency. This phenomenon may help explain why we are now facing the possibility of cataclysmic climate change because we have overstocked the upper atmosphere with CO_2.

5. As explained below, the Sraffa framework greatly facilitates defining sufficient conditions for environmental sustainability and determining under what conditions sustainability is consistent with expanding production. Unfortunately, it is not of much help in another crucial task: estimating the social cost of using nature. However, the Sraffa framework does confirm the value of Pigovian taxes as a policy tool. The failure to charge for emission of GHGs is arguably the single greatest "mistake" in our price system. It has led to a dysfunctional bias in what kinds of new technologies have been adopted and rejected over the past three centuries which is now disrupting the climate at an alarming rate. While there is a great deal of debate about how high the "social cost of carbon" is, we know that it is not zero. Yet when profit maximizing capitalists in some industry j use the criterion: **pa**(j)' + **w**L(j)' + **m**H(j)' < **pa**(j) + **w**L(j) + **m**H(j) to decide whether or not to replace an existing technology {**a**(j), L(j), H(j)} with a new technology {**a**(j)', L(j)', H(j)'} the number they use for m(c) – the price of a metric ton of GHG emissions which is part of the vector of prices for different inputs from nature, **m** – is zero. Estimates of the social price of carbon are hotly disputed. The US Environmental Protection Agency's estimates range from a low of $11 for a metric ton released in 2015 to a high of $105.[7] The Sraffa framework confirms what economists have known dating back to the work of A.C. Pigou in the early twentieth century: If you want capitalists to serve the social interest when making technological choices you need to levy a "Pigovian tax" on GHG emissions equal to the "social cost of carbon." Otherwise they will continue to make socially counterproductive decisions in this regard.

SUFFICIENT CONDITIONS FOR ENVIRONMENTAL SUSTAINABILITY

Assume there are only two primary inputs, homogenous labour, measured in hours, and homogeneous nature, measured in greens. For convenience also assume that the size of the labour force and number of hours worked in every industry remains the same year after year. We assume that nature consists of a certain number of greens, <u>GR</u>, which is initially just sufficient to permit full employment of the labour force. In which case, if production uses up any greens at all it is impossible to define an environmentally sustainable steady state unless nature also regenerates. So, for now we assume that nature regenerates a certain number of greens per year, REG.

The first condition for sustainability is that the number of greens used up as inputs in production during a year, i.e. "nature throughput," cannot exceed the number of greens

---

[7] https://www3.epa.gov/climatechange/Downloads/EPAactivities/social-cost-carbon.pdf.

regenerated during a year. Otherwise there will not be enough greens of nature to allow for production to continue at the same level as the previous year. **N** is our row vector representing the number of greens of nature needed directly and indirectly to make a unit of each produced good in the economy. So, if **x** is the vector of produced outputs, **Nx** represents throughput, the number of greens subtracted from GR because we produced the gross output vector **x** this year. To prevent GR from shrinking, we need **Nx** ≤ REG.[8]

But even if the labour force is not growing what if labour productivity is growing? If labour productivity increases the same number of hours worked in every industry next year will produce a larger **x** than this year. In order to prevent throughput from exceeding regeneration and rendering the economy environmentally unsustainable **N** must decrease. As we have seen, how much **x** rises due to technical changes that increase labour productivity can be represented by a single number, $\rho(l)$, and how much **N** shrinks due to technical changes that increase throughput efficiency can be represented by a single number as well, $\rho(n)$. Provided the number of hours worked does not change, as long as $\rho(n) = \rho(l)$ throughput will not rise but remain constant. In sum:

- The first condition for environmental sustainability is **Nx** ≤ REG. This establishes the level of throughput we must not surpass to maintain environmental sustainability.
- The second condition for sustainability is $\rho(n) = \rho(l)$. This keeps throughput from rising above REG even as labour productivity and the quantities of goods produced increase.

If either condition is violated the economy will become environmentally unsustainable.

IMPLICATIONS FOR STEADY-STATE AND DE-GROWTH ECONOMICS

If we keep discovering new technologies that increase labour productivity output per hour can continue to increase. That's what $\rho(l) > 0$ means. For hundreds of years we have proven capable of finding new technologies that improve our ability to produce more goods per hour. The pace of technological change that increases labour productivity and output may slow down or speed up in the future, as it has in the past, but there is no reason to believe it cannot continue to increase indefinitely.

But will increases in labour productivity prove to be environmentally unstable? Clearly not if we choose to take all increases in productivity in the form of more leisure. If we continue to produce the same vector of outputs, **x**, and simply do so working fewer hours, we do not increase strain on the environment. But what we have just demonstrated is that even if we continue to work the same number of hours as labour productivity grows, and

---

[8] Environmental sustainability is only of concern when nature is no longer infinite in size compared to the throughput a fully employed labour force would produce. We are assuming we have left the "empty" world where **Nx** <<< GR where it would be of no concern if **Nx** exceeds REG.

we therefore produce ever more output, throughput will not increase as long as $\rho(n) \geq \rho(l)$.

Consider a worst-case scenario: Assume that we take none of our increased productivity in the form of leisure, and we increase every component of **x** in the same proportion. So $[1 + \rho(l)]\mathbf{x} = \mathbf{x'} > \mathbf{x}$, and there is no possibility of substituting less throughput intensive goods for more throughput intensive goods. Even so, as long as $\rho(n) = \rho(l)$ **N'x'** will be equal to **Nx**, and therefore **x'** will tread no more heavily on the environment than **x** did. So, at least in theory, it is possible for hours worked to remain constant, labour productivity and output to rise, and throughput to remain constant *provided throughput efficiency rises as fast as labour productivity*. As should now be clear, it all boils down to the relationship between $\rho(n)$ and $\rho(l)$, exactly as those who have focused on "decoupling" claim. Assuming no change in hours worked or the composition of output:

- As long as **Nx** < REG, $\rho(l)$ can exceed $\rho(n)$ until throughput, **Nx**, reaches REG.
- Once **Nx** = REG, $\rho(n) = \rho(l)$ is sufficient to maintain environmental sustainability.
- If **Nx** > REG, $\rho(n) > \rho(l)$ is required to re-achieve environmental sustainability.

Nothing said here should be interpreted to deny that taking more of our productivity increases in the form of leisure rather than additional consumption will be an important part of achieving environmental sustainability in practice. Juliet Schor (1993 and 1999) has done a great deal to draw attention to the astounding fact that on average Americans worked more hours at the end of the twentieth century than we did forty years earlier, even though we were almost twice as productive. Moreover, there is now a great deal of empirical research suggesting that further increases in average consumption in the advanced economies is no longer yielding significant increases in happiness or wellbeing. In which case, social policy should be concentrating on shifting material consumption from those at the top to those at the bottom of the income distribution rather than increasing average consumption. Nor should anything said here be construed to imply that substituting less throughput intensive components for more throughput intensive components in our output vector, **x**, will not be a crucial part of achieving environmental sustainability.

However, if the modelling of environmental sustainability which the Sraffa framework facilitates can help clarify issues, eliminate misperceptions and reduce miscommunications that have plagued debate about the relationship between growth and sustainability it can be useful. Moreover, there are clear strategic and political implications: If lower middle-class workers in the advanced economies come to think environmentalists are telling them that they must abandon hopes for a higher standard of living for their children in order to save the environment, they may be reluctant to become supporters. And if billions living in less developed economies who have yet to enjoy the benefits of economic development come to think environmentalists are telling them that they need to give up all hope of achieving economic development if the environment is to be saved, they may be reluctant to become supporters as well. What the above analysis demonstrates clearly is that environmental sustainability need not be incompatible with producing more goods and therefore increases in what is commonly referred to as "material living standards." In which case, calls for an end to growth or de-growth to save the environment *which give this impression* are misleading and unnecessary as well as politically self-defeating.

APPLYING CONCLUSIONS TO THE REAL WORLD

Of course, the assumption that nature is homogenous and its deterioration can be measured along a single dimension in units of "greens" is grossly unrealistic. Not only is the assumption unrealistic, it prevents us from exploring the beneficial effects of substituting one part of nature that is less scarce for another part that is more scarce to enhance environmental sustainability.[9] What can be said when nature is heterogeneous in meaningful ways?

No doubt many readers have engaged in the exercise which translates one's consumption behaviour into an ecological "footprint" represented as a number of "acres," and then informs you how many "planet earths" would be needed if everyone else should tread on "mother nature" as heavily as you. This may well be a useful tool for raising consciousness. However, precisely because nature is heterogeneous in meaningful ways, the ecological footprint exercise can be grossly misleading if interpreted as a useful guide to policy. Consider three components of nature: she provides sink services for storing greenhouse gases in the upper atmosphere, fresh water and sand. Just as I have a "carbon footprint" we can measure in metric tons of carbon dioxide equivalents, I also have a "fresh water footprint" we can measure in gallons and a "sand footprint" we can measure in tons. However, unless we know whether nature, and therefore humanity, is going to run out of greenhouse gas storage capacity, fresh water, or sand first, we don't know which of my footprints is causing more environmental damage and therefore treading more heavily on the environment. The answer depends on which part of heterogeneous nature is being exhausted more rapidly – the upper atmosphere, water, or sand – as well as on which parts of nature we are more likely to find substitutes for. Moreover, failure to model nature as heterogeneous prevents us from exploring the beneficial effects of substituting one part of nature that is less scarce for another part that is more scarce.

Fortunately, the Sraffa framework can still be helpful in the case of heterogeneous nature, whether it be different "natural resources" or different "natural sink services" that nature provides. It now appears that we may be exhausting nature's ability to store particular material "outputs" of production processes faster than we are using up particular raw material "inputs" from nature. However, there is no reason we cannot include multiple harmful *additions* to the environment (different emissions or pollutants) along with multiple harmful *deletions* from the environment (different raw materials) associated with production processes in our linear production technologies. The easiest way to do this without need to resort to the complication of joint products is to treat the output of p units of a pollutant as an input from a sink service which reduces the storage capacity of the sink by p units.[10]

---

[9] The advantage of being able to rigorously model and measure increases in labour productivity which the unrealistic assumption of homogenous labour makes possible also came at a cost: We cannot use a model with homogeneous labour to explore the beneficial effects of increases in efficiency from technical changes that substitute one kind of labour that is less scarce or more desirable for another kind that is more scarce or less desirable.

[10] Traditionally, economists thought of a "service" from nature as a natural resource "input" like iron ore, but there is no reason the service cannot be storage of an emission like CO_2. As long as iron ore used up (tons deleted from scarce reserves) and CO_2 emissions (metric tons subtracted from available storage capacity in the upper atmosphere) are both listed as inputs in our "recipes"

While the Sraffa framework cannot tell us which parts of nature's many services are at greater risk, and therefore what our policy priorities should be, it provides a way to rigorously measure throughput and increases in throughput efficiency for individual environmental services, at least in theory. As we have seen, simply by turning **H**, the vector of direct nature input coefficients into a matrix with as many different rows as we use different kinds of services from nature, in theory we can measure throughput for individual components of heterogeneous nature and calculate changes in throughput efficiency for each individually. In other words, in the Sraffa framework we can define and rigorously calculate carbon throughput and increases in carbon throughput efficiency, water throughput and increases in water throughput efficiency, and sand throughput and increases in sand throughput efficiency.

Since it is clear that human economic activity is exhausting some parts of nature much faster than others, this is of great practical importance. For example, scientists who have expertise in such matters tell us we need to reduce carbon throughput by more than 90 per cent by 2050 to avoid an unacceptable risk of triggering cataclysmic climate change. While there is good reason to worry about fresh water supplies, most estimate that this problem is not reaching crisis proportions as quickly. In contrast, we can probably increase throughput of sand used to make adobe bricks and concrete for centuries to come. Fortunately, we can measure throughput and increases in throughput efficiency for individual components of heterogeneous nature in the Sraffa framework just as easily as when we pretended that nature was homogeneous and could be measured in greens. In which case "rules" for treating the environment responsibly become:

- For any component of heterogeneous nature (such as sand) for which current levels of throughput are not yet problematic, $\rho(n)$ can fall short of $\rho(1)$ for the time being.
- For any component of heterogeneous nature (perhaps such as water) for which throughput should not rise, $\rho(n)$ must be at least as high as $\rho(1)$.
- For any component of heterogeneous nature (such as greenhouse gas storage capacity) for which throughput must decrease, $\rho(n)$ must exceed $\rho(1)$.[11]

Finally, what can we say about environmental sustainability when we recognize not only that nature is heterogeneous, but also that some of nature's heterogeneous "services" are, for all intents and purposes, non-reproducible, i.e. for these services REG = 0? It is true that we can search for more iron ore, go to greater lengths to extract it from places once thought impenetrable, and make do with lower quality ores. Nonetheless, there is a difference between iron ore and trees which has long been recognized when we classify trees

---

for producing different goods and services, we can calculate throughput and changes in throughput efficiency for both iron ore and CO_2 individually without resort to treating pollutants as joint products. The Sraffa framework is admirably suited to handling joint products which is required for treating fixed capital. But a joint product version of the Sraffa framework is more complicated and is not needed for purposes of measuring throughput, even when throughput takes the form of output "wastes" released into natural sinks.

[11] These are what we might call "worst-case" rules since they assume we take none of our increases in labour productivity as leisure, and we do not change the composition of output to substitute less throughput intensive goods for more throughput intensive goods.

as a reproducible natural resource and iron ore as a non-reproducible natural resource. Strictly speaking the existence of non-reproducible natural resources required for production means that environmental sustainability is impossible. But this does not mean that a sustainable *strategy* is impossible. And we have already begun to describe what such a sustainable "coping strategy" looks like.

Just as we need to search for new technologies that increase environmental throughput efficiency in general, we need to search for new technologies that substitute renewable throughput for non-renewable throughput, and non-renewable throughput that is farther from exhaustion for non-renewable throughput nearing exhaustion. Similarly, just as we need to adjust our output vector **x** to substitute less throughput intensive goods for more throughput intensive goods, we need to adjust **x** to substitute throughputs from nature that regenerate for ones that do not, and we need to adjust **x** to substitute throughputs from nature that may be non-reproducible but are farther from exhaustion for non-reproducible services that are nearing exhaustion.

CONCLUDING OBSERVATIONS

Is all this "shucking and jiving" pointless if sustainability is ultimately impossible? *Not at all!* Even in a worst-case scenario in which there is a non-reproducible service from nature that proves to be impossible for human production activities to do without – i.e. that remains "basic" because no change in technology or adjustment in **x** permits us to do without it altogether – "ultimately" can be a very, very long way off. And in a "better case" scenario we may be able to keep eliminating the need for each non-reproducible service from nature before it is exhausted . . . endlessly.

Moreover, focusing on all this only distracts us from the challenge at hand. Right now, the challenge is to change technologies to increase carbon throughput efficiency and adjust **x** to eliminate the use of all fossil fuels within a few short decades *because we have already exhausted nature's capacity to store greenhouse gases in the upper atmosphere.* The urgency of focusing on this task cannot be over emphasized. Immediate, dramatic, sustained decreases in GHG emissions is of paramount importance.

Hopefully this chapter will have contributed to clarifying the relationship between economic growth and sustainability. But while the primary focus here was not on policy responses to climate change, it would be remiss not to comment on climate policy, even if briefly, in closing.

At a theoretical level the policy solution can appear to be quite simple: Find out what the true social cost of carbon is and charge all emitters a Pigovian tax equal to that amount. And because the Sraffa framework used in this chapter functions at a very high theoretical level, it can lend credence to this response. However, besides the fact that estimating the true social cost of carbon is not an easy task, it would be a tragic mistake to limit our policy response to this single policy tool. Yes, of course we need to "get prices right" in order to prevent predictable dysfunction. But we need to do other things as well or our response will prove insufficient for several reasons.

- Some propose a global "coordinated" carbon tax – set as close to the best estimate of the social price of carbon as can be achieved through international negotiation –

as the most promising approach to international climate change policy (Weitzman, 2013). However, because this policy would impose equal burdens on all countries it would be grossly unfair to countries with lesser responsibility and capability. Moreover, because a uniform global carbon tax would be unfair to many countries, there is good reason to believe they would not implement it. In short, a carbon tax is probably not the best approach for international climate policy. A better approach is to negotiate mandatory national emissions reductions according to differential responsibility and capability, and then allow countries to trade emission reduction credits if they wish (Hahnel, 2012a, 2012b, 2014).
- Real economies do respond to changes in price signals, but the response takes time. Having the wrong price of carbon has caused great harm because the price has been *so* wrong – zero – for *so* long – three hundred years! But we should not expect the correct price, even if imposed tomorrow, to rectify matters immediately. In the case of climate change, we have too little time left to leave matters entirely to "getting prices right." We need other tools as well: Fuel standards for transportation, renewable portfolios for utilities, regulations for energy conservation. And instead of a hands-off government that implements a single tax and then stands aloof, we need activist governments intervening massively to transform our economies from the energy guzzling, fossil fuel addicts they are now to energy efficient economies powered by renewable energy. We need a massive "green new deal" on the scale of how countries transformed their economies during WWII (Pollin, 2015).
- Nobody should put a second-best policy which they can implement on hold because they are waiting for a first-best policy to be implemented at a higher level. If international negotiations break down, national governments must forge ahead anyway. If there is gridlock in Washington, state and cities must forge ahead anyway. If the political process fails to generate strong policies to change incentives, businesses need to forge ahead anyway (Cardwell, 2016). Many who we need to mobilize cannot impose a carbon tax but must contribute in other ways.

Fortunately, all this is what people are now doing more and more. And fortunately, there is no physical reason we cannot respond adequately to the threat of climate change while simultaneously improving economic wellbeing. All that is lacking is the will to do what can and needs to be done.

REFERENCES

Cardwell, D. (2016), 'Apple becomes a green energy supplier, with itself as customer', *New York Times*, Energy and Environment section, 23 August.
Hahnel, R. (2012a), 'Left clouds over climate change policy', *Review of Radical Political Economics*, **44**(2), 141–59.
Hahnel, R. (2012b), 'Desperately seeking left unity on climate change policy', *Capitalism, Nature, Socialism*, **23**(4), 83–99.
Hahnel, R. (2013), 'The growth imperative: beyond assuming conclusions', *Review of Radical Political Economics*, **45**(1), 24–41.
Hahnel, R. (2014), 'An open letter to the climate justice movement', *New Politics*, **56**, 76–83.
Hahnel, R. (2016a), 'A tale of three theorems', *Review of Radical Political Economics*, **46**(1), 125–32.

Hahnel, R. (2016b), 'Environmental sustainability in a Sraffian framework', *Review of Radical Political Economics*, **49**(3), 477–88.
Hahnel, R. (2017), *Income Distribution and Environmental Sustainability in a Sraffian Framework*, New York: Routledge.
Pollin, R. (2015), *Greening the Global Economy*, Cambridge, MA: MIT Press.
Roemer, J. (1981), *Analytical Foundations of Marxian Economic Theory*, Cambridge: Cambridge University Press.
Schor, J. (1993), *The Overworked American: The Unexpected Decline of Leisure*, New York: Basic Books.
Schor, J. (1999), *The Overspent American: Why We Want What We Don't Need*, New York: Harper Perennial.
Weitzman, M. (2013), 'Can negotiating a uniform carbon price help to internalize the global warming externality?', NBER Working Paper No. 19644.

15. Intergenerational altruism: a solution to the climate problem?*
Frikk Nesje and Geir B. Asheim

1. INTRODUCTION

The emissions of greenhouse gases in the atmosphere threaten future climate stability and might undermine the wellbeing of future generations. It is natural to believe that this threat will enhance the concern that people have for their descendants and, thus, strengthen intergenerational altruism. The question that we pose in the present chapter is: Will the increased transfers that such strengthened intergenerational altruism leads to be a blessing for future generations by alleviating the climate problem, or will it aggravate the climate problem?

In a world where the public good problem of controlling the emissions of greenhouse gases is solved by collective action, the possible negative climate externalities of capital accumulation motivated by transfers of consumption potential to future generations will also be internalized. Hence, the dilemma we pose—namely, that the increased intergenerational transfers motivated by the threat of climate change will themselves contribute to increase this threat—only arises if such collective action has not been put in place. So in other words, we ask whether strengthened intergenerational altruism can function as an endogenously emerging second-best substitute if no first-best collective climate action will be undertaken.[1]

We will discuss this dilemma in the context of an intergenerational game played by dynasties. The currently living decision-maker of a dynasty makes choices as to maximize a weighted sum of own utility (of consumption and the stock of greenhouse gases) and the utilities of descendants of the same dynasty, with intergenerational altruism being measured by the weight assigned to the utilities of descendants. The strength of such intergenerational altruism is related to the pure time preference term in the Ramsey rule,

---

* We are grateful for constructive comments by the editor and a reviewer. The chapter is part of the research activities at the Centre for the Study of Equality, Social Organization, and Performance (ESOP) at the Department of Economics at the University of Oslo as well as the Oslo Centre for Research on Environmentally Friendly Energy (CREE). Asheim thanks the support and hospitality of IMéRA at Aix-Marseille University, where part of this work was done.

[1] An alternative motivation for the chapter is the following: Discounting the long-term future has received much attention, most prominently in Nordhaus' discussion of the Stern Review (Nordhaus, 2007; Stern, 2007). As a consequence of the adoption of a low discount rate, the Stern Review effectively advocated for stringent action on climate change. In a second-best setting, however, reducing the discount rate may lower rather than increase welfare due to the public good problem. This point is usually overlooked in the debate on discounting which unrealistically operates in a first-best setting.

so that a relative weight on the utilities of descendants approaching one corresponds to a rate of pure time preference approaching zero.[2]

In this game, factors that motivate the accumulation of capital (and thereby the growth of consumption potential) are the concern for lifetime utility and the weight assigned to the utilities of descendants. The higher the concern for lifetime utility and the heavier the weight assigned to the utilities of descendants, the greater is the motivation for accumulating capital for the purpose of enhancing growth of consumption potential.

Factors that motivate individual control of emissions of greenhouse gases in the atmosphere, provided that dynasties are of positive measure (i.e., that they are able to influence aggregates), are also the concern for lifetime utility and the weight assigned to the utilities of descendants. The higher the concern for lifetime utility and the heavier the weight assigned to the utilities of descendants, the greater is the motivation for individual control of emissions of greenhouse gases. However, due to the public good nature of controlling greenhouse gas emissions, these motivating factors might well be quite weak.

Increased intergenerational altruism increases the weight on the utilities of descendants. We evaluate the desirability of such altruism by the extent to which it facilitates intergenerational transfer without increasing the climate threat: Is increased intergenerational altruism a means for achieving accumulation of consumption potential without accumulation of greenhouse gases? If this is the case, increased intergenerational altruism solves the distributional problem posed by future climate change while alleviating (or at least, without aggravating) the efficiency problem.

We review the models of Jouvet et al. (2000), Karp (2017) and Asheim and Nesje (2016), which should be considered as examples of contributions to the literature.[3] These models are similar in terms of inefficiencies but different in terms of the considered factors motivating the accumulation of capital and the control of greenhouse gas emissions.

Jouvet et al. (2000) consider a simplified overlapping generations model, with lifetime equal to two periods (young and old), in which there is no consumption as young. Karp (2017) considers an overlapping generations model, with expected lifetime constant. Asheim and Nesje (2016) consider a non-overlapping generation model, with lifetime equal to one period. Dynasties are of positive measure in both Jouvet et al. (2000) and Karp (2017) and of zero measure in Asheim and Nesje (2016).

Both Jouvet et al. (2000) and Asheim and Nesje (2016) include motives for the accumulation of capital. In Jouvet et al. (2000) the motive is due to the concern for utility as old

---

[2] It is not related to the other source of a positive consumption discount rate according to the Ramsey rule, namely aversion to inequality/fluctuation under consumption growth.

[3] There are other contributions to this literature that might be relevant. In an early contribution Howarth and Norgaard (1995) made the point that there might be a need for collective action since altruistic agents do not necessarily fully internalize externalities. Rezai et al. (2012) use a non-overlapping generations model which in theory is equivalent to an overlapping generations model with perfect bequest motives. They show that a reduction in the discount rate could lead to accumulation of capital and greenhouse gases in a setting where the public good problem in the control of the emissions of greenhouse gases is severe. While the focus in this chapter is on climate externalities of capital accumulation, there are also other approaches taken in the literature. John and Pecchenino (1994), for example, focus in an overlapping generations model on effects of externalities through consumption. We refer the readers to Jouvet et al. (2000), Karp (2017) and Asheim and Nesje (2016) for an overview of other relevant contributions.

and the welfare of the immediate descendants (which is an aggregate of the utilities of all future descendants). In Asheim and Nesje (2016) it is due to the concern for the welfare of the immediate descendants (which is an aggregate of the utilities of all future descendants). Karp (2017) does not include motives for capital accumulation as the accumulation of capital is exogenous.

Jouvet et al. (2000) and Karp (2017) include motives for control of emissions of greenhouse gases in the atmosphere. In Jouvet et al. (2000) the motive is due to the same factors as above, subject to the limitation imposed by the public good nature of such control. In Karp (2017) it is there due to the concern for lifetime utility and the utility of descendants, subject to the same limitation. Asheim and Nesje (2016) do not include motives for control of emissions of greenhouse gases in the atmosphere, since each dynasty is of zero weight and therefore cannot influence aggregates.

Jouvet et al. (2000) is the only paper with factors motivating both the accumulation of capital and control of emissions of greenhouse gases in the atmosphere. However, there seems to be an inconsistency in how they define the equilibrium of the intergenerational game played by dynasties.[4] As a consequence, we study the accumulation of capital and the control of greenhouse gas emissions separately.

In reviewing the models of Jouvet et al. (2000), Karp (2017) and Asheim and Nesje (2016), we find that transfers to future generations through accumulation of capital might aggravate the climate problem in a second-best setting with insufficient control of greenhouse gas emissions in the atmosphere. In contrast, transfers to the future through control of greenhouse gas emissions will alleviate this problem. Whether increased intergenerational altruism is a means for achieving accumulation of consumption potential without accumulation of greenhouse gases depends on how it affects factors motivating the accumulation of capital and control of greenhouse gas emissions. An argument is provided for why increased intergenerational transfers motivated by the future effects of climate change in fact will increase the threat.

The plan of the chapter is as follows. Section 2 informally presents the models of Jouvet et al. (2000), Karp (2017) and Asheim and Nesje (2016) and clarifies some key concepts. Section 3 discusses factors that motivate the accumulation of capital (and thereby the growth of consumption potential) and the control of greenhouse gas emissions, and asks to what extent increased intergenerational altruism is a solution to the climate problem. Section 4 discusses increased cooperation between dynasties as an alternative solution to the climate problem. Section 5 provides concluding remarks.

The appendix gives a formal presentation of the models of Jouvet et al. (2000), Karp (2017) and Asheim and Nesje (2016) and the results.

2. INFORMAL PRESENTATION OF MODELS

The purpose of this section is to informally present the models of Jouvet et al. (2000), Karp (2017) and Asheim and Nesje (2016), which should be viewed as examples of

---

[4] While dynasties in Jouvet et al. (2000) are of positive weight in the control of the greenhouse gas emissions game, they are of zero weight in the capital accumulation game.

contributions to the literature, and to clarify some key concepts. A formal presentation of these models is given in the appendix. We use the models to facilitate the discussion in Section 3 on whether increased intergenerational altruism is a means for achieving accumulation of consumption potential without accumulation of greenhouse gases in the atmosphere, and thereby is at least a solution to the distributional problem that the threat of climate change poses.

Jouvet et al. (2000)

Consider a discrete time-infinite horizon economy consisting of a finite number of dynasties. Dynasties are interpreted as households of constant population. The structure is a simplified overlapping generations model, with lifetime equal to two periods (young and old), in which there is no consumption as young. What is left of the overlapping generations structure is thus that there is a need for capital accumulation (and capital holdings) in order to transfer wage income earned as young to consumption as old.

As young, a member of a dynasty earns wage income by supplying one unit of labor (exogenously) and receives bequest from the currently living old member. This is all saved for old age. As old, the member allocates its savings between consumption, capital accumulation for bequest and control of greenhouse gas emissions. The member derives utility from consumption and disutility from the stock of greenhouse gases as old. The utility function is strictly concave, with the cross-derivative non-positive.

The currently living old member of a dynasty makes choices as to maximize own welfare, which is a weighted sum of own utility and the welfare of the immediate descendants. Here, intergenerational altruism is non-paternalistic since it takes into account that descendants might care about their descendants. It is of first order since it considers directly immediate descendants only.

The production is a constant-returns-to-scale function of the inputs labor and capital. Thus, output equals units of labor multiplied with a per capita production function, with the capital-labor ratio as argument. The production function is positive, increasing, and strictly concave. Capital fully depreciates after one period. The wage equals the marginal product of labor. Output level and the return on capital are assumed not to be affected by the actions of a single dynasty.[5]

The dynamics of the stock of greenhouse gases depend positively linearly on output level in the current period and the stock of greenhouse gases in the previous period and negatively linearly on the control of greenhouse gas emissions.[6] The dependence of the stock on the control of greenhouse gas emissions can be separated into two additive components, with the first representing the control of a single dynasty and the second representing the aggregate control of all other dynasties.

The equilibrium considered is a symmetric Nash equilibrium. We focus on the case in which capital is accumulated for bequest and there at least is some control of greenhouse

---

[5] Note that this is inconsistent with the assumption of a finite number of dynasties.
[6] While the current stock is a positive function of the stock in the previous period, the change of the stock from the previous period to the current one is a negative function of the stock in the previous period.

gas emissions. Each dynasty can invest either in capital or control of greenhouse gas emissions. Since each dynasty is of positive weight in the control of greenhouse gas emissions game, it is individually rational to invest in some control of greenhouse gas emissions in addition to capital. The equilibrium is solved for by letting the initial member of a dynasty choose its path of capital and control taking as given the corresponding actions by the other dynasties. As a result of the the failure to internalize the climate externality of capital accumulation and a public good problem in the control of greenhouse gas emissions, the equilibrium is inefficient.

Due to the inconsistency in the definition of the equilibrium, where dynasties are assumed to be of positive weight in the control of greenhouse gas emissions game and of zero weight in the capital accumulation game, we assess factors motivating accumulation of capital and control of greenhouse gas emissions in separate.

Karp (2017)

Consider a continuous time-infinite horizon economy consisting of a finite number of dynasties. Here, dynasties are interpreted as same-size countries of constant population. The structure is an overlapping generations model, with expected lifetime and mortality rate constant, and thus, birth rate constant. Since currently living members have identical utility functions and expected lifetime is constant, they are identical.

In this model, accumulation of capital is exogenous (and thus, taken as given), while the control of greenhouse gas emissions is endogenous. The change in the stock of greenhouse gases is assumed to depend positively on aggregate emissions and negatively linearly on the stock of greenhouse gases.

The utility of a member of a dynasty equals its stream of utility flows, discounted by the impatience for own utility and risk-adjusted by the mortality rate. The utility flow depends positively linearly on the logarithm of consumption and negatively linearly on the stock of greenhouse gases.[7] Consumption is Cobb–Douglas in emissions of greenhouse gases, with technology level exogenous. This means that if a dynasty increases its control of greenhouse gas emissions (by reducing its own emissions, and thereby own consumption), it is effectively also increasing the utility of all current and future members of all dynasties.

A representative of the currently living (identical) members of a dynasty, constrained to sharing costs associated with the control of greenhouse gas emissions equally among currently living members, makes choices as to maximize own welfare, which is a weighted sum of own utility and the utilities of descendants. Intergenerational altruism is paternalistic since it does not take into account that descendants might care about their descendants. It is of higher order since it considers directly not only on immediate descendants.

The equilibria considered are stationary symmetric Markov Perfect equilibria (MPE). Here, the action of the currently living representative of a dynasty, which is control of greenhouse gas emissions, depends only on the stock of greenhouse gases. The representatives of each dynasty can invest in control of greenhouse gas emissions. Since each dynasty is of positive weight, it is individually rational to invest in some control of greenhouse gas emissions. An equilibrium is solved for by specifying a decision rule that

---

[7] This is a bit imprecise since it is the representation of the utility flow in equilibrium.

is a best response if and only if it is used by all other representatives, including future representatives of the same dynasty. As a result of a public good problem in the control of greenhouse gas emissions, equilibria are inefficient.

Asheim and Nesje (2016)

Consider a discrete time infinite horizon economy consisting of an uncountably infinite (a continuum) number of dynasties, which, in contrast to Karp (2017), implies that no single dynasty is able to influence aggregates. Dynasties are interpreted as families or tribes of constant population. The structure is a non-overlapping generation model, with lifetime equal to one period. This model therefore does not have concern-for-lifetime-utility motives for accumulation of capital.

Production is a constant-returns-to-scale function using inputs labor and capital. The per capita production function, with the capital-labor ratio as argument, is increasing, and strictly concave. Labor, which is uniformly distributed over dynasties and sum to one, is supplied exogenously.

In this economy there are two types of capital: polluting and non-polluting capital. The productivity of non-polluting capital is a fraction less than one of the productivity of polluting capital. Consumption potential can therefore be accumulated without accumulation of greenhouse gases if only the non-polluting capital is used. This can even be efficient if the reduction in the consumption potential by substitution from polluting to non-polluting capital (due to the lower productivity) is offset by the reduction in pollution that this leads to.

The utility of a member of a dynasty depends positively on consumption, but is adjusted downwards by the stock of greenhouse gases in the atmosphere (as proxied by the aggregate stock of polluting capital). The utility function is increasing, and strictly concave in consumption, while the stock of greenhouse gases adjusts utility downwards in a multiplicative way though a continuous and decreasing adjustment function.

The currently living member of a dynasty maximizes own welfare, which is a weighted sum of own utility and the welfare of the immediate descendants. As in Jouvet et al. (2000) intergenerational altruism is non-paternalistic and of first order.

The equilibrium considered is a symmetric Nash equilibrium. Each dynasty can invest either in polluting or non-polluting capital. Since each dynasty is of zero weight, it is individually rational to invest only in polluting capital since this relaxes its budget constraint without affecting the aggregate stock of polluting capital that adjusts utility for all dynasties downwards. The equilibrium is solved for by letting the initial member of a representative dynasty choose its path of polluting capital taking as given the aggregate path. As a result of the failure to internalize the climate externality of capital accumulation, the equilibrium is inefficient.

3. INTERGENERATIONAL ALTRUISM AS A SOLUTION?

The purpose of this section is to ask whether increased intergenerational altruism is a solution to at least the distributional problem that future climate change poses and to relate the discussion to the properties of the models as presented in Section 2.

Table 15.1 An overview of the considered models

		Jouvet et al.	Karp	Asheim and Nesje
A. The motivating factors				
Accumulation of capital	Lifetime utility	Yes	No	No
	Intergen. altruism	Yes	No	Yes
Contr. of greenh. gas emissions	Lifetime utility	Yes	Yes	No
	Intergen. altruism	Yes	Yes	No
B. Results on incr. intergen. altruism				
Solution to the distributional problem?		It depends	Yes	No
Solution to the efficiency problem?		It depends	Yes	No

To prepare the discussion we first summarize for each model the factors that motivate the accumulation of capital and control of greenhouse gas emissions. Then, we clarify model-by-model whether increased intergenerational altruism can be a solution. A formal presentation of results is given in the appendix. The section is concluded by an argument against the case for increased intergenerational altruism as a second-best substitute if no first-best collective action is undertaken to control of greenhouse gas emissions.

The Motivating Factors

The factors motivating the accumulation of capital and control of greenhouse gas emissions are summarized in Panel A of Table 15.1.

The concern for lifetime utility and the weight assigned to the utilities of descendants are factors that motivate the accumulation of capital. The concern-for-lifetime-utility motive is included in Jouvet et al. (2000) since there is no consumption as young, while the concern-for-the-utilities-of-descendants motive is included in Jouvet et al. (2000) and Asheim and Nesje (2016). There are no motives for the accumulation of capital in Karp (2017) since the accumulation of capital is exogenous.

Provided that dynasties are of positive measure (i.e., that they are able to influence aggregates), the concern for lifetime utility and the weight assigned to the utilities of descendants are also factors that motivate the control of greenhouse gas emissions in the atmosphere. Both the concern-for-lifetime-utility and the concern-for-the-utilities-of-descendants motives are included in Jouvet et al. (2000) and Karp (2017). The concern-for-lifetime-utility motive is included since members care about the stock of greenhouse gases later in life. There are no motives for avoiding accumulation of greenhouse gases in Asheim and Nesje (2016) since dynasties are of zero measure.

Results on Increased Intergenerational Altruism

Increased intergenerational altruism motivated by the future effects of climate change increases the weight on the utilities of descendants in the objective of the currently living decision-maker of a dynasty. The results on whether increased intergenerational altruism is a solution to the distributional problem posed by future climate change while alleviating the efficiency problem are summarized in Panel B of Table 15.1.

Jouvet et al. (2000). In this model, intergenerational altruism affects both the motives for the accumulation of capital for bequest as well as the control of greenhouse gas emissions. Since this is an overlapping generations model and utility depends on consumption and the stock of greenhouse gases while old, this comes on top of the concern-for-lifetime-utility motives for accumulation of capital and the control of greenhouse gas emissions. Increased intergenerational transfers thus takes the form of both bequest and control of greenhouse gas emissions. In effect, consumption potential is accumulated while the effect on accumulation of greenhouse gases is unclear.

If the level of intergenerational altruism is sufficiently low, then capital accumulation, keeping everything else constant, results in decreased greenhouse gas accumulation. Capital accumulation makes available more resources to abate greenhouse gas emissions. The resulting emissions reduction of this effort is larger than the emissions increase due to capital accumulation. If the level of intergenerational altruism is sufficiently high, then capital accumulation, keeping everything else constant, results in increased greenhouse gas accumulation. The reduced emissions resulting from increased effort in greenhouse gas abatement made available by capital accumulation is smaller than the increased emissions due to capital accumulation. Control of greenhouse gas emissions therefore increases. The total effect on the accumulation of greenhouse gases, however, is not investigated in the chapter.

In this model, increased intergenerational altruism therefore does not necessarily solve the distributional problem of future climate change. Since the climate problem can be aggravated, increased intergenerational altruism is not necessarily desirable.

Karp (2017). In this model intergenerational altruism affects the motive for control of greenhouse gas emissions. This comes on top of the concern-for-lifetime-utility motive, since it is an overlapping generations model where the utility flow depends on consumption and the stock of greenhouse gases at later points in life. Increased intergenerational transfers thus takes the form of control of greenhouse gas emissions. In effect, accumulation of greenhouse gases is reduced while capital accumulation is exogenous and thus unaffected. Increased intergenerational altruism can therefore be a solution to the distributional problem posed by future climate change.

Increased intergenerational altruism decreases the steady state value of the stock of greenhouse gases and can in fact lead to a steady state near the steady state that would have been chosen by an infinitely patient social planner. As intergenerational altruism goes toward its upper level, the efficiency problem thus implodes. Since the climate problem is reduced, increased intergenerational altruism is desirable.

Asheim and Nesje (2016). In this model intergenerational altruism affects the motive for the accumulation of (polluting) capital. Since it is a non-overlapping generations model, there are no concern-for-lifetime-utility motives. Increased intergenerational transfers thus takes the form of bequest only. In effect, both consumption potential and greenhouse gases are accumulated. Increased intergenerational altruism therefore does not solve the distributional problem posed by future climate change.

Increased intergenerational altruism increases the steady state value of polluting capital but can lead to a steady state very far from the steady state that would have been chosen by a social planner. As intergenerational altruism goes toward its upper level, the efficiency

318 *Handbook on the economics of climate change*

problem thus explodes. Since the climate problem is increased, increased intergenerational altruism is not desirable.

To summarize, in a second-best setting with insufficient control of greenhouse gas emissions in the atmosphere, transfers to future generations through accumulation of capital might result in accumulation of greenhouse gases, and thereby aggravate the climate problem. In contrast, transfers to the future through control of greenhouse gas emissions will alleviate the climate problem.

Whether increased intergenerational altruism is a means for achieving accumulation of consumption potential without accumulation of greenhouse gases, and thereby is at least a solution to the distributional problem that the effect of climate change poses, depends on how it affects factors motivating the accumulation of capital and the control of greenhouse gas emissions. If increased intergenerational altruism motivates a sufficiently stricter control of greenhouse gas emissions as compared to the enhanced accumulation of capital, then it might be a solution to the distributional problem posed by future climate change—while even alleviating the efficiency problem—since consumption potential can be accumulated without the accumulation of greenhouse gases.

Jouvet et al. (2000) and Asheim and Nesje (2016) include a concern-for-the-utilities-of-descendants motive for the accumulation of capital. However, due to the public good problem in the control of greenhouse gas emissions, climate externalities of capital accumulation will necessarily not be fully internalized. This implies that accumulation of greenhouse gases, as a result of additional capital accumulation, is increasing in intergenerational altruism.

Since it is natural to think of dynasties—which are interpreted as households in Jouvet et al. (2000), same-size countries in Karp (2017), and families or tribes in Asheim and Nesje (2016)—as small relative to the economy at large, the public good problem might in fact be severe. The concern-for-the-utilities-of-descendants motive for the control of greenhouse gas emissions, which is present in Jouvet et al. (2000) and Karp (2017), may therefore be weak, compared to the motive for the accumulation of capital, even if it is the case that intergenerational altruism is high.

Both consumption potential and greenhouse gases will then be accumulated at a higher rate as intergenerational transfers increase. Thus, increased intergenerational altruism will not be a solution to the climate problem.

4. OTHER SOLUTIONS?

The purpose of this section is to clarify whether increased cooperation between dynasties is a solution to the climate problem. This discussion is based on the presentation of the models in Section 2, supported by the results presented in the appendix.

Jouvet et al. (2000) and Asheim and Nesje (2016) do not specifically address the issue of increased but partial cooperation, but rather conditions under which full cooperation can be optimal. In Jouvet et al. (2000) the social planner accumulates less capital and has a higher willingness to pay for control of greenhouse gas emissions. The equilibrium under full cooperation is a symmetric Nash equilibrium, as before, but where capital accumulation for bequest is taxed and control of greenhouse gas emissions is subsidized at appropriate rates. Full cooperation (compared to the case with no cooperation) reduces the motivation for accumulation of capital and increases the motivation for control of greenhouse gas emissions. Asheim and Nesje (2016) consider conditions under which full cooperation can be implemented by requiring that dynasties accumulate non-polluting capital only, and find that this holds if there is a sufficiently small productivity difference between polluting and non-polluting capital. The constrained optimum can be interpreted as a symmetric Nash equilibrium where each dynasty can invest in non-polluting but not polluting capital. The motivation for avoiding greenhouse gas accumulation is thus increased. For both models, this result comes about because the climate externality of capital accumulation is fully internalized when the public good problem in the control of greenhouse gas emissions is solved. Unsurprisingly, full cooperation is a solution to the efficiency problem that future climate change poses since the first-best is obtained.

Karp (2017), on the other hand, also addresses the issue of increased but partial cooperation. Here, the equilibria considered are stationary symmetric MPE, as already defined. Since the number of dynasties can be interpreted as the fragmentation of society into same-size countries of constant population, increased cooperation can be modeled as a reduction in the number of dynasties. Increased cooperation increases the motivation for the control of greenhouse gas emissions because the public good problem becomes less severe. Since cooperation is a means for reducing the accumulation of greenhouse gases without changing the accumulation of consumption potential, it is a solution to the distributional problem imposed by future climate change. Increased cooperation is also a solution to the efficiency problem since the climate problem implodes as cooperation becomes full.

To summarize, increased cooperation between dynasties can be considered a solution to the climate problem.

5. CONCLUDING REMARKS

In this chapter we have argued that increased intergenerational altruism may not function as a second-best substitute in a world threatened by climate change if no first-best collective action is undertaken to control greenhouse gas emissions. In the context of an intergenerational game played by dynasties, the insight is that since dynasties are so small compared to the aggregate economy, the public good problem in the control of greenhouse gases might still be severe. Thus, consumption potential is not likely to be accumulated (through the accumulation of capital) without increasing the climate threat.

REFERENCES

Asheim, G.B. and F. Nesje (2016), 'Destructive intergenerational altruism', *Journal of the Association of Environmental and Resource Economists*, **3**, 957–984.

Golosov, M., J. Hassler, P. Krusell and A. Tsyvinski (2014), 'Optimal taxes of fossil fuel in general equilibrium', *Econometrica*, **82**, 41–88.

Howarth, R.B. and R.B. Norgaard (1995), 'Intergenerational choice under global environment change', in D.W. Bromley (ed), *Handbook of Environmental Economics*, Oxford: Blackwell, pp. 111–138.

John, A. and R. Pecchenino (1994), 'An overlapping generations model of growth and the environment', *Economic Journal*, **104**, 1393–1410.

Jouvet, P.A., P. Michel and J.P. Vidal (2000), 'Intergenerational altruism and the environment', *Scandinavian Journal of Economics*, **102**, 135–150.

Karp, L. (2017), 'Provision of a public good with multiple dynasties', *Economic Journal*, **127**(607), 2641–2664.

Nordhaus, W.D. (2007), 'The *Stern Review on the Economics of Climate Change*', *Journal of Economic Literature*, **45**, 686–702.

Rezai, A., D.K. Foley and L. Taylor (2012), 'Global warming and economic externalities', *Economic Theory*, **49**, 329–351.

Stern, N. (2007), *The Economics of Climate Change: The Stern Review*, Cambridge and New York: Cambridge University Press.

APPENDIX: FORMAL PRESENTATION OF MODELS AND RESULTS

The purpose of the appendix is to formally present the models of Jouvet et al. (2000), Karp (2017) and Asheim and Nesje (2016) and the results discussed in the chapter.

Jouvet et al. (2000)

The case where capital is accumulated for bequest and there is at least some control of greenhouse gas emissions is studied.

At each time t a new generation of $N > 1$ identical members are born into dynasties. They live for two periods.

There are constant-returns-to-scale using inputs L and K: $Y_t = L_t f(k_t)$, with $k_t = K_t/L_t$. f is twice continuously differentiable, positive, increasing, strictly concave and satisfies the Inada conditions. Capital fully depreciates. The factor prices are given by: $w_t = f(k_t) - k_t f'(k_t)$ and $R_t = f'(k_t)$.

Emissions of greenhouse gases at time t is a function of output level, mY_t, with $m > 0$. Let X_t denote the total amount of resources used to control of greenhouse gas emissions. qX_t is the amount abated in period t, with $q > m$. The stock of greenhouse gases depreciates with rate $h \in (0,1]$. The dynamics of the stock is $S_t = mY_t + (1-h)S_{t-1} - qX_t$. For the decision-maker of dynasty i at time t, the dynamics are $S_t = mY_t + (1-h)S_{t-1} - q\bar{X}_t - qx_{it}$, with $\bar{X}_t = X_t - x_{it}$.

Preferences of a member born at time t are represented by the utility function $u(c_{t+1}, S_{t+1})$, where c is consumption. $u_c > 0$, $u_S < 0$, $u_{cc} < 0$, $u_{SS} < 0$, and $u_{cS} \leq 0$ and sufficiently small. Welfare can be expressed in the following manner:

$$\sum_{\tau=t}^{\infty} \alpha^{\tau-t} u(c_{\tau+1}, S_{\tau+1}),$$

with $\alpha \in [0,1)$ denoting intergenerational altruism. Young members supply one unit of labor, earning w_t, inherit z_t and save $s_t = w_t + z_t$. Old members allocate their savings between bequest z_{t+1}, abatement x_{t+1} and consumption $c_{t+1} = R_{t+1}s_t - z_{t+1} - x_{t+1}$.

The symmetric Nash equilibrium is considered. The member of a dynasty born at time t takes as given the actions of other dynasties and chooses c_{t+1}, z_{t+1} and x_{t+1} to maximize own welfare subject to the budget constraints, non-negativity constraints on bequest and abatement, and the dynamics of the stock of greenhouse gases. In equilibrium we have $k_{t+1} = s_t$. Due to symmetry, $X_{t+1} = Nx_{t+1}$.

Assume no bequest and that there exists a unique steady state k^d. We assume that $\alpha > 1/f'(k^d) \equiv \alpha^w$, so that capital will be accumulated for bequest. Define the corresponding steady state $k^\alpha \equiv f'^{-1}(1/\alpha)$. The threshold $\bar{S}^\alpha$, below which no control of greenhouse gas emissions occurs, is defined as the stock of greenhouse gases in the case of capital accumulation for bequest, S^α, that is the solution to

$$0 = u_c(c^\alpha, S^\alpha) + \frac{qu_S(c^\alpha, S^\alpha)}{1 - \alpha(1-h)},$$

with $c^\alpha = f(k^\alpha) - k^\alpha$. In the paper it is proved that $\bar{S}^\alpha$ is decreasing in α.

In the paper it is verified that

$$\frac{dk}{d\alpha} > 0.$$

The change in the stock of greenhouse gases can be described in the following manner: $dS = qNh^{-1}(dc + (1 - (1 - m/q)\alpha^{-1})dk)$. Interpret $1 - m/q$ as the rate of control of greenhouse gas emissions. If $\alpha < 1 - m/q$, accumulating capital, keeping everything else constant, reduces the stock of greenhouse gases. If $\alpha \geq 1 - m/q$, accumulating capital, keeping everything else constant, increases the stock of greenhouse gases.

The following result is proved in the paper:

Proposition 1 *Assume $\alpha > \alpha^w$ and $S^\alpha > \overline{S}^\alpha$. Consumption decreases with the level of intergenerational altruism if $\alpha \geq 1 - m/q$. If $\alpha < 1 - m/q$ and $h < 1$, the effects of an increase in the level of intergenerational altruism on consumption is indeterminate.*

Since consumption decreases in the case with $\alpha \geq 1 - m/q$ and more resources are made available due to capital accumulation, control of greenhouse gas emissions increases. The total effect on the accumulation of greenhouse gases is not investigated in the paper.

Assume that the private and the social discount factors are equal. Consider the following problem of the social planner, which, if decentralized, could be interpreted as the case with cooperation between dynasties:

$$\max_{\{c_t, x_t, P_t, k_{t+1}\}_{t=0}^\infty} \sum_{t=0}^\infty \alpha^t u(c_t, S_t)$$

subject to

$$f(k_t) = c_t + x_t + k_{t+1}; S_t = mNf(k_t) + (1 - h)S_{t-1} - qNx_t;$$
$$x_t \geq 0; k_0, S_{-1} \text{ given.}$$

In the paper it is shown that the social planner accumulates less capital and has a higher willingness to pay for control of greenhouse gas emissions. The social planner solution can be decentralized using two instruments.

Karp (2017)

This paper studies a linear-in-state model, where the consumption function is a simplified version of that proposed by Golosov et al. (2014). We limit our presentation to the case of the stationary symmetric non-limit Markov Perfect equilibria (MPE).

There is a finite number of $N \geq 1$ dynasties, where N represents fragmentation of society. Members of dynasties have lifetime exponentially distributed. The mortality rate is $\theta > 0$. The pure rate of time preference is $r > 0$. Intergenerational altruism can be expressed as $\alpha = 1/(1 + \lambda)$, with $\lambda \leq r$ being the rate of discount of descendants' utilities.

Denote by S_t the stock of greenhouse gases in the atmosphere at time t (which is the state variable), and by x_{it} the greenhouse gas emissions at time t by dynasty i. In the symmetric case, $X_t = Nx_{it}$. The state variable evolves according to

$$\frac{dS}{dt} = BS + X,$$

with $B < 0$.

As N represents fragmentation, the aggregate can be represented by $N = 1$. Let the aggregate utility flow be $u(X_t, S_t; 1) = \ln C(X_t; 1) - \kappa S_t$, where κS_t is loss in consumption due to climate damage. Aggregate consumption is $C(X_t; 1) = A_t X_t^\eta$, with technology level A_t exogenous. Due to symmetry, the flow utility for dynasty i at time t is $u(x_{it}, S_t; N) = (\ln A_t/N) + (\eta/N)\ln x_{it} - (\kappa/N) S_t$.

Denote $U_{it} = \int_{\tau=t}^{\infty} e^{-(r+\theta)(\tau-t)} u(x_{i\tau}, S_\tau; N) d\tau$. Welfare can then be expressed as

$$U_{it} + \theta \int_{\tau=t}^{\infty} e^{-\lambda(\tau-t)} U_{i\tau} d\tau = \int_{\tau=t}^{\infty} D(\tau-t) u(x_{i\tau}, S_\tau; N) d\tau,$$

with $D(\tau - t)$ defined according to

$$D(t) = \frac{\lambda - r}{\lambda - (r+\theta)} e^{-(r+\theta)t} - \frac{\theta}{\lambda - (r+\theta)} e^{-\lambda t}.$$

At each point in time t currently living (identical) members of a dynasty are represented by a representative who shares the chosen cost of control of greenhouse gas emissions equally among the members. Representatives' control depend only on S_t. $\chi(S)$ is an MPE if and only if $x_{it} = \chi(S_t)$ is the best response, for all feasible S, for representatives i, t when all other representatives use this decision rule.

It is shown in the paper that η/κ is the steady state chosen by a social planner with $\alpha = 1$. This steady state is the Green Golden Rule (GGR). It is also verified that there exists an MPE close to the GGR.

Let Φ be the derivative of aggregate emissions with respect to the stock of greenhouse gases at the steady state. Define Y as the steady state stock given by Φ, as a share of GGR.

The following result is proved in the paper:

Proposition 2 *For any $0 \leq \Phi < -B$, increased cooperation or intergenerational altruism (smaller N or larger α) move the MPE steady state closer to GGR: $\frac{dY}{dN} > 0, \frac{dY}{d\alpha} < 0$. For all values of N, there exists an MPE steady state arbitrarily close to the GGR for α close to 1. This steady state is supported by a decision rule corresponding to Φ close to its upper bound ($\Phi = -B$). In contrast, even for full cooperation ($N = 1$), the steady state is bounded away from the GGR for α bounded away from 1.*

Asheim and Nesje (2016)

There is an uncountably infinite number (a continuum) of dynasties, consisting of altruistically linked members each living for one time period.

Production is a constant-returns-to-scale function of capital and labor, so that the per capita production function, f, satisfies the Inada conditions with the additional assumption that per capita consumption is bounded above.

There are two kinds of capital, polluting and non-polluting capital. A consumption stream $_1c = (c_1, c_2, \ldots) \geq 0$ is feasible given a pair of initial capital stocks $(b, g) > 0$ if there exist streams of polluting capital $_0b = (b_0, b_1, b_2...) \geq 0$ and non-polluting capital $_0g = (g_0, g_1, g_2...) \geq 0$ such that $(b_0, g_0) = (b, g)$ and

$$c_t + b_t + g_t = b_{t-1} + g_{t-1} + f(b_{t-1} + (1-\gamma)g_{t-1}) \tag{15A.1}$$

for all $t \in \mathbb{N}$, where $\gamma \in (0,1)$ measures to what extent non-polluting capital is less productive.

Labor is uniformly distributed over a continuum of dynasties i on the unit interval $[0,1]$. Assume that the map from consumption to utility for dynasty i in generation t, in addition to depending on an increasing, strictly concave, and continuously differentiable utility function $u:\mathbb{R}_+ \to \mathbb{R}$, satisfying $u(0) = 0$ and $\lim_{c \to 0} u'(c) = \infty$, also depends on the aggregate amount of polluting capital accumulated by generation $t-1$: $a(b_{t-1})u(\mathbf{c}_t(i))$, where the continuous and decreasing function $a:\mathbb{R}_+ \to \mathbb{R}$, satisfying $a(0) = 1$ and $\lim_{b \to \infty} a(b) = 0$, captures the effect of the climate externalities caused by polluting capital, and where $\mathbf{c}_t(i)$ is the consumption of dynasty i. Refer to $u_t = a(b_{t-1})u(\mathbf{c}_t(i))$ as adjusted utility.

Dynasties have the same u function, they are affected in the same manner by the adjustment, and they have the same non-paternalistic altruistic (NPA) welfare function:

$$(1-\alpha) \sum_{t=0}^{\infty} \alpha^t a(b_t) u(\mathbf{c}_{t+1}(i)),$$

where $\alpha \in (0,1)$ is the per generation factor used to discount future utilities. Each dynasty i can invest in either polluting $\mathbf{b}_t(i)$ or non-polluting $\mathbf{g}_t(i)$ capital. Since each dynasty is of zero weight, it is individually rational for each dynasty to invest in polluting capital only, as this relaxes its budget constraint (15A.1), while not influencing the aggregate stock of polluting capital that adjusts its utility. If the profile of initial ownership to capital is assumed to be uniform, the dynasties will behave in the same manner. This implies that each dynasty i, for all $t \in \mathbb{N}$, chooses $\mathbf{b}_t(i) = b_t$ and $\mathbf{g}_t(i) = 0$ so that the budget constraint (15A.1) is satisfied.

Since u is strictly concave, the analysis can be performed by considering a representative dynasty. The analysis is simplified by considering the case where the initial stock of non-polluting capital, g, is zero, so that only the initial stock of polluting capital, b, is positive. Under this assumption and taking into account that dynasties will choose to accumulate only polluting capital, the set of (polluting) capital streams as a function of the initial stock is $K(b) = \{_0b: b_0 = b \text{ and } 0 \leq b_t \leq b_{t-1} + f(b_{t-1}) \text{ for all } t \in \mathbb{N}\}$. Write $\mathcal{K} = \bigcup_{b \in \mathbb{R}_+} K(b)$. Define $\mathbf{c}(_0b) = (b_0 + f(b_0) - b_1, b_1 + f(b_1) - b_2, ..., b_{t-1} + f(b_{t-1}) - b_t, ...)$ as the consumption stream that is associated with $_0b$. Consumption streams as a function of the initial stock is $C(b) = \{_1c: \text{there is } _0b \in K(b) \text{ s.t. } _1c = \mathbf{c}(_0b)\}$. Say that $_1c \in C(b)$ is efficient if there is no $_1\tilde{c} \in C(b)$ such that $_1\tilde{c} > _1c$.

The symmetric Nash equilibrium is considered. The representative dynasty maximizes the NPA welfare function over all consumption streams $_1c \in C(b)$ while taking the climate externalities caused by the stream of polluting capital, $_0b \in K(b)$, as given. The NPA welfare function $v_\alpha: \mathcal{K} \times \mathcal{K} \to $ defined over capital streams is given by:

$$v_\alpha(_0k, _0b) = (1-\alpha) \sum_{t=0}^{\infty} \alpha^t a(b_t) u(k_t + f(k_t) - k_{t+1}),$$

with $\alpha \in (0,1)$, where $k_0 = b$ and k_t is polluting capital held by the representative dynasty for $t \in \mathbb{N}$. The representative dynasty takes $_0b$ as given when maximizing $v_\alpha(_0k, {_0b})$ over all $_0k \in K(b)$. However, in equilibrium, $_0k = {_0b}$, leading to the following definition: $_0b \in K(b)$ is a NPA equilibrium if

$$v_\alpha(_0b, {_0b}) \geq v_\alpha(_0\tilde{k}, {_0b}) \text{ for all } _0\tilde{k} \in K(b).$$

Define $k_\infty: (0,1) \to \mathbb{R}_+$ by, for all $\alpha \in (0,1)$, $\alpha(1 + f'(k_\infty(\alpha))) = 1$. It follows from the properties of f that k_∞ is well-defined, continuous, and increasing, with $\lim_{\alpha \to 0} k_\infty(\alpha) = 0$ and $\lim_{\alpha \to 1} k_\infty(\alpha) = \infty$. For given $\alpha \in (0,1)$, $k_\infty(\alpha)$ is the capital stock corresponding to the modified golden rule.

The following main result is proved in the paper:

Proposition 3 *Assume $b > 0$ and $g = 0$. Then there is a unique NPA equilibrium, $b^*(b)$, with associated NPA equilibrium consumption stream $c^*(b) = \mathbf{c}(k^*(b))$. Furthermore, $b^*(b)$ is strictly monotone in time, with $\lim_{t \to \infty} b_t^*(b) = k_\infty(\alpha)$, and $c^*(b)$ is efficient, with $\lim_{t \to \infty} c_t^*(b) = f(k_\infty(\alpha))$. Long-term utility adjusted for climate externalities, $\lim_{t \to \infty} a(b_t^*(b)) u(c_t^*(b)) = a(k_\infty(\alpha)) u(f(k_\infty(\alpha)))$, approaches 0 as $\alpha \to 1$.*

A feasible policy is to require the dynasties to accumulate non-polluting capital only, a case which could be interpreted as cooperation between dynasties. It is shown in the paper that such a policy can be efficient, provided that γ, the parameter measuring to what extent non-polluting capital is less productive, is sufficiently small.

16. On intertemporal equity and efficiency in a model of global warming
John M. Hartwick and Tapan Mitra

1. INTRODUCTION

There is increasing concern today over the sustainability of current welfare levels, as one encounters increasing evidence of long-term damage being done to the global environment.

It has been an objective of national income accounting that net national product ought to provide the correct signal of sustainable welfare (see, especially, Weitzman (1976)). Since environmental degradation as well as natural resource depletion have been traditionally omitted in measuring NNP, it has been felt that such measures of NNP have increasingly become erroneous indicators of the level of sustainable welfare. As a result, an aspect of the sustainability issue that has received a lot of attention is "green NNP accounting", which proposes a revision of the concept of net national product to provide a better measure of sustainable welfare.[1]

A different approach, and one that we follow in this chapter, is to explicitly study the nature of paths of sustainable welfare, in the face of global environmental damage. Examples of these damages are depletion of the protective ozone layer covering parts of the atmosphere, and the rise in the average temperature of the Earth's climate system, generally referred to as *global warming*.[2]

The ozone layer prevents most harmful UVB wavelengths of ultraviolet light from passing through the Earth's atmosphere. It is suspected that a variety of biological consequences such as increases in adverse health effects on humans, damage to plants, and reduction of plankton populations in the ocean may result from the increased UV exposure due to ozone depletion. Chlorofluorocarbons (CFCs) and other halogenated ozone depleting substances (ODS) are mainly responsible for man-made chemical ozone depletion.[3]

Global warming is mostly caused by increasing concentrations of greenhouse gases (GHG) and other human activities. Effects of global warming significant to humans include the threat to food security from decreasing crop yields and the abandonment of populated areas due to rising sea levels. The major greenhouse gases are water vapor,

---

[1] See the contributions by Hartwick (1990, 1994), Maler (1991), Asheim (1994), and the references cited in these papers.

[2] Although the increase of near-surface atmospheric temperature is the measure of global warming often reported, most of the additional energy stored in the climate system since 1970 has gone into ocean warming. The remainder has melted ice and warmed the continents and atmosphere. See Rhein et al. (2013) for a comprehensive study.

[3] This led to the adoption of the Montreal Protocol that bans the production of CFCs, halons, and other ozone-depleting chemicals.

carbon dioxide (CO_2), methane, and ozone.[4] Fossil fuel burning has produced about three-quarters of the increase in CO_2 from human activity over the past 20 years. The rest of this increase is caused mostly by changes in land-use, particularly deforestation. Estimates of global CO_2 emissions indicate that burning of coal, oil and gas accounted for most of the emissions.

In this chapter, we wish to focus on the environmental damage caused by global warming, and examine the nature of paths of sustainable welfare in the presence of global warming in a theoretical model of intertemporal resource allocation. Any such theoretical model needs to abstract from reality, and our use of the term "global warming" in the context of the model is necessarily an abstraction, which tries to capture essential aspects of both the causes and effects of global warming which have been described above.

Since a principal cause of global warming is CO_2 emissions from burning of fossil fuels, we model it by postulating a certain law of motion relating the rate of change of (an index of) global warming to the flow use of an exhaustible resource. The particular form of this law of motion will play an important role in our analysis and our conclusions, and we elaborate on these aspects in the following sections.

Let us note that there have been different approaches in the literature to modeling both the pollution accumulation process from CO_2 emissions and the way in which this pollution affects global warming. It is quite common to model the pollution accumulation process as one with a decay rate of carbon accumulation being a linear function of the carbon stock (see Forster (1973)). However, a high carbon stock may destroy the environment's natural self-purification process, as has been pointed out by Forster (1975), Dasgupta (1982) and Comolli (1997) among others, and an inverse U-shaped decay function has been proposed instead. Regarding the effect of this carbon accumulation on global warming, Nordhaus (1991) has modeled it as a two-step process in which the second step operates with a lag: fossil fuel burning slowly increases the carbon stock in the atmosphere, which in turn creates global temperature change after some passage of time. Following Stollery (1998), we model the carbon accumulation process as entirely irreversible, thereby providing a "worst-case" benchmark, and we simplify by ignoring the lags involved in fossil fuel burning producing global warming.

Global warming affects the ability of technology to combine resources to produce current consumption goods, as well as investment goods; investment goods in turn help to produce future consumption. Since consumption is a basic determinant of the welfare level, global warming can affect welfare levels that can be achieved in the present and in the future.[5] But, there can be a more direct effect of global warming on welfare. Given

---

[4] Human activity since the Industrial Revolution has increased the amount of greenhouse gases in the atmosphere, leading to increased radiative forcing from CO_2, methane, tropospheric ozone, CFCs and nitrous oxide. The concentrations of CO_2 and methane have increased by 36% and 148% respectively since 1750. [See EPA (2007), a report on Climate Change by the United States Environmental Protection Agency.] These levels are much higher than at any time during the last 800,000 years, the period for which reliable data has been extracted from ice cores. [See, for example, Lüthi et al. (2008).] Less direct geological evidence indicates that CO_2 values higher than this were last seen about 20 million years ago. [See Pearson and Palmer (2000).]

[5] This production effect of global warming has been studied by Nordhaus (1982, 1991), Sinclair (1992) and Stollery (1998), among others.

a consumption level, the welfare generated by it can be lower at a higher level of global warming; this is a common topic nowadays in discussions of welfare in polluted urban environments.[6] We include both the production effect and the direct welfare effect of global warming in our model.

Apart from these two basic aspects of global warming, we model the dynamic resource allocation process very much along the lines of the seminal work of Solow (1974) and Dasgupta and Heal (1979) on economic growth in the presence of exhaustible resources. Thus, our framework has the simplifications of dealing with only one exhaustible resource, and with a single capital good. The exhaustible resource and the capital good are essential inputs and substitutes in the production of a single good, which can either be invested to augment the stock of the capital good, or consumed.

In this framework, the production effect of global warming is introduced by postulating that (an index of) global warming affects the production of the single capital/consumption good negatively. The welfare effect of global warming is introduced through a utility function. Utility is obtained from the consumption of the single produced good, with more consumption leading to higher utility; in addition, (an index of) global warming has a negative effect on the utility obtained.

We complete the specification of the model by linking the rate of growth of (the index of) global warming to the use of the exhaustible resource. Thus, as mentioned above, in the context of our model, one might equate an index of carbon emissions from the burning of fossil fuels directly to the index of global warming (with no lags), and one views the global warming process itself to be irreversible (with no decay arising from the environment's natural self-purification process).

We explore the nature of equitable paths in this framework where equity means that a constant utility level is maintained over time. Our focus is on equitable paths, which satisfy at least short-run efficiency, so that there are no Pareto gains to be made in the utility stream by reallocation of resources in the short run.[7]

Let us first recall (for ease of comparison) the main observations that emerge from the model of Solow (1974) and Dasgupta and Heal (1979) *without* global warming. The intertemporal equity requirement translates to constant positive consumption over time (with population constant), and one requires sufficient substitutability between man-made capital for the resource in production, as well as sufficient importance of capital in production (in comparison to the exhaustible resource) so as to permit capital accumulation to offset the resource use which must dwindle over time eventually. In the context of the Cobb–Douglas production function, the second condition was precisely characterized by Solow (1974) as the requirement that the capital share in output exceed the resource share in output.[8]

---

[6] This direct welfare effect appears in the work of Forster (1973), Tahvonen and Kuuluvainen (1991), Stollery (1998) and d'Autume, Hartwick and Schubert (2010), among others.

[7] The exact relation between this class of paths, and paths which are both equitable and *long-run* efficient, is not explored in this chapter, but is clearly an issue of considerable interest. Investigation of this issue would also reveal the relation between this class of paths and (non-trivial) *maximin* optimal paths. We hope to report on these connections in future research.

[8] See Cass and Mitra (1991), and Mitra, Asheim, Buchholz and Withagen (2013) for results on more general classes of production functions.

In this framework, short-run efficiency is characterized by *Hotelling's Rule*, which requires that the return on the capital good and on the resource be equalized at each moment of time. This translates to the condition that the price of the resource (in terms of the produced good) grow at the real interest rate determined by the marginal product of capital.

An important insight about this model was the observation by Hartwick (1977) that if the investment in the capital good was equal to resource rents (along a path satisfying Hotelling's Rule), then the path would be equitable.[9] This investment policy has come to be known in the literature as *Hartwick's Rule*. It was later established by Buchholz, Dasgupta and Mitra (2005) that, conversely, any equitable path, which satisfies Hotelling's Rule, *must* follow this investment rule.[10] Thus, Hartwick's Rule captures completely the equity requirement for short-run efficient paths, being both a necessary and sufficient condition.

With global warming affecting production adversely, the capital versus resource importance requirement for intertemporal equity can be expected to be a more stringent one; this is analyzed by Stollery (1998). With global warming affecting utility directly, consumption over time has to grow to keep pace with the global warming on paths of constant utility, and the conditions under which this is possible are explored by d'Autume, Hartwick and Schubert (2010).

We differ from these earlier papers primarily in formalizing the law of motion which describes how global warming is generated by the use of the exhaustible resource. Our description is more general, encompassing the earlier formulations in which the rate of growth of global warming is taken to be proportional to the use of the resource. But, it allows for the possibility that the rate of growth of global warming is proportional to a concave function of the resource use. Denoting the index of (irreversible) global warming (at time t) by $w(t)$, we use the following law of motion to describe how global warming is generated by the use of the exhaustible resource $(r(t))$:

$$\dot{w}(t)/w(t) = \lambda G(r(t)) \text{ for } t \geq 0 \tag{W}$$

where $\lambda > 0$ and G is an increasing, continuous and concave function of resource use.

In the context of models with global warming, the concept of short-run efficiency has not been systematically analyzed, and an extended version of Hotelling's Rule (which we will call EHoR, and is specified formally in Section 3) is typically derived as a necessary condition (along with other necessary conditions) from a dynamic optimization problem involving the maximization of a discounted integral of utilities. This makes it difficult to relate this extended version of Hotelling's Rule to the concept of short-run efficiency, and therefore to see the precise connection of it to the concept of equity. Since our focus is on equitable paths which are at least short-run efficient, we start by formally studying short-run efficiency in our framework (in Section 3). As mentioned above, short-run efficiency

---

[9] An extension of this rule to more general models of capital accumulation was provided by Dixit, Hammond and Hoel (1980).
[10] This result has also been investigated by Withagen and Asheim (1998) and Mitra (2002) in more general models of intertemporal allocation.

means that there are no *Pareto* gains to be made in the utility stream by reallocation of resources in the short run.

Short-run efficiency is seen to lead to a necessary condition, which is more general than the extended version of Hotelling's Rule discussed in the papers by Stollery (1998) and d'Autume, Hartwick and Schubert (2010). If in the formulation of global warming, captured by (W), the function G takes the form:

$$G(r) = r^\mu \text{ for all } r \geq 0 \qquad (G)$$

where $0 < \mu \leq 1$, then this necessary condition is the same as EHoR, discussed in these papers, when $\mu = 1$, but not when $0 < \mu < 1$.

In Sections 4 and 5, we concentrate on the formulation of global warming (W), with G specified as in (G), with $\mu = 1$. As already mentioned, a necessary condition of short-run efficiency is an extended version of Hotelling's Rule, the extension being a reflection of the externality generated by global warming. It is possible to relate this EHoR condition to the standard form of Hotelling's Rule by noting that one can design a (time-dependent) Pigouvian tax such that the *net price* of the resource (in terms of the produced good) grows at the real interest rate determined by the marginal product of capital.[11]

In Section 4, we show that when this EHoR condition holds along a path, it is still true that if the investment in the capital good was equal to resource rents (that is, Hartwick's Rule holds), then the path would be equitable, in the sense of maintaining a constant *utility* level, where the utility depends both on consumption and global warming. And, conversely, we show that when an equitable path satisfies this EHoR condition, then it *must* follow the rule that the investment in the capital good be equal to resource rents. Consequently, any equitable path which is short-run efficient must satisfy Hartwick's Rule. Thus, this investment rule continues to play a key role in the study of equitable paths which are short-run efficient, even in this model of global warming. In fact, our results in this section provide an exact analogue of the three equivalence results of Buchholz, Dasgupta and Mitra (2005) in the corresponding model without global warming, as we also demonstrate that if along an equitable path, Hartwick's Rule holds, then the path must satisfy the extended version of Hotelling's Rule (EHoR).

A major analytical benefit of the results of Section 4 is that in studying paths which are equitable and short-run efficient, one can concentrate on equitable paths which satisfy Hartwick's Rule. This investment rule is clearly easier to work with (analytically) than the extended version of Hotelling's Rule. It has the major simplification, with a Cobb–Douglas production function,[12] that the savings rate is a constant (equal to the share of resource in output).

---

[11] We note a word of caution here. In the model with global warming, the production function is not a concave function in the global warming variable. Consequently, we are not claiming that the extended version of Hotelling's Rule is *sufficient* for short-run efficiency. The fact that we are now dealing with optimization problems on a non-convex feasible set has not been adequately appreciated in this literature, since it has concentrated on deriving *necessary* conditions for optimality, using various forms of the Maximum Principle of optimal control theory.

[12] Throughout, in this chapter, the production function is taken to be of the Cobb–Douglas form.

In Section 5, we provide an explicit solution of an equitable path satisfying Hartwick's investment rule. This is a new result for the Stollery (1998) model. Given our results in Section 4, this is also an equitable path which satisfies the extended version of Hotelling's Rule. Consumption and global warming increase to certain steady state constant levels asymptotically along this equitable path. Thus, one might view Solow's picture of an equitable economy as continuing to be true at least in an asymptotic sense in this framework of global warming.

Our procedure is to first examine the *necessary conditions* for an equitable Hartwick path to exist. These conditions allow us to establish two important *quantitative* implications. The first implication is that the technological restriction of Solow (1974) on the shares of capital (α) and the resource (β) in output:

$$\alpha > \beta$$

must be satisfied. The second implication is that an equitable Hartwick path *must* follow an investment policy:

$$\dot{k}(t) = g(k(t)) \text{ for } t \geq 0$$

where the function g takes the explicit algebraic form:

$$g(x) = \left[\frac{x^\eta}{a + bx^\eta}\right]^{1/(\zeta-1)} \text{ for all } x > 0 \tag{g}$$

where η and ζ are defined in terms of the parameters of the production and utility functions, and where a and b are suitable positive constants.

In order to obtain an explicit solution of an equitable path satisfying Hartwick's investment rule, we choose a particular investment policy function g satisfying (g), by choosing constants a and b appropriately, given the parameters of the model. We then verify that this chosen investment policy function generates a path which is both equitable and satisfies Hartwick's Rule. To illustrate our procedure, we provide an example where the parameters of the model are numerically specified, and we can explicitly solve for both the investment policy function, and the equitable path satisfying Hartwick's Rule.

In Section 6, we specify the law of motion (W) in terms of a *strictly concave* function, G, of resource use; that is, we assume that (G) holds, with $0 < \mu < 1$. Unlike the results obtained in Section 5, the picture of the equitable economy in the long run can be completely altered in this global warming formulation.

In this alternate scenario of global warming, we analyze equitable paths with a constant savings rate. We first examine the *necessary conditions* that must be satisfied by any equitable path on which the savings rate is constant. There are two significant conclusions that emerge from our analysis of the dynamic behavior of investment on such paths.

First, the existence of such a path implies a technological restriction:

$$\alpha > (1/\mu)\beta$$

This can be seen to be a generalization of Solow's technological restriction, which was shown to be an implication of the existence of an equitable Hartwick path in Section 5.

Second, if the investment level on such a path becomes *unbounded* over time (see Section 6.1.5), then the capital stock path *must* exhibit *quasi-arithmetic growth* that is faster than linear growth.

In view of the generalized technological restriction, we formulate an extended version of Hartwick's investment rule (abbreviated as EHR) which requires that investment in the capital good be precisely $(1/\mu)$ times resource rents. When the resource stock is relatively "high" (where "high" is relative to a threshold, determined by the other parameters of the framework), we explicitly construct an equitable path, satisfying the extended Hartwick Rule, to demonstrate that global warming generated can be unbounded along such a path; in fact, along such a path, consumption and global warming are seen to exhibit *quasi-arithmetic growth*. Further, we show that, among equitable paths with a constant savings rate, this path will maximize the constant utility level.[13]

In our model, global warming has a negative effect on welfare levels that can be attained, both because of a direct effect through the utility function, and an indirect effect through the production function. Nevertheless, our results of Sections 4 and 5 show that for equitable paths which are efficient (and therefore satisfy the extended Hotelling Rule), this negative effect by itself does not warrant a savings/investment rate in the capital good in excess of resource rents. In fact, Hartwick's original investment rule continues to be the right investment rule to follow, so long as $\mu = 1$. On the other hand, our results of Section 6 show that, when the elasticity (μ) of the global warming function $G(r)$ is less than 1, then the drag created by global warming warrants a savings/investment rate in excess of resource rents on a path of maximum sustainable utility.

The framework of our analysis is technically more complicated to deal with than the original model of Solow (1974), in which the feature of global warming was absent. Nevertheless, we are able to provide explicit solutions of equitable paths with constant savings rates in Sections 5 and 6. This allows us to directly compare our explicit solutions to the one obtained by Solow (1974), and note the similarities as well as the differences in dynamic behavior along equitable paths in the various models; this is done in Section 7.

A basic property relating to the rates of growth of capital, resource and global warming in all these models is given by the formula:

$$\alpha \frac{\dot{k}(t)}{k(t)} - \beta \left[\frac{-\dot{r}(t)}{r(t)} \right] = (\delta + \rho) \frac{\dot{w}(t)}{w(t)}$$

The right-hand side captures the adverse production (δ) and utility (ρ) effects of (the rate of growth of) global warming. The left-hand side indicates the appropriate rate at which capital must grow to substitute for the declining resource to offset these two effects of global warming and preserve equity.

An asymptotic property which also shows the similarity of these models is that the capital–output ratio eventually exhibits linear growth with time. As a consequence, the real interest rate (given by the marginal product of capital) must exhibit an eventual decline at precisely the reciprocal of time.

In terms of differences of our models of global warming from the model of Solow

---

[13] In particular, our result shows that an equitable path satisfying the standard version of Hartwick's Rule will definitely *not* maximize sustainable utility.

(1974), a basic observation is that along equitable paths with a constant savings rate, the time path of the capital stock is no longer a linear function of time; it is described by a strictly convex function of time. The intuition for this observation is clear. The constant savings rate means that investment must be proportional to consumption, while (irreversible) global warming implies that consumption has to be growing on equitable paths. Thus, investment has to be growing along equitable paths with a constant savings rate.

A second difference is a more subtle one, relating to the rate of decline of resource use compared to the rate of growth of the capital stock along equitable paths with a constant savings rate. In Solow's framework, the rate of decline of resource use along an equitable path with a constant savings rate must always exceed the rate of growth of the capital stock:

$$\left[\frac{-\dot{r}(t)}{r(t)}\right] > \left[\frac{\dot{k}(t)}{k(t)}\right]$$

This result continues to be valid *asymptotically* in the global warming model of Stollery (1998), where μ = 1. When the elasticity (μ) of the global warming function G(r) is strictly between 0 and 1, this result holds when μ is relatively small, but is *reversed* when μ is close to 1. This also indicates that the dynamic behavior observed in Stollery's model *cannot* be viewed as a limiting case of the dynamic behavior observed when 0 < μ < 1.

2. THE FRAMEWORK

Consider a model with one produced good, which serves as both the capital as well as the consumption good, and an exhaustible resource. Labor is assumed to be constant over time. The framework described below is the standard one employed in the literature on intertemporal resource allocation in the presence of an exhaustible resource (see for example, Dasgupta and Heal (1974, 1979), Solow (1974)), modified by the specification of a law of motion of global warming, determined by the use of an exhaustible resource. Such models of global warming have been studied by Stollery (1998) and d'Autume, Hartwick and Schubert (2010), but our specification of the law of motion of global warming differs from these earlier studies, and is described more precisely below.

2.1 Production

Denote by k the stock of the augmentable capital good (which is assumed to be non-depreciating) and by r the flow of the exhaustible resource used. Let $F: \mathbb{R}_+^3 \to \mathbb{R}_+$ denote the gross production function for the capital/consumption good; the production process uses the capital stock k and the exhaustible resource use, r, as inputs, but is in addition affected adversely by the level of global warming, w.[14] The output $Q \equiv F(k,r,w)$ is used

---

[14] In the literature, the notation T has been used for temperature, as the indicator of global warming. Since t is used for time, and almost invariably some finite time horizon is denoted by T, this notation was likely to cause confusion. In view of this, we use w to denote an index of global warming; this should not be difficult to remember.

to augment the capital stock through net investment, $z = \dot{k}$, or to provide consumption, c. Output $F(k, r, w)$ is the only source of flow of consumption or of addition to the stock of capital.

While general forms of the production function are certainly worth analyzing in the context of this model, we assume (as a preliminary first step) that F takes the Cobb–Douglas form:

$$F(k, r, w) = \begin{cases} k^\alpha r^\beta w^{-\delta} & \text{for all } (k, r, w) \in \mathbb{R}^3_{++} \\ 0 & \text{for all } (k, r) \in \mathbb{R}^2_+ \mathbb{R}^2_{++} \text{ and } w \in \mathbb{R}_{++} \end{cases} \quad (2.1)$$

with $\alpha > 0, \beta > 0, \delta > 0$ and $\alpha + \beta < 1$. Thus, both capital and resource are essential for production of positive output. [Note that $F(k, r, w)$ is left undefined for $w = 0$.]

For for all $(k, r, w) \in \mathbb{R}^3_{++}$, we can write:

$$F(k, r, w) = \frac{k^\alpha r^\beta}{w^\delta} = \frac{k^\alpha r^\beta}{\Omega}$$

and interpret Ω as the damage to production because of global warming. Note that we assume $\delta > 0$, thereby allowing for both $0 < \delta \le 1$, and $\delta > 1$. When $\delta > 1$, the damage to production is an increasing, strictly convex function of global warming, w. This makes it comparable to the framework of Nordhaus (2012), who assumes that the damage to production takes the form of a linear-quadratic function of global warming.

2.2 Paths

A *path* from (initial stocks of capital and the exhaustible resource, and an initial level of global warming) (K, M, W) in $\mathbb{R}^2_+ \times \mathbb{R}_{++}$ is a quadruplet of functions $(k(t), r(t), c(t), w(t))$, where $k(\cdot):[0,\infty) \to \mathbb{R}_+$, $r(\cdot):[0,\infty) \to \mathbb{R}_+$, $c(\cdot):[0,\infty) \to \mathbb{R}_+$, $w(\cdot):[0,\infty) \to \mathbb{R}_{++}$ such that $k(t)$, $c(t)$ and $w(t)$ are differentiable functions[15] of t, and $r(t)$ is a continuous function of t, and satisfy:

$$\left. \begin{aligned} &\text{(a)} \quad c(t) = Q(t) - \dot{k}(t) \equiv F(k(t), r(t), w(t)) - \dot{k}(t) \text{ for } t \ge 0 \\ &\text{(b)} \quad \int_0^\infty r(t)\,dt \le M \\ &\text{(c)} \quad \dot{w}(t)/w(t) = \lambda G(r(t)) \text{ for } t \ge 0 \\ &\text{(d)} \quad k(0) = K, w(0) = W \end{aligned} \right\} \quad (2.2)$$

We associate with a path $(k(t), r(t), c(t), w(t))$ from (K, M, W) in $\mathbb{R}^2_+ \times \mathbb{R}_{++}$ the function $M(\cdot):[0,\infty) \to \mathbb{R}_+$, where $M(t)$ is a differentiable function of t, and:

$$M(t) = M - \int_0^t r(s)\,ds \text{ for all } t \ge 0 \quad (2.3)$$

---

[15] Note that along a path $(k(t), r(t), c(t), w(t))$, $k(t)$ is in fact a continuously differentiable function of t, by using (2.2)(a), and $w(t)$ is in fact a continuously differentiable function of t, by using (2.2)(c).

denotes the stock of the resource left over at time t.

The equation (2.2)(c) specifies the law of motion of global warming as a function of resource use:

$$\dot{w}(t)/w(t) = \lambda G(r(t)) \text{ for } t \geq 0 \tag{W}$$

and we specify the parameter λ to be positive and the function $G: \mathbb{R}_+ \to \mathbb{R}_+$ to be a continuous, increasing and concave function on $\mathbb{R}_+$, which is twice continuously differentiable on $\mathbb{R}_{++}$, with $G'(r) > 0$ and $G''(r) \leq 0$ for all $r > 0$. The following function is a special case of G:

$$G(r) = r^\mu \text{ for all } r \geq 0 \tag{G}$$

where the parameter μ is restricted to lie in the half-open interval $(0,1]$. In the limiting case of $\lambda = 0$, we recover the model of Solow (1974). With $\lambda > 0$, and $\mu = 1$, the rate of growth of global warming is proportional to resource use, and this is the formulation used by Stollery (1998) and d'Autume, Hartwick and Schubert (2010). When we specify μ to lie in the open interval $(0,1)$, the rate of growth of global warming is a strictly concave function of resource use.

A path $(k(t), r(t), c(t), w(t))$ from (K, M, W) in $\mathbb{R}_+^2 \times \mathbb{R}_{++}$ is called *interior* if $k(t) > 0, r(t) > 0, c(t) > 0$ for $t \geq 0$.

2.3 Preferences

The welfare of the economy at any moment in time is measured by an instantaneous utility function, $U: \mathbb{R}_+ \times \mathbb{R}_{++} \to \mathbb{R} \cup \{-\infty\}$, continuously differentiable on its domain, which depends on the level of consumption and the level of global warming at that moment of time. We suppose that this utility function takes the form:

$$U(c, w) = u(c/w^\rho) \text{ for all } (c, w) \in \mathbb{R}_+ \times \mathbb{R}_{++} \tag{2.4}$$

where $u: \mathbb{R}_{++} \to \mathbb{R}$ is an increasing, continuously differentiable function on $\mathbb{R}_{++}$, with $u(0) = -\infty = \lim_{x \to 0} u(x)$, and $\rho > 0$ a taste parameter. A concrete example of u with these properties is:

$$u(x) = \begin{cases} \ln x & \text{for } x > 0 \\ -\infty & \text{for } x = 0 \end{cases}$$

With $\rho > 1$, one would accommodate cases in which one requires more than a one percent increase in consumption (c) to compensate for a one percent increase in global warming (w) to leave utility $U(c, w)$ at the same level.

Equitable intertemporal preferences are expressed as requiring that the utility level be a finite constant over time. This notion is made precise by defining a path $(k(t), r(t), c(t), w(t))$ from (K, M, W) in $\mathbb{R}_+^2 \times \mathbb{R}_{++}$ to be *equitable* if there is a real number $\bar{u}$, such that:

$$U(c(t), w(t)) = u(c(t)/w(t)^\rho) = \bar{u} \text{ for all } t \geq 0 \tag{2.5}$$

Note that this definition of equity entails that $(c(t)/w(t)^\rho)$ is itself constant (since u is an increasing function), and this constant is positive, since $u(0) = -\infty$. Since $w(t) > 0$ for $t \geq 0$ along any path, we must therefore have $c(t) > 0$ for all $t \geq 0$, and $c(t)$ must be a continuously differentiable function of t, by (2.2)(c). Further, we must have the rate of change of consumption exactly equal to ρ times the rate of change of (the index) of global warming:

$$\frac{\dot{c}(t)}{c(t)} = \rho \frac{\dot{w}(t)}{w(t)} \text{ for all } t \geq 0 \tag{2.6}$$

3. SHORT-RUN EFFICIENCY

A limiting case of our model, *without* global warming (examined by Solow (1974) and Dasgupta and Heal (1979)), can be obtained by setting $\lambda = 0$, and noting that along any path $(k(t), r(t), c(t), w(t))$ from (K, M, W) in $\mathbb{R}_+^2 \times \mathbb{R}_{++}$, global warming is constant with $w(t) = W$ for all $t \geq 0$. In such a framework, short-run efficiency is characterized by *Hotelling's Rule* (abbreviated as (HoR) in what follows) which requires that the return on the capital good and on the resource be equalized at each moment of time.[16] This translates to the condition that the price of the resource (in terms of the produced good) grows at the real interest rate determined by the marginal product of capital:

$$\frac{\dot{F}_2(k(t), r(t), W)}{F_2(k(t), r(t), W)} = F_1(k(t), r(t), W) \text{ for all } t \geq 0 \tag{HoR}$$

A comparable result for our model *with* global warming has not been investigated in the literature. We try to fill this gap by introducing formally the concept of short-run efficiency for our model, and establishing (in subsection 3.1) the *necessary conditions* for an interior path $(k(t), r(t), c(t), w(t))$ from (K, M, W) in $\mathbb{R}_+^2 \times \mathbb{R}_{++}$ to be short-run efficient.

In our model with global warming, the production function is not a concave function in the global warming variable, w. Consequently, we are not claiming that these derived conditions are *sufficient* for short-run efficiency. The fact that we are now dealing with optimization problems on a non-convex feasible set has not been adequately appreciated in this literature. The reader should note that our cautious statements here, and later in the chapter, with respect to any claim about sufficiency results, are made with good reason.

In subsection 3.2, we concentrate on the formulation of global warming (W), with G specified as in (G), with $\mu = 1$. The necessary condition of short-run efficiency is then shown to be an *extended version of Hotelling's Rule* (abbreviated as (EHoR) in what follows), the extension being a reflection of the externality generated by global warming. Following d'Autume, Hartwick and Schubert (2010), it is possible to relate this EHoR condition to the standard form of Hotelling's Rule by noting that one can design a (time-dependent) Pigouvian tax such that the *net* price of the resource (in terms of the produced good) grows at the real interest rate determined by the marginal product of capital.

---

[16] For a discrete-time version of this model, this result was analyzed in Mitra (1978).

3.1 Necessary Conditions of Short-Run Efficiency

A path $(k(t), r(t), c(t), w(t))$ from (K, M, W) in $\mathbb{R}_+^2 \times \mathbb{R}_{++}$ is said to be *short-run efficient* if, given any S and T, with $0 \leq S < T < \infty$, there is no quadruple of functions[17] $(k'(t), r'(t), c'(t), w'(t))$ with $k'(\cdot):[0,\infty) \to \mathbb{R}_+, r'(\cdot):[0,\infty) \to \mathbb{R}_+, c'(\cdot):[0,\infty) \to \mathbb{R}_+,$ $w'(\cdot):[0,\infty) \to \mathbb{R}_{++}$ for $S \leq t \leq T$, such that $k'(t), c'(t)$ and $w'(t)$ are differentiable functions of t, and $r'(t)$ is a continuous functions of t for $S \leq t \leq T$, and satisfy:

$$
\left.\begin{array}{ll}
\text{(a)} & c'(t) = Q'(t) - \dot{k}'(t) \equiv F(k'(t), r'(t), w'(t)) - \dot{k}'(t) \text{ for } S \leq t \leq T \\
\text{(b)} & \int_S^T r'(t)\,dt \leq M(S) - M(T) \\
\text{(c)} & \dot{w}'(t)/w'(t) = \lambda G(r'(t)) \text{ for } S \leq t \leq T \\
\text{(d)} & k'(S) = k(S),\ w'(S) = w(S) \\
\text{(e)} & w'(T) \leq w(T),\ k'(T) \geq k(T)
\end{array}\right\} \quad (3.1)
$$

and:

$$
\left.\begin{array}{ll}
U(c'(t), w'(t)) \geq U(c(t), w(t)) & \text{for all } S \leq t \leq T \\
U(c'(t), w'(t)) > U(c(t), w(t)) & \text{for some } t \in S, T]
\end{array}\right\} \quad (3.2)
$$

In other words, a path $(k(t), r(t), c(t), w(t))$ from (K, M, W) in $\mathbb{R}_+^2 \times \mathbb{R}_{++}$ is short-run efficient if, given any finite segment of time $[S, T]$, it is not feasible, starting with the same stocks of capital, resource and global warming stocks at S, and ending up with at least as much of the capital and resource stocks at T, and at most as much of the global warming stock at T, to Pareto-wise dominate the utility stream of the path $(k(t), r(t), c(t), w(t))$ on the time interval $[S, T]$.[18]

If an interior path $(k(t), r(t), c(t), w(t))$ from (K, M, W) in $\mathbb{R}_+^2 \times \mathbb{R}_{++}$ is short-run efficient, then it can be checked that, given any S, T satisfying $0 \leq S < T < \infty$, the choice of controls $(r(t), c(t))$ for all $t \in S, T]$ must be such that they solve the following optimal control problem:

$$\text{Max} \int_S^T [F(k'(t), r'(t), w'(t)) - c'(t)]\,dt$$

subject to:

---

[17] Note that unlike $(k(t), r(t), c(t), w(t))$, which is a *path*, the quadruple of functions $(k'(t), r'(t), c'(t), w'(t))$ need not constitute a *path*, since given S and T, it needs to satisfy only the constraints (3.1) and (3.2) which pertain to the finite time interval $[S, T]$, and not the infinite interval $[0, \infty)$.

[18] Note that the second condition in (3.2) holds if and only if the strict inequality holds on an open interval contained in $[S, T]$. This follows from the continuity of $c(t)$ and $w(t)$ as functions of time, and the continuity of the utility function U.

(a) $\dot{k}'(t) = F(k'(t), r'(t), w'(t)) - c'(t)$ for $S \leq t \leq T$
(b) $\dot{M}'(t) = -r'(t)$ for $S \leq t \leq T$
(c) $\dot{w}'(t)/w'(t) = \lambda G(r'(t))$ for $S \leq t \leq T$
(d) $U(c'(t), w'(t)) \geq U(c(t), w(t))$ for $S \leq t \leq T$
(e) $w'(T) \leq w(T), M'(T) \geq M(T)$
(f) $k'(S) = k(S), w'(S) = w(S), M'(S) = M(S)$
(g) $(k'(t), r'(t), c'(t), w'(t)) \geq 0$ for $S \leq t \leq T$

(3.3)

at the values specified by the path $(k(t), r(t), c(t), w(t))$ for all $t \in S, T]$.

Here, we have converted a Pareto optimality problem into a problem of maximizing the terminal capital stock (at T), subject to all the constraints in (3.1) except the terminal capital constraint in (3.1)(e), and including the weak dominance in the utility stream in (3.3)(d). This form is amenable to analysis using the methods of optimal control theory.

Applying a standard result from optimal control theory (see Arrow (1968), p. 90),[19] there exist auxiliary variables $p(t) = (p_1(t), p_2(t), p_3(t))$ (corresponding to the laws of motion of the state variables described in (3.3)(a)–(c)) and Lagrange multipliers $q(t) \geq 0$ (corresponding to the inequality constraints (3.3)(d)) for all $t \in S, T]$ such that the following first-order conditions with respect to the controls hold for all $t \in S, T]$:

(i) $F_2(k(t), r(t), w(t)) + p_1(t) F_2(k(t), r(t), w(t)) + p_2(t)(-1) + p_3(t) w(t) \lambda G'(r(t)) = 0$
(ii) $-1 + p_1(t)(-1) + q(t) U_1(c(t), w(t)) = 0$

(3.4)

and such that the following laws of motion of the auxiliary variables hold for all $t \in S, T]$:

(i) $\dot{p}_1(t) = -F_1(k(t), r(t), w(t)) - p_1(t) F_1(k(t), r(t), w(t))$
(ii) $\dot{p}_2(t) = 0$
(iii) $\dot{p}_3(t) = -F_3(k(t), r(t), w(t)) - p_1(t) F_3(k(t), r(t), w(t))$
       $- p_3(t) \lambda G(r(t)) - q(t) U_2(c(t), w(t))$

(3.5)

Dropping the points of evaluation[20] of the marginal products and marginal utilities, the conditions in (3.4) and (3.5) can be used to obtain the following general necessary condition (abbreviated as (NC)) of short-run efficiency:

$$\left[F_1(\cdot) - \frac{\dot{F}_2(\cdot)}{F_2(\cdot)} \right] = - \left[\frac{G'(r(t)) \lambda w(t) r(t)}{F_2(\cdot) r(t)} \right] \left[F_3(\cdot) + \frac{U_2(\cdot)}{U_1(\cdot)} \right]$$

---

[19] To apply the result, one needs to check a constraint qualification condition; but, this is quite straightforward for this optimal control problem.

[20] For notational ease we drop the arguments of the partial derivatives of the production function and the utility function appearing below. Unless mentioned otherwise, it is understood that all the partial derivatives are evaluated at time t and along the path, that is, at the point $(k(t), r(t), w(t))$, represented by $(\cdot)$, for the production function, and at the point $(c(t), w(t))$, again represented by $(\cdot)$, for the utility function.

$$+ \left[\frac{1}{\{1+p_1(t)\}F_2(\cdot)}\right] p_3(t)w(t)\lambda G''(r(t))\dot{r}(t) NC \tag{1}$$

The technical details leading to (NC) can be found in an appendix to Section 3 (see section 8.1).

3.2 Extended Version of Hotelling's Rule

We illustrate in this subsection the necessary condition (NC) in the case where the function G [in the law of motion of global warming (W)] takes the form, given in (G), with $\mu = 1$. This is the case analyzed in d'Autume, Hartwick and Schubert (2010). In this case $G'(r) = 1$, and $G''(r) = 0$, so that (NC) becomes:

$$\left[F_1(\cdot) - \frac{\dot{F}_2(\cdot)}{F_2(\cdot)}\right] = -\left[\frac{\lambda w(t) r(t)}{F_2(\cdot) r(t)}\right]\left[F_3(\cdot) + \frac{U_2(\cdot)}{U_1(\cdot)}\right]$$

which can be rewritten as as an *extended version of Hotelling's Rule* (abbreviated as (EHoR) in what follows):

$$\left[F_1(\cdot) - \frac{\dot{F}_2(\cdot)}{F_2(\cdot)}\right] = -\left[\frac{\dot{w}(t)}{F_2(\cdot) r(t)}\right]\left[F_3(\cdot) + \frac{U_2(\cdot)}{U_1(\cdot)}\right] \tag{EHoR}$$

In (EHoR), the standard version of Hotelling's Rule (HoR) (noted at the beginning of Section 3) is modified by the right-hand side expression, to take account of the global warming externality in both production and utility.

Notice that the second term on the right-hand side of (NC) in Section 3.1 *does not* vanish when $0 < \mu < 1$, and so (EHoR) is *not* the correct necessary condition in that case.

3.2.1 Pigouvian taxes

Along an interior path $(k(t), r(t), c(t), w(t))$ from (K, M, W) in $\mathbb{R}_+^2 \times \mathbb{R}_{++}$, which satisfies the extended version of Hotelling's Rule, $F_2(k(t), r(t), w(t))$ can be interpreted as the price of the resource in terms of the physical good, and the marginal product of capital $F_1(k(t), r(t), w(t))$ is the real rate of interest.

Then, $F_2(k(t), r(t), w(t))$ is the price of the resource paid by producers of the physical good. If $\tau(t)$ is a Pigouvian carbon tax designed to internalize the global warming externality, then $[F_2(k(t), r(t), w(t)) - \tau(t)]$ is the net resource price received by the owners of the resource. The condition (EHoR) can then be seen equivalently as the condition that this *net* resource price grows at the real interest rate determined by the marginal product of capital, when the time-dependent Pigouvian tax is chosen appropriately.

The choice of the Pigouvian tax $\tau(t)$ is one satisfying:

$$\tau(t) = \int_t^\infty e^{-\int_t^s \rho(u)du}\left[F_3(\cdot) + \frac{U_2(\cdot)}{U_1(\cdot)}\right](-\lambda w(s))ds \tag{PT}$$

where, $\rho(t) \equiv F_1(k(t), r(t), w(t))$, the marginal product of capital, is the real interest rate at time $t \geq 0$, and the partial derivatives are evaluated at time s. The condition (PT) can be seen to be the requirement that the carbon tax at each point in time be equal to the discounted value of future marginal damages of global warming to production and utility, where discounting is done at the real rate of interest $F_1(k(t), r(t), w(t))$.

Let us write the marginal damages (MD) in the (PT) formula, at time s, as:

$$MD(s) = \left[F_3(\cdot) + \frac{U_2(\cdot)}{U_1(\cdot)}\right](-\lambda w(s)) \qquad \text{(MD)}$$

where the partial derivatives are evaluated at time s. Note that these marginal damages are positive because $F_3(\cdot) < 0$, $U_2(\cdot) < 0$, $U_1(\cdot) > 0$, while $(-\lambda w(s)) < 0$. Thus, by (PT), the Pigouvian carbon taxes are also positive.

We now come to the rate of change in Pigouvian taxes. This can be obtained by differentiating (PT) with respect to t:

$$\dot{\tau}(t) = \tau(t)F_1(\cdot) - \left[F_3(\cdot) + \frac{U_2(\cdot)}{U_1(\cdot)}\right](-\lambda w(t)) \text{ for all } t \geq 0$$

which can be rewritten as:

$$\dot{\tau}(t) - \tau(t)F_1(\cdot) = \lambda w(t)\left[F_3(\cdot) + \frac{U_2(\cdot)}{U_1(\cdot)}\right] \text{ for all } t \geq 0 \qquad \text{(PTDOT)}$$

where the partial derivatives are evaluated at time t.

The formula (PTDOT) leads to two results which we now describe.

(i) Proportionate rate of change of Pigouvian tax Since the right-hand side expression in (PTDOT) is negative, we obtain:

$$\dot{\tau}(t) < \tau(t)F_1(k(t), r(t), w(t)) \text{ for all } t \geq 0$$

Thus, we get the important result that the proportionate rate of change of Pigouvian taxes must always be smaller than the real rate of interest:

$$\frac{\dot{\tau}(t)}{\tau(t)} < F_1(k(t), r(t), w(t)) \text{ for all } t \geq 0 \qquad (3.6)$$

Note that Pigouvian taxes might be falling at each point in time (as in Stollery (1998)), in which case (3.6) clearly holds. But, in general, $\dot{\tau}(t) > 0$ might hold over some time-interval, so that Pigouvian taxes would increase over that time-interval. But, even in such cases, its growth rate cannot exceed the rate of interest.

(ii) Proportionate rate of change of Resource Price net of Pigouvian tax Since Pigouvian taxes change over time according to (PTDOT), we get:

$$\left[\frac{\dot{w}(t)}{r(t)F_2(\cdot)}\right]\left[F_3(\cdot) + \frac{U_2(\cdot)}{U_1(\cdot)}\right] = \left[\frac{\dot{w}(t)}{r(t)F_2(\cdot)}\right]\left[\frac{\dot{\tau}(t) - \tau(t)F_1(\cdot)}{\lambda w(t)}\right] = \left[\frac{\dot{\tau}(t) - \tau(t)F_1(\cdot)}{F_2(\cdot)}\right]$$
(3.7)

Thus, if the Extended Hotelling Rule (EHoR) holds, then (3.7) implies:

$$\left[F_1(\cdot) - \frac{\dot{F}_2(\cdot)}{F_2(\cdot)}\right] = -\left[\frac{\dot{\tau}(t) - \tau(t)F_1(\cdot)}{F_2(\cdot)}\right] = \left[\frac{\tau(t)F_1(\cdot)}{F_2(\cdot)} - \frac{\dot{\tau}(t)}{F_2(\cdot)}\right]$$

which can be rewritten as:

$$\left[\frac{\dot{F}_2(\cdot) - \dot{\tau}(t)}{F_2(\cdot)}\right] = \frac{\dot{F}_2(\cdot)}{F_2(\cdot)} - \frac{\dot{\tau}(t)}{F_2(\cdot)} = F_1(\cdot) - \frac{\tau(t)F_1(\cdot)}{F_2(\cdot)} = \frac{F_1(\cdot)[F_2(\cdot) - \tau(t)]}{F_2(\cdot)} \quad (3.8)$$

Transposing terms in (3.8), we get:

$$\left[\frac{\dot{F}_2(\cdot) - \dot{\tau}(t)}{F_2(\cdot) - \tau(t)}\right] = F_1(\cdot) \text{ for all } t \geq 0 \quad (3.9)$$

so that the growth rate of the *net* resource price is the real interest rate.

Remark:
We caution the reader that there is no claim that condition (EHoR) ensures short-run efficiency. Similarly, given appropriate Pigouvian taxes, defined by (PT), there is no claim that condition (3.9) ensures short-run efficiency. The conditions we have derived are necessary conditions of short-run efficiency, and the non-convexity of the feasible set prevents us from making statements about these conditions being sufficient for short-run efficiency.

4. EXTENDED HOTELLING RULE, EQUITY AND HARTWICK'S RULE

As we have noted above, a limiting case of our model, *without* global warming (examined by Solow (1974) and Dasgupta and Heal (1979)), can be obtained by setting $\lambda = 0$, and noting that along any path $(k(t), r(t), c(t), w(t))$ from (K, M, W) in $\mathbb{R}_+^2 \times \mathbb{R}_{++}$, global warming is constant with $w(t) = W$ for all $t \geq 0$.

An important insight about this model was the observation by Hartwick (1977) that if the investment in the capital good was equal to resource rents (along a path satisfying Hotelling's Rule), then the path would be equitable. This investment policy has come to be known in the literature as *Hartwick's Rule*. In such a framework, there is in fact a rich set of equivalence results, which may be described as follows. Consider the following three conditions that a path may satisfy: (i) it is equitable; (ii) it satisfies Hartwick's Rule; (iii) it satisfies Hotelling's Rule. It turns out that if the path satisfies any two of these three conditions, it must also satisfy the third. These equivalence results were established by Buchholz, Dasgupta and Mitra (2005).

In this section, we revisit these results in the context of our current model of global warming, in the case where the function G [in the law of motion of global warming (W)] takes the form given in (G), with $\mu = 1$. An interior path $(k(t), r(t), c(t), w(t))$ from (K, M, W) in $\mathbb{R}_+^2 \times \mathbb{R}_{++}$ is said to satisfy *Hartwick's Rule* (abbreviated as (HR) in what follows) if:

$$\dot{k}(t) = r(t)F_2(k(t), r(t), w(t)) \text{ for all } t \geq 0 \quad \text{(HR)}$$

In this case, we often refer to the path itself as a *Hartwick path*.

It is rather remarkable that the equivalence results continue to hold in this model of global warming, the only modification being that Hotelling's Rule is replaced by (EHoR), the extended version of Hotelling's Rule, discussed in Section 3. Our approach to these equivalence results follows Buchholz, Dasgupta and Mitra (2005) quite closely.

The analogue of Hartwick's original result is already noted in d'Autume, Hartwick and Schubert (2010). The converse result is new in this setting: when an equitable path satisfies the extended version of Hotelling's Rule, then it *must* follow the rule that the investment in the capital good be equal to resource rents. This is an important result because it says that this investment rule is the *only* investment rule consistent with the incontrovertible objectives of equity and efficiency.

4.1 A Useful Identity

We begin our exposition of the three equivalence results by establishing a useful identity. All three results then follow by using this identity at an appropriate point.

Consider any interior path $(k(t), r(t), c(t), w(t))$ from (K, M, W) in $\mathbb{R}_+^2 \times \mathbb{R}_{++}$. Then, we have:

$$\frac{d}{dt}\left[\frac{k(t)}{F_2(\cdot)} - r(t)\right] = \frac{1}{F_2(\cdot)}\frac{dk(t)}{dt} - k(t)\frac{\dot{F}_2(\cdot)}{(F_2(\cdot))^2} - \dot{r}(t) \text{ for } t \geq 0 \quad (4.1)$$

Multiplying through in (4.1) by $F_2(\cdot)$, we obtain:

$$F_2(\cdot)\frac{d}{dt}\left[\frac{k(t)}{F_2(\cdot)} - r(t)\right] = \frac{dk(t)}{dt} - k(t)\frac{\dot{F}_2(\cdot)}{F_2(\cdot)} - F_2(\cdot)\dot{r}(t) \text{ for } t \geq 0 \quad (4.2)$$

Also, differentiating (2.2)(a) with respect to t, we get:

$$\dot{c}(t) = F_1(\cdot)\dot{k}(t) + F_2(\cdot)\dot{r}(t) + F_3(\cdot)\dot{w}(t) - \frac{dk(t)}{dt} \text{ for } t \geq 0 \quad (4.3)$$

Adding (4.2) and (4.3), we obtain the important identity:

$$\dot{c}(t) + F_2(\cdot)\frac{d}{dt}\left[\frac{k(t)}{F_2(\cdot)} - r(t)\right] = \left[F_1(\cdot) - \frac{\dot{F}_2(\cdot)}{F_2(\cdot)}\right]k(t) + F_3(\cdot)\dot{w}(t) \text{ for } t \geq 0 \quad (4.4)$$

Note that this identity is valid for any interior path; equity, Hartwick's Rule or the extended version of Hotelling's Rule are not relevant for its validity. We restrict our analysis to interior paths to make sure that all the relevant partial derivatives are well-defined.

4.2 Equitable Hartwick Paths Satisfy the Extended Hotelling Rule

If we now specialize to an interior path which satisfies Hartwick's Rule, then we obtain from (4.4):

$$\dot{c}(t) = \left[F_1(\cdot) - \frac{\dot{F}_2(\cdot)}{F_2(\cdot)}\right]k(t) + F_3(\cdot)\dot{w}(t) \text{ for } t \geq 0 \quad (4.5)$$

which yields after transposing terms:

$$\left[F_1(\cdot) - \frac{\dot{F}_2(\cdot)}{F_2(\cdot)}\right]k(t) = \dot{c}(t) - F_3(\cdot)\dot{w}(t) \text{ for } t \geq 0 \quad (4.6)$$

Let us now go further and impose the condition of equity on the path; that is, condition (2.5) holds. Consequently, we have:

$$U_1(c(t),w(t))\dot{c}(t) + U_2(c(t),w(t))\dot{w}(t) = 0 \text{ for } t \geq 0$$

which yields:

$$\dot{c}(t) = -\dot{w}(t)\left[\frac{U_2(c(t),w(t))}{U_1(c(t),w(t))}\right] \text{ for } t \geq 0 \quad (4.7)$$

Combining (4.6) and (4.7), we obtain:

$$\left[F_1(\cdot) - \frac{\dot{F}_2(\cdot)}{F_2(\cdot)}\right]k(t) = -\dot{w}(t)\left[F_3(\cdot) + \frac{U_2(c(t),w(t))}{U_1(c(t),w(t))}\right] \text{ for } t \geq 0 \quad (4.8)$$

By Hartwick's Rule we know that $\dot{k}(t) = r(t)F_2(\cdot) > 0$ for all $t \geq 0$ along an interior path. So, the condition (EHoR) follows from (4.8).

4.3 Hartwick Paths Satisfying the Extended Hotelling Rule are Equitable

We now establish the analogue of the result, originally established by Hartwick (1977) in a model without global warming.

Using (EHoR) in the basic identity (4.4), we get for all $t \geq 0$,

$$\dot{c}(t) + F_2(\cdot)\frac{d}{dt}\left[\frac{k(t)}{F_2(\cdot)} - r(t)\right] = \left\{-\left[F_3(\cdot) + \frac{U_2(c(t),w(t))}{U_1(c(t),w(t))}\right]\frac{\dot{w}(t)}{r(t)F_2(\cdot)}\right\}k(t)$$
$$+ F_3(\cdot)\dot{w}(t) \quad (4.9)$$

Now using Hartwick's Rule in (4.9), we obtain for $t \geq 0$,

$$\dot{c}(t) = -\left[F_3(\cdot) + \frac{U_2(c(t),w(t))}{U_1(c(t),w(t))}\right]\dot{w}(t) + F_3(\cdot)\dot{w}(t)$$

$$= \left[-\frac{U_2(c(t),w(t))}{U_1(c(t),w(t))}\right]\dot{w}(t)$$

Rearranging terms yields:

$$U_1(c(t),w(t))\dot{c}(t) + U_2(c(t),w(t))\dot{w}(t) = 0 \text{ for all } t \geq 0 \quad (4.10)$$

which implies that:

$$\frac{dU(c(t),w(t))}{dt} = 0 \text{ for all } t \geq 0$$

Thus, $U(c(t),w(t))$ must be a constant for all $t \geq 0$. Since $(k(t), r(t), c(t), w(t))$ is interior, this constant is not $-\infty$, and so the path is equitable.

4.4 Equitable Paths Satisfying the Extended Hotelling Rule are Hartwick Paths

We come now to the final equivalence result of this section: if an interior path $(k(t), r(t), c(t), w(t))$ from $(K, M, W) \in \mathbb{R}^2_+ \times \mathbb{R}_{++}$ satisfies equity and the extended version of Hotelling's Rule (EHoR), then it must satisfy Hartwick's Rule. If one considers equity and efficiency as the two guiding principles in intertemporal choice, then this result implies that *any* path which satisfies both of these guiding principles must necessarily satisfy Hartwick's Rule. Since the condition (EHoR) is analytically harder to deal with than Hartwick's Rule, this result is a major convenience as it allows us to work with equitable Hartwick paths in Section 5, where we focus on the dynamics of paths which are both equitable and short-run efficient.

The proof of this final equivalence result is more involved than the proofs of the results described in the previous two subsections. To help in the exposition, we break up the proof into four steps. Also, some of the technical details are presented in the appendix to Section 4 (which appears as section 8.2), to make the text more readable.

Let $(k(t), r(t), c(t), w(t))$ be an interior path from $(K, M, W) \in \mathbb{R}^2_+ \times \mathbb{R}_{++}$, which is equitable. Then, $(c(t)/w(t)^\rho)$ must be a positive constant, and let us denote this constant by $\eta > 0$. That is,

$$c(t) = \eta w(t)^\rho \text{ for all } t \geq 0 \tag{4.11}$$

First Step:
We first show, following Solow (1974), that this implies the following restriction on the technological parameters:

$$\alpha \geq \beta \tag{S'}$$

The proof of (S') is a slightly modified version of the proof appearing in Solow (1974, p. 43), and so we present it in the appendix to Section 4 (see section 8.2).

Second Step:
Next we establish, using both equity and condition (EHoR), a differential equation which governs the behavior of the variable:

$$v(t) \equiv \left[\frac{k(t)}{F_2(\cdot)} - r(t)\right] \text{ for all } t \geq 0 \tag{4.12}$$

For this we start with the identity (4.4), and use condition (EHoR) to write for $t \geq 0$,

$$\dot{c}(t) + F_2(\cdot)\frac{d}{dt}\left[\frac{k(t)}{F_2(\cdot)} - r(t)\right] = \left[F_1(\cdot) - \frac{\dot{F}_2(\cdot)}{F_2(\cdot)}\right]k(t) + F_3(\cdot)w(t)$$

$$= \left\{-\left[F_3(\cdot) + \frac{U_2(c(t),w(t))}{U_1(c(t),w(t))}\right]\frac{\dot{w}(t)}{r(t)F_2(\cdot)}\right\}k(t) + F_3(\cdot)\dot{w}(t)$$

$$= (-F_3(\cdot))\dot{w}(t)\left[\frac{k(t)}{r(t)F_2(\cdot)} - 1\right] + \frac{\dot{c}(t)k(t)}{r(t)F_2(\cdot)} \tag{4.13}$$

where in obtaining the last line of (4.13), we have used the fact that the equity condition (2.5) implies:

$$U_1(c(t),w(t))\dot{c}(t) + U_2(c(t),w(t))\dot{w}(t) = 0 \text{ for } t \geq 0$$

Transposing terms in (4.13), we can write, for $t \geq 0$,

$$F_2(\cdot)\frac{d}{dt}\left[\frac{k(t)}{F_2(\cdot)} - r(t)\right] = (-F_3(\cdot))\dot{w}(t)\left[\frac{k(t)}{r(t)F_2(\cdot)} - 1\right] + \dot{c}(t)\left[\frac{k(t)}{r(t)F_2(\cdot)} - 1\right]$$

$$= \frac{(-F_3(\cdot))\dot{w}(t)}{r(t)}\left[\frac{k(t)}{F_2(\cdot)} - r(t)\right] + \frac{\dot{c}(t)}{r(t)}\left[\frac{k(t)}{F_2(\cdot)} - r(t)\right]$$

Thus, dividing through by $F_2(\cdot)$, and using the equity condition (2.6), we get:

$$\dot{v}(t) = \left[\frac{(-F_3(\cdot))\dot{w}(t)}{F_2(\cdot)r(t)} + \frac{pc(t)\dot{w}(t)}{w(t)F_2(\cdot)r(t)}\right]v(t) \text{ for all } t \geq 0 \quad (4.14)$$

Third Step:
The display (4.14) is the basic differential equation governing the dynamic behavior of $v(t)$ for $t \geq 0$, as a consequence of equity and condition (EHoR). Our task is to establish that $v(t) = 0$ for all $t \geq 0$, which is Hartwick's Rule.

Note from (4.14) that if $v(t) > 0$ for some $t = \tau \geq 0$, then by continuity of $v(t)$, there is $\varepsilon > 0$ such that $v(t) > 0$ for all $t \in \tau, \tau + \varepsilon]$. Since the term in square brackets in (4.14) is positive for an interior path, we must have $\dot{v}(t) > 0$ for all $t \in \tau, \tau + \varepsilon]$. Thus, $v(t)$ must be increasing on $[\tau, \tau + \varepsilon]$, and so $v(t) > v(\tau)$ for all $t \in (\tau, \tau + \varepsilon]$.

Claim 1: We claim now that: $v(t) \geq v(\tau)$ for all $t \geq \tau$.
The proof of **Claim 1** is presented in the appendix to Section 4 (see section 8.2). With **Claim 1** established, we have $v(t) \geq v(\tau) \equiv E > 0$ for all $t \geq \tau$.

Using the definition of $v(t)$, we then have:

$$\left[\frac{k(t)}{F_2(\cdot)} - r(t)\right] \geq E \text{ for all } t \geq \tau$$

and so we get, using the fact that $c(t) > 0$ for all $t \geq 0$, and (2.2)(a),

$$\frac{Q(t)r(t)}{\beta Q(t)} = \frac{Q(t)r(t)}{F_2(\cdot)r(t)} = \frac{Q(t)}{F_2(\cdot)} \geq \left[\frac{k(t)}{F_2(\cdot)}\right] \geq r(t) + E \geq E \text{ for all } t \geq \tau$$

This implies that:

$$r(t) \geq \beta E \text{ for all } t \geq \tau$$

But, this clearly contradicts the resource constraint (2.2)(b).
Our analysis has shown that if $v(t) > 0$ for some $t \geq 0$, then this leads us to a contradiction. Thus we can conclude:

$$v(t) \equiv \left[\frac{k(t)}{F_2(\cdot)} - r(t)\right] \leq 0 \text{ for all } t \geq 0 \quad (4.15)$$

Fourth Step:
We are not done, of course. We need now to show that if $v(t) < 0$ for some $t \geq 0$, then we are also led to a contradiction.

Note from (4.14) that if $v(t) < 0$ for some $t = \tau \geq 0$, then by continuity of $v(t)$, there is $\varepsilon > 0$ such that $v(t) < 0$ for all $t \in \tau, \tau + \varepsilon]$. Since the term in square brackets in (4.14) is positive for an interior path, we must have $\dot{v}(t) < 0$ for all $t \in \tau, \tau + \varepsilon]$. Thus, $v(t)$ must be decreasing on $[\tau, \tau + \varepsilon]$, and so $v(t) < v(\tau)$ for all $t \in (\tau, \tau + \varepsilon]$.

Claim 2: We claim now that: $v(t) \leq v(\tau)$ for all $t \leq \tau$.
The proof of **Claim 2** is presented in the appendix to Section 4 (see section 8.2). With **Claim 2** established, we have $v(t) \leq v(\tau) \equiv E < 0$ for all $t \geq \tau$.

Next, we work with (4.14), and given our result, it is convenient to define $V(t) = -v(t)$ for all $t \geq \tau$. Then $V(t) \geq V(\tau) \equiv (-E) \equiv E' > 0$ for all $t \geq \tau$, and:

$$\dot{V}(t) = \left[\frac{(-F_3(\cdot))\dot{w}(t)}{F_2(\cdot)r(t)} + \frac{\rho c(t)\dot{w}(t)}{w(t)F_2(\cdot)r(t)}\right] V(t) \text{ for all } t \geq \tau$$

It will be convenient to rewrite this equation (for $t \geq \tau$) as:

$$\frac{\dot{V}(t)}{V(t)} = \left[\frac{(-F_3(\cdot))\dot{w}(t)}{F_2(\cdot)r(t)} + \frac{\rho c(t)\dot{w}(t)}{w(t)F_2(\cdot)r(t)}\right]$$

$$= \frac{\dot{w}(t)}{w(t)}\left[\frac{(-F_3(\cdot))w(t)}{F_2(\cdot)r(t)} + \frac{\rho c(t)}{F_2(\cdot)r(t)}\right]$$

$$= \frac{\dot{w}(t)}{w(t)}\left[\frac{\delta Q(t)}{\beta Q(t)} + \frac{\rho c(t)}{\beta Q(t)}\right]$$

$$= \frac{\dot{w}(t)}{w(t)}\left[\frac{\delta}{\beta} + \frac{\rho c(t)}{\beta Q(t)}\right]$$

Using (4.15), we have: $k(t) \leq F_2(\cdot)r(t) = \beta Q(t)$, and so by (2.2)(a), we get $c(t) = Q(t) - k(t) \geq Q(t) - \beta Q(t) = (1-\beta)Q(t)$. Using this information above, we get:

$$\frac{\dot{V}(t)}{V(t)} = \frac{\dot{w}(t)}{w(t)}\left[\frac{\delta}{\beta} + \frac{\rho c(t)}{\beta Q(t)}\right] \geq \frac{\dot{w}(t)}{w(t)}\left[\frac{\delta}{\beta} + \frac{\rho(1-\beta)}{\beta}\right] = \frac{\dot{w}(t)}{w(t)}\left[\frac{\delta + \rho(1-\beta)}{\beta}\right] \quad (4.16)$$

Since $\alpha + \beta < 1$, we have $(1 - \beta) > \alpha$; further, by (S') established in the First Step, we have $\alpha \geq \beta$. Thus, $(1 - \beta) > \beta$, and we get from (4.16),

$$\frac{\dot{V}(t)}{V(t)} > \frac{\rho \dot{w}(t)}{w(t)} = \frac{\dot{c}(t)}{c(t)} \text{ for all } t \geq \tau \quad (4.17)$$

where the equality in (4.17) follows from the equity condition (2.6). Since:

$$\frac{d(V(t)/c(t))}{dt} = \frac{V(t)}{c(t)}\left[\frac{\dot{V}(t)}{V(t)} - \frac{\dot{c}(t)}{c(t)}\right] \text{ for all } t \geq \tau$$

we infer from (4.17) that:
$$\frac{d(V(t)/c(t))}{dt} > 0 \text{ for all } t \geq \tau$$

Thus $(V(t)/c(t))$ must be increasing with t, and consequently, there is $q > 0$, such that:
$$V(t) \geq qc(t) \text{ for all } t \geq \tau$$

Using the definition of $V(t)$ and $v(t)$, we obtain for $t \geq \tau$:
$$r(t) \geq \left[\frac{k(t)}{F_2(\cdot)}\right] + qc(t) = \left[\frac{k(t)r(t)}{F_2(\cdot)r(t)}\right] + qc(t)$$

$$= \left[\frac{\{Q(t) - c(t)\}r(t)}{\beta Q(t)}\right] + qc(t)$$

$$= \left[\frac{r(t)}{\beta} - \frac{c(t)r(t)}{\beta Q(t)} + qc(t)\right]$$

$$\geq r(t) - \frac{c(t)r(t)}{\beta Q(t)} + qc(t)$$

This simplifies to:
$$\frac{c(t)r(t)}{\beta Q(t)} \geq qc(t) \text{ for } t \geq \tau$$

which finally yields:
$$r(t) \geq \beta q Q(t) \text{ for } t \geq \tau$$

Using this result, and the equity condition (4.11), we obtain:
$$k(t) = Q(t) - c(t) = Q(t) - \eta w(t)^p \leq (1/\beta q)r(t) - \eta W^p \text{ for all } t \geq \tau$$

Thus, integrating from $t = \tau$ to $t = S > \tau$, we get:
$$k(S) - k(\tau) \leq (1/\beta q) \int_\tau^S r(t)dt - (\eta W^p)(S - \tau) \leq (1/\beta q) M - (\eta W^p)(S - \tau)$$

Then, for S large, we get $k(S) < 0$, a contradiction.
Given the Third and Fourth Steps, we can conclude that $v(t) = 0$ for all $t \geq 0$, which is Hartwick's Rule.

Remark:
It will be observed that in the two equivalence results obtained in Sections 4.2 and 4.3, we use only the general form of the utility function U, and not the particular version specified in (2.4). For the third equivalence result obtained in Section 4.4, we explicitly use the particular version specified in (2.4). It remains an open question whether the third

equivalence result is valid for a more general utility function U than the one specified in (2.4).

5. EQUITABLE HARTWICK PATHS

In this section, we continue our analysis of equitable paths in the context of our model of global warming, in the case where the function G [in the law of motion of global warming (W)] takes the form given in (G), with $\mu = 1$. Thus, all the results of Section 4 continue to be valid. In particular, the last result (Section 4.4) ensures that any efficient equitable path must satisfy Hartwick's investment rule. This provides our principal justification for focusing our attention on equitable Hartwick paths, which we do in this section.

Our objective in this section is to provide an *explicit solution* for an equitable Hartwick path. This result is new for the model of global warming that we are analyzing. To this end, we first examine the *necessary conditions* for an equitable Hartwick path to exist (in subsection 5.1).

These conditions allow us to establish *qualitative* dynamic properties of key variables, namely global warming, consumption, investment and capital stock, and therefore provide useful information about their asymptotic behavior.

We then establish two important *quantitative* implications of the existence of an equitable Hartwick path. The first implication is that the technological restriction of Solow (1974) on the shares of capital and the resource:

$$\alpha > \beta \tag{S}$$

must be satisfied. The second implication is that an equitable Hartwick path *must* follow an investment policy (abbreviated (IP)) of the form:

$$\dot{k}(t) = g(k(t)) \text{ for } t \geq 0 \tag{IP}$$

where:

$$g(x) = \left[\frac{x^\eta}{a + bx^\eta}\right]^{1/(\zeta - 1)} \text{ for all } x > 0 \tag{g}$$

with:

$$\eta \equiv \frac{\alpha}{\beta} - 1 \text{ and } \zeta \equiv \frac{1 + (\delta/\rho)}{\beta}$$

and with a and b positive constants. The advantage of having such an investment policy function is that along an equitable Hartwick path, if we know what the capital stock is at time t, we can read off from the function the appropriate investment (in the augmentable capital good) to undertake at that time t, to continue to stay on the equitable Hartwick path.

In subsection 5.2, we choose a particular function g satisfying (g), by choosing constants a and b appropriately, given the parameters of the model. We then verify that this chosen investment policy function g provides an explicit solution of a path $(k(t), r(t), c(t), w(t))$ from (K, M, W) in $\mathbb{R}_+^2 \times \mathbb{R}_{++}$ which is both equitable and satisfies Hartwick's Rule.

5.1 Equitable Hartwick Paths: Necessary Conditions

If an interior path $(k(t), r(t), c(t), w(t))$ from (K, M, W) in $\mathbb{R}_+^2 \times \mathbb{R}_{++}$ satisfies Hartwick's Rule, then the Cobb–Douglas form of the production function implies that:

$$I(t) \equiv \dot{k}(t) = \beta F(k(t), r(t), w(t)) \equiv \beta Q(t) \text{ for all } t \geq 0 \tag{5.1}$$

and since $c(t) + I(t) = Q(t)$ for all $t \geq 0$, we obtain:

$$c(t) = (1 - \beta) Q(t) \text{ for all } t \geq 0 \tag{5.2}$$

If the path is also equitable, then (2.6) must hold, and consequently using (5.2) and (2.2) (c), we get:

$$\frac{\dot{Q}(t)}{Q(t)} = \frac{\dot{c}(t)}{c(t)} = \rho \frac{\dot{w}(t)}{w(t)} = \rho \lambda r(t) \text{ for all } t \geq 0 \tag{5.3}$$

Using (5.1) and (5.3),

$$\frac{\dot{I}(t)}{I(t)} = \frac{\dot{Q}(t)}{Q(t)} = \frac{\dot{c}(t)}{c(t)} = \rho \frac{\dot{w}(t)}{w(t)} = \rho \lambda r(t) \text{ for all } t \geq 0 \tag{5.4}$$

In other words, the growth rates of consumption, output, and investment (in the capital good) are all equal to each other and depend on the resource use $r(t)$, through the expression $\rho \lambda r(t)$.

Let us denote:

$$R(t) = \int_0^t r(s) ds \text{ for all } t \geq 0 \tag{5.5}$$

Then, (5.4) implies that:

$$w(t) = w(0) e^{\lambda R(t)} \text{ for all } t \geq 0 \tag{5.6}$$

Further, Hartwick's Rule in the form of (5.1), together with the Cobb–Douglas form of the production function, imply that:

$$\dot{k}(t) = \beta k(t)^\alpha r(t)^\beta [w(0) e^{\lambda R(t)}]^{-\delta} \text{ for } t \geq 0 \tag{5.7}$$

To proceed further, let us note that (5.4) and (5.5) imply:

$$I(t) = I(0) e^{\rho \lambda R(t)} \text{ for all } t \geq 0 \tag{5.8}$$

Then (5.6) and (5.8) allow us to write:

$$w(t) = w(0) e^{\lambda R(t)} = \left[\frac{w(0)}{I(0)^{1/\rho}}\right] I(t)^{1/\rho} \text{ for } t \geq 0 \tag{5.9}$$

Also, equation (5.7) implies that:

$$k(t) = \beta k(t)^\alpha \left[\frac{1}{\rho\lambda}\frac{\dot{I}(t)}{I(t)}\right]^\beta \left[\frac{w(0)}{I(0)^{1/\rho}}\right]^{-\delta} I(t)^{-\delta/\rho} \text{ for } t \geq 0 \quad (5.10)$$

by using (5.4) and (5.9), and this can be rewritten as:

$$\dot{I}(t)^\beta k(t)^\alpha = \left[\frac{(\rho\lambda)^\beta}{\beta}\right] \left[\frac{w(0)}{I(0)^{1/\rho}}\right]^\delta k(t)^{1+\beta+(\delta/\rho)} \text{ for } t \geq 0 \quad (5.11)$$

The equation (5.11) gives us a *second-order* differential equation in $k(t)$ that must hold along any equitable path, satisfying Hartwick's Rule, since:

$$\dot{I}(t) \equiv \frac{dk(t)}{dt} \text{ for } t \geq 0$$

To ease the writing, we denote the term (δ/ρ) by δ' in what follows below.

We can write (5.11) equivalently in the form:

$$\dot{I}(t) = \frac{N k(t)^{(1+\beta+\delta')/\beta}}{k(t)^{\alpha/\beta}} \text{ for } t \geq 0 \quad (5.12)$$

where:

$$N = \left[\frac{\rho\lambda}{\beta^{(1/\beta)}}\right] \left[\frac{w(0)}{I(0)^{1/\rho}}\right]^{\delta/\beta} \quad (5.13)$$

Now, we need to understand the nature of the solution to the second-order differential equation (5.12). Fortunately, there is a fairly standard method to proceed in such cases; see, for example, Murphy (1960), pp. 160–61. This involves trying to find a function which determines $\dot{k}$ in terms of k, consistent with the restriction expressed in (5.12). This will itself turn out to be a *first-order* differential equation, but with k rather than t as the independent variable.

Introduce the notation $\pi(k)$ to denote the function we are looking for. Then,

$$\pi(k(t)) = \dot{k}(t) = I(t) \text{ for all } t \geq 0 \quad (5.14)$$

is an identity in t. So, differentiating (5.14) with respect to t, we get:

$$\frac{d\pi(k)}{dk}\dot{k}(t) = \dot{I}(t) \text{ for all } t \geq 0 \quad (5.15)$$

Using (5.12) and (5.15), we can write:

$$\frac{N\pi(k)^{(1+\beta+\delta')/\beta}}{k^{\alpha/\beta}} = \frac{d\pi(k)}{dk}\pi(k) \text{ for all } k > 0 \quad (5.16)$$

Thus, we obtain the differential equation:

$$\frac{d\pi}{dk} = \frac{N\pi^{(1+\delta')/\beta}}{k^{\alpha/\beta}} \text{ for all } k > 0 \quad (5.17)$$

This differential equation can be solved to obtain the form of the function $\pi(k)$.

In fact, (5.17) is a special case of the *Bernoulli* differential equation, a well-known first-order differential equation; see, for example, Murphy (1960), p.26. Note that:

$$\zeta \equiv \frac{1 + \delta'}{\beta} > 1 \quad (5.18)$$

To solve (5.17), make the substitution:

$$z = \pi^{1-\zeta} \quad (5.19)$$

so that:

$$\frac{dz}{dk} = (1 - \zeta)\pi^{-\zeta}\frac{d\pi}{dk} \quad (5.20)$$

and substituting (5.20) in (5.17), we get:

$$\frac{dz}{dk} = (1 - \zeta)\pi^{-\zeta}\frac{d\pi}{dk} = \frac{(1 - \zeta)N\pi^{-\zeta}\pi^{(1+\delta')/\beta}}{k^{\alpha/\beta}} = \frac{N(1 - \zeta)}{k^{\alpha/\beta}} \quad (5.21)$$

Integrating (5.21) from $k(0)$ to $k(t)$, we get:

$$\frac{1}{k(t)^{\zeta-1}} - \frac{1}{k(0)^{\zeta-1}} = \begin{cases} N(1 - \zeta)\left[\dfrac{k(t)^{-(\alpha/\beta)+1}}{-(\alpha/\beta)+1} - \dfrac{k(0)^{-(\alpha/\beta)+1}}{-(\alpha/\beta)+1}\right] & \text{if } (\alpha/\beta) \neq 1 \\ N(1 - \zeta)[\ln k(t) - \ln k(0)] & \text{if } (\alpha/\beta) = 1 \end{cases} \quad (5.22)$$

5.1.1 Dynamic properties of key variables

We first summarize the dynamic qualitative properties of key variables, namely global warming, consumption, investment and capital stock.

For any path $(k(t), r(t), c(t), w(t))$ from (K, M, W) in $\mathbb{R}_+^2 \times \mathbb{R}_{++}$, by (2.2)(b) and (c), and (G) with $\mu = 1$, we must have $W \leq w(t) \leq We^{\lambda M}$ for all $t \geq 0$, and $w(t)$ monotonically increasing over time. Thus, $w(t)$ monotonically converges to a steady state value w^*, where $W \leq w^* \leq We^{\lambda M}$.

Since the path $(k(t), r(t), c(t), w(t))$ from (K, M, W) in $\mathbb{R}_+^2 \times \mathbb{R}_{++}$ is equitable, there is $\sigma > 0$, such that $c(t) = \sigma w(t)^p$ for all $t \geq 0$. This implies that $\sigma W^p \leq c(t) \leq \sigma(w^*)^p$ for all $t \geq 0$, and $c(t)$ must also monotonically increase and converge to a steady state value c^*, where $\sigma W^p \leq c^* = \sigma(w^*)^p$.

Since the path $(k(t), r(t), c(t), w(t))$ from (K, M, W) in $\mathbb{R}_+^2 \times \mathbb{R}_{++}$ is interior and satisfies Hartwick's Rule, we have $\dot{k}(t) = r(t)F_2(k(t), r(t), w(t)) = \beta Q(t)$ for $t \geq 0$, so by (2.2)(a), $c(t) = (1 - \beta)Q(t)$ for all $t \geq 0$, and:

$$I(t) \equiv \dot{k}(t) = r(t)F_2(k(t), r(t), w(t)) = \beta Q(t) = \beta c(t)/(1 - \beta) > 0 \text{ for all } t \geq 0 \quad (5.23)$$

This implies that $\sigma W^p \beta/(1 - \beta) \leq I(t) \leq \sigma(w^*)^p[\beta/(1 - \beta)]$ for all $t \geq 0$, and $I(t)$ must also monotonically increase and converge to a steady state value I^*, where

$$\sigma W^p \beta/(1 - \beta) \leq I^* = \sigma(w^*)^p[\beta/(1 - \beta)]$$

It also follows from (5.23) that we must have $k(t)$ monotonically increasing and becoming *unbounded* over time; that is,

$$k(t) \uparrow \infty \text{ as } t \uparrow \infty \tag{5.24}$$

5.1.2 Solow's technological restriction

Next, we infer that the existence of an equitable Hartwick path $(k(t), r(t), c(t), w(t))$ from (K, M, W) in $\mathbb{R}_+^2 \times \mathbb{R}_{++}$ implies that the technological restriction of Solow (1974) on the shares of capital and the resource (written as (S)):

$$\alpha > \beta \tag{S}$$

must be satisfied.

To see this, note first that in view of (5.23), we can infer from (5.22) that:

$$-\frac{1}{k(0)^{\zeta-1}} \leq \begin{cases} N(1-\zeta)\left[\dfrac{k(t)^{-(\alpha/\beta)+1}}{-(\alpha/\beta)+1} - \dfrac{k(0)^{-(\alpha/\beta)+1}}{-(\alpha/\beta)+1}\right] & \text{if } (\alpha/\beta) \neq 1 \\ N(1-\zeta)[\ln k(t) - \ln k(0)] & \text{if } (\alpha/\beta) = 1 \end{cases} \tag{5.25}$$

Let us examine the first line of (5.25). If $(\alpha/\beta) < 1$, then by (5.24), $k(t)^{-(\alpha/\beta)+1} \to \infty$, and the right-hand side expression would tend to $-\infty$ (since $\zeta > 1$ by (5.18)), leading to a contradiction. Thus, if $(\alpha/\beta) \neq 1$, then it must be the case that $(\alpha/\beta) > 1$, establishing (S).

Now, we examine the second line of (5.25). In this case, by using (5.24), we have $\ln k(t) \to \infty$, and the right-hand side expression would tend to $-\infty$, leading to a contradiction again. Thus, the possibility of the second line of (5.25) cannot arise. That is, $(\alpha/\beta) = 1$ is not possible. This completes our demonstration that (S) must hold.

Using the fact that (S) holds (established above), (5.22) can be simplified to read:

$$\frac{1}{k(t)^{\zeta-1}} - \frac{1}{k(0)^{\zeta-1}} = N(\zeta-1)\left[\frac{1}{[(\alpha/\beta)-1]k(t)^{[(\alpha/\beta)-1]}} - \frac{1}{[(\alpha/\beta)-1]k(0)^{[(\alpha/\beta)-1]}}\right]$$

$$\text{for all } t \geq 0 \tag{5.26}$$

5.1.3 Investment policy function

We come now to the key result of Section 5.1 (already mentioned at the beginning of Section 5), namely that an equitable Hartwick path *must* follow an investment policy of the form:

$$\dot{k}(t) = g(k(t)) \text{ for } t \geq 0 \tag{IP}$$

where g satisfies (g).

To demonstrate this, we proceed as follows. Pick any $T \geq 0$, and fix it. Using (5.26) for $t = T$, we obtain:

$$\frac{1}{k(T)^{\zeta-1}} - \frac{1}{k(0)^{\zeta-1}} = N(\zeta-1)\left[\frac{1}{[(\alpha/\beta)-1]k(T)^{[(\alpha/\beta)-1]}} - \frac{1}{[(\alpha/\beta)-1]k(0)^{[(\alpha/\beta)-1]}}\right] \quad (5.27)$$

Now, let S be an arbitrary point in time satisfying $S > T$. Using (5.26) for $t = S$, we obtain:

$$\frac{1}{k(S)^{\zeta-1}} - \frac{1}{k(0)^{\zeta-1}} = N(\zeta-1)\left[\frac{1}{[(\alpha/\beta)-1]k(S)^{[(\alpha/\beta)-1]}} - \frac{1}{[(\alpha/\beta)-1]k(0)^{[(\alpha/\beta)-1]}}\right] \quad (5.28)$$

Subtract (5.27) from (5.28) to obtain:

$$\frac{1}{k(S)^{\zeta-1}} - \frac{1}{k(T)^{\zeta-1}} = N(\zeta-1)\left[\frac{1}{[(\alpha/\beta)-1]k(S)^{[(\alpha/\beta)-1]}} - \frac{1}{[(\alpha/\beta)-1]k(T)^{[(\alpha/\beta)-1]}}\right] \quad (5.29)$$

Let $S \to \infty$, keeping T fixed. Then, using (5.24) and (S), we have:

$$\frac{1}{[(\alpha/\beta)-1]k(S)^{[(\alpha/\beta)-1]}} \to 0 \text{ as } S \to \infty \quad (5.30)$$

Also, since $I(t) \to I^* = [\beta/(1-\beta)]c^* = [\beta/(1-\beta)]\sigma(w^*)^p$ as $t \to \infty$, we have:

$$\frac{1}{k(S)^{\zeta-1}} \to \frac{1}{[I^*]^{\zeta-1}} \text{ as } S \to \infty \quad (5.31)$$

Using (5.30) and (5.31) in (5.29), we obtain:

$$\frac{1}{[I^*]^{\zeta-1}} - \frac{1}{k(T)^{\zeta-1}} = -\left[\frac{N(\zeta-1)}{[(\alpha/\beta)-1]k(T)^{[(\alpha/\beta)-1]}}\right] \quad (5.32)$$

Denote:

$$a = \frac{N(\zeta-1)}{[(\alpha/\beta)-1]}, \quad b = \frac{1}{[I^*]^{\zeta-1}} \quad (5.33)$$

Then (5.32) can be rewritten as:

$$\frac{1}{k(T)^{\zeta-1}} = b + \frac{a}{k(T)^{[(\alpha/\beta)-1]}} = b + \frac{a}{k(T)^{\eta}} = \frac{a + bk(T)^{\eta}}{k(T)^{\eta}} \quad (5.34)$$

recalling that $\eta \equiv (\alpha/\beta) - 1$. Thus, we obtain:

$$k(T) = \left[\frac{k(T)^{\eta}}{a + bk(T)^{\eta}}\right]^{1/(\zeta-1)} \quad (5.35)$$

Since $T \geq 0$ was arbitrary, we get:

$$k(t) = \left[\frac{k(t)^{\eta}}{a + bk(t)^{\eta}}\right]^{1/(\zeta-1)} \text{ for all } t \geq 0 \quad (5.36)$$

That is, an equitable Hartwick path *must* follow the investment policy given by:

$$k(t) = g(k(t)) \text{ for } t \geq 0 \quad \text{(IP)}$$

where:

$$g(x) = \left[\frac{x^\eta}{a + bx^\eta} \right]^{1/(\zeta - 1)} \text{ for all } x > 0 \quad \text{(g)}$$

5.2 Equitable Hartwick Path: Explicit Solution

In this subsection we provide an explicit solution of an equitable Hartwick path. So, effectively, we show the existence of an equitable Hartwick path by *constructing* it.

Note that the results of Section 5.1 *do not* establish that there is an equitable Hartwick path. What they do imply is that *if* there is an equitable Hartwick path, *then* it must satisfy the investment policy (IP), where g is of the form given by (g) above.

This fact, of course, is an enormous help in providing an explicit solution of an equitable Hartwick path. For, now, given the parameters of the model, it remains to find appropriate values of two parameters a and b appearing in the formula (g), so that when investment policy follows (IP), with g defined in terms of these particular values of a and b, the corresponding path satisfies Hartwick's Rule and is also equitable. To rephrase that statement, we use the parameters of the model to define a function, g, and use that function to provide an explicit solution.

The choice of appropriate values of two parameters, a and b (in the above sense) will involve the values of two key parameters of the model, the initial values of the capital and resource stocks (K,M). Notice that, in contrast, the specific values of these parameters (K,M) play no role at all in our analysis in Section 5.1 (although we do use the information in Section 5.1 that the total use of the resource flow cannot exceed the finite resource stock M).

Let us describe how we choose the appropriate values of a and b, given the model's key parameter values (K,M). This is done by writing two constraints, involving the "unknowns" a and b, and the "known" parameter values (K,M) [as well as other parameters relating to the technology, tastes and the law of motion of global warming].

The first is an *equality* constraint, involving the "unknowns" a and b, and the "known" parameter value K [as well as other parameters relating to the technology, tastes and the law of motion of global warming]. It does not involve the parameter value M. Satisfying this constraint implies that given K [and parameters relating to the technology, tastes and the law of motion of global warming] there is a function ϕ which determines a in terms of b; that is, $a = \phi(b)$.

The second is an *inequality* constraint, involving the "unknowns" a and b, and the "known" parameter values (K,M) [as well as other parameters relating to the technology, tastes and the law of motion of global warming]. When a and b satisfy the two constraints simultaneously, then since $a = \phi(b)$, the value of b must lie in an interval, determined completely by (K,M), and parameters relating to the technology, tastes and the law of motion of global warming.

Once b, and therefore a, are determined satisfying these two constraints, the equitable Hartwick path can be constructed in terms of (a,b) and the parameters of the model.

Choosing a and b to satisfy the two constraints involves considerable algebraic manipulations, which are unavoidable given the nature of the model. So, it is important not to lose sight of the role of these constraints. After we are done with defining the investment policy function, g, we will use it to define a path. Then the first constraint will ensure that the national income accounting condition (2.2)(a) will be satisfied by the path. And, the second constraint will ensure that the condition (2.2)(b) on the total use of the exhaustible resource will also be satisfied by the path. Section 5.2.3 below provides the formal material on the precise role of the two constraints (see displays (5.49) and (5.50) in that subsection).

5.2.1 First constraint
We denote:

$$\eta \equiv \frac{\alpha}{\beta} - 1 \text{ and } \zeta \equiv \frac{1+\delta'}{\beta} = \frac{1+(\delta/\rho)}{\beta} \tag{5.37}$$

and define a function, $f: \mathbb{R}^2_{++} \to \mathbb{R}_{++}$ as follows:

$$f(a,b) = a^\beta \left[\frac{K^\eta}{a+bK^\eta} \right]^{\delta'/(\zeta-1)} \equiv a^{[\beta - \{\delta'/(\zeta-1)\}]} \left[\frac{aK^\eta}{a+bK^\eta} \right]^{\delta'/(\zeta-1)} \text{ for all } (a,b) \in \mathbb{R}^2_{++} \tag{5.38}$$

Note from (5.38) and the fact that (from (S)) $\beta < \alpha < 1 < 1 + \delta'$,

$$0 < \frac{\delta'}{\zeta - 1} = \frac{\delta'\beta}{1+\delta' - \beta} = \beta \left[\frac{\delta'}{(1-\beta) + \delta'} \right] < \beta \tag{5.39}$$

It follows now from (5.38) that, given any $b > 0$, we have $f(a,b) \uparrow \infty$ as $a \uparrow \infty$; this is transparent from the second expression for $f(a,b)$ in (5.38), as both terms increase with a, and the first term goes to infinity as a goes to infinity (using (5.39)). Similarly, given any, $b > 0$, we have $f(a,b) \downarrow 0$ as $a \downarrow 0$.

The first constraint is written as:

$$f(a,b) = \left[\frac{\rho\lambda(\zeta-1)}{\eta} \right]^\beta \left[\frac{W^\delta}{\beta} \right] \equiv S \tag{I}$$

where $f(a,b)$ is given by (5.38). Clearly, given any $b > 0$, there is a *unique* value of a for which the constraint (I) is satisfied. This value of a is denoted as $\phi(b)$. Thus, ϕ is a function from $\mathbb{R}_{++}$ to $\mathbb{R}_{++}$, such that:

$$f(\phi(b),b) = \left[\frac{\rho\lambda(\zeta-1)}{\eta} \right]^\beta \left[\frac{W^\delta}{\beta} \right] \equiv S \text{ for all } b > 0$$

With S defined as in (I), we introduce the notation:

$$S' \equiv S^{(\zeta-1)/\delta'}; \; m \equiv \beta(\zeta-1)/\delta' = [\delta' + (1-\beta)]/\delta' > 1 \tag{5.40}$$

Then the following property of the ϕ function may be noted, as it is useful in what follows:

$$\frac{\phi(b)^m}{b} \to S' \text{ as } b \to \infty \tag{5.41}$$

The proof of (5.41) is given in an appendix to Section 5 (see section 8.3).

5.2.2 Second constraint
For $a > 0$, we define:

$$L(a) = \left[\frac{a\eta}{\rho\lambda(\zeta - 1)}\right]$$

Thus, L is a function from $\mathbb{R}_{++}$ to $\mathbb{R}_{++}$. Using this, we write the second constraint as:

$$\frac{L(a)[a + bK^\eta]^{1/(\zeta-1)}}{b^{\zeta/(\zeta-1)}K^{\eta/(\zeta-1)}[(\alpha/\beta) - 1]K^{(\alpha/\beta)-1}} \leq M \quad \text{(II)}$$

We would like to find $(a, b) \in \mathbb{R}^2_{++}$ which satisfies both (I) and (II). We show now that this can always be done by choosing $b > 0$ sufficiently large (given the parameters of the model), and choosing $a = \phi(b)$, where $\phi(b)$ is the unique value of a for which (I) is satisfied.

The formal demonstration is provided as follows. For any $b > 0$, pick $a = \phi(b)$. Then (I) is satisfied by definition of the function ϕ, and consequently:

$$\left[\frac{S}{a^\beta}\right]^{(1/\delta')} = \left[\frac{K^\eta}{a + bK^\eta}\right]^{1/(\zeta-1)} \quad (5.42)$$

Using this in (II), we can rewrite it as:

$$\frac{\eta a^{(\beta/\delta')+1}}{\rho\lambda(\zeta-1)b^{\zeta/(\zeta-1)}S^{(1/\delta')}[(\alpha/\beta) - 1]K^{(\alpha/\beta)-1}} \leq M \quad (5.43)$$

Note that:

$$\frac{\zeta}{\zeta - 1} = \frac{1 + \delta'}{\beta(\zeta - 1)} = \frac{\delta' + \beta}{\beta(\zeta - 1)} + \frac{(1 - \beta)}{\beta(\zeta - 1)} = \frac{(\delta' + \beta)}{\delta'm} + \frac{(1 - \beta)}{\delta'm} \quad (5.44)$$

since $m = \beta(\zeta - 1)/\delta'$. Using (5.44), we can write:

$$\frac{a^{(\beta/\delta')+1}}{b^{\zeta/(\zeta-1)}} = \frac{[a^m]^{(\delta'+\beta)/\delta'm}}{b^{\zeta/(\zeta-1)}} = \left[\frac{a^m}{b}\right]^{(\delta'+\beta)/\delta'm}\left[\frac{1}{b}\right]^{(1-\beta)/\delta'm} \quad (5.45)$$

Using (5.45) the constraint (5.43) is equivalent to:

$$\left[\frac{a^m}{b}\right]^{(\delta'+\beta)/\delta'm}\left[\frac{1}{b}\right]^{(1-\beta)/\delta'm}\left[\frac{\eta}{\rho\lambda(\zeta - 1)S^{(1/\delta')}[(\alpha/\beta) - 1]K^{(\alpha/\beta)-1}}\right] \leq M \quad (5.46)$$

Thus, given (K, M) and the parameters of the model, if we choose $b > 0$ large enough, and we choose $a = \phi(b)$, the constraint (5.46) will be satisfied because of property (5.41). That is, the pair (a, b) will satisfy both constraints (I) and (II).

5.2.3 Constructing an equitable Hartwick path
We will now construct an equitable Hartwick path $(k(t), r(t), c(t), w(t))$ from $(K, M, W) \in \mathbb{R}^2_+ \times \mathbb{R}_{++}$. For this purpose, we will use a pair of parameters $(a, b) \in \mathbb{R}^2_{++}$ which satisfies constraints (I) and (II) above. [We have already checked above that this can be done.] These parameters will be used to define a real-valued function, g, and the function g will be used to construct the path.

Define $g: \mathbb{R}_{++} \to \mathbb{R}$ as follows:

$$g(x) = \left[\frac{x^\eta}{a + bx^\eta}\right]^{1/(\zeta-1)} \text{ for all } x > 0 \tag{g}$$

Note that g is continuous and strictly positive on $\mathbb{R}_{++}$, and so $[1/g(x)]$ is continuous and positive on $\mathbb{R}_{++}$. In particular, for all $k \geq K$, the function $[1/g(x)]$ is continuous on $[K, k]$, and so its Riemann integral:

$$\int_K^k [1/g(x)] dx$$

exists.[21] Define $H: [K, \infty) \to \mathbb{R}_{++}$ by:

$$H(k) = \int_K^k [1/g(x)] dx \text{ for all } k \geq K \tag{H}$$

Then, we have:

$$H'(k) = \frac{1}{g(k)} > 0 \text{ for all } k \geq K \tag{H'}$$

Since $g(k)$ is increasing in k, we must have $H'(k)$ decreasing in k, and it follows from (g) that $H'(k) \to b^{1/(\zeta-1)} > 0$ as $k \to \infty$. Thus, $H(k) \uparrow \infty$ as $k \uparrow \infty$, and H is one-to-one from $[K, \infty)$ to $[0, \infty)$. Consequently, H has an inverse function, which we denote by h. Then h is a function from $[0, \infty)$ to $[K, \infty)$ satisfying:

$$H(h(t)) = t \text{ for all } t \geq 0 \text{ and } h(H(k)) = k \text{ for all } k \geq K \tag{h}$$

We now define $(k(t), r(t), c(t), w(t))$ as follows:

$$\left. \begin{array}{ll} k(t) = h(t) & \text{for all } t \geq 0 \\ r(t) = \left[\dfrac{g'(k(t))}{\rho\lambda}\right] & \text{for all } t \geq 0 \\ w(t) = We^{\int_0^t [g'(k(s))/\rho] ds} & \text{for all } t \geq 0 \\ c(t) = [(1-\beta)/\beta] g(k(t)) & \text{for all } t \geq 0 \end{array} \right\} \tag{5.47}$$

Note that $(k(t), r(t), c(t), w(t)) \gg 0$ for all $t \geq 0$, since h is a map from $[0, \infty)$ to $[K, \infty)$, and $g(x)$ and $g'(x)$ are both positive for all $x > 0$. Also, clearly, $k(t)$, $c(t)$ and $w(t)$ are differentiable functions of t, and $r(t)$ is a continuous function of t.

We need to check that $(k(t), r(t), c(t), w(t))$ is a *path* from (K, W, M); that is, it satisfies (2.2)(a)–(d). Note that $k(0) = h(0) = h(H(K)) = K$, by using (H) and (h). Also, $w(0) = W$ from the third line of (5.47). Thus, (2.2)(d) is satisfied. Also, from the second and third lines of (5.47), we have:

$$\frac{\dot{w}(t)}{w(t)} = [g'(k(t))/\rho] = \lambda r(t) \text{ for all } t \geq 0$$

---

[21] We will use basic results of Riemann integration theory. These are available in any standard text on mathematical analysis, for example in Goldberg (1964), chapter 7.

so (2.2)(c) is satisfied.

Verification of (2.2)(a):
We proceed next to verify (2.2)(a). Note that by (h), we have:

$$h'(H(k))H'(k) = 1 \text{ for all } k \geq K$$

and so by (H') and (h),

$$\dot{k}(t) = h'(t) = h'(H(h(t))) = h'(H(k(t))) = \frac{1}{H'(k(t))} = g(k(t)) > 0 \text{ for all } t \geq 0 \quad (5.48)$$

Thus,

$$\dot{k}(t) + c(t) = g(k(t)) + [(1-\beta)/\beta]g(k(t)) = g(k(t))/\beta \text{ for all } t \geq 0$$

So, to verify (2.2)(a), we need to show that with the specification of $k(t), r(t), w(t)$ as given in the first three lines of (5.47),

$$F(k(t), r(t), w(t)) = g(k(t))/\beta \text{ for all } t \geq 0 \quad (5.49)$$

It is precisely for this purpose that we chose a and b (in defining the function g) to satisfy the first constraint (I). Thus, the verification of (5.49) involves using (I), together with some algebraic manipulations. This material is presented in an appendix to Section 5 (see section 8.3).

Verification of (2.2)(b):
Given (5.47), in order to verify (2.2)(b), we have to show that:

$$\int_0^\infty \left[\frac{g'(k(t))}{\rho\lambda}\right] dt \leq M \quad (5.50)$$

It is precisely for this purpose that we chose a and b (in defining the function g) to satisfy the second constraint (II). Thus, the verification of (5.50) involves using (II), together with some algebraic manipulations. This material is presented in an appendix to Section 5 (see section 8.3).

5.2.4 Hartwick's Rule

We have verified above that $(k(t), r(t), c(t), w(t))$, defined by (5.47), is a path from (K, W, M). We now verify that the path satisfies Hartwick's Rule. We have checked in (5.48) that:

$$\dot{k}(t) = g(k(t)) \text{ for all } t \geq 0$$

and we have shown in (5.49) that:

$$F(k(t), r(t), w(t)) = g(k(t))/\beta \text{ for all } t \geq 0$$

Thus,

$$k(t) = \beta F(k(t), r(t), w(t)) \text{ for all } t \geq 0$$

which is Hartwick's Rule, given the Cobb–Douglas form of the production function F in (2.1).

5.2.5 Equity

We now proceed to verify that $(k(t), r(t), c(t), w(t))$, defined by (5.47), is an equitable path from (K, W, M). Using the fourth line of (5.47),

$$\frac{\dot{c}(t)}{c(t)} = \frac{g'(k(t))\dot{k}(t)}{g(k(t))} = g'(k(t)) \text{ for all } t \geq 0$$

by using (5.48). Also, from the third line of (5.47),

$$\frac{\dot{w}(t)}{w(t)} = [g'(k(t))/\rho] \text{ for all } t \geq 0$$

Thus, we obtain:

$$\frac{\dot{c}(t)}{c(t)} = \rho \frac{\dot{w}(t)}{w(t)} \text{ for all } t \geq 0$$

This verifies that the path $(k(t), r(t), c(t), w(t))$ from (K, M, W) specified in (5.47) is an equitable path.

5.3 A Numerical Example

We provided an explicit solution of an equitable Hartwick path in (5.47). A crucial part of this solution is the specification of the time path of the capital stock, $k(t)$, given in (5.47) by $h(t)$. This specification $h(t)$ is obtained by starting with the investment policy function $g(x)$, given by the formula (g), integrating the reciprocal of it from the initial stock K to any higher stock k, and then taking the inverse of the resulting function $H(k)$. This procedure, which might appear a bit abstract, can be given a concrete form by working through these steps for an example in which the parameters of the model are numerically specified. We carry out this illustrative exercise below.

5.3.1 Specification of numerical values of the parameters

We specify the following numerical values for the technological parameters of the model. Let $\alpha = (13/24)$, $\beta = (13/36)$, $\delta = (1/12)$, $\rho = 1$. Note that $\alpha > \beta > 0$, and $\alpha + \beta < 1$; in fact, $\alpha + \beta + \delta < 1$. Further, $\delta' = (\delta/\rho) = \delta = (1/12)$. Then, $\eta = (\alpha/\beta) - 1 = (3/2) - 1 = (1/2)$, and $\zeta = (1+\delta)/\beta = (13/12)/(13/36) = 3$, so that $(\zeta - 1) = 2$, and $\delta/(\zeta - 1) = (1/24)$. In addition, let $K = 1$, $W = 1$ and $\lambda = (1/4)$, $\mu = 1$. Note that the numerical values are chosen to simplify as far as is possible the algebraic manipulations involved, since the only purpose of the example is to illustrate the mathematical procedure described in the previous paragraph.

We now have to pick a and b to define the function g.

Pick any $b > 0$. For example, pick $b = 1$. Now, according to constraint (I) (appearing just after (5.39)), a will have to be picked to satisfy the equation:

$$a^\beta \left[\frac{K^\eta}{a + bK^\eta} \right]^{\delta/(\zeta-1)} = \left[\frac{\lambda(\zeta - 1)}{\eta} \right]^\beta \left[\frac{W^\delta}{\beta} \right] \quad \text{(I)}$$

For our numerical choices, (I) can be rewritten as:

$$a^{(13/36)} \left[\frac{1}{a+1} \right]^{(1/24)} = \left[\frac{(1/4)2}{(1/2)} \right]^{(13/36)} \left[\frac{1}{(13/36)} \right]$$

which simplifies to:

$$a^{(13/36)} = (36/13)[a + 1]^{(1/24)} \quad (5.51)$$

There is a unique positive solution a to the equation (5.51). It can be numerically solved to yield:

$$a \approx 24.38111049$$

5.3.2 Defining the investment policy function g

Now, with $b = 1$ and a given as the unique positive solution to (5.51), one can define the function g as:

$$g(x) = \left[\frac{x^\eta}{a + bx^\eta} \right]^{1/(\zeta-1)} = \left[\frac{x^{1/2}}{a + x^{1/2}} \right]^{1/2} \quad \text{for all } x > 0 \quad \text{(gn)}$$

Recall that g has been deliberately defined in such a way that the proposed solution of $k(t)$ in (5.47) is the solution to the differential equation:

$$\dot{k}(t) = g(k(t)) \text{ for all } t \geq 0; k(0) = K \quad \text{(DE)}$$

See display (5.48) in Section 5.2.

With our numerical specifications, the function g takes the special form given in (gn), and with this special form, the differential equation (DE) can be analytically "solved", except that it will be in "inverse form". To see this, use the method of separation of variables to write (DE) as:

$$\int_K^{k(T)} \frac{dk}{g(k)} = \int_0^T dt = T \text{ for all } T > 0 \quad (5.52)$$

Thus, one needs to find the integral on the left-hand side of (5.52). If one can evaluate this integral explicitly, given that g satisfies (gn), then it will be an explicit function of $k(T)$. In other words, (5.52) will express T as a function (say H) of $k(T)$; that is, it will give us the solution that we are seeking in "inverse form". So, by taking the *inverse* of the function H, we will in fact obtain the function (say h) which expresses the solution $k(T)$ in terms of T.

Notice that this corresponds precisely to the general procedure described in Section 5.2.3. That is, we first find the function $H: [K, \infty) \to \mathbb{R}_{++}$ defined by:

$$H(k) = \int_K^k [1/g(x)]dx = \int_K^k \left[\frac{a + x^{1/2}}{x^{1/2}}\right]^{1/2} dx \text{ for all } k \geq K \quad \text{(Hn)}$$

Then, we find its inverse function h, from $[0,\infty)$ to $[K,\infty)$, satisfying:

$$H(h(t)) = t \text{ for all } t \geq 0 \text{ and } h(H(k)) = k \text{ for all } k \geq K \quad \text{(h)}$$

The function $h(t)$ will be the solution to the differential equation (DE) with g specified in (gn).

In Figure 16.1 below, we have a plot of the function H. In Figure 16.2 below, we have a plot of the function h. These are graphical counterparts of the analytics described above.

5.3.3 Evaluating the integral

We can evaluate the integral appearing in (Hn) and therefore find the function H in explicit form. Although the integral in (Hn) is not in a "standard form", it can be reduced to a standard form by making some appropriate substitutions. We provide the technical details, to keep our treatment self-contained.

Our first substitution is obtained by defining:

$$y = x^{1/2} \text{ for all } x > 0 \quad (5.53)$$

Then, by the standard change of variable formula,

$$\int_K^k \left[\frac{a + x^{1/2}}{x^{1/2}}\right]^{1/2} dx = \int_{\sqrt{K}}^{\sqrt{k}} \left[\frac{a + y}{y}\right]^{1/2} 2y\,dy$$

$$= 2\int_{\sqrt{K}}^{\sqrt{k}} [a + y]^{1/2} y^{1/2} dy$$

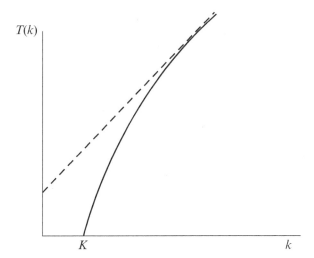

Figure 16.1 Solution of DE in inverse form

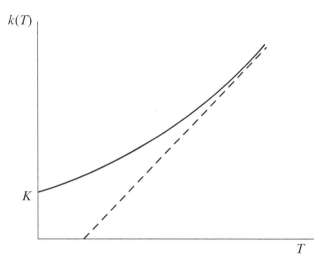

Figure 16.2   Solution of DE

$$= 2\int_{\sqrt{K}}^{\sqrt{k}}[ay + y^2]^{1/2}dy$$

$$= 2\int_{\sqrt{K}}^{\sqrt{k}}\left[\left(y + \frac{a}{2}\right)^2 - \left(\frac{a}{2}\right)^2\right]^{1/2}dy \quad (5.54)$$

Our second substitution is obtained by defining:

$$z = y + \frac{a}{2} \text{ and } A = \frac{a}{2} \quad (5.55)$$

Using (5.54) and the standard change of variable formula,

$$\int_K^k\left[\frac{a + x^{1/2}}{x^{1/2}}\right]^{1/2}dx = 2\int_{\sqrt{K}}^{\sqrt{k}}\left[\left(y + \frac{a}{2}\right)^2 - \left(\frac{a}{2}\right)^2\right]^{1/2}dy$$

$$= 2\int_{A+\sqrt{K}}^{A+\sqrt{k}}[z^2 - A^2]^{1/2}dz \quad (5.56)$$

The integral in (5.56) can be evaluated to yield:

$$2\int_{A+\sqrt{K}}^{A+\sqrt{k}}[z^2 - A^2]^{1/2}dz = 2\left[\frac{z(z^2 - A^2)^{1/2}}{2} - \left(\frac{A^2}{2}\right)\ln\{z + (z^2 - A^2)^{1/2}\}\right]_{A+\sqrt{K}}^{A+\sqrt{k}}$$

$$= \left[z(z^2 - A^2)^{1/2} - A^2\ln\{z + (z^2 - A^2)^{1/2}\}\right]_{A+\sqrt{K}}^{A+\sqrt{k}} \quad (5.57)$$

Thus, we obtain the solution of the differential equation (DE) in "inverse form" [expressing T in terms of k] as:

$$T = (A + \sqrt{k})[(A+\sqrt{k})^2 - A^2]^{1/2} - A^2\ln\{(A+\sqrt{k})+[(A+\sqrt{k})^2 - A^2]^{1/2}\} - C$$

$$\text{for } k \geq K \tag{5.58}$$

where C is a constant, dependent on the initial condition K, given by:

$$C = (A+\sqrt{K})[(A+\sqrt{K})^2 - A^2]^{1/2} - A^2\ln\{(A+\sqrt{K})+[(A+\sqrt{K})^2 - A^2]^{1/2}\} \tag{5.59}$$

Note that the solution (5.58) is in an exact analytical form; it is not an approximation to the solution, unlike a typical numerical simulation.

5.3.4 Plotting the solution

We can plot the solution (5.58) to the differential equation (DE) graphically. To this end, it clearly helps to obtain the derivative of T with respect to k. This is straightforward:

$$\frac{dT}{dk} = \left[1 + \frac{2A}{k^{1/2}}\right]^{1/2} = \left[1 + \frac{a}{k^{1/2}}\right]^{1/2} = \frac{1}{g(k)} \tag{5.60}$$

The equality of the extreme expressions in (5.60) will be recognized to be obvious, as (5.58) is the solution to (DE). The expressions in the middle, obtained by directly differentiating the right-hand expression in (5.58) with respect to k and then simplifying, is a check that the solution to (DE) given in (5.58) is indeed the correct one. For this reason, the steps of the differentiation and simplification are given in an appendix to Section 5 (see section 8.3).

From (5.60), it is clear that:

$$\frac{dT}{dk} > 0 \text{ for } k > 0 \text{ and } \frac{dT}{dk} \downarrow \text{ as } k \uparrow \tag{5.61}$$

Further, $T(k) = 0$ when $k = K$, and:

$$\frac{dT}{dk} \to 1 \text{ as } k \to \infty \tag{5.62}$$

These facts lead us to obtain Figure 16.1, showing that $T(k)$ is a strictly concave, strictly increasing (and continuous) function of k, with $T = 0$ when $k = K$.

The solution of (DE), where k is expressed as a function of T can be obtained directly from Figure 16.1, by viewing Figure 16.1 on its side [this being the graphical equivalent of inverting the function $T(k)$]. This is shown in Figure 16.2.

6. AN ALTERNATE SCENARIO

In this section, we explore an alternate scenario. We specify the law of motion (W) in terms of a *strictly concave* function, G, of resource use; specifically, we assume that (G) holds, with $0 < \mu < 1$. Unlike the results obtained in Section 5, we show that the picture of the equitable economy in the long run can be completely altered in this global warming formulation.

To motivate our analysis, let us review our findings of Sections 4 and 5. While the investment rule ("Invest resource rents" in capital accumulation) originally suggested by Hartwick (1977) was in the context of a model without global warming, Sections 4 and 5 show that the presence of global warming affecting production and utility does not necessarily lead to a different investment rule. Our analysis has shown that in the model of global warming of Stollery (1998) and d'Autume, Hartwick and Schubert (2010), Hartwick's investment rule continues to be the central concept in the study of efficient equitable paths. It not only ensures equity along efficient paths, but it is the only investment rule which does so.

In the alternate scenario of global warming (in which (G) holds with $0 < \mu < 1$), we analyze more generally equitable paths with a constant savings rate (abbreviated as (CSR) below). Recall that given the Cobb–Douglas form of the production function specified in (2.1), an equitable Hartwick path is an equitable path with constant savings rate equal to the resource share (β).

To present our results systematically, we first examine (in Section 6.1) the *necessary conditions* that must be satisfied by any equitable path on which the savings rate is constant. It is worth noting that, even though the focus of our analysis in Section 6 is on the case $0 < \mu < 1$, all our analysis in 6.1.1–6.1.3 applies to both the case $0 < \mu < 1$, and $\mu = 1$, so that the analysis in these subsections can be seen to be a generalization of the corresponding analysis in Section 5.1. There are two significant conclusions that emerge from our analysis of the dynamic behavior of investment on such paths.

First, the existence of such a path implies a parametric restriction:

$$\alpha > \nu\beta = (1/\mu)\beta$$

This can be seen to be a generalization of Solow's technological restriction (S), which was shown to be an implication of the existence of an equitable Hartwick path in Section 5.1.2.

Second, *if* the investment level on such a path becomes *unbounded* over time (see Section 6.1.5), then the capital stock path *must* exhibit *quasi-arithmetic growth* that is faster than linear growth. This is in contrast to both the linear growth of the capital stock in the model of Solow (1974), and the asymptotically linear growth in the model with global warming that we studied in Section 5 (for equitable Hartwick paths). This observation about the growth of the capital stock also implies that investment must exhibit quasi-arithmetic growth, and resource use must exhibit quasi-arithmetic decline.

Note that the possibility of unbounded growth of investment *cannot* occur when $\mu = 1$, for reasons similar to those described in Section 5.1.1; it *can* occur when $0 < \mu < 1$ for reasons given in Section 6.2 below.

Taking our cue from these necessary conditions, we provide an explicit solution in Section 6.2 for an equitable path with a constant savings rate (for an arbitrary initial resource stock). This path is seen to exhibit *quasi-arithmetic growth* of the capital stock, unbounded growth of investment and consumption, and unbounded growth of global warming. Comparing this with the explicit solution of an equitable Hartwick path in Section 5.2, the principal lesson is therefore that the asymptotic dynamic behavior of key variables depends crucially on the particular law of motion of global warming, and only less so on whether there is global warming or not.

Given the explicit solution in Section 6.2, we turn in Section 6.3 to the question of the "best" constant savings rate: one which *maximizes* sustainable utility among all equitable paths with a constant savings rate. We are able to provide the following partial answer to this question. There is a threshold resource stock, $\overline{M}$ (which can be defined precisely in terms of the parameters relating to technology, the utility function, the law of motion of global warming, and the initial stocks of capital and global warming), such that, if the actual resource stock M is *at least as large* as $\overline{M}$, then there is an equitable path with a constant savings rate of $\nu\beta$, and the constant utility level on that path maximizes the constant utility among all equitable paths with a constant savings rate. The constant savings rate of $\nu\beta$ is seen as an extended version of Hartwick's Rule, which coincides with Hartwick's Rule when $\nu = (1/\mu) = 1$ (as in the model of global warming studied in Section 5).

6.1 Equitable Paths with Constant Savings Rate: Necessary Conditions

An interior path $(k(t), r(t), c(t), w(t))$ from (K, M, W) in $\mathbb{R}_+^2 \times \mathbb{R}_{++}$ has a *Constant Savings Rate* (abbreviated as CSR in what follows) if there is $0 < \gamma < 1$, such that:

$$I(t) \equiv \dot{k}(t) = \gamma Q(t) \equiv \gamma F(k(t), r(t), w(t)) \text{ for all } t \geq 0 \quad \text{(CSR)}$$

Given (CSR), (2.2)(a) implies that:

$$c(t) = (1 - \gamma)Q(t) \text{ for all } t \geq 0 \quad (6.1)$$

We focus on the *necessary conditions* that must be satisfied by *equitable paths* which satisfy (CSR). In the first two subsections, we study the behavior of the capital stock and investment along such a path. This leads us to a number of parametric restrictions in the third subsection, culminating in the important restriction that $\alpha > \nu\beta$ (as noted above). The parametric restrictions lead us to a convenient expression for the constant utility in terms of the savings rate in the fourth subsection. Finally, the fifth subsection makes the important observation that if investment is unbounded along such a path, then investment must, in fact, exhibit quasi-arithmetic growth. This last observation provides a useful hint for constructing an explicit solution of an equitable path with a constant savings rate in Section 6.2.

6.1.1 Dynamic behavior of the capital stock

We start with the preliminary result that along any such path, the capital stock must increase monotonically over time and become unbounded as $t \to \infty$. To see this, note that if the path $(k(t), r(t), c(t), w(t))$ is equitable, then there is $\sigma > 0$ such that:

$$c(t) = \sigma w(t)^\rho \text{ for all } t \geq 0 \quad (6.2)$$

By (CSR) and (6.1), it also follows that:

$$\dot{k}(t) = \gamma c(t)/(1 - \gamma) > 0 \text{ for all } t \geq 0 \quad (6.3)$$

Clearly (6.2) and (6.3) imply that:

$$\dot{k}(t) \geq \sigma W^p \gamma/(1 - \gamma) > 0 \text{ for all } t \geq 0 \qquad (6.4)$$

Thus, we must have $k(t)$ monotonically increasing and becoming *unbounded* over time; that is,

$$k(t) \uparrow \infty \text{ as } t \uparrow \infty \qquad (6.5)$$

6.1.2 Dynamic behavior of investment

Next, we proceed to study the dynamic behavior of investment in the capital good ($\dot{k}(t)$) along a (CSR) path which is equitable. For such a path, using (CSR), (6.1) and (6.2),

$$\frac{\dot{I}(t)}{I(t)} = \frac{\dot{Q}(t)}{Q(t)} = \frac{\dot{c}(t)}{c(t)} = \rho \frac{\dot{w}(t)}{w(t)} = \rho \lambda r(t)^\mu \text{ for all } t \geq 0 \qquad (6.6)$$

where $0 < \mu < 1$ (in contrast to $\mu = 1$, assumed in Sections 4 and 5). In other words, the growth rates of consumption, output, and investment (in the capital good) are all equal to each other and depend on the resource use $r(t)$, through the expression $\rho \lambda r(t)^\mu$.

It follows in particular from (6.6) that:

$$w(t) = \left[\frac{I(t)}{I(0)}\right]^{(1/\rho)} w(0) \text{ for } t \geq 0 \qquad (6.7)$$

Using (CSR), (6.6), (6.7) and the Cobb–Douglas form of F,

$$\dot{k}(t) = \gamma k(t)^\alpha \left[\frac{1}{\rho \lambda} \frac{\dot{I}(t)}{I(t)}\right]^{\nu\beta} \left[\frac{I(t)}{I(0)}\right]^{-(\delta/\rho)} w(0)^{-\delta} \text{ for } t \geq 0 \qquad (6.8)$$

where we denote $(1/\mu)$ by ν. This can be rewritten as:

$$\dot{I}(t)^{\nu\beta} \dot{k}(t)^\alpha = \left[\frac{(\rho\lambda)^{\nu\beta}}{\gamma}\right] \left[\frac{w(0)^\delta}{I(0)^{\delta'}}\right] k(t)^{1+\nu\beta+\delta'} \text{ for } t \geq 0 \qquad (6.9)$$

where we denote (δ/ρ) by δ'. We are now in familiar territory, having dealt with a differential equation like (6.9) in Section 5.1.

The equation (6.9) gives us a *second-order* differential equation in $k(t)$ which we can write equivalently in the form:

$$\dot{I}(t) = \frac{N k(t)^{(1+\nu\beta+\delta')/\nu\beta}}{\dot{k}(t)^{\alpha/\nu\beta}} \text{ for } t \geq 0 \qquad (6.10)$$

where:

$$N = \left[\frac{\rho\lambda}{\gamma^{(1/\nu\beta)}}\right] \left[\frac{W^{\delta/\nu\beta}}{I(0)^{\delta'/\nu\beta}}\right] \qquad (6.11)$$

To solve (6.10), we need to find a function which determines $\dot{k}$ in terms of k, consistent with the restriction expressed in (6.10). This will itself turn out to be a *first-order* differential equation, but with k rather than t as the independent variable.

Introduce the notation $\pi(k)$ to denote the function we are looking for. Then,

$$\pi(k(t)) = \dot{k}(t) = I(t) \text{ for all } t \geq 0 \qquad (6.12)$$

is an identity in t. So, differentiating (6.12) with respect to t, we get:

$$\frac{d\pi(k)}{dk}\dot{k}(t) = \dot{I}(t) \text{ for all } t \geq 0 \tag{6.13}$$

Using (6.10) and (6.13), we can write:

$$\frac{N\pi(k)^{(1+\nu\beta+\delta')/\nu\beta}}{k^{\alpha/\nu\beta}} = \frac{d\pi(k)}{dk}\pi(k) \text{ for all } k > 0 \tag{6.14}$$

Thus, we obtain the differential equation:

$$\frac{d\pi}{dk} = \frac{N\pi^{(1+\delta')/\nu\beta}}{k^{\alpha/\nu\beta}} \text{ for all } k > 0 \tag{6.15}$$

This differential equation can be solved to obtain the form of the function $\pi(k)$.
However, before we do that, we need to rule out a parameter configuration.

Parameter Restriction 1:

$$\frac{(1+\delta')}{\nu\beta} \neq 1 \tag{6.16}$$

The proof of (6.16) is provided in an appendix to Section 6 (see section 8.4).

Having established the parameter restriction (6.16), denote $(1+\delta')/\nu\beta$ by ζ, and (noting that $\zeta \neq 1$) make the substitution $z = \pi^{1-\zeta}$, so that:

$$\frac{dz}{dk} = (1-\zeta)\pi^{-\zeta}\frac{d\pi}{dk} \tag{6.17}$$

Substituting (6.17) in (6.15), we get:

$$\frac{dz}{dk} = (1-\zeta)\pi^{-\zeta}\frac{d\pi}{dk} = \frac{(1-\zeta)N\pi^{-\zeta}\pi^{(1+\delta')/\nu\beta}}{k^{\alpha/\nu\beta}} = \frac{N(1-\zeta)}{k^{\alpha/\nu\beta}} \tag{6.18}$$

Integrating (6.18) from $k(0)$ to $k(t)$, we get:

$$\frac{1}{k(t)^{\zeta-1}} - \frac{1}{k(0)^{\zeta-1}} = \begin{cases} N(1-\zeta)\left[\dfrac{k(t)^{-(\alpha/\nu\beta)+1}}{-(\alpha/\nu\beta)+1} - \dfrac{k(0)^{-(\alpha/\nu\beta)+1}}{-(\alpha/\nu\beta)+1}\right] & \text{if } (\alpha/\nu\beta) \neq 1 \\ N(1-\zeta)[\ln k(t) - \ln k(0)] & \text{if } (\alpha/\nu\beta) = 1 \end{cases} \tag{6.19}$$

6.1.3 Parameter restrictions as necessary conditions

We have already noted a parameter restriction (6.16) as a *necessary condition* for the existence of an equitable path with a constant savings rate. We now note three other parametric restrictions, which are also necessary conditions, by using the solution (6.19) on the dynamic behavior of investment on such paths.

We start with the following result.

Parameter Restriction 2:

$$\alpha \neq \nu\beta \tag{6.20}$$

To see this, note first that in view of (6.3), we can infer from (6.19) that:

$$-\frac{1}{k(0)^{\zeta-1}} \leq \begin{cases} N(1-\zeta)\left[\dfrac{k(t)^{-(\alpha/\nu\beta)+1}}{-(\alpha/\nu\beta)+1} - \dfrac{k(0)^{-(\alpha/\nu\beta)+1}}{-(\alpha/\nu\beta)+1}\right] & \text{if } (\alpha/\nu\beta) \neq 1 \\ N(1-\zeta)[\ln k(t) - \ln k(0)] & \text{if } (\alpha/\nu\beta) = 1 \end{cases} \quad (6.21)$$

Let us examine the second line of (6.21). If $(\alpha/\nu\beta) = 1$, then it must be the case that $\zeta = (1 + \delta')/\nu\beta > (\alpha/\nu\beta) = 1$. By using (6.5), we have $\ln k(t) \to \infty$, as $t \to \infty$ and the right-hand side expression would therefore tend to $-\infty$ as $t \to \infty$, leading to a contradiction. Thus, the possibility of the second line of (6.21) cannot arise. That is, $(\alpha/\nu\beta) = 1$ is not possible. This completes our demonstration that the parameter restriction (6.20) must hold.

Using the fact that (6.20) holds, (6.19) can be simplified to read:

$$\frac{1}{k(t)^{\zeta-1}} - \frac{1}{k(0)^{\zeta-1}} = N(\zeta-1)\left[\frac{1}{[(\alpha/\nu\beta)-1]k(t)^{[(\alpha/\nu\beta)-1]}} - \frac{1}{[(\alpha/\nu\beta)-1]k(0)^{[(\alpha/\nu\beta)-1]}}\right]$$

$$\text{for all } t \geq 0 \quad (6.22)$$

This solution (6.22) can be seen to be the basic result on the dynamic behavior of investment. It implies two other parametric restrictions which we now note.

The first restriction is written as follows.

Parameter Restriction 3:

$$\zeta \equiv \frac{(1+\delta')}{\nu\beta} > 1 \quad (6.23)$$

The proof of (6.23) is provided in an appendix to Section 6 (see section 8.4).

The second restriction is written as follows.

Parameter Restriction 4:

$$\alpha > \nu\beta \quad (6.24)$$

To establish (6.24), we use (6.3) and (6.22) to obtain:

$$-\frac{1}{k(0)^{\zeta-1}} \leq N(\zeta-1)\left[\frac{1}{[(\alpha/\nu\beta)-1]k(t)^{[(\alpha/\nu\beta)-1]}} - \frac{1}{[(\alpha/\nu\beta)-1]k(0)^{[(\alpha/\nu\beta)-1]}}\right] \text{ for all } t \geq 0 \quad (6.25)$$

If (6.24) does not hold, then in view of (6.20) we must have $\alpha/\nu\beta < 1$. By using (6.5), we have $k(t) \to \infty$ as $t \to \infty$ and the right-hand side expression in (6.25) would therefore tend to $-\infty$ as $t \to \infty$ (since $\zeta > 1$ by (6.23)), leading to a contradiction. This establishes the parameter restriction (6.24).

On intertemporal equity and efficiency in a model of global warming   369

It is useful to observe that the parameter restriction 4 implies the parameter restriction 3, and therefore also the parameter restrictions 1 and 2. Thus, while it is certainly true that it is obtained by first establishing these weaker parameter restrictions, we do not need to state those weaker restrictions separately, once we state parameter restriction 4. Further, it is clear that the parameter restriction 4 is an extended version of the technological restriction (S) of Solow (1974), which coincides with (S) when ν = 1 (as in Sections 4 and 5).

6.1.4 On the constant utility level

We now establish a convenient expression for the constant utility level that can be attained along any equitable path with a constant savings rate.

Using (6.5), (6.24) and (6.25), and letting $t \to \infty$, we obtain:

$$\frac{1}{k(0)^{\zeta-1}} \geq \frac{N(\zeta - 1)}{[(\alpha/\nu\beta) - 1]k(0)^{[(\alpha/\nu\beta)-1]}} \qquad (6.26)$$

which can be rewritten as:

$$\frac{1}{k(0)^{\zeta-1}} \geq \left[\frac{\rho\lambda}{\gamma^{(1/\nu\beta)}}\right]\left[\frac{W^{\delta/\nu\beta}}{I(0)^{\delta'/\nu\beta}}\right]\frac{(\zeta - 1)}{[(\alpha/\nu\beta) - 1]k(0)^{[(\alpha/\nu\beta)-1]}}$$

by using the expression for N in (6.11). Transposing terms, we obtain:

$$(\alpha/\nu\beta) - 1]K^{[(\alpha/\nu\beta)-1]} \geq k(0)^{\zeta-1}(\zeta - 1)\left[\frac{\rho\lambda}{\gamma^{(1/\nu\beta)}}\right]\left[\frac{W^{\delta/\nu\beta}}{I(0)^{\delta'/\nu\beta}}\right] \qquad (6.27)$$

Denoting:

$$J' \equiv \frac{(\alpha/\nu\beta) - 1]K^{[(\alpha/\nu\beta)-1]}}{(\zeta - 1)\rho\lambda W^{\delta/\nu\beta}}; \quad J = (J')^{\nu\beta/(1-\nu\beta)} \qquad (6.28)$$

we get:

$$I(0)^{(1-\nu\beta)/\nu\beta} = I(0)^{\zeta-1-(\delta'/\nu\beta)} \leq J'\gamma^{(1/\nu\beta)}$$

and:

$$I(0) \leq J\gamma^{1/(1-\nu\beta)} \qquad (6.29)$$

Note that, by (6.2) and (6.3), we have:

$$\sigma = \frac{c(0)}{w(0)^\rho} = \frac{(1 - \gamma)I(0)}{\gamma W^\rho} \qquad (6.30)$$

so, using (6.29), we get an upper bound on the utility along the path in terms of the saving rate γ:

$$u(\sigma) = u\left[\frac{c(0)}{w(0)^\rho}\right] = u\left[\frac{(1-\gamma)I(0)}{\gamma W^\rho}\right] \leq u\left[\frac{(1-\gamma)J\gamma^{\nu\beta/(1-\nu\beta)}}{W^\rho}\right] \qquad (6.31)$$

If we go over the inequalities (6.26), (6.27) and (6.30), (6.31), we note that:
(i) If along the equitable path with constant savings rate, the investment level $k(t)$ goes

to infinity as $t \to \infty$, then equalities hold in all the four inequalities, and the constant utility level is given by:

$$u(\sigma) = u\left[\frac{c(0)}{w(0)^p}\right] = u\left[\frac{(1-\gamma)I(0)}{\gamma W^p}\right] = u\left[\frac{(1-\gamma)J\gamma^{\nu\beta/(1-\nu\beta)}}{W^p}\right] \quad (6.32)$$

(ii) If along the equitable path with constant savings rate, the investment level $k(t)$ remains bounded above as $t \to \infty$, then strict inequalities hold in all the four inequalities, and the constant utility level is given by:

$$u(\sigma) = u\left[\frac{c(0)}{w(0)^p}\right] = u\left[\frac{(1-\gamma)I(0)}{\gamma W^p}\right] < u\left[\frac{(1-\gamma)J\gamma^{\nu\beta/(1-\nu\beta)}}{W^p}\right] \quad (6.33)$$

6.1.5 Unbounded growth of investment

In Section 6.2 below, we will provide an explicit solution of an equitable path with a constant savings rate. That path is seen to also exhibit *quasi-arithmetic growth* in the capital stock that is faster than linear growth, and the investment in the capital good becomes unbounded over time.

In this subsection, we show a result which might be seen as complementing that explicit solution. The result is that if *any* equitable path with a constant savings rate has *unbounded growth of investment*, then the capital stock along that path *must* exhibit quasi-arithmetic growth that is faster than linear growth. Consequently, investment itself must exhibit quasi-arithmetic growth.

The method of demonstrating this result is similar to the one we employed to derive an investment policy function for any equitable Hartwick path in Section 5.1.3. We proceed as follows. Pick any $T \geq 0$, and fix it. Using (6.22) for $t = T$, we obtain:

$$\frac{1}{k(T)^{\zeta-1}} - \frac{1}{k(0)^{\zeta-1}} = N(\zeta-1)\left[\frac{1}{[(\alpha/\nu\beta)-1]k(T)^{[(\alpha/\nu\beta)-1]}} - \frac{1}{[(\alpha/\nu\beta)-1]k(0)^{[(\alpha/\nu\beta)-1]}}\right]$$

Now, let S be an arbitrary point in time satisfying $S > T$. Using (6.22) for $t = S$, we obtain:

$$\frac{1}{k(S)^{\zeta-1}} - \frac{1}{k(0)^{\zeta-1}} = N(\zeta-1)\left[\frac{1}{[(\alpha/\nu\beta)-1]k(S)^{[(\alpha/\nu\beta)-1]}} - \frac{1}{[(\alpha/\nu\beta)-1]k(0)^{[(\alpha/\nu\beta)-1]}}\right]$$

Subtract the first equation from the second to obtain:

$$\frac{1}{k(S)^{\zeta-1}} - \frac{1}{k(T)^{\zeta-1}} = N(\zeta-1)\left[\frac{1}{[(\alpha/\nu\beta)-1]k(S)^{[(\alpha/\nu\beta)-1]}} - \frac{1}{[(\alpha/\nu\beta)-1]k(T)^{[(\alpha/\nu\beta)-1]}}\right]$$

Let $S \to \infty$, keeping T fixed, in this equation. Then, using (6.5) and (6.24), we have:

$$\frac{1}{[(\alpha/\nu\beta)-1]k(S)^{[(\alpha/\nu\beta)-1]}} \to 0 \text{ as } S \to \infty$$

If investment is unbounded along the path, then since $\zeta > 1$ by (6.23), we have:

$$\frac{1}{k(S)^{\zeta-1}} \to 0 \text{ as } S \to \infty$$

Thus, we obtain:

$$\frac{1}{k(T)^{\zeta-1}} = \left[\frac{N(\zeta-1)}{[(\alpha/\nu\beta)-1]k(T)^{[(\alpha/\nu\beta)-1]}}\right] \text{ for all } T \geq 0$$

which can be rewritten as the first-order differential equation:

$$\dot{k}(t) = ak(t)^{\theta'} \text{ for all } t \geq 0$$

where:

$$a = \left[\frac{(\alpha/\nu\beta)-1}{N(\zeta-1)}\right]^{1/(\zeta-1)} \text{ and } \theta' = \frac{(\alpha/\nu\beta)-1}{(\zeta-1)} = \frac{\alpha-\nu\beta}{1+\delta'-\nu\beta}$$

The differential equation can be solved explicitly to obtain the solution:

$$k(t) = [K^{(1-\theta')} + a(1-\theta')t]^{1/(1-\theta')} \text{ for all } t \geq 0$$

Thus, $k(t)$ must exhibit quasi-arithmetic growth, and investment $I(t) = \dot{k}(t)$ must also exhibit quasi-arithmetic growth:

$$I(t) = \dot{k}(t) = a[K^{(1-\theta')} + a(1-\theta')t]^{\theta'/(1-\theta')} \text{ for all } t \geq 0$$

Further, recalling (6.6), resource use $r(t)$ must exhibit quasi-arithmetic decline:

$$r(t) = \left[\frac{1}{\rho\lambda}\frac{\dot{I}(t)}{I(t)}\right]^\nu = \left[\frac{a\theta'}{\rho\lambda}\right]^\nu \left[\frac{1}{K^{(1-\theta')} + a(1-\theta')t}\right]^\nu \text{ for all } t \geq 0$$

6.2 Explicit Solution of an Equitable Path Satisfying CSR

In this subsection, we provide an explicit solution of an equitable path satisfying a constant savings rate. We have seen that the parameter restriction 4 (appearing in Section 6.1.3) is a *necessary condition* for the existence of such a path. So, in this subsection, we impose that parametric restriction throughout, and since it is an extended form of the technological restriction (S) appearing in Section 5, we refer to it in this subsection as (ES):

$$\alpha > \nu\beta \tag{ES}$$

To write the explicit solution, we need to define a few new parameters, in terms of the model's parameters, and the initial conditions (K, W, M). Let us define:

$$\theta = \frac{(\alpha-\nu\beta)}{1-\alpha+\delta'} \text{ where } \delta' = \frac{\delta}{\rho} \tag{6.34}$$

Note that by (ES), we have $\theta > 0$.
Our explicit solution of the capital stock will be of the form:

$$k(t) = (A+Bt)^{1+\theta} \text{ for all } t \geq 0$$

where A and B will be defined in terms of the parameters of the model in what follows.

Thus, the capital stock will exhibit *quasi-arithmetic growth* (compared to linear growth in Solow (1974)).

Clearly, A will be defined by the initial capital stock, K (and θ); that is,

$$K = A^{1+\theta} \tag{6.35}$$

The important choice therefore is B, and this will be dependent on the constant savings rate through a fairly involved formula. We now proceed to describe this formula.

Given any $s > 0$, let us define:

$$B(s) = \left[\frac{sA^{(\theta/\rho)\delta}}{(1+\theta)W^\delta}\right]^{1/(1-\nu\beta)} \left[\frac{\theta}{\rho\lambda}\right]^{\nu\beta/(1-\nu\beta)} \tag{6.36}$$

Notice that $B(s)$ is monotonically increasing in s, and:

$$B(s) \downarrow 0 \text{ as } s \downarrow 0 \text{ and } B(s) \uparrow \infty \text{ as } s \uparrow \infty \tag{6.37}$$

Thus, there is a unique solution $s^* > 0$ such that:

$$\left[\frac{B(s)}{A}\right]^{\nu-1}\left[\frac{\theta}{\rho\lambda}\right]^\nu\left[\frac{1}{\nu-1}\right] = M \tag{6.38}$$

holds if and only if $s = s^*$.

This value of s will be a *candidate* for the savings rate that we are looking for (with (6.38) representing the appropriate version of the resource constraint (2.2)(b)). But, there are two qualifications to this candidacy. First the solution s^* to (6.37) might end up outside the interval $(0, 1)$. [Note that, quite deliberately, we did not restrict s to lie in $(0, 1)$ in defining $B(s)$ in (6.36), and also in writing the equation (6.37) whose solution is s^*.] Second, even if the solution s^* were to lie in $(0, 1)$, we would not pick it as our savings rate if it exceeded $\nu\beta$. For, then, we could choose the lower savings rate of $\nu\beta$ (this would leave some of the resource stock unused), and increase the constant utility along the path. [One can see this from our discussion in Section 6.1.4, but for the sake of clarity, we will elaborate on this point again below.] These informal statements motivate our choice of the savings rate, γ, as:

$$\gamma = \begin{cases} s^* & \text{if } s^* \leq \nu\beta \\ \nu\beta & \text{if } s^* > \nu\beta \end{cases} \tag{6.39}$$

Using this, we then define:

$$B \equiv B(\gamma) = \left[\frac{\gamma A^{(\theta/\rho)\delta}}{(1+\theta)W^\delta}\right]^{1/(1-\nu\beta)} \left[\frac{\theta}{\rho\lambda}\right]^{\nu\beta/(1-\nu\beta)} \tag{6.40}$$

With θ defined by (6.34), A defined by (6.35), γ defined by (6.39) and B defined by (6.40), we are now ready to formally describe a quadruple of functions $(k(t), r(t), c(t), w(t))$ from (K, M, W), which will be an equitable path and will have the constant savings rate γ.

The capital stock path is specified as:

$$k(t) = (A + Bt)^{(1+\theta)} \text{ for all } t \geq 0 \tag{6.41}$$

Next, we specify the path of resource use as:

$$r(t) = \left[\frac{\theta B}{\rho\lambda(A+Bt)}\right]^\nu \text{ for all } t \geq 0 \quad (6.42)$$

The path of global warming is specified as:

$$w(t) = \frac{W(A+Bt)^{\theta/\rho}}{A^{\theta/\rho}} \text{ for all } t \geq 0 \quad (6.43)$$

Finally, the consumption path is specified as:

$$c(t) = \frac{(1-\gamma)(1+\theta)B(A+Bt)^\theta}{\gamma} \text{ for all } t \geq 0 \quad (6.44)$$

With these definitions, we now proceed to verify that $(k(t), r(t), c(t), w(t))$ is a *path* from (K, M, W), as defined in (2.2). For this purpose, we verify condition (2.2)(d), (2.2)(c), (2.2)(a), and then condition (2.2)(b).

By (6.35) and (6.41) we have $k(0) = A^{(1+\theta)} = K$, and by (6.43), we have $w(0) = W$, as required in (2.2)(d).

Condition (2.2)(c):
Given (6.43), we have:

$$\frac{\dot{w}(t)}{w(t)} = \frac{W(\theta/\rho)B(A+Bt)^{(\theta/\rho)-1}}{W(A+Bt)^{(\theta/\rho)}} = \frac{\theta B}{\rho(A+Bt)} = \lambda\left[\frac{\theta B}{\rho\lambda(A+Bt)}\right] = \lambda r(t)^\mu \text{ for all } t \geq 0$$

(6.45)

the last equality in (6.45) following from (6.42). This verifies condition (2.2)(c).

Condition (2.2)(a):
Given (6.41) and (6.44), we have:

$$c(t) = \frac{(1-\gamma)k(t)}{\gamma} \text{ for all } t \geq 0$$

and so:

$$c(t) + k(t) = \frac{k(t)}{\gamma} \text{ for all } t \geq 0$$

Thus, to verify condition (2.2)(a), it is sufficient to show that:

$$\dot{k}(t) = \gamma F(k(t), r(t), w(t)) \text{ for all } t \geq 0 \quad (6.46)$$

Using (6.41), the left-hand side of (6.46) is:

$$\dot{k}(t) = (1+\theta)B(A+Bt)^\theta \text{ for all } t \geq 0 \quad (6.47)$$

Using (6.41), (6.42) and (6.43), the right-hand side of (6.46) is:

$$\gamma F(k(t), r(t), w(t)) = \gamma (A + Bt)^{\alpha(1+\theta)} \left[\frac{\theta B}{\rho\lambda(A + Bt)} \right]^{\nu\beta} \left[\frac{A^{(\theta/\rho)}}{W(A + Bt)^{(\theta/\rho)}} \right]^{\delta}$$

$$= \frac{\gamma A^{(\theta/\rho)\delta} B^{\nu\beta}}{W^{\delta}} \left[\frac{\theta}{\rho\lambda} \right]^{\nu\beta} (A + Bt)^{\alpha(1+\theta) - \nu\beta - (\theta/\rho)\delta}$$

$$= \left[\frac{\gamma A^{(\theta/\rho)\delta} B^{\nu\beta}}{W^{\delta}} \right] \left[\frac{\theta}{\rho\lambda} \right]^{\nu\beta} (A + Bt)^{\theta} \qquad (6.48)$$

the last line in (6.48) following from the definition of θ in (6.34). Thus, in order to verify that (6.46) holds, we need to show that:

$$(1 + \theta) B = \left[\frac{\gamma A^{(\theta/\rho)\delta} B^{\nu\beta}}{W^{\delta}} \right] \left[\frac{\theta}{\rho\lambda} \right]^{\nu\beta}$$

or equivalently:

$$B^{1-\nu\beta} = \left[\frac{\gamma A^{(\theta/\rho)\delta}}{(1 + \theta) W^{\delta}} \right] \left[\frac{\theta}{\rho\lambda} \right]^{\nu\beta}$$

But this clearly holds as it is the defining condition for B in (6.40). This completes the verification of (6.46), and therefore of condition (2.2)(a).

Condition (2.2)(b):
Given (6.42), we have:

$$\int_0^{\infty} r(t) dt = \int_0^{\infty} \left[\frac{\theta B}{\rho\lambda(A + Bt)} \right]^{\nu} dt = \left[\frac{B}{A} \right]^{\nu-1} \left[\frac{\theta}{\rho\lambda} \right]^{\nu} \left[\frac{1}{\nu - 1} \right] \qquad (6.49)$$

Since $\gamma \leq s^*$ (recall (6.39)), we have $B \equiv B(\gamma) \leq B(s^*)$ (recall (6.36)) and so by (6.37),

$$\left[\frac{B}{A} \right]^{\nu-1} \left[\frac{\theta}{\rho\lambda} \right]^{\nu} \left[\frac{1}{\nu - 1} \right] \leq \left[\frac{B(s^*)}{A} \right]^{\nu-1} \left[\frac{\theta}{\rho\lambda} \right]^{\nu} \left[\frac{1}{\nu - 1} \right] = M \qquad (6.50)$$

Clearly, (6.49) and (6.50) verify condition (2.2)(b).

6.2.1 Equity
We now proceed to verify that $(k(t), r(t), c(t), w(t))$ from (K, M, W) is an *equitable* path. Given (6.44), we get:

$$\dot{c}(t) = \frac{(1 - \gamma)(1 + \theta)\theta B^2 (A + Bt)^{\theta-1}}{\gamma} \text{ for all } t \geq 0 \qquad (6.51)$$

Thus, combining (6.44) and (6.51),

$$\frac{\dot{c}(t)}{c(t)} = \frac{(1 - \gamma)(1 + \theta)\theta B^2 (A + Bt)^{\theta-1}}{(1 - \gamma)(1 + \theta) B(A + Bt)^{\theta}} = \frac{\theta B}{(A + Bt)} = \rho \left[\frac{\theta B}{\rho(A + Bt)} \right] = \rho \frac{\dot{w}(t)}{w(t)}$$

$$\text{for all } t \geq 0 \qquad (6.52)$$

the last equality in (6.52) following from (6.45). This verifies that the path $(k(t), r(t), c(t), w(t))$ from (K, M, W) specified in (6.41)–(6.44) is an equitable path.

6.2.2 Global warming

Along the equitable path $(k(t), r(t), c(t), w(t))$ from (K, M, W), specified by (6.41)–(6.44), we will have global warming increase over time with no finite bound. This is immediate from (6.43) which defines $w(t)$. To preserve equity, which we have verified in (6.52) above, consumption (and therefore output and investment) will also increase over time without any finite bound.

This makes the long-run picture under equity substantially different from Solow (1974), Stollery (1998) and d'Autume, Hartwick and Schubert (2010).

6.3 Maximum Sustainable Utility

We have shown in Section 6.2, that for each specification of the parameters of the model relating to technology, tastes, and the law of motion of global warming, and for each specification of the initial conditions (K, M, W), there is an equitable path satisfying a constant savings rate, provided the condition (ES) is satisfied:

$$\alpha > \nu\beta \tag{ES}$$

In Section 6.1, we saw that (ES) was a necessary condition for the existence of an equitable path with a constant savings rate. Thus, its appearance in our existence result of Section 6.2 is only to be expected.

Our existence result in Section 6.2 was constructive: we provided an explicit solution $(k(t), r(t), c(t), w(t))$ from (K, M, W), of the equitable path with a constant savings rate γ, specified in (6.39). Along this path the capital stock exhibits quasi-arithmetic growth, which is faster than linear growth. As a result, investment in the capital good goes to infinity as $t \to \infty$. Thus, we know that the constant utility level specified in (6.32), in Section 6.1.4, is actually attained. Any other equitable path with the *same* constant savings rate γ must attain a lower utility level, according to (6.33) in Section 6.1.4.

Given our results in this section so far, a natural question that arises is the following: among equitable paths with a constant savings rate, which savings rate would *maximize* the constant utility?

If we look at the path $(k(t), r(t), c(t), w(t))$ from (K, M, W) constructed in Section 6.2, we note that it is equitable, with a constant savings rate γ (given by the formula (6.39)), and it satisfies:

$$k(t) \uparrow \infty \text{ as } t \uparrow \infty$$

Thus, the constant utility level associated with it is given by (6.32). That is, the upper bound on the constant utility that can be obtained by *any* equitable CSR path (see (6.31)) $(k'(t), r'(t), c'(t), w'(t))$ from (K, M, W) with the same savings rate γ is actually attained by the path $(k(t), r(t), c(t), w(t))$ from (K, M, W) constructed in Section 6.2.

Let us look at the expression of the utility level in (6.32). This depends on the savings rate γ. Clearly, the utility on the right-hand side of (6.32) will be maximized at that value of γ, at which the function:

$$f(\gamma) \equiv (1 - \gamma)\gamma^{v\beta/(1-v\beta)} \text{ for } \gamma \in [0,1] \quad (6.53)$$

is maximized. Note that $f(0) = f(1) = 0$, while $f(\gamma) > 0$ for all $0 < \gamma < 1$. Differentiating f with respect to γ on $(0,1)$, we obtain:

$$f'(\gamma) = (1 - \gamma)[v\beta/(1 - v\beta)]\gamma^{v\beta/(1-v\beta)-1} - \gamma^{v\beta/(1-v\beta)}$$

$$= \gamma^{v\beta/(1-v\beta)}\left[\frac{(1-\gamma)[v\beta/(1-v\beta)]}{\gamma} - 1\right]$$

Thus,

$$f'(\gamma) 0 \text{ if and only if } (1-\gamma)[v\beta/(1-v\beta)]\gamma$$

which can be rewritten as:

$$f'(\gamma) 0 \text{ if and only if } v\beta\gamma$$

That is, f is increasing in γ on $[0, v\beta]$ and decreasing in γ on $[v\beta, 1]$, and consequently it attains a *unique maximum* when:

$$\gamma = v\beta \quad (6.54)$$

This suggests that the "best" savings rate that emerges from this analysis (in terms of maximizing sustainable utility among equitable paths with constant savings rates) is $v\beta$, which exceeds the savings rate β involved in Hartwick's Rule, since $v = (1/\mu) > 1$ in the present context. [It is also reassuring to note that if in the formula (6.54) we let $v \to 1$, then the best savings rate $\gamma = v\beta$ approaches β, the savings rate involved in Hartwick's Rule.]

The reason that this observation is called a "suggestion" rather than a result is that given the resource stock M, there need not be an equitable path with constant savings rate equal to $v\beta$.

We now proceed to formalize this suggestion into a result. The result is that there is a threshold resource stock, $\overline{M}$, such that if the actual resource stock M is *at least as large* as $\overline{M}$, then there is an equitable path with a constant savings rate of $v\beta$, and the constant utility level on that path does indeed maximize the constant utility among *all* equitable paths with a constant savings rate.

To establish the result, consider the model with a specification of the parameters relating to technology, tastes, and the law of motion of global warming, satisfying:

$$\alpha > v\beta \quad \text{(ES)}$$

and for a specification of the initial conditions (K, W). [Note that the initial resource stock M is not yet specified.]

As in (6.34), define:
$$\theta = \frac{(\alpha - v\beta)}{1 - \alpha + \delta'} \quad \text{where } \delta' = \frac{\delta}{\rho} \tag{6.55}$$

and as in (6.35), define A by:
$$K = A^{1+\theta} \tag{6.56}$$

Now, recalling (6.36), define:
$$B(v\beta) = \left[\frac{v\beta A^{(\theta/\rho)\delta}}{(1+\theta)W^\delta}\right]^{1/(1-v\beta)} \left[\frac{\theta}{\rho\lambda}\right]^{v\beta/(1-v\beta)} \tag{6.57}$$

[obtained by replacing s by $v\beta$ in (6.36)]. Using (6.57), define the threshold resource stock $\overline{M}$ by:
$$\overline{M} = \left[\frac{B(v\beta)}{A}\right]^{v-1} \left[\frac{\theta}{\rho\lambda}\right]^v \left[\frac{1}{v-1}\right] \tag{6.58}$$

To complete description of the parameters of the model, assume that the initial resource stock M satisfies:
$$M \geq \overline{M} \tag{6.59}$$

With θ defined by (6.55), A defined by (6.56), $B \equiv B(v\beta)$ defined by (6.57), and the savings rate defined to be equal to $v\beta$, we now consider the quadruple of functions $(k(t), r(t), c(t), w(t))$ from (K, M, W), defined exactly as in (6.41)–(6.44), but with γ replaced by $v\beta$. As verified in Section 6.2, these functions will satisfy conditions (2.2)(d), (2.2)(c) and (2.2)(a). It remains to verify (2.2)(b) to establish that $(k(t), r(t), c(t), w(t))$ from (K, M, W) is a *path*.

We have $B \equiv B(v\beta)$, and so, using (6.58) and (6.59), we get:
$$\left[\frac{B(v\beta)}{A}\right]^{v-1} \left[\frac{\theta}{\rho\lambda}\right]^v \left[\frac{1}{v-1}\right] = \overline{M} \leq M \tag{6.60}$$

Given (6.42) and (6.60), we have:
$$\int_0^\infty r(t)\,dt = \int_0^\infty \left[\frac{\theta B}{\rho\lambda(A+Bt)}\right]^v dt = \left[\frac{B}{A}\right]^{v-1} \left[\frac{\theta}{\rho\lambda}\right]^v \left[\frac{1}{v-1}\right] \leq M \tag{6.61}$$

verifying the resource constraint (2.2)(b). Thus, $(k(t), r(t), c(t), w(t))$ is a path from (K, M, W). It has a constant savings rate $v\beta$, and it is equitable, by following the procedure used in Section 6.2.1.

The constructed path $(k(t), r(t), c(t), w(t))$ from (K, M, W) satisfies:
$$k(t) \uparrow \infty \text{ as } t \uparrow \infty$$

Thus, the constant utility level associated with it is given by (recall (6.32)):
$$u^* = u\left[\frac{(1-v\beta)J(v\beta)^{v\beta/(1-v\beta)}}{W^p}\right] \tag{6.62}$$

Consider now *any* equitable path with a constant savings rate of $\gamma \neq \nu\beta$. Then the constant utility level is (recalling (6.31)):

$$\tilde{u} \leq u\left[\frac{(1-\gamma)J(\gamma)^{\nu\beta/(1-\nu\beta)}}{W^{\rho}}\right] < u\left[\frac{(1-\nu\beta)J(\nu\beta)^{\nu\beta/(1-\nu\beta)}}{W^{\rho}}\right] = u^* \qquad (6.63)$$

the last inequality in (6.63) following from (6.54).

We have now formally established the result we were after. There is a threshold resource stock, $\overline{M}$ (defined in (6.58)), such that, if the actual resource stock M is at least as large as $\overline{M}$, then there is an equitable path with a constant savings rate of $\nu\beta$, and the constant utility level on that path maximizes the constant utility among *all* equitable paths with a constant savings rate.

7. RATES OF CHANGE OF KEY VARIABLES

In this section, we make some observations about the rates of change of key variables (capital stock, resource use, consumption, investment and global warming) along equitable paths with a constant savings rate. This helps us to see more clearly the effect on the dynamics along equitable paths of (a) introducing global warming (following Stollery (1998)) in the model of Solow (1974), and (b) of varying the Stollery model with alternate formulations of the law of motion of global warming.

In the first subsection, we note some general dynamic properties, which are applicable to this class of models of global warming, including Solow's framework as a special case. They emphasize the *similarities* in dynamic behavior on equitable paths arising from the common structure of these models.

In the second subsection, we examine in detail the specific relation between the rate of growth of the capital stock and the rate of decline of the resource use along equitable paths. The emphasis is on identifying the key *differences* in dynamic behavior along equitable paths, due to the presence of global warming, and the distinct formulations of it.

7.1 General Dynamic Properties

Investment and global warming
Perhaps the most basic dynamic property of equitable paths satisfying (CSR) is that rate of growth of investment in the capital good must be proportional to the rate of growth of the index of global warming.

To see the logic, note that a constant savings rate implies that:

$$\frac{\dot{I}(t)}{I(t)} = \frac{\dot{Q}(t)}{Q(t)} = \frac{\dot{c}(t)}{c(t)} \text{ for all } t \geq 0 \qquad (7.1)$$

while equity demands that:

$$\rho \frac{\dot{w}(t)}{w(t)} = \frac{\dot{c}(t)}{c(t)} \text{ for all } t \geq 0 \qquad (7.2)$$

This gives us:

$$\frac{\dot{I}(t)}{I(t)} = \rho \frac{\dot{w}(t)}{w(t)} \text{ for all } t \geq 0 \tag{7.3}$$

Capital-resource substitution and global warming

Along equitable paths satisfying (CSR), the capital stock increases and resource use declines over time, the former substituting for the latter. In order to maintain equity, this substitution of capital for resource must proceed at a rate to offset both the production and the utility effects of global warming. Our next property provides a precise formulation of this phenomenon.

Using (2.1), we have:

$$\frac{\dot{Q}(t)}{Q(t)} = \alpha \frac{\dot{k}(t)}{k(t)} + \beta \frac{\dot{r}(t)}{r(t)} - \delta \frac{\dot{w}(t)}{w(t)} \text{ for all } t \geq 0 \tag{7.4}$$

On the other hand, constant savings rate and equity yield (by (7.1) and (7.2)),

$$\frac{\dot{Q}(t)}{Q(t)} = \frac{\dot{c}(t)}{c(t)} = \rho \frac{\dot{w}(t)}{w(t)} \text{ for all } t \geq 0 \tag{7.5}$$

Combining (7.4) and (7.5), we can write:

$$\alpha \frac{\dot{k}(t)}{k(t)} - \beta \left[\frac{-\dot{r}(t)}{r(t)} \right] = (\delta + \rho) \frac{\dot{w}(t)}{w(t)} \tag{7.6}$$

The right-hand side of (7.6) captures the production and utility effects of (rate of growth of) global warming. The left-hand side indicates the appropriate rate at which capital must grow to substitute for the declining resource to offset these two effects of global warming and preserve equity.

Remark

Note that (7.3) and (7.6) hold in the model of Stollery (1998), discussed in our Section 5, as well as in the alternate scenario discussed in Section 6. They also hold in the model of Solow (1974), where the right-hand side of (7.3) and (7.6) is zero.

Dynamics of the capital–output ratio

In the model of Solow (1974), along any equitable path satisfying (CSR), we must have both consumption and investment constant over time. This implies two things. First, output must be a constant over time. Second, capital must exhibit *linear growth*. Consequently, the capital–output ratio itself must exhibit linear growth.

In the model of Stollery (1998), or in its variation considered in our Section 6, the situation is more subtle, since output has to grow over time, and capital exhibits different growth patterns depending on the law of motion of global warming. Nevertheless, it is true in all these models that *asymptotically* the capital–output ratio always exhibits linear growth.

To see the logic of this, we first make the preliminary observations that equity demands that consumption is positive and bounded away from zero, while a constant savings rate $(0 < \gamma < 1)$ then implies that investment is positive and bounded away from zero. Thus, we have:

$$k(t) \uparrow \infty \text{ as } t \uparrow \infty \tag{7.7}$$

as already noted earlier in (5.24) and (6.5), and in addition (by (7.3)):

$$\dot{I}(t) > 0 \text{ for all } t \geq 0 \tag{7.8}$$

The result (7.8) leads to the following two logical possibilities:
(a) $I(t) \uparrow \infty$ as $t \uparrow \infty$; or
(b) $I(t)$ increases and converges to some $\bar{I}$, where $0 < I(0) < \bar{I} < \infty$.

Case (a)
If possibility (a) occurs, then we have the situation analyzed in Section 6.1.5, and $k(t)$ must exhibit quasi-arithmetic growth, which is faster than linear growth, and so $I(t) = \dot{k}(t)$ must also exhibit quasi-arithmetic growth. More explicitly, $k(t)$ must satisfy:

$$k(t) = [K^{(1-\theta')} + a(1-\theta')t]^{1/(1-\theta')} \text{ for all } t \geq 0 \tag{7.9}$$

where:

$$a = \left[\frac{(\alpha/\nu\beta) - 1}{N(\zeta - 1)}\right]^{1/(\zeta-1)}, \quad \theta' = \frac{(\alpha/\nu\beta) - 1}{(\zeta - 1)}$$

and:

$$\zeta \equiv \frac{(1+\delta')}{\nu\beta}, \quad N = \left[\frac{\rho\lambda}{\gamma^{(1/\nu\beta)}}\right]\left[\frac{W^{\delta/\nu\beta}}{I(0)^{\delta'/\nu\beta}}\right]$$

Consequently, investment $I(t) = \dot{k}(t)$ must satisfy:

$$I(t) = \dot{k}(t) = a[K^{(1-\theta')} + a(1-\theta')t]^{\theta'/(1-\theta')} \text{ for all } t \geq 0 \tag{7.10}$$

Using (7.9) and (7.10), we see that the capital–output ratio must satisfy:

$$\frac{k(t)}{Q(t)} = \frac{\gamma k(t)}{I(t)} = (\gamma/a)[K^{(1-\theta')} + a(1-\theta')t] \text{ for all } t \geq 0 \tag{7.11}$$

which is linear growth.

Case (b)
If possibility (b) occurs, then given any $\varepsilon > 0$, there is $0 < \tau < \infty$, such that:

$$|I(t) - \bar{I}| < \varepsilon \text{ for all } t \geq \tau$$

Consequently, we have:

$$k(\tau) + (\bar{I} + \varepsilon)(t - \tau) \geq k(t) \geq k(\tau) + (\bar{I} - \varepsilon)(t - \tau) \text{ for all } t \geq \tau$$

and so:

$$\frac{\gamma k(\tau) + (\bar{I} + \varepsilon)(t - \tau)]}{I(t)(\gamma/\bar{I})[K + \bar{I}t]} \geq \frac{k(t)/Q(t)]}{(\gamma/\bar{I})[K + \bar{I}t]} \geq \frac{\gamma k(\tau) + (\bar{I} - \varepsilon)(t - \tau)]}{I(t)(\gamma/\bar{I})[K + \bar{I}t]} \quad \text{for all } t \geq \tau \quad (7.12)$$

Letting $t \to \infty$ in (7.12), we get:

$$\frac{k(t)/Q(t)]}{(\gamma/\bar{I})[K + \bar{I}t]} \to 1 \text{ as } t \to \infty \quad (7.13)$$

This means that the capital–output ratio approaches the linear growth path $(\gamma/\bar{I})[K + \bar{I}t]$ as $t \to \infty$.

Remark
Because the share of capital in output is constant (α) in our framework, the asymptotic linear growth of the capital–output ratio also implies that the real rate of interest (given by the marginal product of capital) must exhibit an eventual decline at precisely the reciprocal of time.

7.2 Growth Rates of Capital and Resource

Solow's framework
In the model of Solow (1974), in which global warming is absent (and which can be treated as a special case of our model, with $\lambda = 0$), the relation (7.6) obtained above yields:

$$\alpha \left[\frac{\dot{k}(t)}{k(t)}\right] = \beta \left[\frac{-\dot{r}(t)}{r(t)}\right] \text{ for all } t \geq 0$$

for any equitable path with a constant savings rate, and therefore for an equitable Hartwick path. Since the existence of a non-trivial equitable path in Solow's framework implies:

$$\alpha > \beta \quad (S)$$

we must have:

$$\left[\frac{-\dot{r}(t)}{r(t)}\right] > \left[\frac{\dot{k}(t)}{k(t)}\right] \text{ for all } t \geq 0 \quad (7.14)$$

That is resource use must decline at a faster rate than the rate of growth of the capital stock.

Intuitively, the logic of this result can be seen as follows. Equity and a constant savings rate imply that consumption, output and investment must be constant over time (see (7.1) and (7.2) above). This means that the capital stock must exhibit precisely linear growth. If resource use decline is at the same rate (or lower), then resource use must exhibit "harmonic" decline or slower (that is, $r(t)$ would be of the order of $(1/t)$ or higher), and this would violate the basic resource constraint that the total resource use has to be finite.

Global Warming with $0 < \mu < 1$:
In the framework of global warming with $0 < \mu < 1$ (discussed in Section 6), since we have an explicit solution of an equitable path, with a constant savings rate $(0 < \gamma < 1)$,

we can directly compare the rate of growth of the capital stock and the rate of decline of resource use.

Recall from (6.41) that the capital stock path in the explicit solution is specified as:

$$k(t) = (A + Bt)^{(1+\theta)} \text{ for all } t \geq 0 \tag{7.15}$$

and from (6.42) that the path of resource use is specified as:

$$r(t) = \left[\frac{\theta B}{\rho\lambda(A + Bt)}\right]^v \text{ for all } t \geq 0 \tag{7.16}$$

Thus, from (7.15), we get:

$$\left[\frac{\dot{k}(t)}{k(t)}\right] = \frac{(1 + \theta)B}{(A + Bt)} \text{ for all } t \geq 0 \tag{7.17}$$

and from (7.16), we get:

$$\left[\frac{-\dot{r}(t)}{r(t)}\right] = \frac{vB}{(A + Bt)} \text{ for all } t \geq 0 \tag{7.18}$$

Consequently,

$$\left[\frac{-\dot{r}(t)}{r(t)}\right] > \left[\frac{\dot{k}(t)}{k(t)}\right] \text{ if and only if } v > (1 + \theta) \tag{7.19}$$

In this characterization result, note that from (6.34), θ itself depends on v. So, in order to make it more useful, one needs to rewrite it so that it involves a comparison of v (equivalently μ) with an expression involving only the rest of the parameters of the model ($\alpha, \beta, \delta, \lambda, \rho$). This is straightforward. Since from (6.34), we have:

$$\theta = \frac{(\alpha - v\beta)}{1 - \alpha + \delta'} \text{ where } \delta' = \frac{\delta}{\rho}$$

we obtain:

$$v > (1 + \theta) \text{ if and only if } v > \frac{1 + \delta'}{1 + \delta' - (\alpha - \beta)} \tag{7.20}$$

Note that by (ES), we have $\alpha > \beta$, and so the expression on the right-hand side of (7.20) is greater than 1. Using (7.20) in (7.19), we can write:

$$\left[\frac{-\dot{r}(t)}{r(t)}\right] > \left[\frac{\dot{k}(t)}{k(t)}\right] \text{ if and only if } v > \frac{1 + \delta'}{1 + \delta' - (\alpha - \beta)} \tag{7.21}$$

Equivalently, one can write:

$$\left[\frac{-\dot{r}(t)}{r(t)}\right] > \left[\frac{\dot{k}(t)}{k(t)}\right] \text{ if and only if } \mu < 1 - \frac{(\alpha - \beta)}{1 + \delta'} \equiv \bar{\mu} \tag{7.22}$$

Note that $\bar{\mu}$, defined in (7.22) satisfies $0 < \bar{\mu} < 1$. When μ is relatively small (precisely, when it is smaller than $\bar{\mu}$), then the result seen in Solow's model will continue to hold. When μ is close enough to 1 (precisely, when it is larger than $\bar{\mu}$), then the traditional result will be *reversed*.

Remarks

As a word of caution, we note that the case $\mu = 1$ *cannot* be treated as part of the above analysis, for then the explicit solution (given in Section 6.2) of an equitable path satisfying (CSR) is no longer valid. To see this, note that with $\mu = 1$, we have $v = 1$ and the resource path specified in (6.42) [and repeated in (7.16) above] will violate the resource constraint, as the total resource use will be infinite.

Global Warming with $\mu = 1$:

In the framework of global warming with $\mu = 1$ (that is, the Stollery (1998) model, discussed in Section 5), since we have an explicit solution of an equitable Hartwick path (which is an equitable path with a constant savings rate of β), we can use that to compare the rate of growth of the capital stock and the rate of decline of resource use.

Compared to the case $0 < \mu < 1$ (discussed above), the investment policy function is analytically less straightforward, and consequently the calculations of the appropriate rates of growth are more involved. Recall from the explicit solution in (5.47) that:

$$r(t) = \frac{g'(k(t))}{\rho\lambda} \text{ for all } t \geq 0 \tag{7.23}$$

while from (5.48),

$$\dot{k}(t) = g(k(t)) \text{ for all } t \geq 0 \tag{7.24}$$

where the function g is given by (recall (g) in Section 5.2):

$$g(x) = \left[\frac{x^\eta}{a + bx^\eta}\right]^{1/(\zeta-1)} \text{ for all } x > 0 \tag{7.25}$$

with:

$$\eta \equiv \frac{\alpha}{\beta} - 1 \text{ and } \zeta \equiv \frac{1 + (\delta/\rho)}{\beta} = \frac{1 + \delta'}{\beta} \tag{7.26}$$

and a and b satisfying the constraints (I) and (II) of Section 5.2.

Using (7.23) and (7.24), it is possible to derive the following formula for all $t \geq 0$:

$$\left[\frac{-\dot{r}(t)}{r(t)}\right] - \left[\frac{\dot{k}(t)}{k(t)}\right] = \frac{(\alpha - \beta) g(k(t))}{\beta \ k(t)}\left[1 - \left\{\frac{(1 + \delta')}{(1 + \delta') - \beta}\right\}\left\{\frac{a}{(a + bk(t)^\eta)}\right\}\right] \tag{7.27}$$

The details of the derivation are included in an appendix to Section 7 (see section 8.5).

As noted in (5.24) [and repeated in (7.7) above], we have $k(t) \uparrow \infty$ as $t \uparrow \infty$, and so the second term in curly brackets in (7.27) goes to zero as $t \to \infty$. Thus, the term in square brackets on the right-hand side of (7.27) approaches 1 as $t \to \infty$, and consequently, for all *large* $t \geq 0$,

$$\left[\frac{-\dot{r}(t)}{r(t)}\right] - \left[\frac{\dot{k}(t)}{k(t)}\right] > 0$$

That is, Solow's result holds in Stollery's model *asymptotically*. Resource use eventually declines at a faster rate than the growth of the capital stock.

Remark

The reader will no doubt have noticed from our above discussion that the behavior of the growth rates of capital versus resource is *discontinuous* as the case $0 < \mu < 1$ approaches the case $\mu = 1$. For $0 < \mu < 1$, and μ close to 1, we have noted that the rate of resource decline must be *smaller than* the rate of growth of capital *for all* $t \geq 0$. At $\mu = 1$, however, the rate of resource decline must be *greater than* the rate of growth of capital at least asymptotically.

REFERENCES

Arrow, K.J. (1968), *Applications of Control Theory to Economic Growth*, Lectures in Applied Mathematics 12 (Mathematics of the Decision Sciences, Part 2), American Mathematical Society, Providence, RI.

Asheim, G.B. (1994), Net national product as an indicator of sustainability, *Scandinavian Journal of Economics*, **96**, 257–65.

d'Autume, A., J.M. Hartwick and K. Schubert (2010), The zero discounting and maximin optimal paths in a simple model of global warming, *Mathematical Social Sciences* **59**, 193–207.

Buchholz, W., S. Dasgupta and T. Mitra (2005), Intertemporal equity and Hartwick's rule in an exhaustible resource model, *Scandinavian Journal of Economics* **107**(3), 547–61.

Cass, D. and T. Mitra (1991), Indefinitely sustained consumption despite exhaustible natural resources, *Economic Theory* **1**, 119–46.

Comolli, P.M. (1997), Pollution control in a simplified general-equilibrium model with production externalities, *Journal of Environmental Economics and Management* **4**, 289–304.

Dasgupta, P.S. (1982), *The Control of Resources*, Harvard University Press, Cambridge, MA.

Dasgupta, P.S. and G.M. Heal (1974), The optimal depletion of exhaustible resources, *Review of Economic Studies*, Symposium on the Economics of Exhaustible Resources **41**, 3–28.

Dasgupta, P.S. and G.M. Heal (1979), *Economic Theory and Exhaustible Resources*, Cambridge University Press, Cambridge, MA.

Dixit, A., P. Hammond and M. Hoel (1980), On Hartwick's rule for regular maximin paths of capital accumulation and resource depletion, *Review of Economic Studies* **47**, 551–6.

EPA (The US Environmental Protection Agency) (2007), *Climate Change Indicators in the United States*, Annual Reports 2007.

Forster, B.A. (1973), Optimal consumption planning in a polluted environment, *Economic Record* **49**(4), 534–45.

Forster, B.A. (1975), Optimal pollution control with a nonconstant exponential rate of decay, *Journal of Environmental Economics and Management* **2**, 1–6.

Goldberg, R.R. (1964), *Methods of Real Analysis*, Blaisdell, Waltham, MA.

Hartwick, J.M. (1977), Intergenerational equity and investing of rents from exhaustible resources, *American Economic Review* **67**, 972–4.

Hartwick, J.M. (1990), Natural resources, national accounting and economic depreciation, *Journal of Public Economics* **43**, 291–304.

Hartwick, J.M. (1994), National wealth and net national product, *Scandinavian Journal of Economics* **96**, 253–6.

Lüthi, D. et al. (2008), High-resolution carbon dioxide concentration record 650,000–800,000 years before present, *Nature* **453**, 379–82.

Maler, K.G. (1991), National accounts and environmental resources, *Environmental and Resource Economics* **1**, 1–15.

Mitra, T. (1978), Efficient growth with exhaustible resources in a neoclassical model, *Journal of Economic Theory* **17**, 114–29.

Mitra, T. (2002), Intertemporal equity and efficient allocation of resources, *Journal of Economic Theory* **107**, 356–76.

Mitra, T., G.B. Asheim, W. Buchholz and C. Withagen (2013), Characterizing the sustainability problem in an exhaustible resource model, *Journal of Economic Theory* **148**, 2164–82.

Murphy, G.M. (1960), *Ordinary Differential Equations and their Solutions*, Van Nostrand, Princeton, NJ.

Nordhaus, W.D. (1982), How fast should we graze the global commons?, *American Economic Review* **72**, 242–6.

Nordhaus, W.D. (1991), To slow or not to slow: the economics of the greenhouse effect, *Economic Journal* **101**, 920–37.

Nordhaus, W.D. (2012), Integrated economic and climate modeling, in Peter Dixon and Dale Jorgenson (eds), *Handbook of Computable General Equilibrium Modeling*, Elsevier, North Holland.

Pearson, P.N. and M.R. Palmer (2000), Atmospheric carbon dioxide concentrations over the past 60 million years, *Nature* **406**, 695–9.

Rhein, M. et al. (2013), Observations: ocean, in T.F. Taylor et al. (eds), *Climate Change 2013: The Physical Science Basis*, Contribution of Working Group I to the Fifth Assessment Report of the Intergovernmental Panel on Climate Change, Cambridge University Press, Cambridge and New York.

Royden, H.L. (1988), *Real Analysis*, 3rd edn, Macmillan, New York.

Sinclair, P.J.N. (1992), High does nothing and rising is worse: carbon taxes should keep declining to cut harmful emissions, *Manchester School* **60**, 41–52.

Solow, R.M. (1974), Intergenerational equity and exhaustible resources, *Review of Economics Studies* Symposium, 29–46.

Stollery, K.R. (1998), Constant utility paths and irreversible global warming, *Canadian Journal of Economics* **31**(3), 730–42.

Tahvonen, O. and J. Kuuluvainen (1991), Optimal growth with renewable resources and pollution, *European Economic Review* **35**, 650–61.

Weitzman, M.L. (1976), On the welfare significance of national product in a dynamic economy, *Quarterly Journal of Economics* **90**, 156–62.

Withagen, C. and G.B. Asheim (1998), Characterizing sustainability: the converse of Hartwick's Rule, *Journal of Economic Dynamics and Control* **23**, 159–65.

386 *Handbook on the economics of climate change*

8. APPENDICES

In this section, we include the technical details that were left out in the text (to ease reading the text, which is already quite mathematical in nature) in Sections 3, 4, 5, 6 and 7. The appendices to these sections keep our exposition self-contained, without making the text unduly cluttered.

In numbering displayed mathematical content, we use A, followed by the Section number for which the Appendix was created, and then the number of the displayed content. Thus, for example, the first display in the Appendix to Section 3 is numbered (A3.1).

8.1 Appendix to Section 3

Denoting $[1 + p_1(t)]$ by $P_1(t)$ for all $t \in S,T]$, and differentiating (3.4)(i) with respect to t, we get:

$$\dot{p}_2(t) = \dot{P}_1(t) F_2(k(t), r(t), w(t)) + P_1(t) \dot{F}_2(k(t), r(t), w(t))$$
$$+ \lambda G'(r(t))[p_3(t)\dot{w}(t) + \dot{p}_3(t)w(t)] + p_3(t)w(t)\lambda G''(r(t))\dot{r}(t) \qquad \text{(A3.1)}$$

Using (3.5)(ii) in (A3.1),

$$\lambda G'(r(t))[p_3(t)\dot{w}(t) + \dot{p}_3(t)w(t)] + p_3(t)w(t)\lambda G''(r(t))\dot{r}(t)$$
$$= -\dot{P}_1(t)F_2(k(t), r(t), w(t)) - P_1(t)\dot{F}_2(k(t), r(t), w(t))$$
$$= P_1(t)F_1(k(t), r(t), w(t))F_2(k(t), r(t), w(t)) - P_1(t)\dot{F}_2(k(t), r(t), w(t))$$
$$= P_1(t)F_2(k(t), r(t), w(t))\left[F_1(k(t), r(t), w(t)) - \frac{\dot{F}_2(k(t), r(t), w(t))}{F_2(k(t), r(t), w(t))}\right] \qquad \text{(A3.2)}$$

where the second line in (A3.2) follows from (3.5)(ii), and the third line in (A3.2) follows from (3.5)(i).

Next, we rewrite (3.5)(iii) as:

$$\dot{p}_3(t) + p_3(t)\frac{\dot{w}(t)}{w(t)} = -[P_1(t)F_3(k(t), r(t), w(t)) + q(t)U_2(c(t),w(t))]$$

so that:

$$\dot{p}_3(t)w(t) + p_3(t)\dot{w}(t) = -[P_1(t)F_3(k(t), r(t), w(t)) + q(t)U_2(c(t),w(t))]w(t)$$

$$= -P_1(t)\left[F_3(k(t), r(t), w(t)) + \frac{q(t)}{P_1(t)}U_2(c(t),w(t))\right]w(t)$$

$$= -P_1(t)w(t)\left[F_3(k(t), r(t), w(t)) + \frac{U_2(c(t),w(t))}{U_1(c(t),w(t))}\right] \qquad \text{(A3.3)}$$

where the last line of (A3.3) follows from (3.4)(ii).

We can now combine (A3.2) and (A3.3) to write:

$$\left[F_1(k(t), r(t), w(t)) - \frac{\dot{F}_2(k(t), r(t), w(t))}{F_2(k(t), r(t), w(t))}\right]$$

$$= \left[\frac{1}{P_1(t)F_2(k(t), r(t), w(t))}\right]\{\lambda G'(r(t))[p_3(t)\dot{w}(t) + \dot{p}_3(t)w(t)] + p_3(t)w(t)\lambda G''(r(t))\dot{r}(t)\}$$

$$= -\left[\frac{\lambda G'(r(t))w(t)}{F_2(k(t), r(t), w(t))}\right]\left[F_3(k(t), r(t), w(t)) + \frac{U_2(c(t), w(t))}{U_1(c(t), w(t))}\right]$$

$$+ \left[\frac{1}{P_1(t)F_2(k(t), r(t), w(t))}\right]p_3(t)w(t)\lambda G''(r(t))\dot{r}(t) \quad (A3.4)$$

8.2 Appendix to Section 4

Proof of First Step (in Section 4.4)

For all $t \geq 0$, we can write:

$$\dot{k}(t) = k(t)^\alpha r(t)^\beta w(t)^{-\delta} - c(t)$$

$$= \frac{k(t)^\alpha r(t)^\beta}{w(t)^\delta} - \eta w(t)^\rho$$

$$\leq Ak(t)^\alpha r(t)^\beta - \eta W^\rho \quad (A4.1)$$

where in the last line of (A4.1), we have used the fact that $w(t) \geq W$ (from (2.2)(c)), and the notation $A = (1/W^\delta)$.

From (A4.1), it follows that:

$$\frac{\dot{k}(t)}{k(t)^\alpha} \leq Ar(t)^\beta = Ar(t)^\beta(1)^{1-\beta} \text{ for all } t \geq 0$$

so that integrating from $t = 0$ to $T > 0$, and applying Holder's Inequality (see, for instance, Royden (1988), p.121),

$$\int_0^T \frac{\dot{k}(t)}{k(t)^\alpha}dt \leq A\int_0^T r(t)^\beta(1)^{1-\beta}dt \leq A\left[\int_0^T r(t)dt\right]^\beta\left[\int_0^T dt\right]^{1-\beta} \leq AM^\beta T^{1-\beta}$$

Integration of the left-hand side of the above display then yields:

$$\frac{k(T)^{1-\alpha} - K^{1-\alpha}}{(1-\alpha)} \leq AM^\beta T^{1-\beta}$$

which can be rewritten as:

$$k(T)^{1-\alpha} \le K^{1-\alpha} + (1-\alpha)AM^{\beta}T^{1-\beta}$$

$$\le K^{1-\alpha}(T+1)^{1-\beta} + (1-\alpha)AM^{\beta}(T+1)^{1-\beta}$$

$$= [K^{1-\alpha} + (1-\alpha)AM^{\beta}](T+1)^{1-\beta}$$

Thus, denoting $[K^{1-\alpha} + (1-\alpha)AM^{\beta}]^{1/(1-\alpha)}$ by B,

$$k(t) \le B(t+1)^{(1-\beta)/(1-\alpha)} \text{ for all } t \ge 0 \tag{A4.2}$$

We now use (A4.2) in (A4.1) to write:

$$\dot{k}(t) \le Ak(t)^{\alpha}r(t)^{\beta} - \eta W^{p}$$

$$\le AB^{\alpha}[(t+1)^{(1-\beta)/(1-\alpha)}]^{\alpha}r(t)^{\beta} - \eta W^{p}$$

$$= AB^{\alpha}[(t+1)^{\alpha/(1-\alpha)}]^{(1-\beta)}r(t)^{\beta} - \eta W^{p}$$

so that integrating from $t = 0$ to $T > 0$,

$$k(T) - k(0) \le AB^{\alpha}\int_{0}^{T}[(t+1)^{\alpha/(1-\alpha)}]^{(1-\beta)}r(t)^{\beta}dt - \int_{0}^{T}\eta W^{p}dt$$

$$\le AB^{\alpha}\left[\int_{0}^{T}(t+1)^{\alpha/(1-\alpha)}dt\right]^{(1-\beta)}\left[\int_{0}^{T}r(t)dt\right]^{\beta}dt - \int_{0}^{T}\eta W^{p}dt$$

$$\le AB^{\alpha}[(1-\alpha)(T+1)^{1/(1-\alpha)} - (1-\alpha)]^{(1-\beta)}M^{\beta} - \eta W^{p}T$$

$$\le AB^{\alpha}M^{\beta}(1-\alpha)(T+1)^{(1-\beta)/(1-\alpha)} - \eta W^{p}T \tag{A4.3}$$

where we again used Holder's Inequality in the second line of (A4.3).

Now if (S') *did not hold*, then $\beta > \alpha$, and so $(1-\beta)/(1-\alpha) \equiv \rho \in (0,1)$. Then, as $T \to \infty$, the right-hand side expression in the last line of (A4.3) must go to $-\infty$, as the (negative) linear term would dominate. Thus, $k(T) < 0$ for large T, a contradiction. This establishes (S').

Proof of Claim 1 (in Third Step of Section 4.4)

For if the claim were false, there would be some $\tau' > \tau + \varepsilon$, for which $v(\tau') < v(\tau)$. Since $v(\tau + \varepsilon) > v(\tau)$, by continuity of $v(\cdot)$, there is some $\tau'' \in (\tau + \varepsilon, \tau')$ for which $v(\tau'') = v(\tau)$. Let $T = \inf\{t \in (\tau + \varepsilon, \tau'): v(t) = v(\tau)\}$. Then, by continuity of

$v(\cdot)$, we must have $v(T) = v(\tau)$, and so $T > \tau + \varepsilon$; further, by definition of T, we must have $v(t) > v(\tau) > 0$ for all $t \in (\tau + \varepsilon, T)$. Thus, by (4.14), $\dot{v}(t) > 0$ for all $t \in (\tau + \varepsilon, T)$, and so v is increasing on $(\tau + \varepsilon, T)$. Thus, by continuity of $v(\cdot)$, we must have $v(T) \geq v(\tau + \varepsilon) > v(\tau)$, which contradicts the conclusion reached earlier that $v(T) = v(\tau)$.

Proof of Claim 2 (in Fourth Step of Section 4.4)
For if the claim were false, there would be some $\tau' > \tau + \varepsilon$, for which $v(\tau') > v(\tau)$. Since $v(\tau + \varepsilon) < v(\tau)$, by continuity of $v(\cdot)$, there is some $\tau'' \in (\tau + \varepsilon, \tau')$ for which $v(\tau'') = v(\tau)$. Let $T = \inf\{t \in (\tau + \varepsilon, \tau') : v(t) = v(\tau)\}$. Then, by continuity of $v(\cdot)$, we must have $v(T) = v(\tau)$, and so $T > \tau + \varepsilon$; further, by definition of T, we must have $v(t) < v(\tau) < 0$ for all $t \in (\tau + \varepsilon, T)$. Thus, by (4.14), $\dot{v}(t) < 0$ for all $t \in (\tau + \varepsilon, T)$, and so v is decreasing on $(\tau + \varepsilon, T)$. Thus, by continuity of $v(\cdot)$, we must have $v(T) \leq v(\tau + \varepsilon) < v(\tau)$, which contradicts the conclusion reached earlier that $v(T) = v(\tau)$.

8.3 Appendix to Section 5

Proof of (5.41) [Property of ϕ function]:
Rewrite the equation (I) as:

$$a^{\beta(\zeta-1)/\delta'} K^\eta = S^{(\zeta-1)/\delta'}[a + bK^\eta] = S'a + S'bK^\eta$$

and then equivalently as:

$$a^m - (S'/K^\eta)a = S'b \tag{A5.1}$$

where S' and m are defined in (5.40).
Next, let us define:

$$\psi(a) = [a^m - (S'/K^\eta)a] \text{ for all } a \geq 0 \tag{A5.2}$$

Note that $\psi(0) = 0$, $\psi(a) > 0$ if and only if $a > \bar{a}$, where $\bar{a} = (S'/K^\eta)^{1/(m-1)}$. Further, for $a > \bar{a}$, we have $\psi' > 0$ and $\psi'' > 0$. Thus, given any $b > 0$, the unique solution to (A5.1), which we denoted by $\phi(b)$, is increasing in b, and in fact $\phi(b) \uparrow \infty$ as $b \uparrow \infty$. Further, since (A5.1) implies:

$$[\phi(b)^{m-1} - (S'/K^\eta)]\left[\frac{\phi(b)}{b}\right] = S'$$

we must have:

$$\frac{\phi(b)}{b} \downarrow 0 \text{ as } b \uparrow \infty \tag{A5.3}$$

Using this information, and rewriting (A5.1) as:

$$\frac{[\phi(b)]^m}{b} - (S'/K^\eta)\frac{\phi(b)}{b} = S'$$

we see that:

$$\frac{[\phi(b)]^m}{b} \to S' \text{ as } b \to \infty$$

which is the desired result (5.31).

Verification of (2.2)(a):
Using (5.48),

$$\frac{dk(t)}{dt} = g'(k(t))k(t) \text{ for all } t \geq 0$$

and so:

$$\frac{d\ln k(t)}{dt} = g'(k(t)) \text{ for all } t \geq 0$$

Integrating, we get:

$$\ln[k(t)/k(0)] = \int_0^t g'(k(s))ds$$

and using (5.48) again,

$$\frac{g(k(t))}{g(k(0))} = e^{\int_0^t g'(k(s))ds} \tag{A5.4}$$

We can use (A5.4) to write:

$$F(k(t), r(t), w(t)) = k(t)^\alpha \left[\frac{g'(k(t))}{\rho\lambda}\right]^\beta \left[\frac{1}{We^{\int_0^t [g'(k(s))/\rho]ds}}\right]^\delta$$

$$= \left[\frac{k(t)^\alpha}{W^\delta}\right]\left[\frac{g'(k(t))}{\rho\lambda}\right]^\beta \frac{g(k(0))^{\delta/\rho}}{g(k(t))^{\delta/\rho}} \tag{A5.5}$$

Thus, to verify (5.50), we need to show that the right-hand side expression in (A5.5) is equal to $[g(k(t)/\beta]$. We proceed now to demonstrate this.

For any $x > 0$, we can calculate:

$$g'(x) = \frac{1}{(\zeta - 1)}\left[\frac{x^\eta}{a + bx^\eta}\right]^{[1/(\zeta-1)]-1}\left[\frac{(a + bx^\eta)\eta x^{\eta-1} - x^\eta b\eta x^{\eta-1}}{(a + bx^\eta)^2}\right]$$

$$= \frac{1}{(\zeta - 1)}\left[\frac{x^\eta}{a + bx^\eta}\right]^{[1/(\zeta-1)]-1}\left[\frac{a\eta x^{\eta-1}}{(a + bx^\eta)^2}\right]$$

$$= \frac{1}{(\zeta - 1)}\left[\frac{a\eta x^{\eta\{[1/(\zeta-1)]-1\}+\eta-1}}{(a + bx^\eta)^{[1/(\zeta-1)]-1+2}}\right]$$

$$= \frac{1}{(\zeta - 1)}\left[\frac{a\eta x^{[\eta/(\zeta-1)]-1}}{(a + bx^\eta)^{[\zeta/(\zeta-1)]}}\right]$$

and:

$$x^\alpha g'(x)^\beta = x^\alpha \left[\frac{1}{(\zeta-1)}\right]^\beta \left[\frac{a\eta x^{[\eta/(\zeta-1)]-1}}{(a+bx^\eta)^{[\zeta/(\zeta-1)]}}\right]^\beta$$

$$= \left[\frac{a\eta}{(\zeta-1)}\right]^\beta \frac{x^{\beta\{[\eta/(\zeta-1)]-1\}+\alpha}}{(a+bx^\eta)^{[\zeta\beta/(\zeta-1)]}}$$

$$= \left[\frac{a\eta}{(\zeta-1)}\right]^\beta \frac{x^{[(1+\delta')\eta/(\zeta-1)]}}{(a+bx^\eta)^{[(1+\delta')/(\zeta-1)]}}$$

$$= \left[\frac{a\eta}{(\zeta-1)}\right]^\beta g(x)^{(1+\delta')} \tag{A5.6}$$

where the simplification in the exponents in the third line comes from noting:

$$\zeta\beta = \frac{(1+\delta')}{\beta}\beta = (1+\delta')$$

and:

$$\beta\left\{\frac{\eta}{\zeta-1} - 1\right\} + \alpha = \frac{\{\eta - (\zeta-1)\}\beta + \alpha(\zeta-1)}{(\zeta-1)}$$

$$= \frac{\{(\alpha/\beta) - 1 - (\zeta-1)\}\beta + \alpha(\zeta-1)}{(\zeta-1)}$$

$$= \frac{\{((\alpha/\beta) - \zeta)\}\beta + \alpha(\zeta-1)}{(\zeta-1)}$$

$$= \frac{\zeta(\alpha-\beta)}{(\zeta-1)} = \frac{\zeta\beta\eta}{(\zeta-1)} = \frac{(1+\delta')\eta}{(\zeta-1)}$$

Using (A5.6) in (A5.5), we obtain:

$$F(k(t), r(t), w(t)) = \left[\frac{k(t)^\alpha}{W^\delta}\right]\left[\frac{g'(k(t))}{\rho\lambda}\right]^\beta \frac{g(k(0))^{\delta'}}{g(k(t))^{\delta'}}$$

$$= \left[\frac{g(K)^{\delta'}}{W^\delta}\right]\left[\frac{a\eta}{(\zeta-1)\rho\lambda}\right]^\beta g(k(t))$$

$$= \left[\frac{K^\eta}{a+bK^\eta}\right]^{\delta'/(\zeta-1)} \frac{1}{W^\delta}\left[\frac{a\eta}{(\zeta-1)\rho\lambda}\right]^\beta g(k(t))$$

$$= g(k(t))/\beta \tag{A5.7}$$

the last line in (A5.7) following from the fact that:

$$\left[\frac{K^\eta}{a+bK^\eta}\right]^{\delta'/(\zeta-1)}\frac{1}{W^\delta}\left[\frac{a\eta}{(\zeta-1)\rho\lambda}\right]^\beta=\frac{1}{\beta} \tag{A5.8}$$

since (A5.8) is precisely constraint (I), after transposing appropriate terms. This verifies (5.50) and therefore (2.2)(a).

Verification of (2.2)(b):
For $x > 0$, we can calculate (as we did in verifying (2.2)(a) above),

$$\left[\frac{g'(x)}{\rho\lambda}\right]=\left[\frac{a\eta}{\rho\lambda(\zeta-1)}\right]\left[\frac{x^{\{[\eta/(\zeta-1)]-1\}}}{(a+bx^\eta)^{[\zeta/(\zeta-1)]}}\right]$$

$$=\left[\frac{a\eta}{\rho\lambda(\zeta-1)}\right]\left[\frac{x^\eta}{(a+bx^\eta)}\right]^{\zeta/(\zeta-1)}\left[\frac{1}{x^{\alpha/\beta}}\right] \tag{A5.9}$$

where in the second line of (A5.9), the simplification in the exponent comes from the observation that:

$$\left[\frac{\eta}{\zeta-1}-1\right]-\frac{\eta\zeta}{\zeta-1}=\frac{\eta[1-\zeta]}{\zeta-1}-1=-(1+\eta)=-\frac{\alpha}{\beta}$$

Thus, we have:

$$\int_0^\infty\left[\frac{g'(k(t))}{\rho\lambda}\right]dt\le\left[\frac{a\eta}{\rho\lambda(\zeta-1)}\right]\left[\frac{1}{b}\right]^{\zeta/(\zeta-1)}\int_0^\infty\left[\frac{1}{k(t)^{\alpha/\beta}}\right]dt \tag{A5.10}$$

We now need to work on an upper bound for the integral in (A5.10). Note that by (5.48), we have $\dot k(t) > 0$ for all $t \ge 0$, and so $g(k(t)) > g(K)$ for all $t \ge 0$. Using this information, we can write (using (5.48)),

$$k(t)=K+\int_0^t\dot k(s)ds=K+\int_0^t g(k(s))ds\ge K+\int_0^t g(K)ds=[K+g(K)t]$$

and so:

$$\int_0^\infty\left[\frac{1}{k(t)^{\alpha/\beta}}\right]dt\le\int_0^\infty\frac{1}{[K+g(K)t]^{\alpha/\beta}}dt=\frac{1}{g(K)K^{(\alpha/\beta)-1}[(\alpha/\beta)-1]} \tag{A5.11}$$

Thus,

$$\int_0^\infty\left[\frac{g'(k(t))}{\rho\lambda}\right]dt\le\left[\frac{a\eta}{\rho\lambda(\zeta-1)}\right]\left[\frac{1}{b}\right]^{\zeta/(\zeta-1)}\left[\frac{1}{g(K)K^{(\alpha/\beta)-1}[(\alpha/\beta)-1]}\right] \tag{A5.12}$$

Recall that $(a,b)\in\mathbb{R}^2_{++}$ have been chosen to satisfy constraint (II), which is:

$$\left[\frac{a\eta}{\rho\lambda(\zeta-1)}\right]\frac{[a+bK^\eta]^{1/(\zeta-1)}}{b^{\zeta/(\zeta-1)}K^{\eta/(\zeta-1)}[(\alpha/\beta)-1]K^{(\alpha/\beta)-1}}\le M$$

Using this in (A5.12), we get:

$$\int_0^\infty\left[\frac{g'(k(t))}{\rho\lambda}\right]dt\le M$$

verifying (5.51) and therefore (2.2)(b).

Verification of (5.60):
Differentiating (5.58) with respect to k, we obtain:

$$\frac{dT}{dk} = (A + \sqrt{k})(1/2)[(A + \sqrt{k})^2 - A^2]^{-(1/2)}2(A + \sqrt{k})(1/2)k^{-(1/2)}$$

$$+ [(A + \sqrt{k})^2 - A^2]^{(1/2)}(1/2)k^{-(1/2)}$$

$$- \frac{A^2(1/2)k^{-(1/2)} + (1/2)[(A+\sqrt{k})^2 - A^2]^{-(1/2)}2(A+\sqrt{k})(1/2)k^{-(1/2)}\}}{(A + \sqrt{k}) + [(A + \sqrt{k})^2 - A^2]^{(1/2)}} \quad (A5.13)$$

So, we get:

$$2k^{1/2}\frac{dT}{dk} = (A + \sqrt{k})^2[(A + \sqrt{k})^2 - A^2]^{-(1/2)} + [(A + \sqrt{k})^2 - A^2]^{(1/2)}$$

$$- \frac{A^2\{1 + [(A + \sqrt{k})^2 - A^2]^{-(1/2)}(A + \sqrt{k})\}}{(A + \sqrt{k}) + [(A + \sqrt{k})^2 - A^2]^{(1/2)}} \quad (A5.14)$$

This in turn yields:

$$[(A + \sqrt{k})^2 - A^2]^{(1/2)} 2k^{1/2} \frac{dT}{dk} = (A + \sqrt{k})^2 + [(A + \sqrt{k})^2 - A^2]$$

$$- \frac{A^2[(A + \sqrt{k})^2 - A^2]^{(1/2)} + (A + \sqrt{k})\}}{(A + \sqrt{k}) + [(A + \sqrt{k})^2 - A^2]^{(1/2)}}$$

$$= 2[(A + \sqrt{k})^2 - A^2] \quad (A5.15)$$

This can be simplified to read:

$$\frac{dT}{dk} = \frac{[(A + \sqrt{k})^2 - A^2]^{(1/2)}}{k^{1/2}} = \left[1 + \frac{2A}{k^{1/2}}\right]^{1/2} = \left[1 + \frac{a}{k^{1/2}}\right]^{1/2} \quad (A5.16)$$

as claimed in (5.60), by using $a = 2A$ from (5.55).

8.4 Appendix to Section 6

Proof of Parameter Restriction 1 (Display (6.16)):
Suppose, contrary to (6.16), we have:

$$\frac{(1 + \delta')}{\nu\beta} = 1 \quad (A6.1)$$

Then we also have:

$$\nu\beta = (1 + \delta') > 1 > \alpha \quad (A6.2)$$

Using (A6.1) in the differential equation (6.15), we obtain the solution:

$$\pi(k) = \pi(K)\frac{e^{N'k^{\xi}}}{e^{N'K^{\xi}}} = Ce^{N'k^{\xi}} \text{ for all } k \geq K \qquad (A6.3)$$

where $\xi = [1 - (\alpha/\nu\beta)] > 0$, $N' = N/\xi$ and $C = \pi(K)/e^{N'K^{\xi}}$.

Using (2.2)(a), we can write for all $t \geq 0$

$$\dot{k}(t) = [k(t)^{\alpha}r(t)^{\beta}/w(t)^{\delta}] - c(t)$$

$$\leq [k(t)^{\alpha}r(t)^{\beta}/W^{\delta}] \qquad (A6.4)$$

since $w(t) \geq 0$ by (2.2)(c), and $c(t) \geq 0$ for all $t \geq 0$. Since $\dot{k}(t) > 0$ for all $t \geq 0$ by (6.3), we have $k(t) \geq K$ for all $t \geq 0$. Thus, we obtain, by using (A6.3) and (A6.4),

$$r(t)^{\beta} \geq \frac{W^{\delta}\dot{k}(t)}{k(t)^{\alpha}} = \frac{W^{\delta}\pi(k(t))}{k(t)^{\alpha}} = \frac{W^{\delta}\pi(k(t))}{k(t)^{\alpha}} = \frac{W^{\delta}Ce^{N'k(t)^{\xi}}}{k(t)^{\alpha}} \text{ for all } t \geq 0 \quad (A6.5)$$

Recall from (6.5) that $k(t) \uparrow \infty$ as $t \uparrow \infty$. So, the right-hand side expression in (A6.5) goes to ∞ as t goes to infinity. This implies that there is $0 < T < \infty$, such that:

$$r(t)^{\beta} \geq 1 \text{ for all } t \geq T$$

and this clearly violates the resource constraint (2.2)(b). This contradiction establishes the parameter restriction (6.16) in the text.

Proof of Parameter Restriction 3 (Display 6.23)):

Suppose (6.23) does not hold. Then, in view of (6.16), we must have $\zeta < 1$. It follows that:

$$\frac{\alpha}{\nu\beta} < \frac{(1+\delta')}{\nu\beta} < 1 \qquad (A6.6)$$

We now rewrite (6.22) as:

$$k(t)^{1-\zeta} - k(0)^{1-\zeta} = N(1-\zeta)\left[\frac{k(t)^{[1-(\alpha/\nu\beta)]}}{[1-(\alpha/\nu\beta)]} - \frac{k(0)^{[1-(\alpha/\nu\beta)]}}{[1-(\alpha/\nu\beta)]}\right] \text{ for all } t \geq 0 \ (A6.7)$$

By using (6.5) and (A6.6), we have $k(t)^{[1-(\alpha/\nu\beta)]} \to \infty$, as $t \to \infty$, and so (A6.7) yields:

$$\frac{k(t)^{1-\zeta}}{k(t)^{[1-(\alpha/\nu\beta)]}} \to \frac{N(1-\zeta)}{[1-(\alpha/\nu\beta)]} > 0 \text{ as } t \to \infty \qquad (A6.8)$$

Using (6.5) and (A6.8), we can find $0 < T < \infty$, such that:

$$\left.\begin{array}{l}(i) \quad k(t) > 1 \text{ for all } t \geq T \\ (ii) \quad \dfrac{k(t)^{1-\zeta}}{k(t)^{[1-(\alpha/\nu\beta)]}} \geq C \text{ for all } t \geq T\end{array}\right\} \qquad (A6.9)$$

where:

$$C = \frac{N(1-\zeta)}{2[1-(\alpha/\nu\beta)]} \quad (A6.10)$$

Using (A6.6) and (A6.9), we obtain:

$$k(t) \geq C^{1/(1-\zeta)}k(t)^{[1-(\alpha/\nu\beta)]/(1-\zeta)} \geq C^{1/(1-\zeta)}k(t) = C'k(t) \text{ for all } t \geq T \quad (A6.11)$$

Using (A6.4), we also have:

$$k(t) \leq \frac{k(t)^\alpha r(t)^\beta}{W^\delta} \text{ for all } t \geq 0 \quad (A6.12)$$

Combining (A6.11) and (A6.12), and using (A6.9)(i), we obtain:

$$W^\delta C' \leq W^\delta C' k(t)^{1-\alpha} \leq r(t)^\beta \text{ for all } t \geq T$$

and this clearly violates the resource constraint (2.2)(b). This contradiction establishes the parameter restriction (6.23) in the text.

8.5 Appendix to Section 7

Proof of the formula in (7.27):
We begin with the general dynamic property established in (7.6), and write it as:

$$\beta\left[\frac{-\dot{r}(t)}{r(t)}\right] = \alpha\frac{\dot{k}(t)}{k(t)} - (\delta + \rho)\frac{\dot{w}(t)}{w(t)}$$

$$= \alpha\frac{\dot{k}(t)}{k(t)} - (\delta + \rho)\lambda r(t)$$

Thus, we get:

$$\left[\frac{-\dot{r}(t)}{r(t)}\right] = \frac{\alpha \dot{k}(t)}{\beta k(t)} - \frac{(\delta + \rho)}{\beta}\lambda r(t) \text{ for all } t \geq 0$$

This yields the basic expression of comparison of the growth rates of capital and resource:

$$\left[\frac{-\dot{r}(t)}{r(t)}\right] - \frac{\dot{k}(t)}{k(t)} = \frac{(\alpha-\beta)}{\beta}\frac{\dot{k}(t)}{k(t)} - \frac{(\delta+\rho)}{\beta}\lambda r(t) \text{ for all } t \geq 0 \quad (A7.1)$$

We now use the expressions for $r(t)$ and $k(t)$ in (7.13) and (7.14) respectively to obtain:

$$\left[\frac{-\dot{r}(t)}{r(t)}\right] - \left[\frac{\dot{k}(t)}{k(t)}\right] = \frac{(\alpha-\beta)}{\beta}\left[\frac{g(k(t))}{k(t)}\right] - \frac{(\delta+\rho)}{\beta}\lambda\left[\frac{g'(k(t))}{\rho\lambda}\right] \text{ for all } t \geq 0 \quad (A7.2)$$

As verified in (A5.9) (in the appendix to Section 5), we have:

$$\left[\frac{g'(k(t))}{\rho\lambda}\right] = \left[\frac{a\eta}{\rho\lambda(\zeta-1)}\right]\left[\frac{k(t)^\eta}{a+bk(t)^\eta}\right]^{\zeta/(\zeta-1)}\left[\frac{1}{k(t)^{\alpha/\beta}}\right] \quad (A7.3)$$

and manipulating the expression on the right-hand side of (A7.3), we have:

$$\begin{bmatrix} \dfrac{g'(k(t))}{\rho\lambda} \end{bmatrix} = \begin{bmatrix} \dfrac{a\eta}{\rho\lambda(\zeta-1)} \end{bmatrix}\begin{bmatrix} \dfrac{k(t)^\eta}{a+bk(t)^\eta} \end{bmatrix}^{[1+1/(\zeta-1)]}\begin{bmatrix} \dfrac{1}{k(t)^{\alpha/\beta}} \end{bmatrix}$$

$$= \begin{bmatrix} \dfrac{a\eta}{\rho\lambda(\zeta-1)} \end{bmatrix} g(k(t))\begin{bmatrix} \dfrac{k(t)^\eta}{a+bk(t)^\eta} \end{bmatrix}\begin{bmatrix} \dfrac{1}{k(t)^{\alpha/\beta}} \end{bmatrix}$$

$$= \begin{bmatrix} \dfrac{a\eta}{\rho\lambda(\zeta-1)} \end{bmatrix}\begin{bmatrix} \dfrac{g(k(t))}{k(t)} \end{bmatrix}\begin{bmatrix} \dfrac{1}{a+bk(t)^\eta} \end{bmatrix}$$

$$= \begin{bmatrix} \dfrac{\alpha-\beta}{\rho\lambda(1+\delta'-\beta)} \end{bmatrix}\begin{bmatrix} \dfrac{g(k(t))}{k(t)} \end{bmatrix}\begin{bmatrix} \dfrac{a}{a+bk(t)^\eta} \end{bmatrix} \quad (A7.4)$$

where we used (7.26) in the third and fourth lines of (A7.4). Now, using (A7.4) in (A7.2), we obtain:

$$\begin{bmatrix} \dfrac{-\dot{r}(t)}{r(t)} \end{bmatrix} - \begin{bmatrix} \dfrac{\dot{k}(t)}{k(t)} \end{bmatrix} = \dfrac{(\alpha-\beta)}{\beta}\begin{bmatrix} \dfrac{g(k(t))}{k(t)} \end{bmatrix}\left[1 - \left\{\dfrac{(\delta+\rho)}{\rho(1+\delta'-\beta)}\right\}\left\{\dfrac{a}{a+bk(t)^\eta}\right\}\right] \text{ for all } t \geq 0$$

(A7.5)

which gives us (7.27) after making the substitution $\delta' = (\delta/\rho)$.

17. Transformational change: parallels for addressing climate and development goals*
Penny Mealy and Cameron Hepburn

1. INTRODUCTION

Climate change and poverty alleviation are, as Stern (2016) has coined, 'the twin defining challenges of our century'. Historically, efforts to address these two challenges have been conflicted. Adverse impacts of climate change are likely to hit the poorest of this world hardest, but traditional industrial routes out of poverty are dangerously emissions-intensive. Such tensions have been major sticking points in earlier climate negotiations and largely underpinned the failure of the 2009 Copenhagen COP to reach a global climate agreement (Nordhaus, 2010).

However, two important developments suggest a new global readiness to move beyond historical conflicts and instead take advantage of key commonalities and collective interests. The 2015 adoption of the Sustainable Development Goals (SDGs) demonstrated an acute awareness that any plan to advance living standards of present and future generations must address the inseparable links between people, the planet and prosperity (UN, 2015; Griggs et al., 2013; Brown, 2015). Further, the Paris Agreement (UNFCCC, 2015), which has been ratified by an overwhelming majority of countries, provides a promising new international platform to progress a unique collective framework for global climate cooperation. The confluence of these global agendas represents an historic opportunity to marry efforts on climate and development fronts and drive significant progress on sustainable development.

Against this encouraging backdrop, this chapter draws attention to a somewhat under-appreciated, but profoundly important commonality in the twin climate and development challenges: both require societies to navigate and manage system-wide transformative change.

Unfortunately, transformational change is a concept that is not yet well defined or understood. To our knowledge, a generally accepted definition of transformational change does not as yet exist (Mersmann et al., 2014). There is also limited consensus on how it should be meaningfully measured – particularly as transformational change often involves dynamic processes occurring at multiple scales and dimensions (Geels et al., 2016; Turnheim et al., 2015). Further, while traditional economic modelling frameworks are well-suited for studying marginal changes over short-term horizons, they are poor tools

---

* We would like to gratefully acknowledge support from *Partners for the New Economy* and the Oxford Martin School project on *Sensitive Intervention Points to the Post-Carbon Transition*. We would also like to thank Sam Fankhauser and Frank Jotzo, and Berthold Herrendorf, Richard Rogerson and Akos Valentinyi for their kind permission to reproduce their work in Figure 17.1 and Figure 17.2 of this chapter.

for analysing dis-equilibrium dynamics, non-linear feedbacks and emergent properties that commonly characterize transformational change processes. As a better understanding of the process of transformational change could catalyse progress on both climate and development fronts, this chapter explores parallel efforts in respective fields.

First, we examine the nature and importance of transformational change processes in climate and development contexts. Unprecedented changes in both the low-carbon landscape and global economic environment have taken many by surprise. While these unfolding dynamics are invalidating traditional analytical approaches based on assumed patterns of incremental change and challenging long-held notions about growth and development, they are also offering much needed alternative possibilities for achieving climate and development goals. Indeed, given the sheer magnitude of climate and development challenges – and the rate at which change must occur to mitigate the worst climate impacts, being able to both navigate and *drive* the process of transformational change is now seen as a critical policy imperative.

Although significant efforts are underway to understand transformative change processes in respective climate and development fields, we argue that there are also key advantages from better integration across domains. For example, by considering the dynamics of economic development processes, climate policy can better account for mitigation opportunities, particularly as countries shift from energy intensive manufacturing activities towards services. Similarly, by taking advantage of the present impetus towards low-carbon futures, developing countries could seize an unprecedented opportunity to not only accelerate their growth, but also attain a much higher *quality* of sustainable, inclusive and resilient development.

Second, we explore empirical patterns of transformational change in climate and development contexts. We find that climate and development fields share a number of things in common. Aggregate changes (e.g. in GDP/capita or emissions) are often broken down into between-sector and within-sector changes, and key driving forces of change invariably relate to technology, preferences and their endogenous evolution. While historical stylized facts associated with the development process (such as the shift from agriculture to manufacturing to services) have been well documented, empirical patterns associated with low-carbon transformations are still emerging, and also likely to vary more across countries with different productive structures.

Third, we compare and contrast four alternative methodological tools for analysing and modelling transformational change: Network analysis, computable general equilibrium (CGE) models, macro-econometric/input–output models, and agent-based models. Each has key strengths and weaknesses, which are important to bear in mind when applying them to policy questions relating to transformational change. Moreover, as models influence behaviour in the real world, getting these right, or at least less wrong, is not a mere academic curiosum but could be vital for making simultaneous progress on climate and development goals.

This chapter proceeds as follows: In section 2 we examine the importance of transformational change in climate and development contexts and highlight key areas where more amalgamation across domains could be advantageous. Section 3 examines existing approaches to measure transformational change and identifies commonalities across climate and development frameworks. Section 4 reviews existing modelling methodologies and reflects on their ability to appropriately model transformational change. Section 5 concludes.

2. TRANSFORMATIONAL CHANGE IN CLIMATE AND DEVELOPMENT CONTEXTS

2.1 The Importance of Transformational Change in Achieving Climate and Development Goals

The past decades have seen unprecedented transformations across both economic and low-carbon fronts. Rapid technological progress, population growth, urbanization and changing global market structures are combining to shape many countries' development trajectories in new and interesting ways (Fankhauser and McDermott, 2016). Equivalently, startling reductions in renewable energy costs and mounting international cooperation to curb emissions are drastically changing the contours of the low-carbon landscape (Trancik, 2014). Such changes have prompted a re-think of long-held notions about growth and development. They also invalidate commonly used traditional analytical approaches based on assumed patterns of incremental change.

From a policy perspective, many scholars are now stressing that without radical, non-marginal change, limiting global warming to 2°C will be extremely difficult (Burch et al., 2014; Fankhauser and Stern, 2016; Reid et al., 2010). The magnitude of emissions reductions required in the rapidly diminishing time frame necessitates much more than incremental change along existing developmental and technological paradigms (Perez, 2015; Zenghelis, 2015). At the same time, developing countries are encountering a significant paradigm shift of their own. With traditional industrial development routes now potentially less viable due to impinging automation and globalization forces (Rodrik, 2016), development agencies and policy makers are facing the prospects of paving profoundly different pathways to prosperity in the 21st century.

While significant research efforts are underway to understand transformative change processes in respective climate and development fields, there are also key advantages from better integration across domains. In what follows, we outline key areas where climate policy could benefit from better accounting for economic transformation processes and equally, where development policy could benefit from the present impetus to transition towards low-carbon futures.

2.2 Why Climate Mitigation Efforts Should Better Account for Economic Transformation Processes

The Paris Agreement (UNFCCC, 2015) represented a significant step forward in achieving universal consensus and commitment to move towards a sustainable future. However, there is still a significant gap between country Nationally Determined Contributions (NDCs) and the emissions reductions required by 2030 to plausibly remain on a 2°C pathway (Rogeli et al., 2016). In order to attain a reasonable chance of avoiding a rise in global average temperatures by more than 2°C, policy makers will need to harness as many emission reduction opportunities as possible. Better accounting for non-linear dynamics associated with low-carbon technology learning curves, uptake rates and shifts in consumer behaviour and societal norms is a critical first step, and improved understanding of what drives these tipping points could help us speed them up (Aghion et al., 2014; Russil and Nyssa, 2009; Boyd et al., 2015).

However, understanding emissions implications of future economic structural shifts is also key. One of the biggest drivers of errors in historical energy and emissions projections is the failure to anticipate macroeconomic changes – particularly as a country shifts from energy intensive manufacturing activities towards services (Grubb et al., 2015). Most climate-economy models take the structure of the economy as fixed and rarely incorporate macroeconomic dynamics (Zenghelis, 2015). This is particularly problematic for countries like China, whose future emissions trajectory will have important climate consequences, but is also undergoing a process of rapid socio-economic transformation (Green and Stern, 2016). For example, Grubb et al. (2015) recently reviewed projections of 89 scenarios from 12 different models for China's emissions through to 2030. They found that not only is the range of projections extremely large indicating a high degree of uncertainty, most scenarios do not account for the economy's macroeconomic structure and the potential to shift away from its currently high share of manufacturing. Given the consistent historical trend for countries to transition towards less energy-intensive service-based activities as incomes rise, incorporating structural change dynamics in climate-economy models could depict very different climate implications of China's future development path.

While this chapter is primarily focused on drawing parallels between economic transformation and the low-carbon transformation, it is worth noting that climate *adaptation* efforts can also benefit from a better understanding of the economic transformation process. Measures to lessen adverse climate change impacts have traditionally been approached as a static concept, with efforts to reduce vulnerability in developing countries often focusing on protecting existing structures and livelihoods, such as safeguarding agricultural output. However, as many developing countries are undergoing a process of socio-economic transformation, overly static adaptation plans could hamper development progress by placing too much emphasis on sectors that are likely to become less important over time (Bowen et al., 2016; Kocornick-Mina and Fankhauser, 2015). In light of the important links between climate risks and the development process, there has recently been increased interest in understanding what might constitute 'climate-resilient development' (Denton et al., 2014). This approach seeks to better account for the dynamism of the development process and calls for adaptation efforts to become more transformational (Fankhauser and McDermott, 2016).

2.3 Why Development Should Better Account for Low-Carbon Transformation Processes

When it comes to the key development objective of lifting people out of poverty, climate change has traditionally been seen as a major stumbling block, with the world's poorest and most vulnerable likely to shoulder the biggest burden of climate impacts (Hallegatte et al., 2015; Brown, 2015; Collier et al., 2008; Althor et al., 2016). Further, well-worn emissions-intensive industrial pathways that led today's advanced economies to prosperity are now recognized as being incompatible with the rapidly diminishing global carbon budget (Stern, 2007; IPCC, 2007). Consequently, the ability of developing countries to 'catch-up' to advanced countries' income levels by following well established manufacturing routes (Rodrik, 2012) might now be less feasible. This leaves many developing countries with the challenging prospect of scouting out new, un-trodden and uncertain paths to prosperity.

However, with new momentum building in the low-carbon space, particularly on the technological and international negotiation fronts, climate change may now present developing countries with an unprecedented opportunity for growth acceleration (ECA, 2016). Never in history have developing countries been able to leverage such compelling global pressure and financial assistance to adopt new technologies, develop better infrastructure and protect natural capital. Further, as development agencies and policy makers are now increasingly calling for new modes of sustainable, inclusive and resilient development, the 21st century could see developing countries attaining a much higher *quality* of development than their predecessors (UNIDO, 2016; Garnaut et al., 2013; Mlachila et al., 2014).

A particularly important case in point relates infrastructure development. The current stock of infrastructure presently accounts for 60 per cent of global greenhouse gas emissions and if existing, long-lived energy and transportation assets are not stranded, they are projected to commit the world to a significant degree of warming (Davis et al., 2010; Guivarch and Hallegatte, 2011; Pfeiffer et al., 2016). However, with the number of people living in cities projected to increase from 3.5 billion to 6.5 billion by 2050 (Fankhauser and Stern, 2016) and the overwhelming majority of this increase likely to occur in Africa and Asia (UN, 2014), developing countries have a one-time opportunity to shape cleaner, compact and coordinated cities and urban structures (Floater et al., 2014; NCE, 2016). Getting infrastructure right has an enormous potential to curb emissions and lock rapidly developing countries in to a low-carbon growth trajectory (Bhattacharya et al., 2015). Further, as noted by Fuller and Romer (2014), 'nothing else will create as many opportunities for social and economic progress'.

3. EMPIRICAL PATTERNS OF TRANSFORMATIONAL CHANGE IN CLIMATE AND DEVELOPMENT CONTEXTS

Despite its widely recognized importance in achieving climate and development goals, transformational change is a concept that is not yet well defined or understood. Like notions of 'world peace' and 'living happily ever after', transformational change often inspires broad endorsement and significant approval, but is challenging to definitively pin down. To our knowledge, a generally accepted definition does not as yet exist. Nevertheless, development and climate economists have examined empirical patterns of system-wide transformational change within their respective fields. Although there are distinct differences, we highlight important commonalities that could stimulate shared learning.

3.1 Empirical Patterns of Structural Change on the Development Process

When examining system-wide transformative change in the context of development, economists often distinguish between changes arising from two key processes. The first relates to the structural shifts *between* sectors, which usually occurs as resources are allocated from low productivity to higher productivity activities. The second relates to changes occurring *within* sectors, which are often related to economy-wide productivity increases due to improvements in key economic 'fundamentals', such as institutions or human capital (Rodrik, 2013).

In examining structural shifts, economists have found that over the last 200 years, economic growth and development has generally been associated with successive transitions across three broad sectors. Early in a country's development phase, most people and resources are engaged in subsistence agriculture, where labour productivity is very low (Herrendorf et al., 2014). However, as productivity improves, agricultural workers are released to engage in higher value-add industrial activities. Employment and value added shares in manufacturing begin rising as agriculture shares decline. As the economy advances further, the manufacturing share reaches a peak and begins declining, as the service sector becomes much more dominant. These empirical trends are illustrated in Figure 17.1, where Herrendorf et al. (2014) have plotted the empirical decline in agriculture, inverted U-shape in manufacturing and rise in services experienced by ten advanced countries.

To understand what drives the observed structural changes, researchers have focused on two key factors. The first relates to technological progress. As agricultural workers adopt new technologies and techniques, the increase in agricultural productivity allows surplus labour to then be released to other industrial or services activities that pay higher wages (Lewis, 1954). Further, since rates of technological change differ across sectors, some sectors will generate higher improvements in growth and living standards than others (Chenery, 1986). The second factor relates to how consumer preferences change with rising income. As people become richer, they tend to spend a lower relative share of their income on food and more on durable (manufacturing) goods and services (Chenery, 1979; Cypher and Dietz, 2009). This increases domestic demand for new non-agricultural industries and economic activity.

3.2 Empirical Patterns in the Transition Towards a Low-Carbon Economy

When examining shifts towards a low-carbon economy, researchers have also examined empirical patterns in how countries' emissions change as the economy evolves. A sizeable body of empirical research has examined the 'Environmental Kuznets Curve' (EKC) hypothesis, which postulates an inverted U-shape relationship between pollutants and per capita income (Cole et al., 1997; Stern et al., 1996). One explanation for this posited relationship relates to the evolution of economic structure – as a nation progresses from relatively clean agricultural activities to dirty industrial manufacturing activities to cleaner service-based activities, one would expect emissions to rise and subsequently fall. A second explanation relates to the evolution of preferences – as societies become richer, the general expectation is that they are more likely to have a greater preference for environmental quality (Dinda, 2004). Empirical evidence strongly supports the upward sloping part of the EKC, with many countries tending to experience an increase in emissions in the early phases of their development. However, there has been relatively inconclusive evidence to support the EKC's downward sloping section (Kaika and Zervas, 2013). A number of studies have found that urban or local air quality indicators that directly impact human health follow the hypothesized U relationship with income (Grossman and Krueger, 1995; Seldon and Song, 1994; Stern and Common, 2001). For other pollutants, the trend observed among higher income countries tends to be quite mixed, with emissions tending to depend more on country-specific conditions, technologies and policies than its prosperity (Ota, 2017; Fankhauser and Jotzo, 2017).

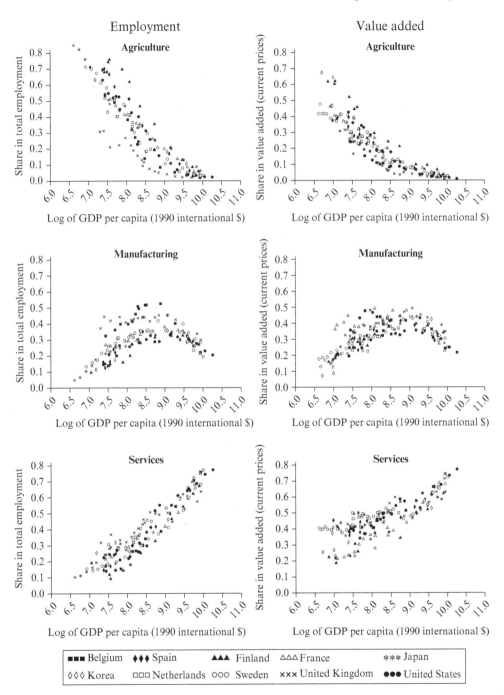

Figure 17.1 Sectoral shares of employment and value added: selected developed countries, 1800–2000

The Kaya identity offers a different lens to view system-wide changes in countries' emissions (Raupach et al., 2007; Rosa and Dietz, 2012). This identity is specified as follows, where C relates to emissions from human sources, E relates to energy consumption, and Y relates to economic output.

$$C \equiv \frac{C}{E} \cdot \frac{E}{Y} \cdot Y$$

This formulation allows progress towards the low-carbon economy to be analysed in terms of reductions in carbon intensity of energy (C/E), or reductions in the energy intensity of economic output (E/Y), or both. Fankhauser and Jotzo (2017) recently illustrated how countries' carbon intensity and energy intensity have evolved over time. As shown in Figure 17.2, energy intensity has been steadily decreasing in all countries over the 1990–2011 period. However, only high-income countries are managing to decarbonize their energy.

In a similar fashion to the study of empirical patterns in development economics, aggregate changes in carbon intensity and emissions intensity are also often broken down into structural shifts and system-wide efficiency improvements (Lenzen, 2016). However, unlike the relatively consistent structural dynamics and drivers observed in the economic development process, patterns in emissions and energy consumption across countries and sectors are more mixed. This in part is due to studies employing a range of different

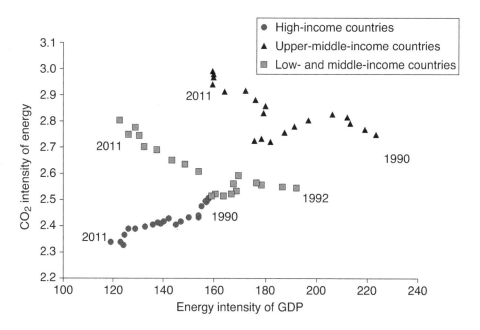

Source: Reproduced with permission from Fankhauser and Jotzo (2017).

Figure 17.2 Long-term trajectory of energy intensity and carbon intensity of energy, 1990/92 to 2011

decomposition methodologies, industrial sector classifications and applying analysis to a wide diversity of regions and time periods (Su and Ang, 2012). But it is also likely that factors driving emissions are different for different economies (Yao et al., 2014).

3.3 Key Commonalities

In our brief overview of empirical frameworks commonly applied to understand patterns of change in development and emissions-related areas, three commonalities are apparent. First, both fields seek to understand observed dynamics in similar ways. Across the development and emissions-related literature, 'transformative change' is recognized as a phenomenon involving between sector changes and within sector changes. Key aggregates of interest are consequently decomposed into variation arising from sectoral shifts and system-wide efficiency improvements.

Second, both domains have significant commonalities in their key objective. While development economists are generally interested in what drives increases in GDP and climate economists are concerned with what reduces emissions, both tend to view system-wide progress through the lens of a productivity measure – for development, improving labour productivity is what matters, while for climate change, improving carbon productivity is key. However, it is this commonality that drives significant conflict. Traditionally, economies have increased their labour productivity by increasing their use of energy. Due to the emissions intensity of traditional energy sources, this in turn has generated greater emissions (Taylor, 2009).

The extent to which these twin objectives can be pursued together ultimately depends on our ability to decouple emissions from the development process (Hepburn and Bowen, 2013). As shown in Figure 17.2, many advanced economies are moving in the right direction, particularly as increasing value add is derived from digitally enhanced material-light, intelligence-heavy activities (Baptist and Hepburn, 2012). However, current rates of progress are presently not sufficient to plausibly keep the planet on a 2°C pathway (NCE, 2014). A further critical sticking point remains for developing countries. As no advanced economy has ever become rich without undergoing an emissions-intensive industrial phase (Felipe et al., 2014), it is presently unknown whether development on the basis of 'industries without smoke stacks' is feasible (Page, 2014).

Such challenges underscore the pressing need to understand and leverage drivers of transformational change towards low-carbon development pathways. Incremental change along existing paradigms will clearly fall short (Perez, 2015). This brings us to the third commonality across climate and development research domains: in both economic and low-carbon transformation studies, key driving forces of change generally relate to either technological improvements (which alter labour or carbon productivity rates) or demand-side factors relating to consumer preferences.

In relation to technological change, a key focus in the climate literature has been to better understand the non-linear dynamics associated with cost improvements in renewable energy technologies (Koh and Magee, 2008; Schilling and Esmundo, 2009). Research has also investigated the potential for policy to influence the pace of change through stimulating greater production and making directed investments in R&D (Farmer and Lafond, 2015; Fischer and Newell, 2008; Menanteau et al., 2003). The nature of knowledge spillovers is a further important factor, with research showing that low-carbon innovation tends to have greater positive benefits for the local economy than innovation incumbent, carbon-intensive technologies (Dechezleprête et al., 2013; 2014). Many studies have also examined the significant potential for developing countries to take advantage of new decentralized forms of energy distribution platforms (such as mini-grids), and leapfrog the traditional centralized energy distribution systems that characterize most developed countries (Alstone et al., 2015; Levin and Thomas, 2016).

However, technological *diffusion* is also a critical facilitating (and limiting) factor in both development and low-carbon transitions. In development research, the difficulty transferring productive knowledge from advanced countries to less developed countries is a common explanation for why some countries remain poor (Keller, 2004; Hidalgo et al., 2007; Hausmann et al., 2014; Bahar et al., 2014). Understanding the likelihood of successful technological transfer and its underpinning drivers is profoundly relevant for both development and climate policy makers seeking to facilitate transformative change. There are a number of important and encouraging developments in this field, which we will return to in the following section.

The nature of demand-side influences, particularly as they relate to underlying consumer preferences has received substantially less research attention in both climate and development fronts. In an extensive review of over 900 articles on structural change, Silva and Teixeira (2008) noted that only a small minority of articles examined demand-side factors. While there is a growing body of research relating to behavioural aspects of energy consumption and climate change awareness (Gowdy, 2008; Steg, 2008; Dietz, 2009; Gillingham et al., 2009) surprisingly little research seeks to address climate implications of how preferences may evolve over extensive time periods. The long-time scales that development and climate policy are often concerned with tends to invalidate standard approaches that take preferences as given and assume future preferences will look like the past. Further, as Mattauch and Hepburn (2016) argue, failing to account for preference evolution and the extent to which preferences can be shaped by policy can significantly overestimate the cost of climate mitigation. The same is also likely to be true for many cost–benefit analyses applied in development contexts, such as large-scale infrastructure investments. The nature of preference formation and its co-evolution with policy choices is an important area for future research that will likely have important implications for understanding transformative change in both development and climate change related areas.

4. METHODOLOGICAL TOOLS FOR MODELLING AND UNDERSTANDING TRANSFORMATIONAL CHANGE IN CLIMATE AND DEVELOPMENT CONTEXTS

A wide range of analytical frameworks and decision-making tools are employed within development and climate domains. This section focuses on four key approaches and considers their relative strengths and weaknesses for helping researchers and policy makers better understand and manage the process of transformational change.

4.1 Network Analysis

Network analysis, which has become a widely embraced analytical framework within a diverse range of disciplines, offers important avenues for understanding transformational change. While traditional methods have tended to study transformational change in terms of sectoral shifts in resource allocation and within-sector productivity improvements, network analysis allows analysis of the *connections across sectors* and provides a means of estimating the probability of transitioning from one sector to another.

Hidalgo et al.'s (2007) application of network analysis to global trade data is one of the most notable contributions to the development context. Their 'Product Space' network, where traded products are represented as nodes linked to each other if they are more likely to require similar production capabilities (or know-how), has received significant attention from both scholars and policy makers as it provides a new lens to both visualize countries' productive structures and analyse feasible development pathways. Hidalgo et al.'s (2007) work is based on a theory advanced in the economic complexity and economic geography literature, which proposes that economic development is *path dependent* due to the underlying knowledge accumulation process. Just as it is easier to become good at making shirts if you already now how to make trousers, Hidalgo et al. (2007) used network analysis to demonstrate that it is easier for countries to become competitive in new products that require similar capabilities (or production know-how) to what they already do.

This style of analysis has also been applied to regional and industry data (O'Clery et al., 2016; Neffke et al., 2011; Boschma et al., 2013), labour-flows data (Neffke and Henning, 2013) and input–output data (Radebach et al., 2016), and offers a number of important insights for developmentally oriented transformational change. First, not all product specializations (and productive capabilities) are equal in terms of their knowledge spillover benefits and future diversification opportunities. The Product Space has a distinct core-periphery structure with manufacturing products (such as metals, machinery and chemicals) occupying the densely connected core of the network, and other products (such as agriculture and mining) tending to locate in the sparsely connected periphery. This finding further re-iterates the importance of manufacturing in the developmental process. Not only is manufacturing associated with unconditional convergence in labour productivity (Rodrik, 2012), which means that industries starting at lower labour productivity levels experience more rapid labour productivity growth. Manufacturing sectors also tend to have higher connectivity (in terms of shared capabilities and knowledge spillover opportunities) with other industries. As such, manufacturing has a unique ability to expand the set of 'adjacent possible' industries that are relatively easy to diversify into (Hausmann and Chauvin, 2015).

Second, the position of countries' exports in the Product Space can help identify *specific* new development opportunities that countries could more easily transition towards. Hidalgo et al. (2007) showed that countries were much more likely to diversify into products that were 'nearby' to their existing exports in the Product Space. By developing a measure (known as 'proximity density') that captures how similar a country's current exports are to a given undeveloped product in terms of their requisite production capabilities, they showed that while the probability of moving to exports that were far away (proximity density = 0.1) was almost zero, the probability of transitioning increases to

0.15 if a country's export basket contains closer products (proximity density = 0.8). The proximity density measure is particularly useful for industry players or policy makers seeking to identify 'adjacent possible' development opportunities that are more likely to be successful. However, when it comes to driving radical transformations in countries' productive structures, one of the key challenges is that it invariably involves big, difficult jumps in knowledge accumulation across the Product Space.

Recently, work has also sought to apply this type of analysis to examine transitions towards the green economy. By applying network analysis to a dataset of traded environmental products, Mealy and Teytelboym (2017) examined countries' green productive capabilities and their capacity to develop new green industries in the future. They showed that network-based measures, such as their Green Complexity Potential measure, were significantly predictive of future increases in countries' green trade competitiveness. Moreover, by exploring countries' positions in the 'green product space' and investigating how countries' competitiveness in green products evolved over time, Mealy and Teytelboym (2017) found that green diversification is strongly path-dependent: countries gaining early success in green capabilities tend to have much greater opportunities for diversifying into new product markets.

While recent efforts to incorporate network analysis into development and climate related research is yielding promising results, considerable future work awaits – particularly in better understanding the dynamic nature of networks. Although existing efforts have illuminated cross-country differences in productive structures, we are yet to understand how and why they change in particular ways. Further, very little research has examined the extent to which a nation's productive structure may be shaped by policy.

4.2 CGE Models

The overarching paradigm currently used to conduct policy analyses in *both* fields is computable general equilibrium (CGE) modelling. CGE models are large-scale numerical models that aim to simulate how the economy might evolve in response to changes in policy, technology or other exogenous events (such as a drought or flood). Often described as an exercise of 'theory with numbers' (Wing and Balistreri, 2018) they encompass two key components: detailed data on the structure of the economy and a theoretically derived system of equations dictating how the economy is likely to respond to particular changes. The underlying data used to calibrate CGE models are based on input–output tables or social accounting matrices. Aiming to provide a 'snapshot' of the economy at a particular point in time, these datasets include transaction values between key economic sectors (including government and representative households) and econometrically estimated parameters (or elasticities) capturing how different economic actors respond to changes in relative prices. The data are linked to a system of equations based on general equilibrium theory. These equations are numerically solved to determine how the simulated economy transitions to a new equilibrium by re-balancing supply and demand in different markets (Wing, 2004).

The ability of CGE models to incorporate data on the current state of the economy and provide theoretically grounded projections of its future evolution has made them popular tools within development and climate policy analysis. In development policy, CGE models have been frequently applied to examine the impacts of trade liberalization migration, industrial policy and pro-poor growth strategies (Ackerman and Gallagher, 2008; Devarajan and Robinson, 2013). In climate policy, CGE models are commonly applied to examine the cost and benefits associated with the introduction of a carbon tax, emission-trading schemes and subsidy-based schemes such as feed in tariffs (Wing, 2009; Adams and Parmenter, 2013). They form the core component of some Integrated Assessment Models (Bosetti et al., 2006; Hasegawa et al., 2017; Scrieciu et al., 2013), which are frequently used by policy makers to understand inter-relationships between economic, energy and climate systems (Farmer et al., 2015).

When applied to reasonably short-term horizons, CGE models have key strengths. Their capacity to explicitly capture interrelationships between markets for final goods, intermediate goods, government expenditure and households enable policy analysts to quantitatively study how the impacts of a particular change may filter through the economy and directly or indirectly impact different sectors and households (Wing, 2004; Chi et al., 2014). Unlike input–output models (discussed below), which are limited in their ability to incorporate actors' behavioural responses to prices and consequently tended to overestimate impacts associated with a given change, CGE models assign a more important role to prices and supply-side constraints (West, 1995). By explicitly specifying household utility functions, CGE models are also able to provide an estimate of the aggregate income and welfare impacts associated with a particular change. The ability to provide a quantitative estimate for impacts on 'winners' and 'losers' of proposed policies have made CGE models significantly influential in many policy debates (Hughes et al., 2016; Devarajan and Robinson, 2005).

However, when CGE models are applied to longer time horizons, or to the context of transformational change, CGE models encounter a number of issues. First, many CGE models encompass a relatively *static* framework, meaning that the modelled producers and consumers make optimizing decisions in a single period (Ahmed and O'Donoghue, 2005). In these models, policy analysis is based on a comparison of two alternative future equilibrium states of the economy – one with the policy change and one without. As these models do not explicitly represent the process of adjustment from one equilibrium to another, any economic impacts associated with transitional dynamics are unable to be accounted for (Ackerman and Gallagher, 2008; Scrieciu, 2007).

An alternative class of *dynamic* CGE models aims to better trace model variables over the projected time path – usually at yearly intervals. These models can be divided into two types: recursive dynamic models and forward-looking models (Devarajan and Robinson, 2013). Unfortunately, neither of these options provide convincing frameworks for realistically capturing agent behaviours or structural dynamics associated with transformational change. Recursive dynamic CGE models assume that agents are completely myopic and make optimizing decisions only on the basis of current and past prices and other model variables. Agents' lack of consideration about the future prevents meaningful analysis of any inter-temporal investment and savings decisions (Babiker et al., 2009).

In contrast, forward-looking models do encompass agents that incorporate the future expectations into their decisions (Pratt et al., 2013). However, these models are usually characterized by agents who are unrealistically perfectly rational and omnisciently able to solve inter-temporal optimization problems over all modelled periods (Richiardi, 2015; Babiker et al., 2009). While dynamic stochastic general equilibrium (DSGE) models enable agents to explicitly incorporate uncertainty about future states of the world, these models are usually so computationally intensive that much of the important economic sectoral detail (which makes these models useful in the first place) needs to be significantly simplified (Devarajan and Robinson, 2013). In addition, projections produced by most forward-looking models are, by construction, smooth, efficient adjustments along a balanced equilibrium path, which are hardly appropriate to characterize the turbulent, out-of-equilibrium dynamics often associated with societies experiencing transformational change (Rezai et al., 2013; Nordhaus, 2010).

A second issue complicating the application of CGE models to transformational change is their heavy reliance on equilibrium outcomes to characterize dynamics and future projections. Almost all CGE models assume capital and labour markets clear – usually instantaneously at each modelled time period. Not only is this at odds with historical observations in which economic activity was commonly subject to long, unanticipated recessions, persistent unemployment (Blanchard and Summers, 1987), 'secular stagnations' (Hansen, 1939) and institutional-induced capital (Harberger, 1959). It also misses essential characteristics of the development and low-carbon transition process that are important for policy makers to account for (DeCanio, 2003). For example, Chenery (1979) emphasizes that as developing economies are often characterized by the persistence of surplus labour and under-utilized capital, models should be designed to allow for the existence of disequilibrium rather than exclude it by assumption. Similarly, a number of scholars have argued that assuming a first-best world where all resources are optimally employed negates the potential economic benefits that green stimulus industrial policies could potentially provide to stagnant or depressed economies (Hasselman and Kovalevsky, 2013; Wolf et al., 2013; Barker et al., 2012). While recent developments in new Keynesian style DSGE models have incorporated frictions in price and wage adjustments and even the possibility of involuntary unemployment (Smets and Woulters, 2007; Christiano et al., 2010; Kemfert, 2003), few have been applied to climate or development policy (Scrieciu et al., 2014; Fagiolo and Roventini, 2016).

4.3 Macro-Econometric and Input–Output Simulation Models

Macro-econometric and input–output simulation models offer an alternative modelling framework that involves less restrictive assumptions about optimization and equilibrium. These are similar to CGE models in that they are based on detailed economic sectoral data (such as input–output or social accounting matrices) and are able to capture production and consumption flows across different industries. Their dynamics are also built on a system of equations. However, unlike CGE models, they are less strictly tied to neoclassical general equilibrium theory.

Macro-econometric and input–output simulation models instead draw on a more diverse range of economic fields (such as behavioural, ecological and evolutionary economics (Scrieciu et al., 2013)) and alternative economic paradigms (such as the Keynesian or neo-Keynesian framework, where demand is the key driving force of growth and supply adjusts to meet demand, subject to supply constraints (Boulanger and Brechet, 2005; Rezai et al., 2013)). Instead of making optimization assumptions, they draw on historical data and econometrics to characterize key behavioural parameters and are said to implicitly characterize a form of bounded rationality (Barker et al., 2016). Macro-econometric simulation models also differ to CGE models in their price formation. Rather than determining prices by imposing market-clearing assumptions, they employ a markup on unit costs, which depend on the level of competition in each sector (Cambridge Econometrics, 2014).

Macro-econometric and input–output simulation models are generally estimated using reasonably long stretches of time-series data and have the advantage that they can account for out-of-equilibrium dynamics that may have characterized historical observations. An important implication is that, unlike CGE models, they can capture the existence of unemployment and underutilized resources, which are particularly relevant considerations when assessing the *benefits* of industrial policies or green stimulus measures (Barker et al., 2016; IRENA, 2016). These types of models are also well suited to capture technological detail underpinning different sectors, as well as direct and indirect effects of endogenous technological change (Wiebe and Lutz, 2016; Lutz et al., 2007).

On the one hand, macroeconometric models are less bounded by the 'straight-jacket' of strong general equilibrium theoretical impositions. However, on the other hand, they have to be much more intimately tied to historical data. This heavy reliance on past data to estimate future projections can be particularly problematic in the context of transformational change, as by definition, such a process entails a future state that could look significantly different from historical trends (Scrieciu, 2007; Köhler et al., 2006). Moreover, the fairly rigid format of input–output tables makes it challenging to consider how the industrial structure of an economy may evolve in response to climate or development policies.

4.4 Agent-Based (Multi-Agent) Models

Agent-based (or multi-agent) models offer yet another modelling paradigm. They aim to simulate the behaviour of social or economic systems by explicitly representing heterogeneous, networked agents that interact and make decisions through prescribed behavioural rules (Farmer and Foley, 2009; Bonabeau, 2002). Agents do not need to necessarily make optimizing decisions, but can be programmed to make decisions in accordance with a number of behavioural typologies that are usually informed by empirical studies (Valbuena et al., 2008; Smajgl et al., 2011). A key advantage of this modelling paradigm, which is not accessible in CGE or macro-econometric simulation models, is its ability to explicitly model how macro-level dynamics arise from the (often probabilistic) interaction of these heterogeneous entities (Mercure et al., 2016; Vespignani, 2012; Holland and Miller, 1991). This emergence attribute is commonly examined in studies investigating social and collective phenomena, such as traffic jams and stock market crashes (Farmer and Foley, 2009). However, it is also known to be an important characteristic underpinning the 'tipping-point' dynamics associated with technological adoption and product diffusion rates (Dosi, 1982; Silverberg et al., 1988), social opinions (Watts and Dodds, 2007); and climate change awareness (Russil and Nyssa, 2009), which each play a key role in development and low-carbon transition processes.

A further important point is that while agent-based models (ABMs) do not explicitly enforce market clearing, equilibrium prices and quantities often do emerge as the result of agent's buying and selling decisions (Gintis, 2007). This framework consequently allows researchers and policy makers to analyse conditions under which equilibrium outcomes do and don't occur over different time horizons. In addition, in the presence of multiple equilbria, ABMs are powerful tools to understand the probability and process by which economies end up in different states (such as clean vs dirty energy outcomes) (Farmer et al., 2015; Arthur, 2006).

In comparison to CGE and macro-econometric approaches, agent-based modelling is still a relatively young and emerging analytical framework. However, its flexibility and intuitive framework for understanding change processes occurring within complex systems has seen it attract increasing interest, particularly within the climate-related literature. Many studies, which compare the relative merits of different approaches for analysing sustainability and climate policy-making, consistently provide strong appraisals and attest to its potential for offering important complementary insights to existing analytical tools (Boulanger and Brechet, 2005; Bassi, 2014; Safarynsky et al., 2012; Farmer et al., 2015; Balint et al., 2016).

However, there are two key challenges currently hindering faster progress and wider dissemination of ABMs within the policy landscape. First, there is a present lack of commonly accepted modelling standards within the ABM community. As a result, the ABM literature is replete with diverse and bespoke implementations, which often use a variety of different programming languages and design structure (Müller et al., 2013; Richiardi et al., 2006). Further, as models (and their output) are often complicated and poorly documented (Angus and Hassani-Mahmooei, 2015; Lee et al., 2015), replicating results or applying existing models to other contexts can be difficult and time-consuming (Richiardi, 2015). Second, estimation, calibration and validation procedures are often particularly challenging within ABMs (Faglio et al., 2007). These procedures aim to ensure the model and its parameters are as scientifically robust and defensible as possible. As ABMs often involve a large number of parameters and allow macro-level dynamics to emerge from the interactions and behaviours of micro-level agents, their estimation and calibration can be significantly more involved than other modelling approaches discussed in this chapter.[1]

5. CONCLUSION

In light of the pressing need to better understand transformational change processes, particularly as they relate to climate and development contexts, this chapter has sought to draw these fields together, highlighting key commonalities and shared learning opportunities.

It is clear that better integration across climate and development domains is paramount – and there are clear advantages for both research *and* policy. In relation to research, climate and development economists have traditionally studied the process of transformative change in separate fields and with differing emphases. However, identifying key commonalities in respective change processes may not only improve

---

[1] That said, calibration and validation of CGE and DSGE models are also subject to criticism. CGE models are often applied to policy analysis without being exposed to any soft of validation process (van Dijk, Philippidis and Wolter, 2016; Beckman, Hertel and Tyner, 2011). Calibrating and validating DSGE models is both more challenging and more controversial, as models often require a number of ad-hoc tweaks or 'frictions' to enable them to fit the data (Faglio and Roventin, 2012; Grauwe, 2010).

shared learning outcomes, it could also illuminate a more generalized theory of transformational change. For policy, a lack of integration in climate and development initiatives can lead to outcomes that are at best, myopic and at worst, detrimental to their intended objectives.

In terms of methodological tools for analysing and modelling transformational change, this chapter has reviewed four different approaches that have been used in both climate and development contexts. Network analysis provides a useful framework to investigate relationships across economic sectors, and allows scholars and policy makers to better understand technological diffusion and industrial transition possibilities. However, there has been relatively little work on the dynamism of economic networks and how industrial structure may be shaped by policy. CGE models are the standard policy analysis tool in both climate and development fields, and while these models have key strengths when applied to short-term horizons, our review casts doubt on their ability to appropriately capture dis-equilibrium dynamics and emergent attributes of the transformative change process. While macro-econometric models deal better with out-of-equilibrium settings, their strong reliance on historical data weakens their ability to predict fundamentally new and different economic structures and dynamics.

In contrast, the flexibility of agent-based models make them a promising framework for analysing the formation of new economic arrangements and 'tipping point' dynamics emerging from complex interactions between heterogeneous agents. If appropriate standards and validation procedures can be progressed and adopted, these models could offer researchers and policy makers an important tool for better navigating and driving the process of transformational change to make simultaneous progress on climate and development goals.

REFERENCES

Ackerman, F. and K.P. Gallagher (2008), 'The shrinking gains from global trade liberalization in computable general equilibrium models: a critical assessment', *International Journal of Political Economy*, 37(1), 50–77.
Adams, P.D. and B.R. Parmenter (2013), 'Computable general equilibrium modeling of environmental issues in Australia', in P.B. Dixon and D. Jorgenson (eds), *Handbook of Computable General Equilibrium Modeling*, Vol. 1, Amsterdam: North-Holland, pp. 553–657.
Aghion, P., C. Hepburn, A. Teytelboym and D. Zenghelis (2014), 'Path dependence, innovation and the economics of climate change', Centre for Climate Change Economics and Policy/Grantham Research Institute on Climate Change and the Environment Policy Paper and Contributing Paper to New Climate Economy.
Ahmed, V. and C. O'Donoghue (2005), 'Using CGE and microsimulation models for income distribution analyses: a survey', Working Paper No. 0089, Department of Economics, National University of Ireland, Galway.
Alstone, P., D. Gershenson and D.M. Kammen (2015), 'Decentralized energy systems for clean electricity access', *Nature Climate Change*, 5(4), 305–14.
Althor, G., J.E. Watson and R.A. Fuller (2016), 'Global mismatch between greenhouse gas emissions and the burden of climate change', *Scientific Reports*, 6, 20281.
Angus, S.D. and B. Hassani-Mahmooei (2015), '"Anarchy" reigns: a quantitative analysis of agent-based modelling publication practices in JASSS, 2001–2012', *Journal of Artificial Societies and Social Simulation*, 18(4), 16.
Arthur, W.B. (2006), 'Out-of-equilibrium economics and agent-based modeling', in K. Judd and L. Tesfatsion (eds), *Handbook of Computational Economics*, Vol. 2, Amsterdam: North-Holland, pp. 1551–64.
Babiker, M., A. Gurgel, S. Paltsev and J. Reilly (2009), 'Forward-looking versus recursive-dynamic modeling in climate policy analysis: a comparison', *Economic Modelling*, 26(6), 1341–54.
Bahar, D., R. Hausmann and C.A. Hidalgo (2014), 'Neighbors and the evolution of the comparative advantage of nations: evidence of international knowledge diffusion?', *Journal of International Economics*, 92(1), 111–23.

Balint, T., F. Lamperti, A. Mandel, M. Napoletano, A. Roventini and A. Sapio (2016), 'Complexity and the economics of climate change: a survey and a look forward', *Ecological Economics*, **138**, 252–65.

Baptist, S. and C. Hepburn (2013), 'Intermediate inputs and economic productivity', *Philosophical Transactions of the Royal Society A: Mathematical, Physical and Engineering Sciences*, **371**(1986), 20110565.

Barker, T., A. Anger, U. Chewpreecha and H. Pollitt (2012), 'A new economics approach to modelling policies to achieve global 2020 targets for climate stabilisation', *International Review of Applied Economics*, **26**(2), 205–21.

Barker, T., E. Alexandri, J.F. Mercure, Y. Ogawa and H. Pollitt (2016), 'GDP and employment effects of policies to close the 2020 emissions gap', *Climate Policy*, **16**(4), 393–414.

Bassi, A. (2014), 'Using simulation models for green economy policy making: a comparative assessment', *Review of Business and Economics Studies*, (1), 88–99.

Beckman, J., T. Hertel and W. Tyner (2011), 'Validating energy-oriented CGE models', *Energy Economics*, **33**(5), 799–806.

Bhattacharya, A., J. Oppenheim and N. Stern (2015), 'Driving sustainable development through better infrastructure: key elements of a transformation program', Brookings Global Working Paper Series.

Blanchard, O.J. and L.H. Summers (1987), 'Hysteresis in unemployment', *European Economic Review*, **31**(1–2), 288–95.

Bonabeau, E. (2002) 'Agent-based modeling: methods and techniques for simulating human systems', *Proceedings of the National Academy of Sciences*, **99**(suppl 3), 7280–87.

Boschma, R., A. Minondo and M. Navarro (2013), 'The emergence of new industries at the regional level in Spain: a proximity approach based on product relatedness', *Economic Geography*, **89**(1), 29–51.

Bosetti, V., C. Carraro, M. Galeotti, E. Massetti and M. Tavoni (2006), 'WITCH – a world induced technical change hybrid model', *The Energy Journal*, **27**(Special issue), 13–37.

Boulanger, P.M. and T. Bréchet (2005), 'Models for policy-making in sustainable development: the state of the art and perspectives for research', *Ecological Economics*, **55**(3), 337–50.

Bowen, A., C. Duffy and S. Fankhauser (2016), '"Green growth" and the new Industrial Revolution', Policy Brief, Grantham Research Institute on Climate Change and the Environment, Global Green Growth Institute, January 2016.

Boyd, R., F. Green and N. Stern (2015), 'The road to Paris and beyond: policy paper', Centre for Climate Change Economics and Policy, Grantham Research Institute on Climate Change and the Environment.

Brown, A. (2015), 'Climate change and Africa', *Nature Climate Change*, **5**, 811.

Burch, S., A. Shaw, A. Dale and J. Robinson (2014), 'Triggering transformative change: a development path approach to climate change response in communities', *Climate Policy*, **14**(4), 467–87.

Cambridge Econometrics (2014), 'E3ME Technical Manual, Version 6.0', available at https://www.camecon.com/wp-content/uploads/2016/09/E3ME-Manual.pdf.

Chenery, H.B. (1986), 'Growth and transformation', in H. Chenery, S. Robinson and M. Syrquin (eds), *Industrialization and Growth: A Comparative Study*, New York: OUP, pp. 13–36.

Chenery, H.B. and H. Elkington (1979), *Structural Change and Development Policy*, New York: Oxford University Press for the World Bank.

Chi, Y., Z. Guo, Y. Zheng and X. Zhang (2014), 'Scenarios analysis of the energies' consumption and carbon emissions in China based on a dynamic CGE model', *Sustainability*, **6**(2), 487–512.

Christiano, L.J., M. Trabandt and K. Walentin (2010), 'Involuntary unemployment and the business cycle', European Central Bank Working Paper Series No 1202, available at https://www.ecb.europa.eu/pub/pdf/scpwps/ecbwp1202.pdf?d807a66f1e6b8186917361f60f6911e9.

Cole, M.A., A.J. Rayner and J.M. Bates (1997), 'The environmental Kuznets curve: an empirical analysis', *Environment and Development Economics*, **2**(4), 401–16.

Collier, P., G. Conway and T. Venables (2008), 'Climate change and Africa', *Oxford Review of Economic Policy*, **24**(2), 337–53.

Cypher, J.M. and J.L. Dietz (2009), *The Process of Economic Development*, 3rd edn, London: Routledge.

Davis, S.J., K. Caldeira and H.D. Matthews (2010), 'Future CO_2 emissions and climate change from existing energy infrastructure', *Science*, **329**(5997), 1330–33.

DeCanio, S.J. (2003), *Economic Models of Climate Change: A Critique*, New York: Palgrave Macmillan.

Dechezleprêtre, A., R. Martin and M. Mohnen (2013), 'Knowledge spillovers from clean and dirty technologies: a patent citation analysis', Grantham Research Institute on Climate Change and the Environment Working Paper No. 135.

Dechezleprêtre, A., R. Martin and M. Mohnen (2014), 'Policy brief: clean innovation and growth', London: Imperial College, available at https://spiral.imperial.ac.uk/handle/10044/1/17753.

Denton, F., T.J. Wilbanks, A.C. Abeysinghe, I. Burton, Q. Gao, M.C. Lemos, T. Masui, K.L. O'Brien and K. Warner (2014), 'Climate-resilient pathways: adaptation, mitigation, and sustainable development', in Field, C.B., V.R. Barros, D.J. Dokken, K.J. Mach, M.D. Mastrandrea, T.E. Bilir, M. Chatterjee, K.L. Ebi, Y.O. Estrada, R.C. Genova, B. Girma, E.S. Kissel, A.N. Levy, S. MacCracken, P.R. Mastrandrea and L.L. White (eds), *Climate Change 2014: Impacts, Adaptation, and Vulnerability. Part A: Global and Sectoral Aspects. Contribution of Working Group II to the Fifth Assessment Report of the Intergovernmental Panel on Climate*, Cambridge and New York: Cambridge University Press, pp. 1101–31.

Devarajan, S. and S. Robinson (2002), 'The influence of computable general equilibrium models on policy', TMD discussion papers 98, International Food Policy Research Institute (IFPRI).

Devarajan, S. and S. Robinson (2013), 'Contribution of computable general equilibrium modeling to policy formulation in developing countries', in P.B. Dixon and D. Jorgenson (eds), *Handbook of Computable General Equilibrium Modeling*, Vol. 1, Amsterdam: North-Holland, pp. 277–98.

Dietz, T., G.T. Gardner, J. Gilligan, P.C. Stern and M.P. Vandenbergh (2009), 'Household actions can provide a behavioural wedge to rapidly reduce US carbon emissions', *Proceedings of the National Academy of Sciences*, **106**(44), 18452–6.

Dinda, S. (2004), 'Environmental Kuznets curve hypothesis: a survey', *Ecological Economics*, **49**(4), 431–55.

Dosi, G. (1982), 'Technological paradigms and technological trajectories: a suggested interpretation of the determinants and directions of technical change', *Research Policy*, **11**(3), 147–62.

ECA (Economic Commission for Africa) (2016), 'Greening Africa's industrialisation', *Economic Report on Africa*, Addis Ababa, available at http://www.un.org/en/africa/osaa/pdf/pubs/2016era-uneca.pdf.

Fagiolo G. and A. Roventini (2012), 'Macroeconomic policy in DSGE and agent-based models', Working Paper 2012-17, Economix.

Fagiolo, G. and A. Roventini (2016), 'Macroeconomic policy in DSGE and agent-based models redux: new developments and challenges ahead', available at SSRN: https://ssrn.com/abstract=2763735 or http://dx.doi.org/10.2139/ssrn.2763735.

Fagiolo, G., A. Moneta and P. Windrum (2007), 'A critical guide to empirical validation of agent-based models in economics: methodologies, procedures, and open problems', *Computational Economics*, **30**(3), 195–226.

Fankhauser, S. and F. Jotzo (2017), 'Economic growth and development with low-carbon energy', Centre for Climate Change Economics and Policy Working Paper No. 301, Grantham Research Institute on Climate Change and the Environment, Working Paper No. 267.

Fankhauser, S. and T.K. McDermott (eds) (2016), *The Economics of Climate-Resilient Development*, Cheltenham, UK and Northampton, MA, USA: Edward Elgar Publishing.

Fankhauser, S. and N. Sternab (2016), 'Climate change, development, poverty and economics', Centre for Climate Change Economics and Policy Working Paper No. 284, Grantham Research Institute on Climate Change and the Environment, Working Paper No. 253.

Farmer, J.D. and D. Foley (2009), 'The economy needs agent-based modelling', *Nature*, **460**(7256), 685–6.

Farmer, J.D. and F. Lafond (2016), 'How predictable is technological progress?', *Research Policy*, **45**(3), 647–65.

Farmer, J.D., C. Hepburn, P. Mealy and A. Teytelboym (2015), 'A third wave in the economics of climate change', *Environmental and Resource Economics*, **62**(2), 329–57.

Felipe, J., A. Mehta and C. Rhee (2014), 'Manufacturing matters... but it's the jobs that count', ADB Economics Working Paper Series, No. 420, available at https://www.adb.org/sites/default/files/publication/149984/ewp-420.pdf.

Fischer, C. and R.G. Newell (2008), 'Environmental and technology policies for climate mitigation', *Journal of Environmental Economics and Management*, **55**(2), 142–62.

Floater, G., P. Rode, A. Robert, C. Kennedy, D. Hoornweg, R. Slavcheva and N. Godfrey (2014), 'Cities and the new climate economy: the transformative role of global urban growth', NCE Cities Paper 01, LSE Cities, London School of Economics and Political Science.

Fuller, B. and P. Romer (2014), 'Urbanization as opportunity', World Bank Policy Research Working Paper (6874).

Garnaut, R., C. Fang, and L. Song (2013), 'China's new strategy for long-term growth and development', in R. Garnaut, C. Fang and L. Song (eds), *China: A New Model for Growth and Development*, Canberra, Australia: Australian National University E Press, pp. 1–16.

Geels, F.W., F. Berkhout and D.P. van Vuuren (2016), 'Bridging analytical approaches for low-carbon transitions', *Nature Climate Change*, **6**(6), 576.

Gillingham, K., R.G. Newell and K. Palmer (2009), 'Energy efficiency economics and policy', *Annual Review of Resource Economics*, **1**(1), 597–620.

Gintis, H. (2007), 'The dynamics of general equilibrium', *The Economic Journal*, **117**(523), 1280–309.

Gowdy, J.M. (2008), 'Behavioral economics and climate change policy', *Journal of Economic Behavior & Organization*, **68**(3), 632–44.

Grauwe, P. (2010), 'The scientific foundation of dynamic stochastic general equilibrium (DSGE) models', *Public Choice*, **144**(3), 413–43.

Grazzini, J. and M. Richiardi (2015), 'Estimation of ergodic agent-based models by simulated minimum distance', *Journal of Economic Dynamics and Control*, **51**, 148–65.

Green, F. and N. Stern (2017), 'China's changing economy: implications for its carbon dioxide emissions', *Climate Policy*, **17**(4), 1–15.

Griggs, D., M. Stafford-Smith, O. Gaffney, J. Rockström, M.C. Öhman, P. Shyamsundar, W. Steffen, G. Glaser, N. Kanie and I. Noble (2013), 'Policy: sustainable development goals for people and planet', *Nature*, **495**(7441), 305.
Grossman, G.M. and A.B. Krueger (1995), 'Economic growth and the environment', *The Quarterly Journal of Economics*, **110**(2), 353–77.
Grubb, M., F. Sha, T. Spencer, N. Hughes, Z. Zhang and P. Agnolucci (2015), 'A review of Chinese CO_2 emission projections to 2030: the role of economic structure and policy', *Climate Policy*, **15**(sup1), S7–S39.
Guivarch, C. and S. Hallegatte (2011), 'Existing infrastructure and the 2°C target', *Climatic Change*, **109**(3–4), 801–5.
Hallegatte, S., M. Bangalore, M. Fay, T. Kane and L. Bonzanigo (2015), 'Shock waves: managing the impacts of climate change on poverty', Washington, DC: World Bank Publications.
Hansen, A.H. (1939), 'Economic progress and declining population growth', *American Economic Review*, **29**(1), 1–15.
Harberger, A.C. (1959), 'Using the resources at hand more effectively', *American Economic Review*, **49**(2), 134–46.
Hasegawa, T., S. Fujimori, A. Ito, K. Takahashi and T. Masui (2017), 'Global land-use allocation model linked to an integrated assessment model', *Science of the Total Environment*, **580**, 787–96.
Hasselman, K. and D. Kovalevsky (2013), 'Simulating animal spirits in actor-based environmental models', *Environmental Modelling & Software*, **44**, 1042.
Hausmann, R. and J. Chauvin (2015), 'Moving to the adjacent possible: discovering paths for export diversification in Rwanda', Faculty Research Working Paper Series, 15-022.
Hausmann, R., C.A. Hidalgo, S. Bustos, M. Coscia, A. Simoes and M.A. Yildirim (2014), *The Atlas of Economic Complexity: Mapping Paths to Prosperity*, Cambridge, MA: MIT Press.
Hepburn, C. and A. Bowen (2013), 'Prosperity with growth: Economic growth, climate change and environmental limits', in R. Fouquet (ed.), *Handbook on Energy and Climate Change*, Cheltenham: Edward Elgar Publishing, pp. 617–38.
Herrendorf, B., R. Rogerson and Á. Valentinyi (2014), 'Growth and structural transformation', in P. Aghion and S. Durlauf (eds), *Handbook of Economic Growth*, Vol. 2B, Amsterdam: North-Holland, pp. 855–941.
Hidalgo, C.A., B. Klinger, A.L. Barabási and R. Hausmann (2007), 'The product space conditions the development of nations', *Science*, **317**(5837), 482–87.
Holland, J.H. and J.H. Miller (1991), 'Artificial adaptive agents in economic theory', *American Economic Review Papers and Proceedings*, **81**(2), 365–70.
Hughes, A., T. Caetano and H. Trollip (2016), 'Modelling a more equitable future', Mitigation Action Plans & Scenarios, Cape Town, available at http://mapsprogramme.org/outputs/modelling-equitable-future/.
IPCC (2007), 'Climate change 2007: mitigation of climate change. Contribution of Working Group III to the Fourth Assessment Report of the Intergovernmental Panel on Climate Change', S. Solomon, D. Qin, M. Manning, Z. Chen, M. Marquis, K.B. Averyt, M. Tignor and H.L. Miller (eds), Cambridge and New York: Cambridge University Press.
IRENA (2016), 'Renewable energy benefits: measuring the economics', IRENA, Abu Dhabi, available at http://www.irena.org/DocumentDownloads/Publications/IRENA_Measuring-the-Economics_2016.pdf.
Kaika, D. and E. Zervas (2013), 'The environmental Kuznets curve (EKC) theory – Part A: concept, causes and the CO_2 emissions case', *Energy Policy*, **62**, 1392–402.
Keller, W. (2004), 'International technology diffusion', *Journal of Economic Literature*, **42**(3), 752–82.
Kemfert, C. (2003), 'Applied economic-environment-energy modeling for quantitative impact assessment', *Puzzle-solving for Policy II*, **1**, 91.
Kocornick-Mina, A. and S. Fankhauser (2015), 'Climate change adaptation in dynamic economies: the cases of Columbia and West Bengal, Grantham Research Institute on Climate Change and the Environment', Global Green Growth Institute, September 2015.
Koh, H. and C.L. Magee (2008), 'A functional approach for studying technological progress: extension to energy technology', *Technological Forecasting and Social Change*, **75**(6), 735–58.
Köhler, J., T. Barker, D. Anderson and H. Pan (2006), 'Combining energy technology dynamics and macro-econometrics: the E3MG model', *The Energy Journal*, **27**(Special issue), 113–33.
Lee, J.S., T. Filatova, A. Ligmann-Zielinska, B. Hassani-Mahmooei, F. Stonedahl, I. Lorscheid, A. Voinov, G. Polhill, Z. Sun and D.C. Parker (2015), 'The complexities of agent-based modeling output analysis', *Journal of Artificial Societies and Social Simulation*, **18**(4), 4.
Lenzen, M. (2016), 'Structural analyses of energy use and carbon emissions: an overview', *Economic Systems Research*, **28**(2), 119–32.
Levin, T. and V.M. Thomas (2016), 'Can developing countries leapfrog the centralized electrification paradigm?', *Energy for Sustainable Development*, **31**, 97–107.
Lewis, W.A. (1954), 'Economic development with unlimited supplies of labor', *Manchester School of Economic and Social Studies*, **22**, 139–91.

Lutz, C., B. Meyer, C. Nathani and J. Schleich (2007), 'Endogenous innovation, the economy and the environment: impacts of a technology-based modelling approach for energy-intensive industries in Germany', *Energy Studies Review*, **15**(1), 1–17.
Mattauch, L. and C. Hepburn (2016), 'Climate policy when preferences are endogenous – and sometimes they are', *Midwest Studies in Philosophy*, **40**(1), 76–95.
Mealy, P. and A. Teytelboym (2017), 'Economic complexity and the green economy', available at SSRN: https://ssrn.com/abstract=3111644.
Menanteau, P., D. Finon and M.L. Lamy (2003), 'Prices versus quantities: choosing policies for promoting the development of renewable energy', *Energy Policy*, **31**(8), 799–812.
Mercure, J.F., H. Pollitt, A.M. Bassi, J.E. Viñuales and N.R. Edwards (2016), 'Modelling complex systems of heterogeneous agents to better design sustainability transitions policy', *Global Environmental Change*, **37**, 102–15.
Mersmann, F., K.H. Olsen, T. Wehnert and Z. Boodoo (2014), 'From theory to practice: understanding transformational change in NAMAs', accessed at https://epub.wupperinst.org/frontdoor/index/index/docId/5700.
Mlachila, M.M., R. Tapsoba and M.S.J.A. Tapsoba (2014), 'A Quality of Growth Index for developing countries: a proposal (No. 14-172)', International Monetary Fund Working Paper.
Müller, B., F. Bohn, G. Dreßler, J. Groeneveld, C. Klassert, R. Martin, M. Schlüter, J. Schulze, H. Weise and N. Schwarz (2013), 'Describing human decisions in agent-based models – ODD+D, an extension of the ODD protocol', *Environmental Modelling & Software*, **48**, 37–48.
Neffke, F. and M. Henning (2013), 'Skill relatedness and firm diversification', *Strategic Management Journal*, **34**(3), 297–316.
Neffke, F., M. Henning and R. Boschma (2011), 'How do regions diversify over time? Industry relatedness and the development of new growth paths in regions', *Economic Geography*, **87**(3), 237–65.
New Climate Economy (NCE) (2014), 'Better growth, better climate: charting a new path for low-carbon growth and a safer climate, The Global Report', New Climate Economy, London, available at https://newclimateeconomy.report/2016/wp-content/uploads/sites/2/2014/08/NCE-Global-Report_web.pdf.
New Climate Economy (NCE) (2016), 'The sustainable infrastructure imperative, 2016 Report', New Climate Economy, London, available at https://newclimateeconomy.report/2016/.
Nordhaus, W.D. (2010), 'Economic aspects of global warming in a post-Copenhagen environment', *Proceedings of the National Academy of Sciences*, **107**(26), 11721–6.
Nordhaus, W.D (2013), 'Integrated economic and climate modeling', in P.B. Dixon and D. Jorgenson (eds), *Handbook of Computable General Equilibrium Modeling*, Vol. 1, Amsterdam: North-Holland, pp. 1069–131.
O'Clery, N., A. Gomez-Lievano and E. Lora (2016), 'The path to labor formality: urban agglomeration and the emergence of complex industries', Center for International Development Working Paper (No. 78).
Ota, T. (2017), 'Economic growth, income inequality and environment: assessing the applicability of the Kuznets hypotheses to Asia', *Palgrave Communications*, **3**, 1–23.
Page, J. (2014), 'Rediscovering structural change: manufacturing, natural resources, and industrialization', *Oxford Handbooks Online*, accessed 31 October 2016 at http://ezproxy-prd.bodleian.ox.ac.uk:2067/view/10.1093/oxfordhb/9780199687107.001.0001/oxfordhb-9780199687107-e-019.
Perez, C. (2015), 'Capitalism, technology and a green global golden age: the role of history in helping to shape the future', *The Political Quarterly*, **86**(S1), 191–217.
Pfeiffer, A., R. Millar, C. Hepburn and E. Beinhocker (2016), 'The "2 C capital stock" for electricity generation: committed cumulative carbon emissions from the electricity generation sector and the transition to a green economy', *Applied Energy*, **179**, 1395–408.
Pratt, S., A. Blake and P. Swann (2013), 'Dynamic general equilibrium model with uncertainty: uncertainty regarding the future path of the economy', *Economic Modelling*, **32**, 429–39.
Radebach, A., J.C. Steckel and H. Ward (2016), 'Patterns of sectoral structural change: empirical evidence from similarity networks', available at SSRN: https://ssrn.com/abstract=2771653.
Raupach, M.R., G. Marland, P. Ciais, C. Le Quéré, J.G. Canadell, G. Klepper and C.B. Field (2007), 'Global and regional drivers of accelerating CO_2 emissions', *Proceedings of the National Academy of Sciences*, **104**(24), 10288–93.
Reid, W.V., D. Chen, L. Goldfarb, H. Hackmann, Y.T. Lee, K. Mokhele, E. Ostrom, K. Raivio, J. Rockström, H.J. Schellnhuber and A. Whyte (2010), 'Earth system science for global sustainability: grand challenges', *Science*, **330**(6006), 916–17.
Rezai, A., L. Taylor and R. Mechler (2013), 'Ecological macroeconomics: an application to climate change', *Ecological Economics*, **85**, 69–76.
Richiardi, M. (2015), 'The future of agent-based modelling', LABORatorio R. Revelli Working Paper Series 141, LABORatorio R. Revelli Centre for Employment Studies.
Richiardi, M.G., R. Leombruni, N.J. Saam and M. Sonnessa (2006), 'A common protocol for agent-based social simulation', *Journal of Artificial Societies and Social Simulation*, **9**(1).

Rodrik, D. (2012), 'Unconditional convergence in manufacturing', *The Quarterly Journal of Economics*, **128**(1), 165–204.
Rodrik, D. (2013), 'Structural change, fundamentals and growth: an overview', Institute for Advanced Study, available at https://pdfs.semanticscholar.org/d540/b22af021344573476a67ada49e89471faef3.pdf.
Rodrik, D. (2016), 'Premature deindustrialization', *Journal of Economic Growth*, **21**(1), 1–33.
Rogelj, J., M. Den Elzen, N. Höhne, T. Fransen, H. Fekete, H. Winkler, R. Schaeffer, F. Sha, K. Riahi and M. Meinshausen (2016), 'Paris Agreement climate proposals need a boost to keep warming well below 2°C', *Nature*, **534**(7609), 631–9.
Romer, P. (2016), 'The trouble with macroeconomics', Working Paper, available at https://paulromer.net/wp-content/uploads/2016/09/WP-Trouble.pdf.
Rosa, E.A. and T. Dietz (2012), 'Human drivers of national greenhouse-gas emissions', *Nature Climate Change*, **2**(8), 581.
Russill, C. and Z. Nyssa (2009), 'The tipping point trend in climate change communication', *Global Environmental Change*, **19**(3), 336–44.
Safarzyńska, K., K. Frenken and J.C. van den Bergh (2012), 'Evolutionary theorizing and modeling of sustainability transitions', *Research Policy*, **41**(6), 1011–24.
Schilling, M.A. and M. Esmundo (2009), 'Technology S-curves in renewable energy alternatives: analysis and implications for industry and government', *Energy Policy*, **37**(5), 1767–81.
Scrieciu, S.S. (2007), 'The inherent dangers of using computable general equilibrium models as a single integrated modelling framework for sustainability impact assessment: a critical note on Böhringer and Löschel (2006)', *Ecological Economics*, **60**(4), 678–84.
Scrieciu, S., A. Rezai and R. Mechler (2013), 'On the economic foundations of green growth discourses: the case of climate change mitigation and macroeconomic dynamics in economic modeling', *Wiley Interdisciplinary Reviews: Energy and Environment*, **2**(3), 251–68.
Scrieciu, S.Ş., V. Belton, Z. Chalabi, R. Mechler and D. Puig (2014), 'Advancing methodological thinking and practice for development-compatible climate policy planning', *Mitigation and Adaptation Strategies for Global Change*, **19**(3), 261–88.
Selden, T. and D. Song (1994), 'Environmental quality and development: is there a Kuznets curve for air pollution emissions?', *Journal of Environmental Economics and Management*, **27**, 147–62.
Silva, E.G. and A.A. Teixeira (2008), 'Surveying structural change: seminal contributions and a bibliometric account', *Structural Change and Economic Dynamics*, **19**(4), 273–300.
Silverberg, G., G. Dosi and L. Orsenigo (1988), 'Innovation, diversity and diffusion: a self-organisation model', *The Economic Journal*, **98**(393), 1032–54.
Smajgl, A., D.G. Brown, D. Valbuena and M.G. Huigen (2011), 'Empirical characterisation of agent behaviours in socio-ecological systems', *Environmental Modelling & Software*, **26**(7), 837–44.
Smets, F. and R. Wouters (2007), 'Shocks and frictions in US business cycles', *American Economic Review*, **97**(3), 586–606.
Steg, L. (2008), 'Promoting household energy conservation', *Energy Policy*, **36**(12), 4449–53.
Stern, D.I. (1998), 'Progress on the environmental Kuznets curve?', *Environment and Development Economics*, **3**, 175–98.
Stern, N. (2007), 'The economics of climate change', *American Economic Review*, **98**(2), 1–37.
Stern, N. (2009), 'Managing climate change and overcoming poverty: facing the realities and building a global agreement', Policy Paper, Centre for Climate Change Economics and Policy Grantham Research Institute on Climate Change and the Environment.
Stern, N. (2016), 'Economics: current climate models are grossly misleading', *Nature*, **530**(7591), 407–9.
Stern, D.I. and M.S. Common (2001), 'Is there an environmental Kuznets curve for sulfur?', *Journal of Environmental Economics and Management*, **41**(2), 162–78.
Stern, D.I., M.S. Common and E.B. Barbier (1996), 'Economic growth and environmental degradation: the environmental Kuznets curve and sustainable development', *World Development*, **24**(7), 1151–60.
Su, B. and B.W. Ang (2012), 'Structural decomposition analysis applied to energy and emissions: some methodological developments', *Energy Economics*, **34**(1), 177–88.
Taylor, L. (2009), 'Energy productivity, labor productivity, and global warming', in J.M. Harris and N.R. Goodwin (eds), *Twenty-first Century Macroeconomics: Responding to the Climate Challenge*, Cheltenham, UK and Northampton, MA, USA: Edward Elgar Publishing, Chapter 6.
Trancik, J.E. (2014), 'Renewable energy: back the renewables boom', *Nature*, **507**(7492), 300–302.
Turnheim, B., F. Berkhout, F. Geels, A. Hof, A. McMeekin, B. Nykvist and D. van Vuuren (2015), 'Evaluating sustainability transitions pathways: bridging analytical approaches to address governance challenges', *Global Environmental Change*, **35**, 239–53.
United Nations (2014), 'World urbanization prospects: the 2014 revision, highlights', Department of Economic and Social Affairs, Population Division, United Nations (ST/ESA/SER.A/352).
United Nations (2015), 'Transforming our world: the 2030 Agenda for Sustainable Development', Resolution

adopted by the General Assembly on 25 September 2015, available at http://www.un.org/ga/search/view_doc.asp?symbol=A/RES/70/1&Lang=E.

UNFCCC (United Nations Framework Convention on Climate Change) (2015), 'Adoption of the Paris Agreement', Report No. FCCC/CP/2015/L.9/ Rev.1, available at http://unfccc.int/resource/docs/2015/cop21/eng/l09r01.pdf.

UNIDO (United Nations Industrial Development Organisation) (2016), 'Industrial Development Report 2016: the role of technology and innovation in inclusive and sustainable industrial development', Vienna, available at https://www.unido.org/sites/default/files/2015-12/EBOOK_IDR2016_FULLREPORT_0.pdf.

Valbuena, D., P.H. Verburg and A.K. Bregt (2008), 'A method to define a typology for agent-based analysis in regional land-use research', *Agriculture, Ecosystems & Environment*, **128**(1), 27–36.

van Dijk, M., G. Philippidis and G. Woltjer (2016), 'Catching up with history: a methodology to validate global CGE models', FOODSECURE Technical paper (No. 9), LEI Wageningen UR.

Vespignani, A. (2012), 'Modelling dynamical processes in complex socio-technical systems', *Nature Physics*, **8**(1), 32–9.

Watts, D.J. and P.S. Dodds (2007), 'Influentials, networks, and public opinion formation', *Journal of Consumer Research*, **34**(4), 441–58.

West, G.R. (1995), 'Comparison of input–output, input–output + econometric and computable general equilibrium impact models at the regional level', *Economic Systems Research*, **7**(2), 209–27.

Wiebe, K.S. and C. Lutz (2016), 'Endogenous technological change and the policy mix in renewable power generation', *Renewable and Sustainable Energy Reviews*, **60**, 739–51.

Wing, I.S. (2004), 'Computable general equilibrium models and their use in economy-wide policy analysis: everything you ever wanted to know (but were afraid to ask)', MIT Joint Program on the Science and Policy of Global Change, Technical Note no. 6, available at http://web.mit.edu/globalchange/www/MITJPSPGC_TechNote6.pdf.

Wing, I.S. (2009), 'Computable general equilibrium models for the analysis of energy and climate policies', in J. Evans and L.C. Hunt (eds), *International Handbook on the Economics of Energy*, Cheltenham, UK and Northampton, MA, USA: Edward Elgar Publishing, pp. 332–66.

Wing, I.S. and E.J. Balistreri (2018), 'Computable general equilibrium models for policy evaluation and economic consequence analysis', in S.-H. Chen, M. Kaboudan and Y.-R. Du (eds), *The Oxford Handbook of Computational Economics and Finance*, Oxford: Oxford University Press, pp. 139–203.

Winkler, H., R. Spalding-Fecher and L. Tyani (2002), 'Comparing developing countries under potential carbon allocation schemes', *Climate Policy*, **2**(4), 303–18.

Wolf, S., S. Fürst, A. Mandel, W. Lass, D. Lincke, F. Pablo-Martí and C. Jaeger (2013), 'A multi-agent model of several economic regions', *Environmental Modelling & Software*, **44**, 25–43.

Xu, X.Y. and B.W. Ang (2013), 'Index decomposition analysis applied to CO_2 emission studies', *Ecological Economics*, **93**, 313–29.

Yao, C., K. Feng and K. Hubacek (2014), 'Driving forces of CO_2 emissions in the G20 countries: an index decomposition analysis from 1971 to 2010', *Ecological Informatics*, **26**, 93–100.

Zenghelis, D. (2015), 'Decarbonisation: innovation and the economics of climate change', *The Political Quarterly*, **86**(S1), 172–90.

18. Less precision, more truth: uncertainty in climate economics and macroprudential policy*
Cameron Hepburn and J. Doyne Farmer

1. INTRODUCTION

Earth's climate is a complex system. Our experience with weather forecasts attests that it is very difficult to predict what will emerge from the trillions of interactions between molecules in the atmosphere, oceans and on land. Modelling the climate—which can be understood roughly as "30-year average weather"—requires vast computing power (Palmer, 2011). At best, such models produce imperfect probability estimates of potential outcomes.[1]

Climate forecasting is both easier and harder than weather forecasting. On the one hand, it is much easier to predict the long-run average of a variable than an individual draw—the long-run average of the weather in a given location is easier to predict than the specific weather on a particular date in the future. On the other hand, much greater computing power is required over much longer time scales to predict the climate; over 30 years, the climate system will be subjected to many more shocks and larger external "forcings" than is possible over, say, five days. These forcings include the human emissions of greenhouse gases (see section 3).

As human impact on the climate increases—largely through increased concentrations of CO_2 and other greenhouse gases in the atmosphere and greater heat retention on Earth—the climate system moves further and further from current conditions. As this process continues, the data series and information sets that underpin our ability to understand and predict weather and climate systems become less reliable. In short, the relevant domain incorporates more "unknown unknowns"—a cause for concern given the existence of tipping points in the system.

The global financial system is also a complex system. Indeed, the climate and financial systems exhibit several similarities. Among these are:

- Significant feedbacks;
- Thresholds, tipping points and non-linearities, sometimes with irreversibilities;

---

* The authors would like to thank, without implicating, Tera Allas, Myles Allen, Eric Beinhocker, Howard Covington, David Frame, David Hendry, Tim Palmer, Matthew Scott, Natalie Seddon and participants at a workshop at the Bank of England on 2 April 2014, particularly David Spiegelhalter and Misa Tanaka. We are very grateful for valuable input from an anonymous referee and we thank Alexander Teytelboym and Penny Mealy for comments on this chapter.

[1] Frame and Stone (2012) make the point that the 1990 Intergovernmental Panel on Climate Change (IPPC) forecasts of warming have turned out to be "roughly right" once various surprise events are factored into account.

- Fat-tailed distributions, where there is a higher probability of events far from the mean;
- Non-equilibrium system dynamics, often chaotic, that do not necessarily settle down; and
- Emergent properties that are highly sensitive to initial conditions.

Furthermore, both systems have experienced the gradual build-up of pressures of various kinds (e.g. emissions in the atmosphere, greater financial market interconnectedness and leverage) that shifted the systems into more unchartered territory. Given the prima facie similarities between the two systems, it is worth asking whether some of the various techniques employed by physicists, mathematicians and economists to understand climate systems might be relevant to the economics of financial systems, including macroprudential policy.

This chapter explores and synthesises what might be learned for macroprudential financial policy from climate science and economics. There is much to be learned in both directions. And caution is required, for there are also significant differences between the systems, not least the reflexivity in the financial system that is absent in the climate system (see section 4).

This chapter is structured as follows. In section 2, we consider our poor track record in estimating the uncertainty in predictions of key scientific constants. Our historical overconfidence at estimating fundamental constants of nature is remarkable. And these constants are stationary, unlike the emergent properties of the climate and the financial system. Section 3 considers modelling within climate change science and economics. This includes an overview of climate system modelling, a brief review of current integrated assessment models (IAMs), recent critiques of such models, and the theoretical developments that have been stimulated by their failings. Section 4 briefly, and very partially, attempts to draw possible lessons for macroprudential regulation. It starts from the recognition that human systems should be conceived as complex adaptive dynamic systems. We offer five possible lessons. Section 5 briefly considers the interactions between the two systems and section 6 concludes.

2. CAUTIONARY TALES OF HUMAN OVERCONFIDENCE

In a widely cited paper, Svenson (1981) found that over 90% of American and 69% of Swedish drivers rate themselves as better than the median driver in their country. A propensity for overconfidence is widely present in human judgements (Lichtenstein and Fischhoff, 1977; Harvey, 1997; Pallier et al., 2002). Unfortunately, scientists are also humans and overconfidence emerges in various forms as a (dangerous) feature of even some of the most careful scientific studies.

The important overconfidence aspect relevant to this chapter is how we model and treat systemic uncertainty. Do models exclude or ignore events that are possible and important? Are parameter estimates reported with error bars that are too narrow? Do we even have the right model for the system we are attempting to understand? We briefly review the relevant sources of uncertainty, before turning to examine some recent scientific history.

Sources of Uncertainty

There are a variety of classifications for the sources and types of uncertainty in descriptive (i.e. not normative[2]) parameters. For the purposes of this chapter, we employ a simple two-part division into parameter and model uncertainty.[3]

Parameter uncertainty arises where, even if we have the correct model of the structure of the system, the data only allows parameter estimation within specific uncertainty bounds. There are two sorts of error that can arise in parameter estimation: *systematic error* where the sample/model mean is a biased estimate of the real thing (perhaps because of faulty equipment, inaccurate data, flaws in procedure, etc.); and *random error*, which is a problem when there are only limited data/measurements—with enough data/measurements, random error goes to zero.

In contrast, model uncertainty arises where the structure of the world is either unknown or changing in an unknown way, or for some reason it is impossible to be certain that we have the right model. Models are, by definition, simplifications of reality,[4] and for complex systems such as we are interested in here it is almost always the case that there is substantial model uncertainty.

In statistics, model uncertainty is called *bias* and parameter uncertainty is called *variance*. There is typically an inherent trade-off between the two: A complicated model with many free parameters may suffer from too much variance, while a simple model with only a few parameters may suffer from too much bias. Good modelling practice strikes a balance between the two. In situations with a high degree of Knightean uncertainty (many unknown unknowns) one may end up with both problems at once. The danger of this happening becomes more severe as one moves from simple to complex systems, and as one moves from physical science to social science. One must be vigilant about overconfidence in these situations

Parameter Uncertainty

Fundamental physical constants should be relatively easy to estimate. Physical constants derive from the laws of nature, and under standard theories in physics do not change over time or space. Barring a shift in the laws of nature, constants such as the speed of light, Planck's constant (which relates the energy of a photon to its frequency), and the mass of an electron should be just that—constants—and should not change.

---

[2] It is normally not appropriate to think of uncertainty in empirical parameters in the same manner as variation in parameters about value judgements. Such value parameters include those that represent preferences (such as intergenerational discount rates, or values of life), or decision (or control) variables, selected by humans to achieve some "best" outcome. Rather than estimating probability distributions, parametric sensitivity analyses can be conducted on such variables to understand the relationship between choices, values and outcomes (Morgan and Henrion, 1990).

[3] This ignores various other sources and categorisations of uncertainty, see Morgan and Henrion (1990).

[4] Following Rissanen (1978) a model can be viewed in terms of information compression. A model makes valid predictions only to the extent that it is capable of compressing information. The best model is one for which the data plus the model has the minimum description length. See also the summary by Bais and Farmer (2008).

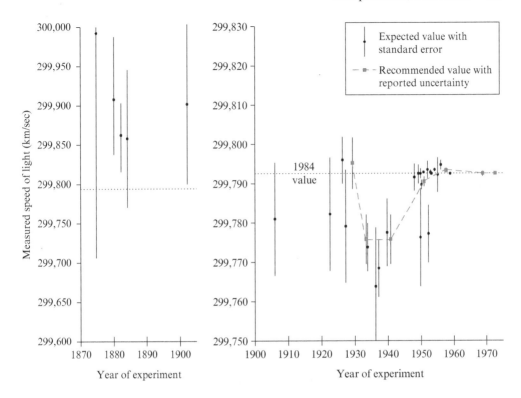

Source: Henrion and Fischhoff (1986).[5]

Figure 18.1 Estimated values for the speed of light, 1870–1973

The history of the measurement of fundamental physical constants provides a useful cautionary tale. Scientists have frequently been remarkably overconfident in the accuracy of their measurements. Figure 18.1 from Henrion and Fischhoff (1986) shows how estimated values of the speed of light have changed from 1870 to 1973. If the errors are normally distributed, then one expects the standard error to contain the measurement 70% of the time; instead they are contained only 50% of the time. More troubling, there is a period during the 1930s and 40s where all six experiments performed during this period yielded measurements that were systematically low by roughly two standard deviations.

Figure 18.2 shows similar estimates for Planck's constant from 1952 to 1973, where this problem is even more severe. All estimates are systemically low, with the true value far outside of the error estimate.

Finally, to underscore that this is not an isolated problem, Figure 18.3 presents the recommended values for the mass of an electron, showing the same problem once again.[6]

---

[5] Representation due to David M. Hassenzahl (2004).
[6] The interested reader is encouraged to look at Henrion and Fischhoff (1986) for further examples of scientific overconfidence, including estimates of Avagadro's number, the charge of an electron and the inverse fine structure constant.

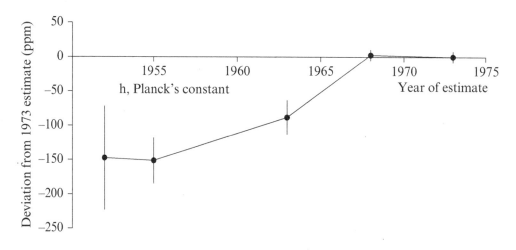

Source: Henrion and Fischhoff (1986).

Figure 18.2 Recommended values for Planck's constant, 1952–1973

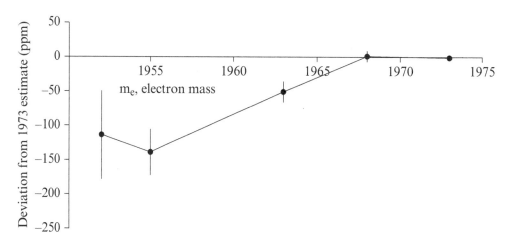

Source: Henrion and Fischhoff (1986).

Figure 18.3 Recommended values for the mass of an electron, 1952–1973

The implications for estimating parameters in far more complex systems (like the climate or financial system) are sobering.

In considering the reasons for the result, Henrion and Fischhoff (1986) note that "some of the apparent overconfidence reflects a deliberate decision to ignore the harder-to-assess sources of uncertainty", and further cite Taylor et al. (1969) in noting that individual differences in attitude of investigators play a role: "variation in attitude, although out of keeping with scientific objectivity, is nevertheless unavoidable so long as scientists are human beings". Properly estimating error bars is at least partially a subjective matter, and it is often difficult to think of the reasons why one's error bars might be too small.

Another source of error in measuring fundamental constants is that there can be a difference between the quantity of interest (e.g. the speed of light in a vacuum) and the quantity actually measured (e.g. the speed of light in a near vacuum). In other words, we are overconfident about the ability of the quantity actually measured to proxy for the quantity of interest. Overconfidence is very difficult to eliminate in measurements of physical constants, partly because it is much easier to underestimate the scale of little-known or unknown sources of error than to overestimate them, precisely because they are little-known or unknown.

Another example of systematic error in parameter uncertainty is the use of past data for forecasting future performance. (This can also be a source of model uncertainty.) The standard errors emerging from a regression on past data are informative about the fit of the model to history, but they may or may not provide information about the future. Extrapolating from the past to the future involves an additional form of uncertainty—the fact that the future may not be like the past. This error is not captured by statistical measures of the model's fit to past data. Estimating how likely the past is to be a good indicator of the future is highly likely to involve subjective judgement. The error involved in this subjective judgement, while often entirely ignored, is likely to be considerably larger than the errors involved in fitting a model to a historical time series. The problem, again, is that even when it is not ignored, this error is underestimated. False precision is the result.

The problems of overconfidence in estimates of uncertainty in fundamental physical constants are likely to be magnified as we move into the domain of policy analysis, especially in the climate and financial systems. This is because the quantities of interest are far less precisely known, they may not be stationary, and subjective judgements of systematic error are likely to be even more important.[7] Furthermore, the consequences of overconfidence in estimating important variables in climate and financial policy could be catastrophic. Indeed, one can plausibly argue that such failures have already been catastrophic.

Model Uncertainty

Every model is a simplification of reality, and the choice between models is essentially a choice between what is captured and what is omitted by the model. Because all models are false, it is inappropriate to attempt to state a probability that a particular model is the "right" model. Nonetheless, one model can be "better" than another, in the sense that it is more likely to capture the variables of interest, and hence deliver insight and forecasts that are more useful.

The distinction between model uncertainty and parameter uncertainty is not always as clear as it might seem. Sometimes a more general model can be created which encompasses two specific models, so that the choice between models becomes a choice between parameter values in the more general model.

---

[7] See also the separate claim of Ioannidis (2005) that "most research findings are false".

Table 18.1 Likelihood scale from Mastrandrea et al. (2010)

Term	Likelihood of outcome (%)
Exceptionally unlikely	0–1
Very unlikely	0–10
Unlikely	10–33
About as likely as not	33–66
Likely	66–100
Very likely	90–100
Extremely likely	95–100
Virtually certain	99–100

Communicating Uncertainty

Facing up to this uncertainty requires a humility that tends to be hard won. We are confident that climate scientists have made more progress in this domain than financial economists. The most recent report of the United National Intergovernmental Panel on Climate Change (IPCC) uses two different concepts for communicating the degree of certainty or uncertainty in its findings: the "confidence" in the claim and the "likelihood" of the claim (Mastrandrea et al., 2010). Confidence can be "very low", "low", "medium", "high" or "very high". Confidence is related to the quality of the evidence (limited, medium or robust) and the level of agreement between experts (low, medium or high). Likelihood is given a quantitative definition, as shown in Table 18.1. It was concluded, for instance, with "high confidence" that for a "likely" chance of meeting the 2°C target, substantial cuts in emissions are required by mid-century (IPCC, 2014). But it was only with "low confidence" that it was concluded that some mitigation measures raise prices for some energy services (IPCC, 2014).

One conclusion from comparing the two fields is that conclusions about the likely state and future evolution of the financial system might be expressed with similar markers of confidence and likelihood that are built on a deeper humility of what is known and indeed knowable.

3. UNCERTAINTY IN CLIMATE SCIENCE AND ECONOMICS

Weather forecasting provides one of the best practical success stories in which a complex system that was once viewed as unpredictable can now be forecast with much greater accuracy. The story of the development of meteorology over the last 150 years is a remarkable one, ably told by Edwards (2010) in *The Vast Machine*. He notes (p. 431) that we are now, finally, in a position where:

> Computer models assimilate observations in near-real time from a far-flung network of sensors on land, at sea, in the upper air, and in outer space. Global weather forecasts and analyzed data zoom around the world in minutes. Refashioned and interpreted by national weather centers and commercial forecast services, they serve countless human ends, from agriculture, shipping, insurance, and war to whether you are going to need an umbrella in the morning. Weather forecasting

today is woven tightly into the fabric of everyday life. It's an infrastructure: ubiquitous, reliable (within limits), widely shared, and transparent. You can get a forecast for any place on the planet.

The process of developing this remarkable "global knowledge infrastructure" (Edwards, 2010) did not happen overnight. On the contrary, the processes by which this system has developed have involved a myriad of different inventions and innovations over time, and enormous investment in the "knowledge infrastructure" that serves as its backbone.

Climate modelling piggybacks on top of weather modelling. Global circulation models for the climate are essentially weather models that are run for long periods of time under counterfactual scenarios such as increased greenhouse gases. Because the underlying models are validated everyday by making weather forecasts we can have some confidence in them, at least as long as the effects of the counterfactual scenarios are properly understood (e.g. because we understand the properties of greenhouse gases). Without the data and modelling infrastructure that supports weather prediction, climate science as we know it would not be possible. It is hard enough as it is, and some scientists are calling for the next grand phase of scientific development, at vast expense—a CERN for the climate (Palmer, 2011).

Layering the social sciences on top of the physical sciences to develop models that can inform climate policy makes life even tougher. Climate policy requires an understanding of the impact of changes in long-run climate variables on human well-being. This requires blending different disciplines, from physics and chemistry to philosophy, economics and politics, and then arguably also sociology and psychology to fully understand the climate-economy feedbacks.

In an attempt to address this challenge, "integrated assessment models" (IAMs), have been designed to try to incorporate and integrate insights from different fields. Such models have been used to support cost–benefit analysis of climate change, and to determine values such as the "social cost of carbon"—the marginal damage done by the release of a tonne of carbon dioxide into the atmosphere.[8]

However, IAMs have tended to treat the connection between the economic and climate systems through a linear chain of reasoning, as shown in Figure 18.4. IAMs start with the assumption that economic activity produces emissions of greenhouse gases, particularly CO_2, which leads to increased concentrations of such gases in the atmosphere,[9] trapping heat on Earth. The additional heat can be stored in the oceans—expanding the volume of the oceans and leading to sea level rise—or in the atmosphere, raising surface temperatures, particularly at the poles, and melting glaciers (leading to more sea level rise). This has several effects—on plant and animal life, on ice coverage, plant growth and death, and ultimately changing the climate itself. As the climate changes, there are physical impacts on human societies, in the form of changes in storm patterns, droughts, floods, desertification, and so on. Damages are often represented by a simple "damage function" that describes the percentage of GDP lost as a simple function (often exponential or

---

[8] Examples of IAMs include the PAGE model (Hope, 2013), the DICE model (Nordhaus, 2010, 2013) and the FUND model (Anthoff and Tol, 2012).
[9] Some of the CO_2 is also stored in the oceans, increasing acidity and leading to shifts in marine ecosystems, such as the expected elimination of coral reefs and the ecosystems associated with them by 2040.

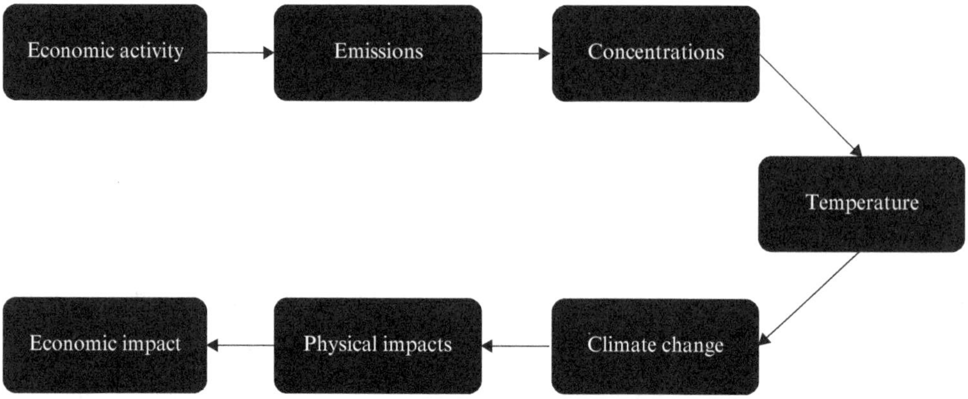

Figure 18.4 Stylised schematic of an integrated assessment model of climate change

quadratic) of temperature. Impacts are assumed to be economically damaging on balance, with more damage as we move further from the current climate. The monetary valuation of these impacts, using the assumed damage function, is the benefit of avoiding these impacts by reducing greenhouse gas emissions, and can be compared with the costs of abatement.

Some of the IAMs—particularly the DICE model of Nordhaus (2016) and the FUND model of Tol (2018)—conclude that a gradual policy response is all that is called for. "Optimum" levels of greenhouse gases, according to these models, should rise above levels not seen for millions of years, when humans didn't exist, and Earth's climate and physical geography was very different. For instance, estimated "optimal" warming is found to be 3.5°C in 2100 and rising by Nordhaus (2016). This level of warming implies serious risks of altering the physical geography of Earth, with the likelihood of increased migrations of humans and other species towards the poles and to higher ground (Oppenheimer, 2013). There is evidence that recent climate change has already impacted the timing of spring events, and the distribution and behaviour of animals and plants from across the globe, on the land and in the oceans (Parmesan and Yohe, 2003; Parmesan, 2006). Allen and Frame (2007) observe: "Once the world has warmed by 4°C, conditions will be so different from anything we can observe today (and still more different from the last ice age) that it is inherently hard to say when the warming will stop." In our view it is implausible for warming of 3.5°C, with the associated risks, to be considered "optimal" (see also Dietz and Stern (2015)).

What, if anything, is wrong with such models? Plenty, according to Schneider (1997), Ackerman et al. (2009), Pindyck (2013), Stern (2013) and Farmer et al. (2015). The question is helpful, not least because it might provide financial system modellers with some reminders of how *not* to model complex adaptive systems. We argue here that there are four main challenges for climate system modellers, not all of which are well-managed by IAMs. The four problems are inadequate treatment of: (1) structural uncertainty, (2) feedbacks, (3) thresholds; and finally (4) reliance on increasingly irrelevant past data. We consider each in turn.

First, because the climate system is a complex system, there will be features of the system that are inherently unpredictable. For instance, the risks associated with extreme climate change remain poorly understood. This is largely a result of deep structural uncertainty, rather than a problem of inadequate scientific effort, although doubtless more time and money would deliver improved knowledge. A related challenge is that, while the paleoclimate record shows a relationship between CO_2 and temperature, there are indications that the causal relationship may be bi-directional (Shakun et al., 2012).

Some of this structural uncertainty can be reduced by using ensembles of models. Consider the derivation of estimates for an important parameter called the climate sensitivity, which describes the equilibrium increase in temperature following a doubling of atmospheric concentrations of CO_2 (e.g. from 280 ppm pre-industrial to 560 ppm).[10] Basic calculations by Arrhenius in 1896 led him to conclude it was likely to take a value of around 5–6°C. Over the last century, considerable effort has been invested to pin this value down. Indeed, depending on the methods employed, the mean estimate derived from one method can be in the tail of another estimated distribution. Examining the various distributions collected by Knutti and Hegerl (2008) gives a picture of the full potential range of the climate sensitivity parameter as being from not much above 1°C to as much as 10°C.

However, it would be inappropriate to combine these distributions as if they were independent draws into a meta-distribution and, by taking the spread of the forecast of models, conclude that the uncertainty in the standard error is the spread of the models. The problem with this is that the (human) scientists designing the models are influenced by each other, and make similar errors. So each model is not a completely independent draw, and our confidence in the results is stronger than it should be. (See, for example, the systemic low estimates for the speed of light in the 1930s and 40s.) How much larger could the real error be? Section 3 gives reason to be humble, especially given that we have evidence of positive feedbacks in the system (see below) and are increasingly in unknown territory.

So how should structural uncertainty, or "Knightian uncertainty" or "ambiguity", be accounted for in climate modelling? This question has motivated an applied literature within environmental economics. Various authors (Li, 2004; Weitzman, 2007a, 2007b, 2009, 2013; Millner et al., 2012) have observed that incorporating ambiguity and/or extreme risks into economic analysis would increase the weight placed on future damages, sometimes significantly, motivating higher carbon prices and stronger climate policy.[11] This is indeed the conclusion of the minority of IAMs that do include such extreme risks. For instance, Hope (2013) finds that the top 1% of runs contribute around 20% to the mean value of the social cost of carbon.

Second, in the climate system, an initial temperature increase might trigger positive feedback processes that lead to further temperature increases. Physical feedbacks could include the following:

---

[10] Equilibrium of the climate system is reached over several hundred years, so that some physicists call into question the relevance of the climate sensitivity and would prefer instead that focus is on the transient climate system response (Allen and Frame, 2007).

[11] See also more generally Hansen (2007) and Hansen and Sargent (2010).

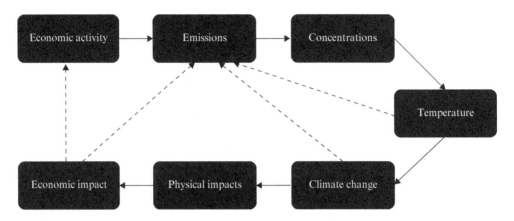

Figure 18.5 Stylised schematic of an integrated assessment model of climate change including selected feedbacks

- collapse of Greenland (or Antarctic) ice sheets, which reduces the reflection of heat back into space, thus increasing warming;
- release of methane (a powerful GHG) from the deep oceans or from the thawing of the permafrost;
- die-back of the Amazon rainforest or impacts on other ecosystems, reducing CO_2 uptake and increasing warming.

There may also be socioeconomic feedbacks (both positive and negative in the system). Hotter temperatures will require more cooling (and less heating), thermal energy generating plants may become less efficient, and so on.

When these feedbacks are included, the effective climate sensitivity increases. Previdi et al. (2013) argue that incorporating both fast and slow feedbacks might increase an initial climate sensitivity of around 3°C towards closer to 4–6°C. Figure 18.5 provides a simple schematic, suggesting some of the first steps that might be taken to adjust the linear IAM structure, set out in Figure 18.4 above, into more suitable form for the system at hand.

Third, it is clear from research that in addition to uncertainty and feedbacks, there may be "tipping points" in the Earth system that send it towards an altogether different state (Lenton et al., 2008). These sorts of outcomes, while hopefully highly unlikely, cannot be ruled out (Stainforth et al., 2005). Here, the analogy with the financial sector is imperfect—while the odds of catastrophic outcomes may be extremely low, there is no possibility of a planetary bailout. Weitzman (2009) argues that policy depends almost entirely on how high-temperature damages and tail probabilities are conceptualised and modelled.[12]

Finally, past data are increasingly less useful for modelling the current climate. Direct measurements of historical GHG captured in ice cores suggests that historical concentrations of GHGs over the last 800,000 years have ranged between approximately 180 ppm

---

[12] Nordhaus (2011) replies that if we can learn and act in a timely manner, this situation is not so worrisome.

and 300 ppm (Lüthi et al., 2008). Pre-industrial concentrations were 280 ppm. We are now above 400 ppm and rising at 2–3 ppm per annum. No human within history of civilisation over the last 10,000 years has experienced the conditions we are experiencing today. It is more difficult to understand the likely response of a complex system to a shock that takes it completely out of sample and beyond historical or human experience.

4. IMPLICATIONS FOR MACROPRUDENTIAL REGULATION

What does all this mean for macroprudential regulation? Unlike for the climate system, where a global system of measurement and modelling allows disaggregated forecasts to be made, at present no individual entity has a good grasp of what lies within the overall global financial system. Each individual bank may (or may not) have a fair understanding of its own open positions and risk limits, but each bank manages its individual position largely in ignorance of the overall system. This, of course, was one of the factors behind the 2008 financial crisis (Haldane and May, 2011).

Ultimately, it is up to financial market regulators and central bankers to determine whether anything can be learned from experience with understanding and modelling the climate system. Nevertheless, in this section, we will tentatively offer what we consider to be the top six implications that might emerge.

Implications

First, as implied above, the most valuable lesson from climate system modelling relates to the benefits of *en mass* data capture. Data over long time series are very helpful in understanding complex systems, indeed they are essential. In the climate system, paleoclimatology data from ice cores, tree rings, corals, oceans and lake sediments have been vital in understanding the sorts of states that are semi-stable and the types of transitions that are possible.[13] Data at many scales are required: it is impossible to properly understand a complex system with only aggregate data. In the climate system, as noted, we have both aggregate data and many time series at specific spatial locations—but even this is not enough, data models are required to get a smooth and complete dataset for the globe that can be used for subsequent modelling.

It requires effort to acquire the necessary data. Just as the troposphere does not reveal its secrets without human intrusion and examination, banks will not hand over commercially critical valuable information unless they are forced to do so. In part, they have good reason for this reluctance—such data could potentially be used to reverse engineer the operations of banks and large profits could potentially be made. More importantly, financial system data could be extremely dangerous in the wrong hands—deliberately triggering a financial market collapse for military, corporate or personal gain could not be ruled out.

In fact, some central banks, such as the Bank of England and the European Central Bank, already have substantial data of this type, but we are just beginning to make use

---

[13] http://www.ncdc.noaa.gov/data-access/paleoclimatology-data.

432 *Handbook on the economics of climate change*

of it. Hansen (2012) notes that the Office of Financial Research, in the US Treasury Department, was established following the Dodd-Frank bill with a remit to compile appropriate data and to identify measures of systemic risk. It remains to be seen whether this particular initiative will succeed on the scale required—and even if it does, coordination with other financial systems would be required to deliver truly successful financial system modelling. Nevertheless, it is a start.

Second, less precision in reporting model results is likely to mean more truth.[14] We have seen in section 2 that overconfidence is pervasive, and one should not expect estimates pertaining to the climate and financial systems to be immune, not least given that the relevant outputs are not fundamental physical constants, and hence could be wildly wrong. It is extremely difficult to predict some of the emergent properties of such systems.

Third, a wide range of potential scenarios should be explored by central bankers and regulators. Within the climate science community, a set of plausible scenarios are developed for use in climate modelling and research, and these are updated periodically, most recently in 2008.[15] We must do our best to think through the failure modes, whether it is the forests of the Amazon or Lehman Brothers. We recognise that resources are limited everywhere, and one can speculate all day about what might happen. Even qualitative scenario analysis can be useful. What are the impacts on the UK financial system from a financial crisis in China? Or a major new conflict in Asia, Eastern Europe or the Middle East? Or a pandemic? Or systemic implications from climate policy and impacts? What would be the impacts on the financial system? What are appropriate macroprudential preventative actions, and what are appropriate responses should these sorts of events occur?

Fourth, resilience should serve as an important concept, alongside efficiency.[16] There may well be an efficiency loss in a more resilient system, which may have greater redundancy during normal conditions. There is great social value in digging up and burning many fossil fuels, but enormous risks and costs too. Allowing fossil fuel use to continue for too long would be very dangerous for society at large. Similarly, there is social value in allowing banks to have leveraged balance sheets, but enormous risks too. Allowing the use of greater leverage to continue for too long would be very dangerous to society at large. However, to date, short-run "efficiency" has comprehensively trumped resilience in both systems, and this needs to change.

Fifth, a precautionary approach that achieves an appropriate balance between resilience and efficiency is required in a complex system with unknown thresholds, non-linearities and tipping points. Analysis can identify areas where there is reason to believe there may be a system boundary, or cliff edge, and create policy to ensure that the system does not drive off one of those cliff edges. An inevitable challenge is that we will not know precisely

---

[14] Accuracy is defined as the proximity of a measurement to its true value. Precision is defined as the degree to which repeated measurements show the same results, and is commonly expressed in statistics as the reciprocal of the variance.

[15] For instance, the most recent scenarios include four "representative concentration pathways" (RCPs), which describe plausible but very different greenhouse concentration trajectories, and hence four different climate futures. The four RCPs are named according to the radiative forcing in the year 2100, and hence are called RCP2.6, RCP4.5, RCP6 and RCP8.5 (IPCC, 2008).

[16] There is more than one definition of resilience, but it is broadly the capacity of an ecosystem or other system to respond to a shock with relatively minimal damage and with a rapid recovery.

(or even vaguely) where the cliff edge stands. And the location of the cliff edge may move over time. The point is to allow an adequate buffer zone such that the risk of driving over the cliff is minimal. In climate change, it now appears likely that the cumulative emission of 1 trillion tonnes of carbon might serve as one such system—indeed planetary—boundary (Rockström et al., 2009).

One might complain that it is almost impossible to set appropriate precautionary buffers if we do not know where the thresholds lie. However, in the climate system we now know enough to know that there are a lot of dangerous events that might happen. The feedbacks noted above for instance, such as the thawing of the permafrost releasing methane or the collapse of the ice sheets, appear "too big to fail"—the latter could lead to metres of sea level rise, and once gone, would not return before a substantial cooling of Earth towards ice age temperatures.[17] The underlying problem in both systems is non-linear positive feedback loops and that can enormously amplify modest effects. Modelling everything precisely is both impossible and less important than having an order-of-magnitude idea of the key risks.

Finally, given that such systems are constantly evolving, policy must continually evolve and adapt too. Knowing this, the policy interventions can be designed to ensure the flexibility and optionality of responding in different ways is preserved. While sometimes it is extremely important to signal forward commitment to markets through policy announcements, in order to shape expectations in a socially useful manner, it is also important to avoid commitments that are potentially ill-suited to an evolving system.

Limitations to the Analogy

Despite the apparently useful analogy between the climate system and the financial system, there are also some important differences. Modelling the financial system is arguably more difficult, because the agents in the financial system are intelligent, while atoms in the climate system are not intelligent. Atoms do not behave on the basis of their subjective expectations about likely future states of the world. Agent intelligence implies that the financial system will function as a *complex **adaptive** system*, not merely a complex system, as agents adjust their strategies and move into different niches seeking higher levels of fitness, e.g. as measured by profitability in the financial context.

Moreover, agents in the financial system do not merely behave on the basis of expectations about future states of nature. They develop strategies based on their own internal models of how other agents behave, and how those other agents believe other agents (including themselves) will behave. Beinhocker (2013) refers to this a "complex **reflexive** system"—merging the notion of a complex adaptive system with the notion of reflexivity (Soros, 2013). This might lead to even more dangerous feedback dynamics, or a greater variety of attractors, and greater system instability than in a system where reflexivity is not present.[18]

---

[17] When Greenland was 3 to 5°C warmer than today, a large proportion of the ice sheet had melted (Velicogna, 2009).

[18] Bronk (2013), for instance, argues that the reflexivity of markets coupled with social networks and contagion of ideas and emotions may lead to "shared narratives and analytical homogeneity in markets", where market participants go along with the crowd to save themselves

The core idea is that humans are embedded into the social systems that we are seeking to understand, and the process of gaining insight and understanding into the system may well change the way in which the system functions. For instance, although value at risk (VaR) has been used as a regulatory measure of choice, it does not take important systemic effects into account: the use of VaR introduces its own dynamics into a financial system, and can generate extreme risks even though each agent is individually acting in good faith.[19]

So, the problem of modelling and managing risk and uncertainty in financial systems is even more challenging than modelling and managing risk for the climate system. Because of the additional reflexivity in the financial system, and because we do not have a century of collaborative science and data collection behind us, understanding and managing risk in the financial system may be more difficult than in the climate system.

5. FINANCIAL AND CLIMATE SYSTEM INTERACTIONS

The focus of this chapter has been upon the usefulness of analogies between modelling uncertainty in the climate and financial systems, rather than upon the interaction between the two systems. However, given that the two systems can potentially shock each other in interesting ways, we briefly consider the inter-relationships here. We structure our reflections in two parts.

First, the physical impacts of climate change—storms, floods, droughts, fires, heatwaves etc.—will damage assets, change asset values, and have an impact on the financial sector through the insurance and banking systems. Overall, these impacts are unlikely to be severe in the short-term, notwithstanding the gradual increase in natural disasters and losses that are being absorbed by the insurance and reinsurance industry. Providing premiums rise accordingly, these losses should be manageable. However, as the climate system moves further outside historical experience, and past data become a more and more unreliable guide to future impacts, assessing probabilities, and hence providing insurance products, becomes more and more difficult. One might expect some insurance markets to break down, and for government to need to intervene, in a much greater fashion than as has happened in the recent United Kingdom negotiations about Flood Re. In the medium to longer term, there is a significant chance that we will realise at least a 3–4°C increase in global average temperatures, with likely dramatic shifts in climate and habitats for humans and others species. It would be altogether surprising if such a shift in the physical geography of the planet did not have significant impacts on the economic system, lowering economic growth and affecting various macroeconomic variables, which would also, needless to say, have more than a mild impact on the financial system.

---

the hassle of thinking for themselves. With insufficient cognitive diversity and/or heterogeneity of beliefs, Bronk (2013) argues that actors will not identify and address anomalies and novelties, increasing the potential for market instability.

[19] See Aymanns and Farmer (2015) and Aymanns et al. (2016) for models that illustrate how the use of VaR based on moving averages of historical risk levels and high leverage can result in bulk oscillations of the financial system that closely resemble the slow run up in prices leading to the crisis of 2008.

Second, in the short run (by which we mean the next five years or so), steps taken by governments to reduce the worst impacts of climate change will change the returns to different asset classes. Much economic activity is bound up with production and consumption that has so far only been economically viable because of the implicit subsidy of allowing firms to dump carbon pollution into the common atmosphere for free. When this implicit subsidy is removed, and an appropriate price is put upon carbon, economic activity will shift. There will be winners and losers. In the medium run, it is well-known that there are more fossil resources, even reserves, than can be permitted to be emitted into the atmosphere (Hepburn et al., 2014) and a surplus of fossil-fuelled assets above ground (Pfeiffer et al., 2016). Some of these assets will certainly see write downs in their valuations. It is far from clear that this could lead to a financial system collapse, but if the affected firms in the oil and gas, power, and mining industries are highly connected, and/or highly leveraged, there would appear to be at least a small risk of a trigger that could cause a shock to the financial system.

6. CONCLUSION

Appropriate modelling of a global complex system, such as the climate system or the financial system, requires good data collection at multiple scales and wise models that both shape and employ those data. This requires the existence of suitable organisations and a degree of international collaboration. These elements, hard won for meteorological and climatological modelling, are not yet present in financial system modelling. Furthermore, debate and discussion in financial market regulation, given the significance of reflexivity and participants' mental models, is even more valuable than in the climate context. Even with such data, it is extremely difficult to predict some of the emergent properties of complex systems, and even with wise models, history suggests we will get it wrong again, and by a margin that lies well outside estimated error bars.

The other lesson is that financial system regulation should incorporate analysis of "what if" scenarios that explore the consequences of specific (or combined) failures at various nodes in the network. Point estimates, often without uncertainty ranges (see Tol (2018) and subsequent Editorial Note (2015)), are manifestly inadequate and dangerously misleading. Based on the scenarios, and a better understanding of the system, policy should focus on resilience as well as efficiency (whether conceptualised as a high relative coefficient of risk and prudence or in some other form). Once considerations of resilience are more prevalent, it is natural to adopt a precautionary approach that is respectful of (potentially unknown) thresholds, non-linearities and tipping points to reduce the risk of going over a system boundary to tolerably low levels. Finally, it may be fruitful to map this approach onto conventional economic notions of risk aversion and precautionary behaviour, and then develop other areas of the economic literature, such as instrument choice (e.g. prices vs quantities) under ambiguity, rather than just risk. Such policy choices need to be made across most regulatory domains, including those with complex systems.

REFERENCES

Ackerman, F., S.J. DeCanio, R.B. Howarth and K. Sheeran (2009), 'Limitations of integrated assessment models of climate change', *Climatic Change*, **95**(3–4), 297–315.
Allen, M. and D. Frame (2007), 'Abandon the quest! Uncertainty in climate sensitivity: the long-term warming response to doubling carbon dioxide may not be a problem after all, unless the politicians choose to turn it into one', *Science*, **318**, 582–83.
Anthoff, D. and R.S.J. Tol (2012), 'Climate damages in the FUND model: a comment', *Ecological Economics*, **81**, 42.
Aymanns, C. and J.D. Farmer (2015), 'Dynamics of the leverage cycle', *Journal of Economic Dynamics and Control*, **50**, 155–79.
Aymanns, C., F. Caccioli, J.D. Farmer and V. Tan (2016), 'Taming the Basel leverage cycle', *Journal of Financial Stability*, **27**, 263–77, ISSN 1572-3089.
Bais, F.A. and J.D. Farmer (2008), 'The physics of information', in P. Adriaans and J. Van Benthem (eds), *Philosophy of Information*, Amsterdam: Elsevier, pp. 609–84.
Beinhocker, E.D. (2013), 'Reflexivity, complexity and the nature of social science', *Journal of Economic Methodology*, **20**(4), 330–42.
Bronk, R. (2013), 'Reflexivity unpacked: performativity, uncertainty and analytical monocultures', *Journal of Economic Methodology*, **20**(4), 343–49.
Crutchfield, J.P. (2009), 'The hidden fragility of complex systems: consequences of change, changing consequences', in G. Ascione, C. Massip, and J. Perello (eds), *Cultures of Change: Social Atoms and Electronic Lives*, Barcelona: ACTAR D Publishers, pp. 98–111.
Dietz, S. and N. Stern (2015), 'Endogenous growth, convexity of damage and climate risk: how Nordhaus' framework supports deep cuts in carbon emissions', *The Economic Journal*, **125**(583), 574–620.
Edwards, P. (2010), *A Vast Machine: Computer Models, Climate Data, and the Politics of Global Warming*, Cambridge, MA: MIT Press.
Farmer, J.D., C. Hepburn, P. Mealy and A. Teytelboym (2015), 'A third wave in the economics of climate change', *Environmental and Resource Economics*, **62**(2), 329–57.
Frame, D.J. and D.A. Stone (2012), 'Assessment of the first consensus prediction on climate change', Letters, *Nature Climate Change*, **3**(4), 357.
Haldane, A.G. and R.M. May (2011), 'Systemic risk in banking ecosystems', *Nature*, **469**(7330), 351.
Hansen, L.P. (2007), 'Beliefs, doubts and learning, valuing macroeconomic risk', *American Economic Review*, **97**(2), 1–30.
Hansen, L.P. (2012), 'Challenges in identifying and measuring systematic risk', NBER Working Paper 18505.
Hansen, L.P. and T.J. Sargent (2010), 'Fragile beliefs and the price of uncertainty', *Quantitative Economics*, **1**(1), 129–62.
Harvey, N. (1997), 'Confidence in judgment', *Trends in Cognitive Science*, **1**, 78–82.
Hassenzahl, D.M. (2004), 'Understanding uncertainty: definitions and tools for risk analysts', accessed on 29 March 2014 at https://faculty.unlv.edu/dmh/ratl/Lectures/ppts/Uncertainty.ppt.
Henrion, M. and B. Fischhoff (1986), 'Assessing uncertainty in physical constants', *American Journal of Physics*, **54**, 791–8.
Hepburn, C., E. Beinhocker, J.D. Farmer and A. Teytelboym (2014), 'Resilient and inclusive prosperity within planetary boundaries', *China & World Economy*, **22**(5), 76–92.
Hope, C. (2008), 'Optimal carbon emissions and the social cost of carbon over time under uncertainty', *Integrated Assessment*, **8**(1), 107–22.
Hope, C. (2013), 'How high should climate change taxes be?', in R. Fouquet (ed.), *Handbook on Energy and Climate Change*, Cheltenham, UK and Northampton, MA: Edward Elgar Publishing, pp. 403–14.
Hope, C. and K. Schaefer (2016), 'Economic impacts of carbon dioxide and methane released from thawing permafrost', *Nature Climate Change*, **6**(1), 56–9.
Ioannidis, J.P.A. (2005), 'Why most published research findings are false', *PLoS Medicine*, **2**(8), e124, 0696–0701.
IPCC (2008), 'Towards new scenarios for analysis of emissions, climate change, impacts, and response strategies', Geneva: Intergovernmental Panel on Climate Change, p. 132.
IPCC (2014), 'Summary for policymakers', in O. Edenhofer, R. Pichs-Madruga, Y. Sokona, E. Farahani, S. Kadner, K. Seyboth, A. Adler, I. Baum, S. Brunner, P. Eickemeier, B. Kriemann, J. Savolainen, S. Schlömer, C. von Stechow, T. Zwickel and J.C. Minx (eds), *Climate Change 2014: Mitigation of Climate Change. Contribution of Working Group III to the Fifth Assessment Report of the Intergovernmental Panel on Climate Change*, Cambridge and New York: Cambridge University Press.
Knutti, R. and G.C. Hegerl (2008), 'The equilibrium sensitivity of the Earth's temperature to radiation changes', *Nature Geoscience*, **1**, 735–43.

Lenton, T.M., H. Held, E. Kriegler, J. Hall, W. Lucht, S. Rahmstorf and H.J. Schellnhuber (2008), 'Tipping elements in the Earth's climate system', *Proceedings of the National Academy of Science*, **105**(6), 1786–93.

Li, A. (2004), 'Investigating the effect of risk and ambiguity aversion on the social cost of carbon', Dissertation for MSc in Environmental Change and Management, University of Oxford, Oxford, accessed on 29 March 2014 at http://www.cameronhepburn.com/wp-content/uploads/2012/10/Li2004.pdf.

Lichtenstein, S. and B. Fischhoff (1977), 'Do those who know more also know more about how much they know?', *Organizational Behavior and Human Performance*, **20**, 159–83.

Lüthi, D., M. Le Floch, B. Bereiter, T. Blunier, J.-M. Barnola, U. Siegenthaler, D. Raynaud, J. Jouzel, H. Fischer, K. Kawamura and T.F. Stocker (2008), 'High-resolution carbon dioxide concentration record 650,000–800,000 years before present', *Nature*, **453**(7193), 379–82.

Mastrandrea, M.D., C.B. Field, T.F. Stocker, O. Edenhofer, K.L. Ebi, D.J. Frame, H. Held, E. Kriegler, K.J. Mach, P.R. Matschoss, G.K. Plattner, G.W. Yohe and F.W. Zwiers (2010), 'Guidance note for lead authors of the IPCC fifth assessment report on consistent treatment of uncertainties', Intergovernmental Panel on Climate Change, Jasper Ridge, CA.

Millner, A., S. Dietz and G. Heal (2012), 'Scientific uncertainty and climate policy', *Environmental and Resource Economics*, **55**(1), 21–46.

Morgan, M.G. and M. Henrion (1990), *Uncertainty: A Guide to Dealing with Uncertainty in Quantitative Risk and Policy Analysis*, New York: Cambridge University Press.

New, M., D. Liverman, H. Schroder and K. Anderson (2011), 'Four degrees and beyond: the potential for a global temperature increase of four degrees and its implications', *Philosophical Transactions of the Royal Society A*, **369**, 6–19.

Nordhaus, W.D. (2010), 'Economic aspects of global warming in a post-Copenhagen environment', *PNAS*, **107**(26), 11721–6.

Nordhaus, W.D. (2011), 'The economics of tail events with an application to climate change', *Review of Environmental Economics and Policy*, **5**(2), 240–57.

Nordhaus, W.D. (2013), *The Climate Casino: Risk, Uncertainty, and Economics for a Warming World*, New Haven, CT: Yale University Press.

Nordhaus, W.D. (2016), 'Projections and uncertainties about climate change in an era of minimal climate policies' (No. w22933), National Bureau of Economic Research.

Oppenheimer, M. (2013), 'Climate change impacts: accounting for the human response', *Climatic Change*, **117**(3), 439–49.

Pallier, G., R. Wilkinson, V. Danthiir, S. Kleitman, G. Knezevic, L. Stankov and R.D. Roberts (2002), 'The role of individual differences in the accuracy of confidence judgments', *The Journal of General Psychology*, **129**(3), 257–99.

Palmer, T. (2011), 'A CERN for climate change', *Physics World*, **24**, 14–15.

Parmesan, C. (2006), 'Ecological and evolutionary responses to recent climate change', *Annual Reviews of Ecology, Evolution and Systematics*, **37**, 637–69.

Parmesan, C. and G. Yohe (2003), 'A globally coherent fingerprint of climate change impacts across natural systems', *Nature*, **421**, 37–42.

Pfeiffer, A., R. Millar, C. Hepburn and E. Beinhocker (2016), 'The "2 C capital stock" for electricity generation: committed cumulative carbon emissions from the electricity generation sector and the transition to a green economy', *Applied Energy*, **179**, 1395–408.

Pindyck, R.S. (2013), 'Climate change policy: what do the models tell us?', *Journal of Economic Literature*, **51**(3), 860–72.

Previdi, M., B.G. Liepert, D. Peteet, J. Hansen, D.J. Beerling, A.J. Broccoli, S. Frolking, J.N. Galloway, M. Heimann, C. Le Quéré, S. Levitus, and V. Ramaswamy (2013), 'Climate sensitivity in the Anthropocene', *Quarterly Journal of the Royal Meteorological Society*, **139**, 1121–31.

Rissanen, J. (1978), 'Modeling by the shortest data description', *Automatica*, **14**, 465–71.

Rockström, J., W. Steffen, K. Noone, Å. Persson, F.S. Chapin, III, E.F. Lambin, T.M. Lenton, M. Scheffer, C. Folke, H.J. Schellnhuber, B. Nykvist, C.A. de Wit, T. Hughes, S. van der Leeuw, H. Rodhe, S. Sörlin, P.K. Snyder, R. Costanza, U. Svedin, M. Falkenmark, L. Karlberg, R.W. Corell, V.J. Fabry, J. Hansen, B. Walker, D. Liverman, K. Richardson, P. Crutzen and J.A. Foley (2009), 'A safe operating space for humanity', *Nature*, **476**(7263), 472–5.

Schneider, S.H. (1997), 'Integrated assessment modeling of global climate change: transparent rational tool for policy making or opaque screen hiding value-laden assumptions?', *Environmental Modeling and Assessment*, **2**(4), 229–49.

Shakun, J.D., P.U. Clark, F. He, S.A. Marcott, A.C. Mix, Z. Liu, B. Otto-Bliesner, A. Schmittner, and E. Bard (2012), 'Global warming preceded by increasing carbon dioxide concentrations during the last deglaciation', *Nature*, **484**, 49–54.

Soros, G. (2013), 'Fallability, reflexivity and the human uncertainty principle', *Journal of Economic Methodology*, **20**(4), 309–29.

Stainforth, D.A., T. Aina, C. Christensen, M. Collins, N. Faull, D.J. Frame, J.A. Kettleborough, S. Knight, A. Martin, J.M. Murphy, C. Piani, D. Sexton, L.A. Smith, R.A. Spicer, A.J. Thorpe and M.R. Allen (2005), 'Uncertainty in predictions of the climate response to rising levels of greenhouse gases', *Nature*, **433**(7024), 403–6.

Stern, N. (2013), 'The structure of economic modelling of the potential impacts of climate change: grafting gross underestimation of risk onto already narrow science models', *Journal of Economic Literature*, **51**(3), 838–59.

Svenson, O. (1981), 'Are we less risky and more skillful than our fellow drivers?', *Acta Psychologica*, **47**, 143–51.

Taylor, B.N., W.H. Parker and D.N. Langenburg (1969), *The Fundamental Constants and Quantum Electrodynamics*, New York: Academic Press.

Tol, R. (2018), 'The economic impacts of climate change', *Review of Environmental Economics and Policy*, **12**(1), 4–25.

Tripati, A.K., C.D. Roberts and R.A. Eagle (2009), 'Coupling of CO_2 and ice sheet stability over major climate transitions of the last 20 million years', *Science*, **326**(5958), 1394–7.

Velicogna, I. (2009), 'Increasing rates of ice mass loss from the Greenland and Antarctic ice sheets revealed by GRACE', *Geophysical Research Letters*, **36**(19).

Weitzman, M.L. (2007a), 'A review of the *Stern Review on the Economics of Climate Change*', *Journal of Economic Literature*, **45**(3), 703–24.

Weitzman, M.L. (2007b), 'Subjective expectations and asset-return puzzles', *American Economic Review*, **97**, 1102–30.

Weitzman, M.L. (2009), 'On modeling and interpreting the economics of catastrophic climate change', *Review of Economics and Statistics*, **91**(1), 1–19.

Weitzman, M.L. (2013), 'Tail-hedge discounting and the social cost of carbon', *Journal of Economic Literature*, **51**(3), 873–82.

Index

Acemoglu, D. 84, 271
adaptation 400
adaptation strategies in European coastal cities 167–8
  conclusions 196–7
  construction of adaptation infrastructure 185, 190, 192–6
  damage costs in main cities 181–4, 185, 186–7
  Glasgow case study 184–5, 188–96
  methods 169–81
additive damages, *see* economic growth and social cost of carbon: additive versus multiplicative damages
Africa, historical 263
agent-based models (ABMs) 411–12
agents 433
agriculture and agricultural productivity 260–61, 262, 264, 265, 273, 402, 403
air pollution
  deaths caused by 13–17, 41–2, 45–6, 47, 48, 66–7
  health costs and impacts of 13–18, 20
  modeling accumulation process 327
  sources of 15–16
  see also policy evaluation (implementation of Paris Agreement), China
air quality co-benefits 13–21, 28–9
Alexeev, V.A. 127, 131, 132
altruism, *see* intergenerational altruism
analytical tractability 128, 158–9
anthropic principle 274
ARD (Acemoglu/Robinson/Dunning) model 84
Asheim, G.B. 311–12, 315, 316, 317–18, 319, 323–5
assets 434, 435
Australia, costs of sea-level rise 167–8
autocracy, and host–MNC relations in resource-rich countries 87, 90–92

'backward-looking' considerations, setting aside 233–4
bargaining game/theory 76–9, 93–5, 96, 104, 129
  global commons, 'bargaining to lose' 106–11
  hold-up problem 96–101, 102, 104–5
  host country government (HC) 79–80
  incentive structures 79–81
  multilateral development agencies (IFIs) 80–81
  obsolescing bargain model 73, 96, 101–2
  in permeability context 81–2
Barrett, S. 245, 247
Bass model 240–41, 242
behavioral tipping points, *see* tipping and reference points in climate change games
beliefs 272
bias 422
bilateral investment treaties (BITs) 85, 99–101, 102
border tax adjustments (BTAs) 235–6
BRICs (Brazil, Russia, India and China) 270
British Empire 262–4
Bronk, R. 433–4
Brown, S. 168
Bubb, R.J. 99
Buchholz, W. 252
bureaucratic growth, misdirection of 78–9
bureaucrats 80
business cycles 283
Byzantium 261

Cao, Jing 33
cap-and-dividend policies 26
cap-and-giveaway-and-trade policies 26
cap-and-spend policies 26
capital, dominance of 275
capital–output ratio 332, 379–81
capitalism 275, 276
carbon cycle 204, 221
carbon dividend policies 26–8
carbon footprint 305
carbon-free economy, timing of switch to 201, 207–8, 211, 214, 215, 217
carbon intensity of energy/output 220, 221, 404, 405
carbon leakage 235–6
carbon markets 151
carbon permits 21, 26, 151–3
carbon pricing 21–2, 32, 308
  benefits of 34
  China 24, 25
  distributional incidence (incidence analysis) 49–57
  support for 116

439

US (United States) 22, 24, 25
  see also optimal global climate policy, and regional carbon prices
carbon rent allocation 21–3, 24, 29
  efficiency-equity trade-off 28
  payers 23–5
  receivers 26–8
carbon stock 327
carbon tax 21–3, 26
  China 33–4, 37, 43–8, 51–4, 55–7
  coordinated global 307–8
  intertemporal equity and efficiency (global warming model) 339–41
  optimal 129, 140–43, 148–51, 201–2, 208, 215–16
  power generation 38
  support for 116
  see also economic growth and social cost of carbon: additive versus multiplicative damages
cascades 272–3
Castruccio, S. 135
CETA (Carbon Emissions Trajectory Assessment) model 114
CFCs (chlorofluorocarbons) 242
CGE (computable general equilibrium) models 408–10, 412
chaotic past climate episodes 273–4
Chichilnisky, G. 109, 129, 151, 152, 153, 224, 225, 268
Chichilnisky-Heal, N. 106, 107, 108, 109, 110
China
  air pollution 13–14, 15, 16
  carbon pricing 24, 25
  future development path 400
  historical 260, 262
  see also policy evaluation (implementation of Paris Agreement), China
clean development mechanism 234
climate change policy under spatial heat transport and polar amplification 127–9
  climate externality price and 'safety first' utility 154–8, 165–6
  conclusions and future research 158–60
  cross latitude effects 143–8
  heat transport and climate change policy 140–48
  optimal policies 148–53
  temperature dynamics and heat transport 130–34
  temperature paths and polar amplification 145–7, 164–5
  welfare maximization under heat transfer 134–40

climate policy 307–8; *see also* distributional issues in climate policy; optimal global climate policy, and regional carbon prices; policy evaluation (implementation of Paris Agreement), China
climate sensitivity 429, 430
climate system, *see* uncertainty, systems
CO_2 emissions 327, 427
  China 33–4, 37–9, 40–41, 42–5, 47–8, 65
  factors 49–50
  lack of incentives to reduce 301–2
  Paris Agreement pledges 32
  transient climate response to cumulative CO_2 emissions (TCRCE) 128
  see also greenhouse gas (GHG) accumulation and cyclical growth; Integrated Assessment Models (IAMs)
Co-Pollutant Cost of Carbon (CPCC) 16–17, 20
coal and coal excise, China 37, 40, 41–2, 43, 45–6, 47, 48, 57, 65, 66
coalition formation games 245–6, 250, 251–2
coastal flooding, *see* sea-level rise
collapse, *see* Malthus, collapse and climate change
common but differentiated responsibilities (CBDR) 232–4, 243
community benefit funds 21
complex systems, *see* uncertainty, systems
Condorcet, Nicolas de 266, 269
Conference of the Parties (COP) negotiations 239, 254–5
confidence and likelihood 426
contracts, hidden 108
cooperation 267–8, 270–71
Copernicus 266
Corn Laws 262
corruption 71–2, 77, 78, 83, 98
costs of production, and carbon tax 54–7
cumulative carbon emissions 127–8
cyclical growth, *see* greenhouse gas (GHG) accumulation and cyclical growth

Daniel, P. 95
Dannenberg, A. 247, 252
Dasgupta, P.S. 328
data capture 431–2
de-growth 296, 304
deaths caused by air pollution 13–17, 41–2, 45–6, 47, 48, 66–7
delegation 247–8
demand-side influences 406
democracy 271, 272
  democratic accountability, and permeability 68–9, 76, 77–9

and host–MNC relations in resource-rich countries 86–7, 89, 90–92
Desmet, K. 128–9
development goals, *see* transformational change (climate and development goals)
DICE (Dynamic Integrated Climate-Economy) model 114–16, 122–3, 226, 428
Dietz, S. 122–3
diffusion 240–41, 242
discount rate 12, 117, 122–3, 157–8, 310, 311
Disraeli, Benjamin 262
distribution of climate change impacts 121–2, 128, 141–3
distributional issues in climate policy 12–13
  air quality co-benefits 13–21
  carbon rent allocation 21–8
  conclusions 28–9
  incidence analysis, carbon pricing 49–57
DSGE (dynamic stochastic general equilibrium) models 409–10, 412
dynamic CGE (computable general equilibrium) models 409
dynasties 310–11, 313–15, 318–19, 321–5

early adopters 242
Earth, uniqueness of 274–5
ecological footprint 305
economic goods 296–7
economic growth 147–8; *see also* greenhouse gas (GHG) accumulation and cyclical growth; growth and sustainability
economic growth and social cost of carbon: additive versus multiplicative damages 199–202
  conclusions 216–18
  functional forms, calibration and computational implementation 209, 220–23
  Integrated Assessment Model (IAM) of Ramsey growth and energy transitions 203–8
  policy simulation and optimization 208–16
economic underdevelopment 70–71, 74
Edwards, P. 426–7
efficiency, *see* intertemporal equity and efficiency (global warming model)
electricity, China 35, 38, 43, 44
electricity-using capital, increasing efficiency of 39, 43, 44
electron, recommended values for mass of 423–4
emissions, CO_2, *see* CO_2 emissions
emissions/carbon permits 21, 26, 151–3
emissions reductions 13
  air quality co-benefits 18–21

Paris Agreement pledges 32
requirement for 399
emissions tax, *see* carbon tax
emissions trading systems (ETS) 224, 225–6, 234–5, 237
  China 37–8, 43–4, 45–6, 47, 48, 51, 57
energy balance model (EBM) 130
energy intensity 40, 50, 54–5, 56, 282, 284, 288, 404, 405
energy prices 212–14, 228–9
  China 41, 42, 51–5
Enkhbayar, Nambar 71–2
Enlightenment 265–7
Environmental Kuznets Curve (EKC) 402
Environmental Protection Agency (EPA) 116
'environmental threshold concerns' 252–3
environmental throughput 296–7
equity 232–4
  and air quality co-benefits 19
  efficiency-equity trade-off, carbon rent allocation 28
  intergenerational 12
  see also intertemporal equity and efficiency (global warming model)
equity weights (EMUC) 120
ethnic fragmentation 72, 73, 75
European Union (EU), Thematic Strategy on Air Pollution (TSAP) 20
evolutionary game theory (EGT) models 244
expectations, environmental negotiations 249–50
Expected Shortfall (ES) 175, 176–7, 180, 182–8, 190, 192, 193–6
experiments, climate change 246–8
exports 55, 407–8
externalities
  climate externality price and 'safety first' utility 154–8, 165–6
  optimal taxes 140–43, 148–51
extinctions 273

fairness
  as concept 270
  and other-regarding preferences 249–52
feedback response time (FRT) 120–21
feedbacks 429–30, 433–4
financial system 420–21, 431–5
First World War 263
Fischhoff, B. 423, 424
'forward-looking' considerations 233–4
fossil fuels, and heat transport, *see* climate change policy under spatial heat transport and polar amplification
framing 253
Francis (Pope) 264–5

Full Social Cost of Carbon (FSCC) 18
FUND (Climate Framework for Uncertainty, Negotiation and Distribution) model 114–16, 123, 224, 226, 428

games, *see* bargaining game/theory; tipping and reference points in climate change games
generation cost elasticities 65
Glasgow, sea-level rise and adaptation strategies 178, 179–80, 184–5, 188–96
global commons, 'bargaining to lose' 106–11
global warming
   and climate sensitivity 429, 430
   feedbacks 429–30, 433
   'optimal' 428
   Paris Agreement pledges 32
   see also intertemporal equity and efficiency (global warming model)
global warming damages, *see* economic growth and social cost of carbon: additive versus multiplicative damages
Glorious Revolution 271
Golosov, M. 201, 204, 206–7, 208, 217
government, host country 79–80
Great Britain
   Empire 262–4
   Glorious Revolution 271
green products 408
greenhouse gas (GHG) accumulation and cyclical growth 281–2
   accumulation 283–4
   conclusions 291
   functional forms and parameterization 293–5
   impacts on labor 291
   macroeconomic relationships 282–3
   steady-states 285–90
greenhouse gases (GHGs) 326–7, 430–31; *see also* Integrated Assessment Models (IAMs)
Gribbin, J. 274
growth and sustainability 296–7
   applying conclusions to real world 305–7
   conclusions 307–8
   implications for steady-state and de-growth economics 303–4
   Sraffa model 297–9
   sufficient conditions for environmental sustainability 302–3
   technical change, labor productivity and throughput efficiency 299–302
Grubb, M. 400
Grüning, C. 251

Gsottbauer, E. 253
Guzman, A.T. 99

Hadjiyiannis, C. 249
Hallegatte, S. 177, 181
Hartwick's Rule 329, 330–32, 341–8, 387–9
   equitable paths 348–63, 389–93
Hawking, S. 267
Heal, G. 100, 224, 225, 242, 328
health costs and impacts of air pollution 13–18, 20, 66–7
heat transport, *see* climate change policy under spatial heat transport and polar amplification
hegemonic power and leadership 268, 269–70
Henrion, M. 423, 424
Hidalgo, C.A. 407
hold-up problem 96–101, 102, 104–5
Hope, Chris 114, 115
host–MNC relations in resource-rich countries 83–5
   conclusions 101–2
   framework for understanding relationship between host and MNC 93–6, 104
   hold-up problem 96–101, 102, 104–5
   investment in resource-rich LDCs 101
   model 85–92
   see also permeability approach to post-transition resource extraction
Hotelling rent 212–13, 214
Hotelling's Rule (and extension) 110, 206, 329–30, 336, 339, 341–5, 387–9

ice melt 145–6
incidence analysis 49–57
income elasticities
   electricity-using products 64
   road fuel 65
India
   air pollution 13–14, 15, 16
   historical 263, 264
inequality-aversion (IA) 250–51
infrastructure development 401
input–output simulation models 410–11
institutions, formation of 271
insurance 434
Integrated Assessment Models (IAMs) 127–9, 203–8, 224, 226–7, 427–31
   flourishing research 122–3
   influence 115–16
   origins 114–15
   regional responsibilities and impacts 121–2
   representative results under uncertainty 118–21
   scrutiny 116–18

intergenerational altruism 310–12
  conclusions 319
  factors motivating capital accumulation and emissions control 316
  formal presentation of models and results 321–5
  increased cooperation between dynasties as solution 318–19
  informal presentation of models 312–15
  as solution 315–18
intergenerational equity 12
International Environmental Agreements (IEAs)
  'environmental threshold concerns' 252–3
  and game theory 245–6, 249–52
International Monetary Fund (IMF) 116
intertemporal allocation of resources model 107–11
intertemporal equity and efficiency (global warming model) 326–33
  alternate scenario (strictly concave function in specification of law of motion) 363–78, 393–5
  equitable Hartwick paths 348–63, 389–93
  equity, extended Hotelling's Rule and Hartwick's Rule 341–8, 387–9
  framework 333–6
  rates of change of key variables 378–84, 395–6
  short-run efficiency 336–41, 386–7
investment, *see* greenhouse gas (GHG) accumulation and cyclical growth; host–MNC relations in resource-rich countries; intertemporal equity and efficiency (global warming model); permeability approach to post-transition resource extraction
IPCC (Intergovernmental Panel on Climate Change) 169, 426
İriş, D. 248, 252–3
irreversibility 240

Jackson, C.H. 127
Jouvet, P.A. 311–12, 313–14, 316, 317, 318–19, 321–2
Jury Theorem 269, 271–4

Kahneman, D. 252
Karp, L. 311, 312, 314–15, 316, 317, 318, 319, 322–3
Kaya identity 283, 404
Kindleberger, C. 269
Kolstad, C.K. 251–2
Kopp, R.E. 173–4, 177
Kőszegi, B. 254
Kunreuther, H. 100, 242

Kyoto Protocol 234, 243

labor and labor productivity
  and economic development 402, 405
  greenhouse gas accumulation and cyclical growth 282, 284, 287, 288, 289, 291
  growth and sustainability 297–302, 303–4, 305
Lange, A. 250–51, 252
Langen, P.L. 127
lead-free fuel 242
leaders, and framing 253
legal environments 72
Leibniz, Gottfried 267
Levy, J. 253
likelihood and confidence 426
living standards 304
Logos and *Mythos* 264–9, 274
loss-aversion 252–4
low-carbon transformation process 400–401, 402, 404–6, 410

macroeconometric models 410–11
macroeconomic changes 400
macroprudential financial policy/regulation 421, 431–4
Malthus, collapse and climate change 260–65
  beliefs and Condorcet's Jury Theorem 271–4
  conclusions 275–6
  guardians of 'common home' 274–5
  Logos and *Mythos* 264–9, 274
  prisoners' dilemma, cooperation and morality 269–71
manufacturing, and development process 402, 403, 407
markets
  and climate change 275
  volatility of 272–3
Markowitz, H. 253
Marx, Karl 283
Mastrandrea, M.D. 426
mathematics 266, 267
Mealy, P. 408
Middle East, historical 263
Milinski, M. 246–7, 248
mitigation scenarios, and cyclical growth, *see* greenhouse gas (GHG) accumulation and cyclical growth
Mlodinow, L. 267
model uncertainty 422, 425
Mongolia
  mining agreements and host–MNC relations 92, 95, 96, 108

see also permeability approach to post-transition resource extraction
mortality, from air pollution 13–17, 41–2, 45–6, 47, 48, 66–7
multinational corporations (MNCs) and multilaterals
 and underdevelopment of resource-rich nations 68–9, 73, 74, 75–9, 80–82, 108
 see also host–MNC relations in resource-rich countries
multiplicative damages, *see* economic growth and social cost of carbon: additive versus multiplicative damages
Mythos and *Logos* 264–9, 274

National Academy of Sciences (NAS) 123
National Energy Modelling System (NEMS) 33
national income accounting/net national product 326
nationalism 72, 73
Nationally Determined Contributions (NDCs) 236, 399
nature, services from 305–7
Nesje, F. 311–12, 315, 316, 317–18, 319, 323–5
Net Present Value (NPV) 118–19, 120
network analysis 407–8
New International Economic Order (NIEO) 75–6
Newton, Isaac 266
NICE (Nested Inequalities Climate-Economy) model 224–5, 226–31
non-overlapping generations model 311–12, 315, 316, 317–18, 319, 323–5
Nordhaus, William 12, 114, 115, 117, 224, 226, 310, 327, 428
North, G.R. 130–31, 134
Nyborg, K. 250

obsolescing bargain model 73, 96, 101–2
OECD (Organisation for Economic Co-operation and Development), and air pollution 13–15, 16
oil producing countries 83
optimal control theory 337–8
optimal global climate policy, and regional carbon prices 224–6
 conclusions 236–7
 discussion 232–6
 modeling 226–9
 results 229–32, 233
other-regarding preferences 249–52
overconfidence 421–6

overlapping generations models 311, 313–15, 316, 317, 318–19, 321–3
ozone layer 326

PAGE (Policy Analysis of the Greenhouse Effect) model 114–17, 123, 224
PAGE09 (Policy Analysis of the Greenhouse Effect) model 118–21
parameter uncertainty 422–5
Parfit, D. 274–5
Paris Agreement (2015) 32, 239–40, 243, 244, 254–5, 397, 399; *see also* policy evaluation (implementation of Paris Agreement), China
Parry, Ian 47, 64, 65, 66, 67
past data, using for forecasting 425, 430–31
paths, intertemporal equity and efficiency (global warming model) 334–5
permeability approach to post-transition resource extraction 68–9, 106
 analysis of incentive structures of actors in bargaining game 79–81
 bargaining game in permeability context 81–2
 bargaining theory overview 76–9
 case comparison, Mongolia and Zambia 70–76
 conclusions 82
 extension of approach (global commons) 106–11
 resource curse 69–70
 see also host–MNC relations in resource-rich countries
Peters, W. 251
Pigouvian taxes 302, 307, 330, 339–41
Pindyck, R.S. 117
Planck's constant, recommended values for 423, 424
polar amplification, *see* climate change policy under spatial heat transport and polar amplification
policy evaluation (implementation of Paris Agreement), China 32–4
 analytical framework 34–9, 61–7
 baseline projections 40–42
 conclusions 57–8
 energy sectors 35–7
 incidence analysis 49–57
 model equations 61–4
 parameterization 64–7
 policies 37–9
 policy comparison 42–8
 results 40–48
political parties 253
political underdevelopment 68–9, 71–3, 76–9

pollution, *see* air pollution
Pope Francis 264–5
population and population growth 260, 261, 262, 264
poverty alleviation 397
power sector, China 35, 38, 41–2, 43–4, 64–5, 66
precipitation, damages from 135
preferences 245–6, 406
   intertemporal equity and efficiency (global warming model) 335–6
   non-standard 248–54
   other-regarding 249–52
   pure time preference (PTP) rate 120, 310–11
   reciprocal 249–50
   reference-dependent 252–4
price-based policies 19–20
price elasticities
   of electricity 64–5
   of fuel 47, 65–6
principal–agent problem 80, 87
prisoners' dilemma 99, 239, 249, 250, 267, 269–70
private property rights 109
probability density function (PDF) 242, 244
Product Space 407–8
production
   costs, and carbon tax 54–7
   intertemporal equity and efficiency (global warming model) 333–4
   know-how 407
productivity 401–2, 405; *see also* agriculture and agricultural productivity; labor and labor productivity
prospect theory 252, 253
proximity density 407–8
public goods games 243–4, 246–8, 251–2
pure time preference (PTP) rate 120, 310–11

Rabin, M. 250, 254
Radical Enlightenment 265–7
Ramsey growth model 203–8
random error 422
rationality 265–7
reciprocal preferences 249–50
reference-dependent preferences 252–4
reference points, *see* tipping and reference points in climate change games
religion 264–6, 274
renewable energy, transition to, *see* economic growth and social cost of carbon: additive versus multiplicative damages
renewable power generation, China 38

rent allocation, *see* carbon rent allocation
representative concentration pathways (RCPs) 169, 170–72, 173, 174, 175, 180–85, 188, 196, 432
resilience 432
resource-rich countries/resource curse, *see* host–MNC relations in resource-rich countries; permeability approach to post-transition resource extraction
RICE (Regional Integrated Climate-Economy) model 224, 226–7
Rio Tinto 80–81, 108
risk
   attitude 253
   perception 246–7
   see also adaptation strategies in European coastal cities
road fuel, China 39
road transport, China 35, 65
Robinson, J. 84, 271
Rome (Roman Empire) 260–61, 264
Rose-Ackerman, S. 99
Rossi-Hansberg, E. 128–9

S-shaped diffusion curve 240–41, 242
Sandler, T. 252
Sanjagav, Bayartsogt 72
Sata, Michael 71, 72
savings, intertemporal equity and efficiency (global warming model) 332, 333, 364–5, 369–72, 375–80, 381–2, 383
$SCCO_2$ (social cost of CO_2), *see* social cost of carbon
Schofield, N. 266, 269, 272
sea-level rise
   assessing damage costs and adaptation needs in coastal cities 177–8, 179–80
   construction of adaptation infrastructure 185, 190, 192–6
   costs of 167–8
   damage costs in main European coastal cities 181–4, 185, 186–7
   estimating local (LSLR) 172–4, 175
   Glasgow case study 184–5, 188–96
   and polar amplification 145
   risk calculation with Monte Carlo simulation 178, 180–81
   risk measures 174–7
   stochastic approach to modeling 169–72, 173, 174
sectoral permit/tax systems 21
services sector, and structural change 402, 403
Sheeran, K. 129, 151, 152, 153
slave trade 262

social cost of carbon 17, 18, 302, 307; *see also* economic growth and social cost of carbon: additive versus multiplicative damages; Integrated Assessment Models (IAMs)
social networks 80–81
social norms 267–8
social price
  of climate externality 154–6, 165–6
  of fossil fuels, and heat transport 144–5, 163–4
social welfare function, and additive and multiplicative damages 200, 203, 209, 211, 214–15
Solow, R.M. 328, 331, 332–3, 352, 379, 381
Solow–Swan growth model 283, 287, 290, 295
soul 267
South Africa, historical 263
Spanish empire 261–2
spatial heat transport, *see* climate change policy under spatial heat transport and polar amplification
speed of light, estimated values for 423, 425
Sraffa model 297–9, 302, 304, 305–6
stability agreements 77, 95–6
state, permeability of 107–11
Steffen, W. 167–8
Stern, N. 40, 122–3, 200, 214–15, 217, 229, 230
Stern Review on the Economics of Climate Change 12, 115, 116–17, 310
Stollery, K.R. 327, 329, 333, 379, 383
structural change, and development process 401–2, 403, 405–6
Suez Canal 262–3, 263–4
sunk costs 96
Sunley, E.M. 95
Sunstein, C.R. 15
sustainability and growth, *see* growth and sustainability
Sustainable Development Goals (SDGs) 397
systematic error 422

Tavoni, A. 247, 252–3
taxes
  border tax adjustments (BTAs) 235–6
  Pigouvian 302, 307, 330, 339–41
  see also carbon tax
technical/technological change 402, 405–6
  and labor productivity 299–300, 303
  throughput and throughput efficiency 300–302
technological transfer 406
temperature dynamics and heat transport 130–34
Teytelboym, A. 408

threshold public goods games (TPGs) 243–4, 246–8
throughput and throughput efficiency 296, 300–302, 303, 304, 306, 307
tipping and reference points in climate change games 239–40
  climate change experiments 246–8
  coalition formation games 245–6, 250, 251–2
  discussion 254–5
  key features of dangerous climate change 242–3
  non-standard preferences 248–54
  regime shifts and catastrophic climate change 240–42
  threshold public goods games (TPGs) 243–4, 246–8
tipping points 430, 432–3
Tol, Richard 114, 115, 200
trading ratios 21
tragedy of the commons 267
transformational change (climate and development goals) 397–8
  conclusions 412–13
  economic transformation processes 399–400
  empirical patterns 401–6
  importance of 399
  low-carbon transformation processes 400–401
  methodological tools for modeling and understanding 406–12
transient climate response (TCR) 120, 121, 123
transient climate response to cumulative CO_2 emissions (TCRCE) 128
transparency, and host–MNC relations in resource-rich countries 84–5, 86, 90, 92
Turing, A. 267
Tversky, A. 252

UN Framework Convention on Climate Change (UNFCCC) 232–4, 243; *see also* Paris Agreement (2015); policy evaluation (implementation of Paris Agreement), China
uncertainty 118–21, 168, 243, 245, 247, 268–9
uncertainty, systems 421
  climate science and economics 426–31
  communicating 426
  conclusions 435
  feedbacks 429–30, 433–4
  financial and climate system interactions 434–5
  implications for macroprudential regulation 431–4
  model uncertainty 425
  overconfidence 421–6

parameter uncertainty 422–5
reliance on past data 430–31
sources of 422
structural uncertainty 428–9
tipping points 430, 432–3
underdevelopment, and resource wealth, *see* permeability approach to post-transition resource extraction
US (United States)
air pollution 13–14, 15–17, 18–20
carbon dividend policy 27–8
carbon pricing 22, 24, 25
climate policy options 21
co-pollutant intensity and impacts 18–19, 20, 21
coastal extreme events 167
equity 19
as hegemonic power 270
historical 262, 263–4
religion 266

Value-at-Risk (VaR) 175–7, 180, 182–8, 192, 196, 434
value of a statistical life (VSL) 14–15, 16–17
values 270, 422
van den Bergh, J.C. 253
variance 422
Vasconcelos, V.V. 244

violent protest 72–3
Vogt, C. 250–51

Walpole, Robert 262
war 275
weather forecasting 420, 426–7
Weitzman, M.L. 117, 202, 211, 242
welfare effects
China 46–8
formulas for gains of policies 64
global warming 327–8
regional carbon prices 231–2, 233–4, 235, 236–7
welfare function (social), and additive and multiplicative damages 200, 203, 209, 211, 214–15
welfare maximization, under heat transfer 134–40
Wells, S. 260
working hours 303–4
World Bank 69, 80–82, 98, 108, 116
World Health Organization (WHO) 13

Zambia
mining agreements and host–MNC relations 96
see also permeability approach to post-transition resource extraction